AF361716

Microstructure and Corrosion
Behavior of Advanced Alloys

Microstructure and Corrosion Behavior of Advanced Alloys

Editor

Marián Palcut

MDPI • Basel • Beijing • Wuhan • Barcelona • Belgrade • Manchester • Tokyo • Cluj • Tianjin

Editor
Marián Palcut
Slovak University of Technology
in Bratislava
Slovakia

Editorial Office
MDPI
St. Alban-Anlage 66
4052 Basel, Switzerland

This is a reprint of articles from the Special Issue published online in the open access journal *Materials* (ISSN 1996-1944) (available at: https://www.mdpi.com/journal/materials/special_issues/microstructure_corrosion_behavior_advanced_alloys).

For citation purposes, cite each article independently as indicated on the article page online and as indicated below:

LastName, A.A.; LastName, B.B.; LastName, C.C. Article Title. *Journal Name* **Year**, *Volume Number*, Page Range.

ISBN 978-3-0365-3044-4 (Hbk)
ISBN 978-3-0365-3045-1 (PDF)

© 2022 by the authors. Articles in this book are Open Access and distributed under the Creative Commons Attribution (CC BY) license, which allows users to download, copy and build upon published articles, as long as the author and publisher are properly credited, which ensures maximum dissemination and a wider impact of our publications.

The book as a whole is distributed by MDPI under the terms and conditions of the Creative Commons license CC BY-NC-ND.

Contents

About the Editor

Marián Palcut (b. 1981) graduated with honors in physical chemistry at the Comenius University in Bratislava, Slovakia. He obtained his PhD in Materials Science and Technology at the Norwegian University of Science and Technology (NTNU) in Trondheim in 2007. He completed his postdoctorate at the Technical University of Denmark and the University of Oslo. At present, he holds a faculty position at the Slovak University of Technology. He defended his habilitation thesis in May 2019. Assoc. Prof. Marián Palcut has published 39 articles indexed in Web of Science and Scopus. His papers have been cited more than 500 times (excluding self-citations). His current h index is 14. Dr. Palcut is a regular editor of Advances in Materials Science and Engineering and a frequent reviewer of international peer-reviewed journals in Materials Science and Technology. He has participated in several national and international scientific projects focused on investigating the corrosion resistance of advanced metallic materials and alloys. His current research interests include corrosion studies of complex alloys, high entropy alloys, and lead-free solders.

Preface to "Microstructure and Corrosion Behavior of Advanced Alloys"

In many industrial applications, metallic materials are exposed to harsh operating conditions. Due to a combination of chemical and thermal stresses, the constructional and functional materials are degraded and their utility properties are lost. These undesirable events are of a physicochemical nature and are commonly known as 'corrosion'. Several internal and external factors influence the corrosion behavior of materials. Internal factors include the chemical composition, microstructure, and phase constitution of the materials. External factors that can significantly contribute to an increase in corrosion resistance include surface treatments and the use of various protective coatings. In this Special Issue Book, 3 reviews and 18 original research papers focused on the complex relationships between the microstructure, phase constitution, and corrosion behavior of metallic materials are collected. Both high temperature and low temperature corrosion studies are included as they investigate the physicochemical processes at the material interfaces. Furthermore, possibilities for increasing the corrosion resistance of metallic materials are studied by means of surface modification and application of protective layers. This Special Issue Book, Microstructure and Corrosion Behavior of Advanced Alloys, displays the diversity and complexity of modern corrosion research. It is hoped that it will become a valuable source of reference for corrosion scientists.

I am thankful to Ms. Hellen Hu from MDPI for providing valuable support during Special Issue editing.

This work was supported by the Slovak Research and Development Agency under the Contract no. APVV-20-0124.

Marián Palcut
Editor

Review

Duplex Steels Used in Building Structures and Their Resistance to Chloride Corrosion

Mariusz Maslak [1],*, Marek Stankiewicz [1] and Benedykt Slazak [2]

[1] Faculty of Civil Engineering, Cracow University of Technology, Warszawska 24, 31-155 Cracow, Poland; goziolko@cyfronet.pl

[2] DAIKO SRL Welding Materials, Viale Felissent 84/D, 31100 Treviso, Italy; bslazak@gmail.com

* Correspondence: mmaslak@pk.edu.pl; Tel.: +48-501546577

Abstract: Welded structures made of duplex steels are used in building applications due to their resistance to local corrosion attack initiated by chlorides. In this paper, the material and technological factors determining the corrosion resistance are discussed in detail. Furthermore, recommendations are formulated that allow, in the opinion of the authors, to obtain a maximum corrosion resistance for welded joints. The practical aspects of corrosion resistance testing are also discussed, based on the results of qualification tests. This work is of a review character. The conclusions and practical recommendations are intended for contractors and investors of various types of structures made of the duplex steel. The recommendations concern the selection and use of duplex steels, including the issues of metallurgy, welding techniques, and corrosion protection.

Keywords: duplex steels; welded joints; corrosion resistance; corrosion tests; chloride environment

Citation: Maslak, M.; Stankiewicz, M.; Slazak, B. Duplex Steels Used in Building Structures and Their Resistance to Chloride Corrosion. *Materials* **2021**, *14*, 5666. https://doi.org/10.3390/ma14195666

Academic Editor: Marián Palcut

Received: 29 July 2021
Accepted: 23 September 2021
Published: 29 September 2021

Publisher's Note: MDPI stays neutral with regard to jurisdictional claims in published maps and institutional affiliations.

Copyright: © 2021 by the authors. Licensee MDPI, Basel, Switzerland. This article is an open access article distributed under the terms and conditions of the Creative Commons Attribution (CC BY) license (https://creativecommons.org/licenses/by/4.0/).

1. Introduction

Modern two-phase ferrite-austenitic duplex stainless steels demonstrate an excellent combination of high strength and relatively high corrosion resistance [1]. Due to their fair ductility, the risk of catastrophic brittle fracture initiation is considerably limited [2]. Nowadays, the particular impulse amplifying the development of new types of duplex stainless steel is driven by increasing demand for crude oil mining from under the seabed. It translates into the need for the design of specific corrosion-resistant oil transmission installations both at offshore platforms and oil tankers. The newest, fourth generation of duplex steels has been designed recently for this purpose [3]. These materials design actions were taken to balance their microstructure and to equalize the ferrite and austenite corrosion resistance.

The corrosion resistance testing of duplex steels has become the subject of many research programs. Various research computational methods were used and led to the exploration of possible applications [4–16]. However, the knowledge, despite its extensive discussion in the scientific community, has not always been effectively translated into practical recommendations for investors and designers. The authors of this article have repeatedly encountered a lack of understanding of basic issues related to the application and technology of duplex steels, which are currently also one of the most expensive products of the metallurgical industry. Therefore, we saw the need to present in a comprehensive, though simplified manner, a practical guide for the rational use of individual duplex steel grades. We hope that this paper will allow designers to optimally use the basic advantages of duplex steels.

The authors' observations on the progress of pitting corrosion in duplex steels are presented in this work. The tests were performed as part of the Welding Procedure Qualification Record (WPQR) certificate [17]. The testing procedure was in accordance with the guidelines of ASTM G48-11 (method A) [18] with simultaneous consideration of the acceptance criteria of the NORSOK M601: 2008 [19] and ASTM A923-03 [20] standards.

1

2. General Characteristics of Duplex Steel

Duplex steels can be divided into five basic categories, depending on the percentage of alloying elements (Cr, Mo, Ni, Mn, Cu, and N—Figure 1). These groups are as follows:

1. Lean Duplex Stainless Steel (LDSS);
2. Standard Duplex Stainless Steel with 22% Cr (DSS 22% Cr);
3. High Alloyed Standard Duplex Stainless Steel with 25% Cr (DSS 25% Cr);
4. Super Duplex Stainless Steel (SDSS);
5. Hyper Duplex Stainless Steel (HDSS).

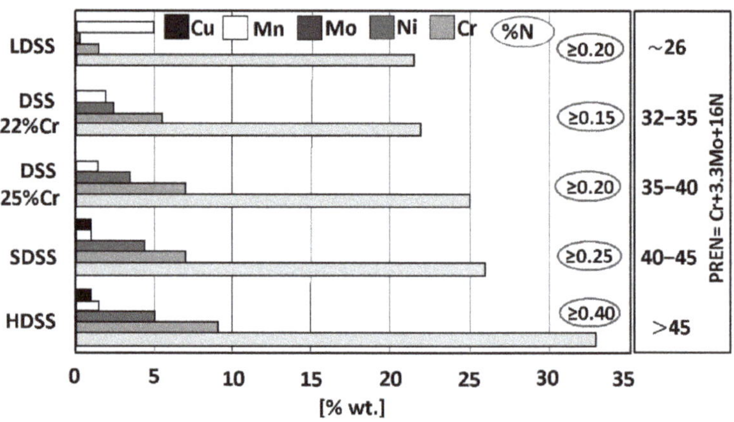

Figure 1. Chemical compositions and Pitting Resistance Equivalent Numbers (PREN) for various categories of duplex steels—according to [21].

A typical microstructure of properly balanced duplex stainless steel is given in Figure 2. It includes ferritice (dark) and austeniteic grains (light). The chemical composition of duplex stainless steel is limited by thermodynamic stability of austenite and ferrite and also by nitrogen solubility limit. Stability areas of duplex steel with respect to combined Cr + Mo mass fraction are given in Figure 3. Below 20% Cr + Mo, there is a risk of martensitic transformation of austenite. At above 35% Cr + Mo a δ-ferrite instability and formation of harmful secondary phases may occur. Moreover, a variation of nitrogen content in HDSS may influence the phase stability. Such difficulties are, in general, caused by high nitrogen vapor pressure.

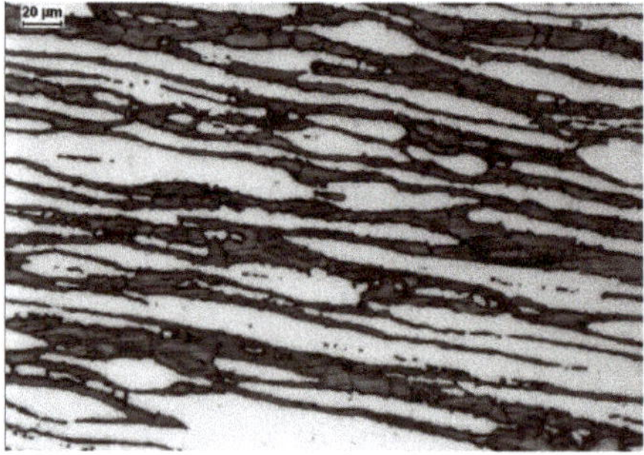

Figure 2. A properly balanced duplex steel microstructure containing approximately 50% of each δ-ferrite (dark areas) and γ-austenite (light areas)—according to [21].

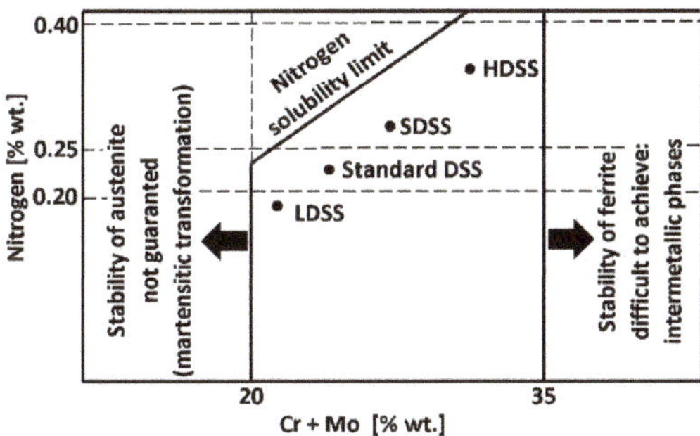

Figure 3. Stability areas of conventional duplex steels—according to [21].

Balanced duplex stainless steels are theoretically located at a 50% ferrite content line in Schaeffler–DeLong diagrams (Figure 4). However, in practice in properly balanced steel microstructure, the ferrite—austenite proportions range are wider, i.e., $\frac{ferrite}{austenite} \approx \frac{50\%-10\%}{50\%+10\%}$.

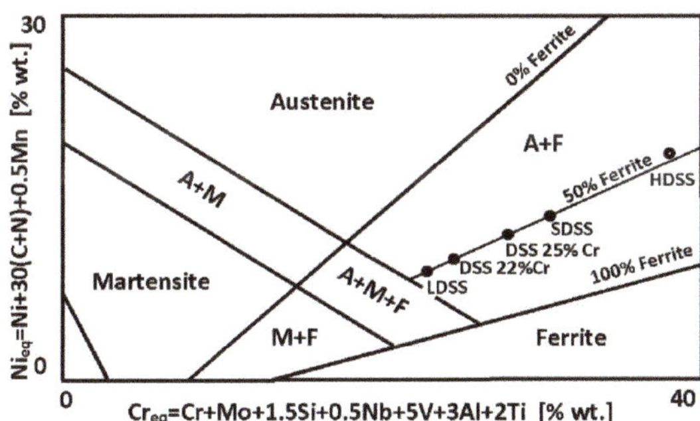

Figure 4. Location of individual categories of duplex steels in the Schaeffler–DeLong diagram—according to [21].

The appropriate use of duplex steel demands the welding engineers perform thoughtful actions based on well-established engineering knowledge and experience. A commonly applied routine-based approach to welding admittedly leads to obtaining the expected mechanical properties of welding joints. Nevertheless, it does not guarantee the concurrent obtaining of required corrosion resistance for the joints. It is known that the corrosion resistance of joints in the chloride environment impact zone reaches only 50–80% of the parent material corrosion resistance [22].

3. PREN Chloride Corrosion Resistance Index

In the right of Figure 1, the values of Pitting Resistance Equivalent Number (PREN) have been given for each group of duplex steels. This index is used to evaluate the pitting corrosion resistance of the steels in a chloride environment. The PREN indicator refers to the thermodynamically stable steels, i.e., the steels after their final heat treatment [23]. In the case of duplex steels, the following formula given by Herbsleb is used to calculate the value [24]:

$$PREN = Cr + 3.3Mo + 16N \tag{1}$$

In this equation Cr, Mo, and N are weight percentages of the corresponding elements. For SDSS and HDSS, which contain W or Cu, different formulae may be used [24]. These are as follows:

- Okamoto formula:

$$PRENW = Cr + 3.3(Mo + 0.5W) + 16N \tag{2}$$

- Heimgartner formula:

$$PRENCu = Cr + 3.3Mo + 15N + 2Cu \tag{3}$$

- Extended formula:

$$PRENEXT = Cr + 3.3(Mo + 0.5W) + 2Cu + 16N \tag{4}$$

In these formulae, the content of a particular alloying element is given in its mass percentage (%wt.).

The PREN values range from 26 (for LDSS steels with average pitting corrosion resistance) up to above 45 (for HDSS steels with high corrosion resistance). In both cases, the resistance is significantly higher when compared to conventional steel grades. However, the values of PREN should be treated as comparative data. The final selection of steel for a given application should be based on tests carried out in a given corrosive medium. The usefulness of the PREN indicator for estimating the analogous resistance of welding consumables is limited due to different welding techniques that can be used and hence the different levels of nitrogen introduction into the welded melt metal. The terms "high corrosion resistance" and "average corrosion resistance" should be therefore interpreted as relative measures.

4. Alloying Elements Influence on Duplex Steels Corrosion Resistance

Duplex steels usually crystallize from a liquid in the form of δ-ferrite. During alloy cooling, the lattice structure stresses increase because of Fe replacement by elements with larger radii, e.g., Ni.

In the areas with a higher concentration of austenite-forming elements, the A2-type ferrite lattice structure is transformed to an A1-type lattice (with 25% larger lattice parameter) at δ-solvus temperature (Figure 5a). This phenomenon is accompanied by a decrease in tension and a simultaneous decrease in inter-granular borders energy. This is an enhancing factor for $\delta \rightarrow \delta + \gamma$ transition. Due to the presence of alloying elements, there is an increase in Cr, Si, Mo, W, P concentration in A2-type ferrite lattice and Ni, N, Cu, Mn, C in A1-type austenite lattice. The transition appears to be diffusion-limited [25]. As a consequence, the newly formed austenite assumes a lamellar (island) structure. Out of homogeneous δ-ferrite, a two-phase structure $\delta + \gamma$ is formed, with its components varying from one another in terms of corrosion resistance. Chrome and molybdenum (ferrite formers) strongly increase the electrochemical potential. Consequently, the corrosion resistance is concentrated in ferrite. In austenite, only an interstitial solution of nitrogen significantly can increase the electrochemical potential of steel (Table 1; Figure 5b–d).

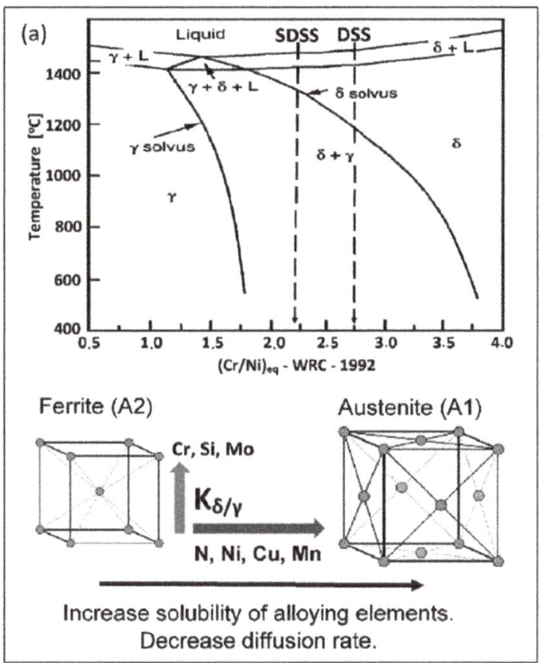

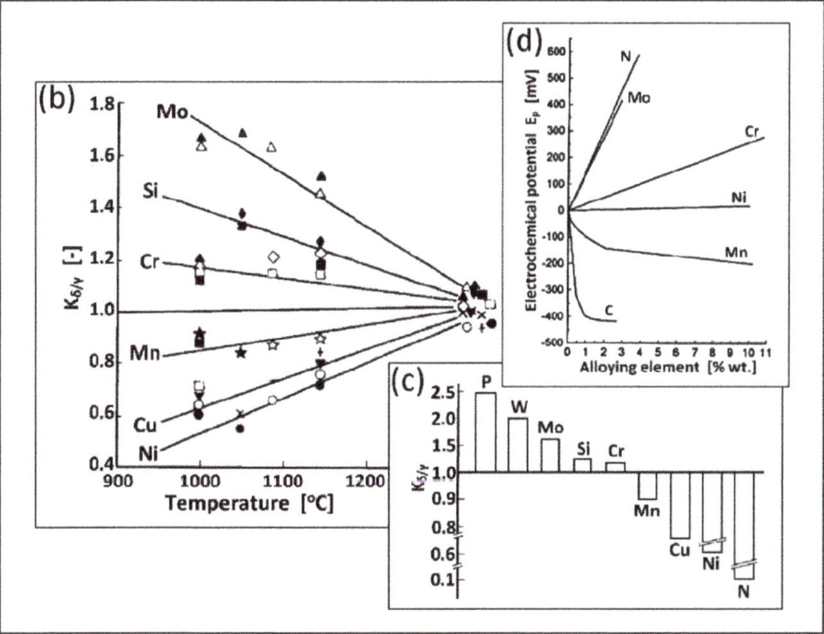

Figure 5. Metallurgical transformations in duplex steels: (**a**) phase equilibrium diagram and austenitic transformation δ → δ + γ (according to [27]), (**b**) distribution coefficient of alloy additions $K_{\delta/\gamma}$ as a function of temperature (according to [28]), (**c**) typical values of the partition coefficient of alloying elements $K_{\delta/\gamma}$ for supersaturated and water-cooled steels (according to [28]), (**d**) influence of alloying elements on the size of electrochemical austenite potential in stainless steels 18/8 (according to [29]).

Table 1. Maximum solubility of alloy additives in ferrite and austenite (according to [26]).

Alloy Additive	Solubility (%wt.)		Crystal Lattice
	Ferrite	Austenite	
W	35	4.7	A2
Mo	31	1.7	A2
Mn	3.5	100	A1
Cr	100	12.5	A2
Cu	2.1	12	A1
Ni	6	100	A1
Si	11	1.7	A4
C	0.03	2.1	-
N	0.1	2.8	-

The influence of typical alloying elements on duplex steel corrosion resistance is presented in Table 2. In general, Mn and Ni reduce the corrosion resistance. However, these elements are required due to the strengthening effect of Mn and the need for Ni to initiate a ferrite decomposition and form a two-phase structure.

The process of δ-ferrite decomposition and formation of the $\delta + \gamma$ two-phase structure takes place at temperatures of 800–1200 °C. Due to its diffusive nature, the kinetics of the transformation depends on the cooling rate (Figure 6). A slow cooling causes a formation of γ-austenite in the amount close to thermodynamic equilibrium. The distribution of alloying elements between the matrix components is also close to equilibrium in this situation. The rapid cooling, on the other hand, creates a metastable structure with a lower austenite content. It increases the risk of harmful secondary phases precipitation from ferrite supersaturated with alloying additives. The formation of harmful precipitates only within ferrite is a consequence of several dozen times higher diffusion coefficients and many times lower solubility of interstitial elements (C, N) in ferrite compared to austenite [25].

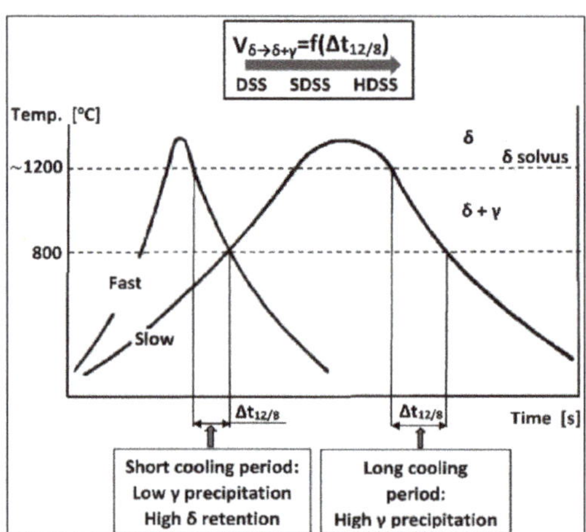

Figure 6. Influence of cooling rate on the content of δ-ferrite in duplex steels (according to [30]). Left-hand frame: fast cooling, high δ-ferrite content; right-hand frame: slow cooling, high content of γ-austenite precipitates.

To obtain a higher amount of γ-austenite in the final steel microstructure, it is desirable to cool it slowly in the temperature range of the two-phase structure. It can be ensured by a sufficiently high linear energy of welding. At a temperature below 1050 °C, the situation is reversed, and a higher cooling rate is necessary to counterbalance the high tendency to secondary phase precipitations so that the cooling line does not cross the upper Time-Temperature-Transformation (TTT) curve (Figure 7).

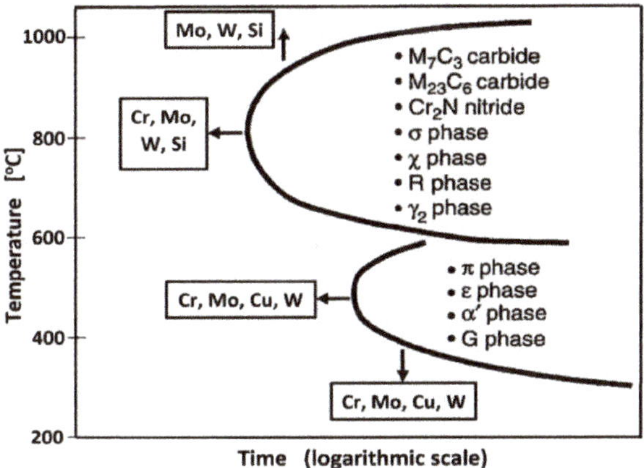

Figure 7. Schematic diagram of TTT (Time-Temperature-Transformation) specified for the typical duplex steels (according to [31]).

Table 2. Influence of various alloy additions and steel microstructure on duplex steel pitting and slotted corrosion resistance (according to [32,33]).

Alloying Element	Effect	Cause	Formal Limitations
C	Negative	A surplus of C causes secretion of chromium carbides with associated chromium-depleted zones	Up to approximately 0.03% of the content
Si	Positive	Stabilizes the passive top layer	Up to approximately 2% due to the negative influence of Si on the stability of the structure and on the Nitrogen solubility
Mn	Negative	Mn-rich sulfides can initiate pitting. Mn can also destabilize the passive surface layer	The content of over 2% of Mn increases the risk of precipitation of harmful intermetallic phases
S	Negative	Sulfides are a strong initiator of pitting corrosion	High resistance to pitting corrosion is only possible with a content of less than 0.003% S
Cr	Positive	Stabilizes the passive top layer	25% to 28% maximum depending on the Mo content. Higher Cr content increases the risk of precipitation of harmful intermetallic phases
Ni	Negative	Increased Ni content lowers the PREN index of austenite	The Ni content is limited to the amount necessary to form about 50% of austenite
Mo	Positive	Stabilizes the passive top layer and its subsurface metal substrate	About 4–5% Mo depending on the Cr content. Mo increases the risk of precipitation of harmful intermetallic phases
N	Positive	Significantly increases the PREN index of austenite	About 0.15% in LDSS steels. About 0.3% in SDSS steels and slightly over 0.4% in 25% Cr alloys, with high Mo and Mn content
W	Positive	It probably works in the same way as Mo	Increases the tendency to release harmful intermetallic phases
Cu	Uncertain	Marginal effect	Maximum content up to approximately 2.5%

5. The TTT Diagram, CPT Temperature, and the Ferrite Number Determined for Duplex Steels

In high-alloy duplex steels of older generations (SDSS and HDSS), the precipitation start time is short [34]. Exceeding the activation energy of the secondary phase precipitations, expressed by the intersection of the cooling line with the TTT curve increases the sensitivity to accumulation of heat exposure effects resulting from welding the steel. It is because the energy of the precipitation reaction is always lower than the activation energy of this process [35].

The formation of secondary phases in duplex steels, which are hard and brittle and thus harmful, takes place in two temperature ranges (Figure 7). The upper curve, corresponding to the range of 600–1050 °C, shows the precipitation of nitrides, carbides, and intermetallic phases as a result of prolonged thermal exposure of steel due to insufficiently rapid cooling. Thus, the welding with high linear energies facilitates the transformation of δ to $\delta + \gamma$, which is favorable but also increases the probability of the secondary phase formation within the ferrite at insufficiently fast cooling. The consequence of this is the need to strictly adhere to the recommended linear energies of welding and to control the cooling process of the joint during welding. The lower TTT curve, corresponding to the temperature of 300–550 °C, shows the remaining secondary changes in the steel microstructure. The most important is the change of δ-ferrite to acicular secondary α'-ferrite, significantly reducing the ductility and toughness of steel. The lower limit of the occurrence of the unfavorable α'-ferrite determines the highest temperature of the long-term thermal exposure, which is about 300 °C.

The advancement of harmful changes in the microstructure, lower ductility, and corrosion resistance depends on the total heat exposure time to both critical ranges on the TTT curves (Figure 8a,b). The exposure of the lowest alloyed LDSS to temperatures between 800–1000 °C may exceed 10 h without the release of harmful phases. However, this time for the standard DSS 22% Cr steel is reduced to about 30–60 min, and for the HDSS steel is even 5–10 min [36]. The presence of 0.5% secondary phases at the boundaries of δ-ferrite causes a dramatic decrease in the breaking work (Figure 8a). The same volume of the harmful and easily released Cr_2N phase lowers the critical pitting temperature (CPT) by up to 20 °C (Figure 8b).

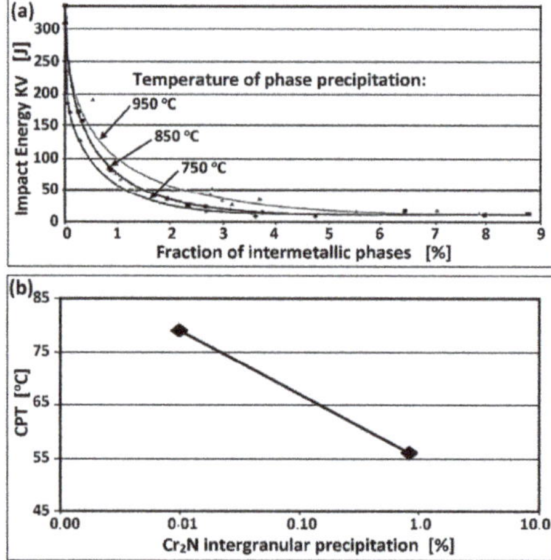

Figure 8. Changes in the microstructure lowering ductility and corrosion resistance of duplex steels, including (**a**) breaking energy KV as a function of intermetallic phase content in duplex steels (according to [37]), (**b**) CPT (Critical Pitting Temperature) index as a function of the content of intercrystalline Cr_2N precipitates (according to [38]).

Nitrogen in duplex steel increases the kinetics of austenite formation (Figure 9). With increasing nitrogen content, the high-temperature equilibrium $\delta + \gamma$ area expands towards lower Ni concentrations, and the temperature of austenite formation increases. It may increase even to the temperature of liquids, which means crystallization of a small part of austenite directly from the liquid metal. A significant increase in the transformation rate $\delta \rightarrow \delta + \gamma$ occurs, which, with sufficiently rapid cooling, makes it possible to avoid the effects of shifting the TTT curves on the time axis towards the origin of the coordinate system. For this reason, nitrogen-rich steels can be cooled in the temperature range of 800–1200 °C faster, without the fear of exceeding the final ferrite content above the permissible level (70%). The addition of N_2 to the forming and shielding gases in Tungsten Inert Gas Arc Welding (TIG, 141) is of fundamental importance for obtaining a properly balanced weld microstructure.

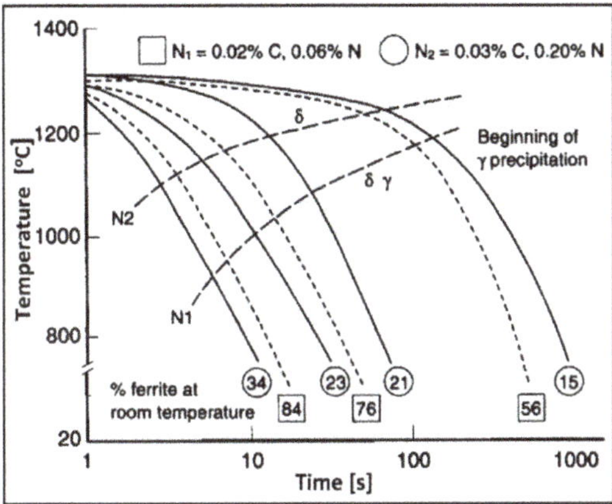

Figure 9. TTT diagram for DSS 25% Cr duplex steels with different alloyed nitrogen content (according to [28]).

The balanced microstructure of duplex steel is obtained in a two-stage supersaturation heat treatment. The first stage is homogenizing annealing. It is most effective at the temperature of maximum nitrogen solubility in solid solution, which, depending on the steel grade, is 1050–1150 °C (Figure 10a,b). At this temperature, harmful secondary precipitates in δ-ferrite dissolve, and the released atomic nitrogen diffuses into austenite since its solubility in the interstitial solution is many times greater (see Table 1). This applies especially to thick-walled elements made of the SDSS and HDSS steels, where the practical cooling rate in the critical temperature range is too low to avoid the precipitation of secondary phases. The second stage is rapid cooling of the steel in water, limiting the possibility of re-separation. As a result, the steel microstructure is balanced, i.e., the proportion of both components equals $\frac{ferrite}{austenite} \approx \frac{50\%^{-10\%}}{50\%^{+10\%}}$ and the resistance to pitting corrosion and also the KV impact strength is increased.

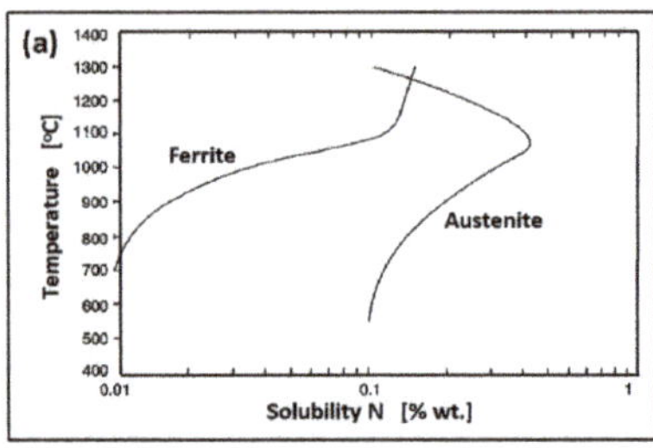

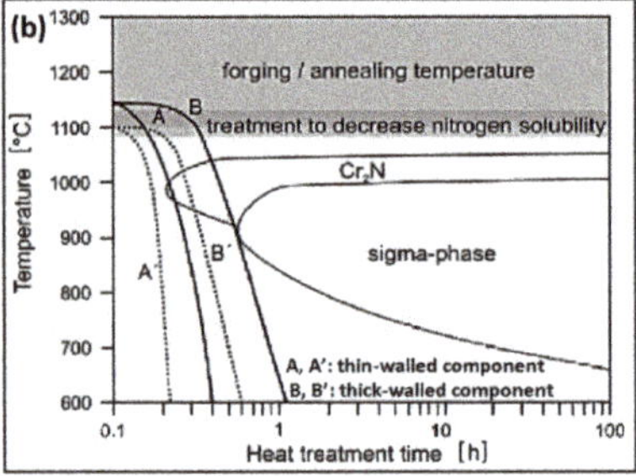

Figure 10. Heat treatment of duplex steel, including (**a**) solubility of Nitrogen in ferrite and austenite (according to [38]), (**b**) heat treatment of the thin-walled and thick-walled duplex steel elements (according to [23]).

The volumetric fraction of δ-ferrite in the structure can be verified experimentally either by microscopic examination (image analysis) or by magnetic methods [4,5]. Values of the ferrite number (FN) are converted into the volumetric fraction of δ-ferrite (%δ) based on the following relationship [39]:

$$\%\delta = 0.7\text{FN} \tag{5}$$

In this equation, %δ (%) is a volumetric fraction of δ-ferrite in the steel microstructure. If it is not possible to perform the appropriate tests, the estimation of the volumetric fraction of %δ-ferrite in duplex steel can be made using the data from the metallurgical certificates in conjunction with the following relationships [36]:

$$\%\delta = 4.01\text{Cr}_{eq} - 5.6\text{Ni}_{eq} + 0.016\text{T} - 20.93 \ (\%\text{wt.}) \tag{6}$$

$$\text{Cr}_{eq} = \text{Cr} + 1.73\text{Si} + 0.88\text{Mo} \ (\%\text{wt.}) \tag{7}$$

$$\text{Ni}_{eq} = \text{Ni} + 24.55\text{C} + 21.75\text{N} + 0.4\text{Cu} \ (\%\text{wt.}) \tag{8}$$

In the equations, T is the homogenizing annealing temperature, and the remaining factors are element weight fractions.

The area of the weld with the lowest resistance to pitting corrosion is the heat-affected zone (HAZ). The HAZ is particularly endangered by the time of impact near the fusion line of the temperature exceeding the δ-solvus level when the microstructure of duplex steel is a single-phase and the free growth inhibitory factor δ of the ferrite grains is missing. Lowering the temperature below the level of δ-solvus again, when the joint is cooling down, activates the δ → δ + γ reaction, which, as mentioned earlier, is a diffusion-controlled transformation. The larger the original grain size, the longer it takes to reach thermodynamic equilibrium. It is difficult to re-achieve a microstructure with a balanced proportion of both matrix phases from the developed ferrite grains. The cumulative effect of the thermal cycles of welding successive weld beads leads to an increase in the ferrite content in the HAZ at the expense of austenite. Furthermore, it leads to an increased tendency to form secondary phases from the ferrite and thus to lower both the ductility and resistance to pitting corrosion. The increased content of austenite-forming Ni in the weld and the absorption of strongly austenitic-forming N from the shielding gases intensify the kinetics of the $\delta \to \delta + \gamma$ transformation, thereby preventing the reduction in austenite content in the weld microstructure. Therefore, the selection of the parent material with the highest possible austenite content, within the limits of the correct balance of the initial duplex steel microstructure, is important for the subsequent resistance of the HAZ to pitting corrosion and its KV impact strength.

6. Pitting Corrosion of Stainless Steels Initiated When Exposed to Chloride Impact

Steel pitting corrosion is an electrochemical process that takes place in halide-containing electrolytes [40]. The corrosion of this type attacks the material locally and quickly perforates the metal, causing the loss of tightness. It is considered to be one of the most severe forms of corrosion. It has the highest intensity in stationary solutions due to their uneven saturation with depolarizers, i.e., oxygen or hydrogen ions, and the resulting formation of the so-called concentration cells. The intensity of pitting corrosion increases with the temperature of the electrolyte. Mixing the solution equalizes the depolarizers' concentration at the metal surface, thereby slowing the progress of corrosion [41]. A permanent polarization of the cell, i.e., blocking the anode or cathode reaction, may stop the progress of pitting corrosion.

A simplified diagram of the steel pitting corrosion mechanism is shown in Figure 11. At structural defects, such as ferrite/austenite interfaces, non-metallic inclusions (especially sulfide), secondary phase precipitates (including carbides and nitrides), the passive layer can be easily punctured by aggressive chloride anions. A local, short-circuited corrosion micro-cell is created with a small anode area and a large cathode area. A consequence of the disproportion of the anode surface to the cathode surface ratio is a high anodic current density. Fe^{2+} ions dissolve into solution and are consequently captured by Cl^- anions to form iron chloride ($FeCl_2$). Around the pit, a cathodic depolarization reaction takes place (oxygen-type, hydrogen-type, or both), which maintains the operation of the electrochemical cell by balancing the electric charge.

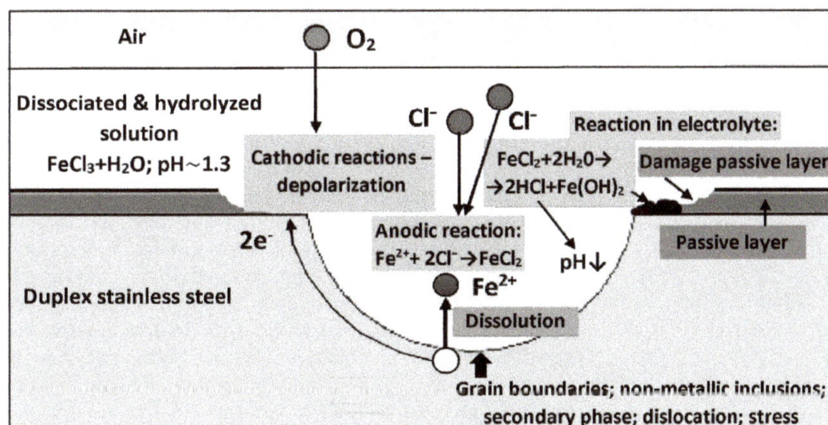

Figure 11. Pitting corrosion mechanism of duplex steel when exposed to chloride attack (according to [21]).

The pitting corrosion is autocatalytic, self-accelerating the growth of pits due to lowering the pH of the corrosive solution inside the growing pit [40]. In the initial stage, the pitting corrosion is of inter-crystalline character [42]. A balanced duplex steel microstructure, within the limits of $\frac{ferrite}{austenite} \approx \frac{50\% - 10\%}{50\% + 10\%}$ reduces the tendency to harmful precipitations of carbides and nitrides at the intergranular boundaries. This simultaneously reduces the risk of developing inter-crystalline corrosion.

7. Passive Layer and Corrosion Protection Mechanisms Identified as Being Specific to Duplex Steels

The corrosion protection of Cr-Ni type stainless steels is based on a durable and tight passive layer. The passive layer should be chemically inert to an aggressive environment. The passive layer in the oxidizing environment forms itself spontaneously and reaches a thickness of 2–4 nm [43]. It has an ability to self-rebuild (re-passivation ability). The most important role is played by Cr, which forms a complex oxide (Fe, Cr)$_2$O$_3$ on the steel surface. The minimum content of Cr in Cr-Ni type steels to form the passive layer is 10.5%. The higher the Cr fraction in (Fe, Cr)$_2$O$_3$ is, the tighter and more corrosion resistant the passive layer is. Mo and N are incorporated into the passive layer in a lesser content. Molybdenum has a nobler electrode potential compared to Cr. It can form MoO$_2$ and MoO(OH). Nitrogen may be present in the passive layer in the form of anions. As such, it may inhibit surface adsorption of negative Cl$^-$ anions. A synergic activity of Mo and N is observed. The synergic activity increases the duplex steel pitting resistance to a greater extent than it would be expected from the cumulative content of both elements separately [44].

The oxidation of steel surface during welding is accompanied by migration of Cr and Mo to the surface layer. This, in turn, lowers the electrochemical potential of the metal under the passive layer and increases the risk of corrosion previously initiated. The chemical etching and passivation of the welded joints surface are an effective way to counteract this threat. Nitrogen dissolved in the steel reacts with H$^+$ to form NH$_4$$^+$, thus partially neutralizing the acidic pH of the corrosive medium within the pits, limiting their growth [44]. It is believed that this may be the mechanism of active corrosion protection of austenite over which a thinner passive layer forms. The effectiveness of this protection is evidenced by the value of the nitrogen weighting factor in the PREN formula determined for pure austenite [23]:

$$PREN = Cr + 3.3Mo + 30N \tag{9}$$

A nitrogen deficiency in the passive layer of duplex steel limits the effectiveness of the passive layer over the austenite grains. In Figure 12a, the pits formed in the filling of the gas tungsten welded joint (TIG, 141) are shown in detail. The fillings were laid without nitrogen in the shielding gas. The joint in question was subjected to a laboratory pitting corrosion resistance test when exposed to FeCl$_3$ solution. A fragment of extensive corrosion pitting was selected for analysis. The alloying elements' distribution in the part of the pitting surface indicated by a white arrow revealed a relationship between the distribution of Cr in the surface layer and Cl absorbed from the corrosive solution. In places with a high Cr concentration (the surface layer above the ferrite grains), the content of absorbed Cl on the surface was low. On the other hand, in places with a low concentration of Cr (the surface layer above the austenite grains), the amount of absorbed Cl was high. It is important to note that the high concentration of absorbed Cl occurred at places with the highest corrosion intensity. In the case considered here, the SDSS weld metal had low N content and, therefore, an insufficiently balanced microstructure. The preferred corrosion attack was on the subsurface austenite grains and then, due to the presence of harmful secondary phases, on the adjacent grain boundaries. Subsequently, the grains of both basic phases of the metallic matrix were etched.

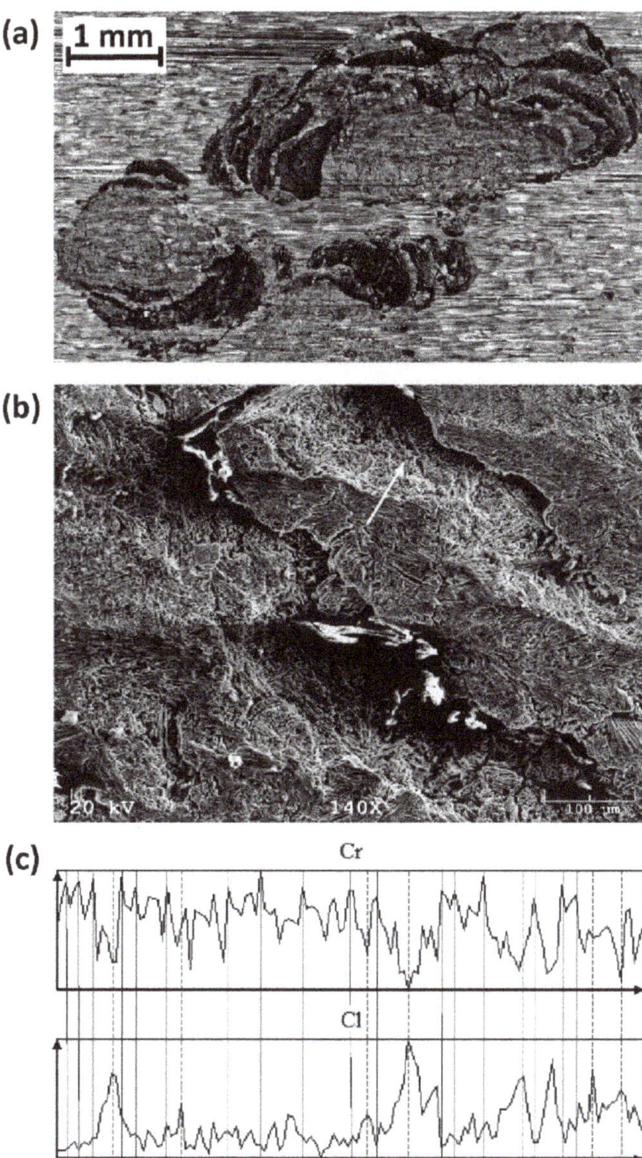

Figure 12. Chloride pits identified in the SDSS weld metal (1.4501, F55). Joint filling obtained after automatic TIG welding in Ar shielding gas without N_2. In particular: (**a**) magnified optical photography, (**b**) a photo taken with a scanning microscope, (**c**) Cl and Cr contents measured along the white arrow marked in Figure 12b. The following locations are marked by vertical lines in Figure 12c: solid lines—local maxima of the Cr content in the surface layer and the corresponding local minima of the absorbed Cl content, dashed lines—local minima of the Cr content in the surface layer, and the corresponding local maxima of the absorbed Cl content.

8. Recommendations on the Chemical Compositions of Duplex Steel Grades with Improved Corrosion Resistance

Because of the different physico-chemical properties of the passive layer over the ferrite and austenite areas, the balanced microstructure reduces the risk of pitting corrosion. Nowadays, the metallurgical industry offers duplex steel products with a microstructure balance within the limits of $\frac{ferrite}{austenite} \approx \frac{50\%-10\%}{50\%+10\%}$. However, due to the risk of excessive ferritization of the heat-affected zone (HAZ) on the welded joints, the parent material should be used with the δ-ferrite content not exceeding 55% [45–49].

In each type of steel, the presence of carbon and sulfur precipitates with high surface energy facilitates the initiation of corrosion. For this reason, the selection of materials with the lowest C and S contents possible is recommended. The presence of atomic oxygen, dissolved in the solid steel solution, supports electrochemical corrosion due to its role in the cathodic reaction. It is therefore advisable to deeply deoxidize the liquid metal [50].

It is recommended to use the fourth-generation duplex steel. Welding of the high-alloy duplex steels (both SDSS and HDSS) of the previous generations was associated with technological difficulties to ensure an appropriate cooling time ($\Delta t_{12/8}$) necessary for the proper balance of the weld microstructure. Increased kinetics of austenite formation and reduced sensitivity to selective corrosion of the δ-ferrite in the fourth generation steel softens the welding technological regime necessary to obtain the required resistance to pitting corrosion of such welded joints [22].

Duplex steels should not be welded without additional material due to the risk of excessive ferritization of the weld, resulting from melting of the parent material. The weld metals should have a Ni content higher by 2–4% compared to the parent material in order to increase the kinetics of austenite formation $\delta \rightarrow \delta + \gamma$ as well as to compensate for the lack of subsequent balancing of the welded joints microstructure by heat treatment. For this reason, it is recommended to weld the lower grades of duplex steels with the weld metal of the composition corresponding to the higher grade (e.g., DSS steels should be welded with the weld metal of the chemical composition of SDSS steels). This applies in particular to the weld root beads exposed to direct contact with the corrosive medium. Examples of auxiliary materials dedicated to welding various grades of duplex steels are given in [51].

Nitrogen present in most duplex steel grades enhances the kinetics of austenite formation. The relationship between the nitrogen content and δ-ferrite content in the weld is linear (Figure 13).

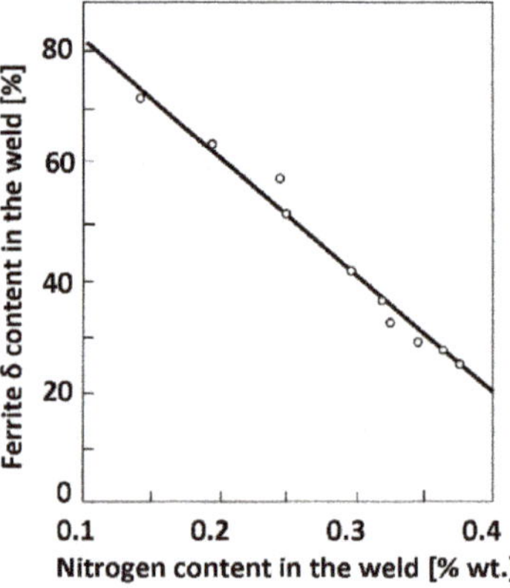

Figure 13. Influence of nitrogen content in a weld made with the TIG method on the volume fraction of δ-ferrite in the steel microstructure (according to [52]).

The high vapor pressure of nitrogen at the welding temperature causes its migration from the weld pool to the surrounding environment and, consequently, reduces the share of austenite in the weld structure, thereby increasing the risk of losing corrosion resistance and lowering the impact strength.

The nitrogen-poor weld contains up to 80% δ-ferrite. Since the electric arc does not transfer the electrically neutral nitrogen atoms, N is not present in arc welding consumables. Therefore, the only option to increase the N content in the weld metal during the welding is the addition of N_2 to the shielding and forming gases. The necessary amount of N_2 depends on its solubility limit in duplex steel and increases with increasing Ni concentration in the steel. Nitrogen content in shielding gases for GTAW welding (TIG, 141) should be in the range of 1–1.2% for DSS 22% steel and 2–2.5% for DSS 25%, both SDSS and HDSS steels.

The weld root is usually the area with the greatest risk of corrosion. Therefore, it is important to use N_2-rich mixture as forming gas. However, the use of shielding gases with an excessively high N_2 concentration may result in exceeding the N solubility limit in solid solution and the appearance of weld porosity, especially in thick-walled joints. Since the corrosion resistance of stainless steel is primarily determined by the properties of the surface layer, the beads of the multi-run welds can be welded in pure argon to avoid porosity. In such cases, the pitting corrosion resistance tests should not include the weld filling.

The addition of 20–40% of helium to shielding gas increases the thermal energy supplied to the weld, and this allows increasing the welding efficiency with the GTAW method (TIG, 141). Furthermore, the full control of the O_2 content in welding gases prevents its absorption in the weld pool, as well as a harmful O increase in the solid solution. In addition, it allows a reduction in the thickness of the oxide layer above the welded joint and thus the depth of the depletion of the steel surface layer in Cr and Mo. Therefore, it is recommended to use welding gases with O_2 content below 200 ppm for duplex steels and to flush the pipes from the weld root side with forming gas in order to reduce the O_2 content as much as possible. It is suggested to limit the O_2 concentration to 25 ppm.

9. Recommended Welding Technologies

Expensive duplex steels are primarily used because of their high corrosion resistance in chloride environments. The correct welding technology selected for these steels should therefore ensure sufficient corrosion resistance of at least those areas of welded joints that remain in contact with the aggressive medium. In the case of single-sided welding, e.g., pipelines, small vessels, and containers, it is usually the weld root with the adjacent heat-affected zone that is most susceptible to corrosion. Whenever possible double-sided welds should be designed since balancing the microstructure and achieving the required level of resistance to pitting corrosion of the weld face is much easier than with the weld root. In single-sided joints with an accessible weld root, the backing weld of the root may be used to improve the low corrosion resistance. To each bead of the weld, it is necessary to introduce the appropriate amount of thermal energy, limit the access of oxygen, provide the necessary time for decomposition of δ-ferrite, and for the formation of an optimal amount of γ-austenite. This is done by slow cooling between the 1200 °C and 1050 °C, and the release of harmful secondary phases is prevented by quick cooling between the 1050 °C and 300 °C. The temperature range of 1050 °C to 850 °C requires a particularly intensive cooling of the steel. In our opinion, the optimal heat amount, introduced into the weld to ensure the expected cooling rate above and below 1050 °C, seems to be the fundamental issue in welding duplex steels with classic arc methods. Moreover, the welding of subsequent weld beads may not cause adverse effects in the microstructure of the preceding weld beads, especially in the areas of the weld root, which are in direct contact with the corrosive environment. Therefore, to ease the control and improve the uniformity of the heat transfer, it is recommended to use mechanized welding instead of manual welding.

Limiting the access of oxygen to the root of the weld requires the use of low-oxygen welding methods. In Figure 14a, the relationship between the breaking energy KV of the weld metal and its oxygenation is shown in detail. The lowest degree of oxygenation in the weld metal is achieved by using the GTAW method (TIG, 141) and PAW (Plasma Arc Welding) method (151), which is a GTAW method extension.

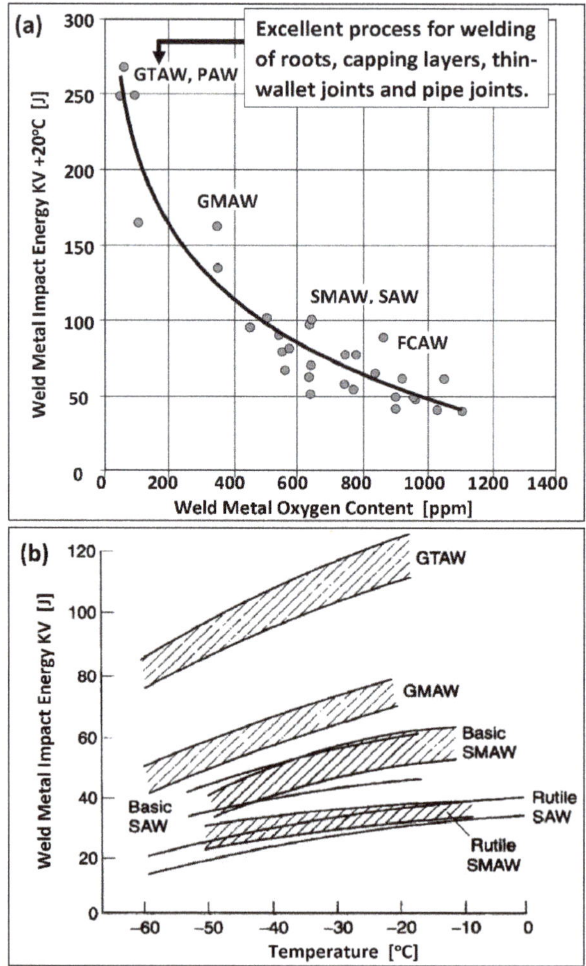

Figure 14. Breaking energy of the weld metal: (**a**) breaking energy KV determined at 20 °C as a function of oxygen content in the duplex weld metal, (**b**) breaking energy KV determined at 0–(−60) °C, specified for various welding methods (according to [28]). The meaning of individual abbreviations is explained in the Abbreviations of this paper.

The ease of GTAW (TIG, 141) made this method the primary choice. For the pipe connections, a modified GMAW-STT method (Surface Tension Transfer, MIG-STT, 131-STT) may be used. It provides a three to four times higher welding efficiency in rela-tion to the conventional GTAW and gives a comparable resistance to pitting corrosion. Additionally, it guarantees a satisfactory ductility of the material down to −40 °C [53]. It should be noted that the use of high-oxygen, slag arc welding methods for this type of welds, such as SMAW (111), SAW (121), or FCAW (114), reduces the corrosion resistance of the joints and also reduces the breaking energy of the weld metal. At the same time, it increases the lower operating temperature threshold for welded joints use (Figure 14b). This is a consequence of the high content of atomic oxygen dissolved in the solid solution and the presence of oxide inclusions at the grain boundaries.

As most standards do not require pitting tests of the whole weld cross-section, any flux-type welding (high-oxygen) methods can be applied for inner layers of thicker multi-run joints. However, particular caution should be taken while welding SDSS and HDSS steels due to their high yield strength and a stronger tendency to brittle fracture. In such a situation, an application of the low-oxygen welding methods (both GTAW and PAW) at the whole cross-section of the weld and precise balancing both of the weld microstructure and the whole heat-affected zone (HAZ) is required for obtaining reasonable impact resistance.

The regulation of γ-austenite formation kinetics in duplex steel welds is dependent on the proper shaping of the weld groove. The shapes of the grooves should be in general analogous to those formed in acid-resistant austenitic steels. Nevertheless, minor discrepancies are possible in the current situation. Examples of typical grooves recommended for welding duplex steels are presented in [38]. For single-sided welding, the grooves should be shaped to obtain a wider root gap, lower root face, and wider groove angle (bevel) [46]. The wider root gap and lower root face limit the weld metal and the parent material mixing rate, which reduces the Ni content in the weld metal of the weld root. The root bead should be massive enough to counteract the nitrogen deficiency at this welding step by extending the cooling time in the austenite formation temperature range. The weld root should be welded using high linear energy, within limits recommended by the weld metal manufacturer. Under-heating of the weld root bead accelerates the cooling in the austenite formation temperature range. Nevertheless, excessive overheating lengthens the cooling time and stimulates the precipitation of harmful secondary phases. The consequence is a reduction in pitting corrosion resistance and impact strength. The following filling beads are often called "cold" runs. They should be welded with the linear energy reduced by up to 75% and should not be massive so as not to cause changes in the microstructure of the root run and in the HAZ, reaching directly under the passive layer. The following filling runs are to be welded with recommended increased heat input energy, which in the face layer reaches up to 150% of the heat input used for the weld root [54]. The thermal effects of the successive layers of the weld, lying above the "cold" run, must in no way affect the microstructure and properties of the weld root run.

The most problematic for maintaining the proper microstructure and sufficient pitting resistance seems to be the single- and two-runs of the thin-walled welds. Such problems appear in the seal welds of shell-and-tube heat exchangers [55]. Delicate girth welds in-between massive perforated bottom and thin-walled pipe are often welded with an intense mixing rate of weld and parent material of the pipe, and additionally, the cooling rate is higher due to massive perforated bottom. The content of δ-ferrite usually exceeds the limiting value of 70% even when using recommended welding material. In such cases, the use of austenitic weld metal with a high Mo content allows obtaining ferrite amount slightly smaller than 70%, which means a limited resistance to pitting corrosion. Furthermore, due to the cumulative effects of heat exposure of duplex steel and precipitation of secondary phases, it is not advisable to cut the materials thermally. In order to avoid a heat accumulation during welding, it is not recommended to preheat the steel, apart from drying the surface at a temperature not exceeding 100 °C. For the same reason, the inter-pass weld temperature should be strongly limited.

The recommendations given above allow obtaining welded joints resistant to pitting corrosion for standard duplex DSS 22% Cr steels. However, with increasing Cr content, the necessary cooling rate must be controlled. An insufficient austenite content in the weld can occur if the cooling rate is too high. A release of harmful secondary phases can be observed if the cooling rate is too low. The welding of high Cr duplex steels can be facilitated by using a combined welding method, for example, GTAW with assisted cooling of the weld by a micro-jet injector and argon as a refrigerant [56,57]. The rapidly expanding gas quickly removes heat from the weld, allowing for 2–3 times higher intensification of the cooling process. The introduction of micro-jet cooling allows increasing the welding heat.

It ensures an increase in the share of austenite in the microstructure and thus increases the pitting corrosion resistance of the joint. Moreover, the controlled and appropriately rapid cooling below 1000 °C avoids the formation of harmful secondary phases.

The implementation of the high-temperature heat treatment after the welding, carried out to properly balance the microstructure of duplex steel joints, is possible only for small objects that can be fully homogenized annealed at 1050–1150 °C and then supersaturated in water. Local heat treatment with the use of heating mats cannot be used due to degradation of the steel microstructure at the edges and due to the impossibility of rapid cooling. The large objects can be stress-relieved by tempering at temperatures below 300 °C for about 10 h so as not to initiate the microstructural changes within the lower TTT curve, as shown previously in Figure 7.

The corrosion resistance of welded joints made of duplex steel can be improved by chemical etching. The etching removes the oxide layer formed above the weld and heat-affected zone as a result of welding and re-establishes a more compact passive layer. The original oxide layer on the joint and in HAZ can be thick up to 100 nm [54]. The layer is enriched with Fe_2O_3 and thus has a low resistance to pitting. The large depletion layer of Cr and Mo further facilitates the development of pitting corrosion. The etched surfaces are usually treated with highly oxidizing nitric acid, hydrofluoric acid, or peroxide [58]. The chemically formed passive layer is tighter, and the concentration of Cr_2O_3, MoO_2, and MoO(OH) in the layer is higher.

10. Additional Requirements for Welding Quality Control

Ensuring the corrosion resistance of duplex steel welded joints requires extending the conventional routine quality control activities with a few additional measures. The first is the need to control the O_2 content in shielding and purge gases. The concentration should be kept below 200 ppm O_2. This can be achieved by sufficient gas purging flow of the inside area. The second requirement is the need for continuous and accurate monitoring of the weld inter-pass temperature during the welding. In duplex steels welds, this temperature is usually significantly lower than in other steels [59]. An overheating may cause the precipitation of harmful secondary phases. Moreover, it is necessary to constantly monitor the ferrite content with a ferritometer. This is especially important in the case of single- and double-run of thin-walled welds, as they have an increased tendency to excessive ferritization.

The changes in the microstructure of duplex steel joints are usually accompanied by a decrease in the breaking energy KV measured in the impact test. Low values of the toughness KV of the considered weld, or the whole HAZ, tested according to the standard ASTM A923 (Method B) [20], may indicate a possible lack of sufficient corrosion resistance. In this case, it is recommended to perform specialized pitting corrosion resistance tests in the chloride environment. The final control of the passivity of such welded joints, both in installations and in structures made of stainless steel, may be performed after the final etching and passivation using portable testers such as, for example, the Oxyliser 3 probe [60].

11. Corrosion Resistance Tests of Welded Joints Made of Duplex Steel, Carried out for the Chloride Environment

The American standard ASTM G48 [18] is the leading standard for corrosion resistance testing of duplex alloys and their welded joints. It contains several fundamental test procedures for assessing the resistance of stainless steels and related alloys in a ferric chloride solution. However, in the ASTM G48 standard, the criteria for evaluating the test results obtained after the experiments are not explicitly defined. Therefore, the results should be interpreted in conjunction with other guidelines taken, for example, from the Norwegian standard NORSOK M601 [19] or American standard ASTM A923 [20]. The corrosive environment in these tests is an oxygenated aqueous 6% $FeCl_3$ solution. This salt partially hydrolyzes in water. The temperature of the solution increases the degree of hydrolysis, which results in a more acidic solution. The $FeCl_3$ salt does not introduce foreign metal cations into the corrosive environment. The solution is not oxidizing. As such, it does not passivate the metal surfaces and has a high penetration capacity for the surface micro-damages [61].

The results of $FeCl_3$ tests are affected by temperature and autocatalytic course of pitting corrosion. According to the ASTM G48-method A standard procedures, the recommended test temperature for duplex DSS 22% Cr steels is 22 ± 2 °C and for SDSS steels is 35 ± 2 °C. The use of thermostatic water baths with temperature stabilization at ± 0.2 °C is recommended. If there is no consensus on the temperature conditions of tests, a deviation from recommendations of the ASTM G48-method A standard is permissible, provided that all the other elements of the standardized test procedure are followed. This deviation is allowed since in less corrosive environments, such as, for example, NaCl solution, the test temperature may be correspondingly higher. Due to the autocatalytic nature of the pitting corrosion, the extension of the test duration is accompanied by an increase in the average daily mass loss. Then, there is a gradual blurring of differences in corrosion losses between materials of different resistance, and the probability of obtaining an unreliable test result increases. For these reasons, the test time originally proposed in ASTM G48-method A (72 h) has been reduced to 24 h in the NORSOK M601 and ASTM A923 standards. The standards recommend using flat samples with dimensions of 25 mm × 50 mm or sections of tubular surfaces which are equivalent to these sizes. Any unevenness caused by machining should be smoothed and sharp edges rounded. Moreover, efforts should be made to minimize the side surfaces of the samples. In the case of thick samples, taken, for example, from multi-pass joints, cutting a thin sample from the weld root or weld face layer can be a good option as it is responsible for the corrosion resistance of the entire joint. Ideally, the exposed surface should be representative of the corrosion risks within the joint. It must therefore encompass the joint itself, HAZ, and base material. It is also advisable to mirror the surface roughness of the welded joint. The root and face weld surfaces should not be mechanically polished [62]. In the comparative tests of pitting corrosion resistance of the basic materials, a maximum standardization of the shape, dimensions, and surface conditions of samples must be guaranteed. It is also necessary to round the sample edges.

The NORSOK M601 standard supplements the requirements with preliminary etching in HNO_3 and HF solutions. By etching, a thick and leaky passive layer above the weld and heat-affected zone is removed. A re-passivation under free oxidation in the air requires at least 24 h to obtain a sufficiently thick and tight passive layer. The NORSOK M601 standard suggests maintaining the time interval between the sample preparation and the test itself. Direct corrosion testing immediately after etching may result in uniform corrosion without visible pitting, exceeding the limit of the allowable weight loss. In such a case, it is necessary to repeat the corrosion test by doing the preparations again and keeping a 24 h interval between the HNO_3 + HF pre-etch operation and the main $FeCl_3$ test.

It is permissible and beneficial to replace the manual sample washing with an ultrasonic bath. This ensures more effective removal of corrosion products from the sample surface and positively affects the reliability of mass measurements. The test should be performed in a stationary medium with free air access to the $FeCl_3$ solution. Cutting off the access of oxygen and solution stirring causes the polarization of corrosion cells and the electrochemical processes between the sample and the solution may cease.

The results of the pitting resistance test according to ASTM G48-method A are assessed based on the weight loss measurements and visual inspection. Due to the small weight loss of the samples during the test, analytical balances with a measurement accuracy of 1 mg should be used. The permissible maximum weight loss, according to NORSOK M601, is $4 \text{ g}/(\text{m}^2 \text{ per day})$. It corresponds to a corrosion rate of ~0.2 mm/year. In the tests associated with ASTM A901; however, a weight loss of $1 \text{ g}/(\text{m}^2 \text{ per day})$ is allowed, which corresponds to a uniform corrosion rate of ~0.05 mm/year. In practice, these thresholds are achieved at the time of few tiny pit appearances on the sample surface, noticeable at 20× magnification. The occurrence of a single pitting on the surface of a sample, noticeable 20× magnification, qualifies the test result as negative.

In Figures 15 and 16, typical visual assessment results are shown [21]. The tests were carried out according to procedures recommended for use in the ASTM G48-method A standard. The photos presented in Figure 15 show the classic DSS 22% Cr steel, grade 2205 (1.4462, F51), hand-welded with the GTAW method (TIG, 141), shielded with Ar + N_2 mixture, with 2209 wire. A well-balanced parent material with δ-ferrite content of ~51% was used for welding. As there is a relatively long initiation time of secondary phase precipitation, it is possible to extend the thermal exposure time in the range of austenite forming temperature by introducing more heat to the weld. As a result, ~42% δ-ferrite content has been obtained in the root of the weld, and ~55% in its face. The thermal welding cycle increased the ferrite content in the HAZ to ~62%. Due to the relatively low temperature, the amount was still within safe limits. The cooling rate below 1050 °C was sufficiently high, and no harmful secondary phases were released at the ferrite grain boundaries. The corrosion resistance test performed according to the ASTM G48-method A procedure, in conditions typical for the DSS duplex steel, showed a weight loss of less than 1 g/(m^2 per day).

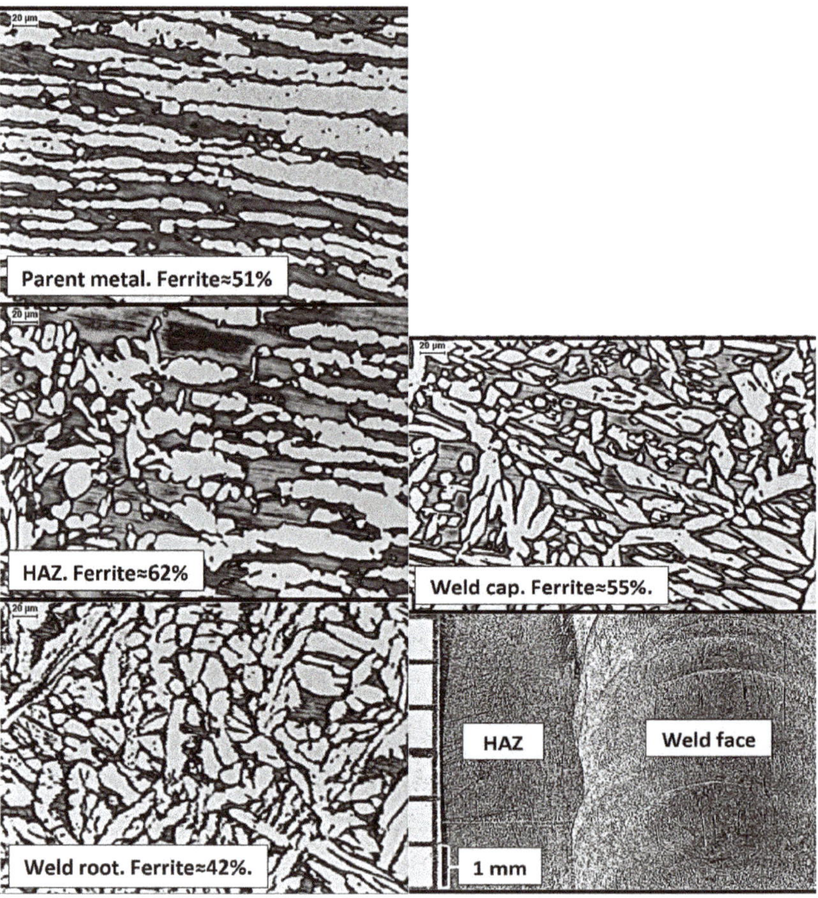

Figure 15. Authors' analysis of welded connection made of duplex DSS steel with high resistance to pitting corrosion identified for the environment of chlorides.

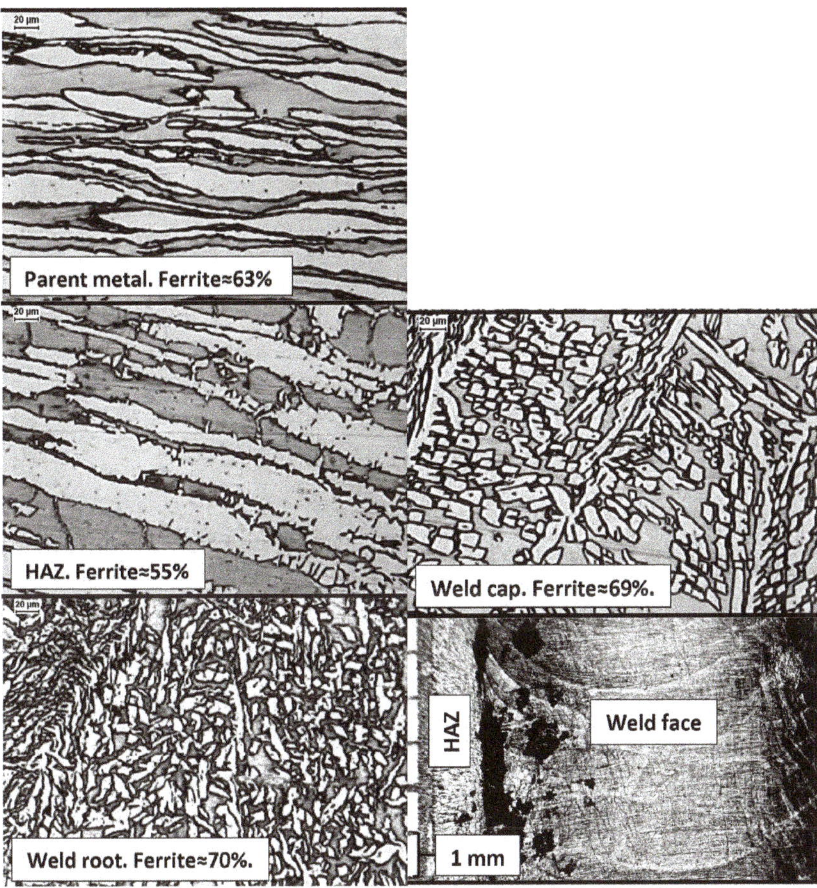

Figure 16. Authors' analysis of welded connection made of super duplex SDSS steel with a low resistance to pitting corrosion identified for the environment of chlorides.

The photograph of a representative sample surface, taken with a $20\times$ magnification (Figure 15), does not reveal any signs of pitting corrosion. Similarly, no corrosion changes were found on the weld root side. It shows that in the case of DSS 22% steel, the compliance with the material and technological recommendations allows an average experienced welder to obtain a joint with the required level of pitting corrosion resistance.

The opposite situation is shown in Figure 16. The SDSS duplex steel, grade 2507 (1.4410, F53), was subjected to manual GTAW welding (TIG, 141), shielded with Ar + N_2 mixture, with 2509 wire. Nevertheless, either an insufficiently balanced or heterogeneous base material was used for welding with increased δ-ferrite content, amounting to ~63% in the base material and to ~55% in the HAZ. It can be presumed that the heat introduced into the joint was insufficient to form an optimal amount of austenite. This is indicated by the high content of δ-ferrite, both in the root and in the face of the weld, amounting to ~70%, and by the small size of austenite dendrites. The representative photo of a sample surface, taken from the weld face side with $20\times$ magnification, reveals intense pitting corrosion in the weld face at the fusion line (Figure 16). It may be a result of a too-short cooling time, low Ni content in the welding wire, use of shielding gas with low N_2 content, or a manual error in welding, causing excessive local mixing of the parent material and weld metal. The corrosion resistance test, carried out according to the ASTM G48-method A procedure, showed a weight loss of over 10 g/(m^2 per day). This example shows the sensitivity of SDSS steels to welding technology errors, resulting in a lack of resistance to pitting.

12. Concluding Remarks

This work was meant to be a practical guide for designers, contractors, and investors of various types of structures made of duplex steel. It provides a set of recommendations. The suggestions presented here result from many years of professional experience of the authors in the field. Our observations show that many mistakes are still made, mainly due to the incorrect selection of steel or an inappropriately selected welding technology.

The basic criterion for assessing the suitability of a given steel grade was its resistance to pitting corrosion. Sufficient corrosion resistance can be achieved by:

- careful selection of the parent material with the lowest possible content of ferrite-forming elements, the lowest possible content of C and S in the melt and properly balanced steel microstructure within limits $\frac{\text{ferrite}}{\text{austenite}} \approx \frac{50\%^{-10\%}}{50\%^{+10\%}}$;
- use of the low-oxygen welding methods, protection of the weld pool against oxygen absorption from the atmosphere;
- use of a dedicated additional material with an increased Ni content (2%–4%), and ensuring the enrichment of the weld metal with austenitic nitrogen by using forming and protective gases with an appropriate N_2 content;
- regulating the kinetics of phase transformations during welding by appropriately shaping the welding groove, controlling the amount of heat introduced into the weld and massiveness of subsequent beads;
- avoiding harmful thermal exposure at temperatures exceeding 300 °C, including adherence to the recommended inter-pass temperature during welding and its continuous monitoring;
- monitoring the level of δ-ferrite content in the joints;
- chemical etching and passivation of both the weld area and adjacent HAZ.

The application of a joint cooling method, with the use of a micro-jet injector in mechanized welding, may support the regulation of phase changes kinetics in the high alloyed DSS 25% Cr, SDSS, and HDSS steels. It enables welding with high linear energies of the arc, which positively influences the δ-ferrite decomposition and formation of austenite in the weld. It provides a fast cooling of the joint within the safe temperature range and outside the harmful phase transformations of both the steel and duplex weld metal.

The pitting corrosion resistance tests in chlorides should be carried out in environmental conditions that do not exceed the limiting resistance of the tested materials. The tests should be conducted on those areas of the welded joint that are essential for corrosion protection in the specific installation of the welded structure.

Author Contributions: Conceptualization, M.M., M.S., and B.S.; methodology, M.S. and B.S.; validation, M.S.; formal analysis, M.S. and B.S.; resources, M.S. and B.S.; data curation, M.S. and B.S.; writing—original draft preparation, M.M. and M.S.; writing—review and editing, M.M. and M.S.; visualization, M.M., M.S., and B.S.; supervision, M.M. All authors have read and agreed to the published version of the manuscript.

Funding: This research received no external funding.

Institutional Review Board Statement: Not applicable.

Informed Consent Statement: Not applicable.

Data Availability Statement: Not applicable.

Conflicts of Interest: The authors declare no conflict of interest.

Abbreviations

WPQR	Welding Procedure Qualification Record
LDSS	Lean Duplex Stainless Steel
DSS	(Standard) Duplex Stainless Steel
SDSS	Super Duplex Stainless Steel
HDSS	Hyper Duplex Stainless Steel
PREN	Pitting Resistance Equivalent Number
$K_{\frac{\delta}{\gamma}}$	Partition coefficient of alloying elements between δ-ferrite and γ-austenite
$\Delta t_{12/8}$	Temperature range between 1200 °C and 800 °C
%wt.	Weight percent
TTT	Time-Temperature-Transformation
FN	Ferrite Number
GTAW	Gas Tungsten Arc Welding
PAW	Plasma Arc Welding
GMAW	Gas Metal Arc Welding
SMAW	Shielded Metal Arc Welding
SAW	Submerged Arc Welding
FCAW	Flux-Cored Arc Welding
HAZ	Heat Affected Zone
STT	Sufrace Tension Transfer
CPT	Critical Pitting Temperature
VOD	Vacuum Oxygen Decarburization

References

1. Kahar, S.D. Duplex stainless steels—An overview. *Int. J. Eng. Res. Appl.* **2017**, *7*, 27–36. [CrossRef]
2. Marrow, T.J.; Sunoog, K.; Oh, J.T. The effect of microstructure on brittle fracture and stress corrosion cracking in duplex stainless steel. In Proceedings of the Stainless Steel World 99 Conference on Corrosion-Resistant Alloys, The Hague, The Netherlands, 16–18 November 1999.
3. Levkov, L.; Shurygin, D.; Dub, V.; Kosyrev, K.; Balikoev, A. New generation of super duplex steels for equipment gas and oil production. In Proceedings of the International Conference on Corrosion in the Oil and Gas Industry, St. Petersburg, Russia, 22–24 May 2019.
4. Perren, R.A.; Sutter, T.A.; Uggowitzer, P.J.; Weber, L.; Magdowski, R.; Böhni, H.; Speidel, M.O. Corrosion resistance of super duplex stainless steel in chloride ion containing environments: Investigations by means of a new microeletrochemical method. I. Precipitation-free states. *Corros. Sci.* **2001**, *43*, 707–726. [CrossRef]
5. Perren, R.A.; Sutter, T.A.; Solenthaler, C.; Gullo, G.; Uggowitzer, P.J.; Böhni, H.; Speidel, M.O. Corrosion resistance of super duplex stainless steel in chloride ion containing environments: Investigations by means of a new microeletrochemical method. II. Influence of precipitates. *Corros. Sci.* **2001**, *43*, 727–745. [CrossRef]
6. Singh Raman, R.K.; Siew, W.H. Role of nitrite addition in chloride stress corrosion cracking of a super duplex stainless steel. *Corros. Sci.* **2010**, *52*, 113–117. [CrossRef]
7. Vignal, V.; Zhang, H.; Delrue, O.; Heintz, O.; Popa, I.; Peultier, J. Influence of long-term ageing in solution containing chloride ions on the passivity and the corrosion resistance of duplex stainless steels. *Corros. Sci.* **2011**, *53*, 894–903. [CrossRef]
8. Kim, S.-T.; Lee, I.-S.; Kim, J.-S.; Jang, S.-H.; Park, Y.-S.; Kim, K.-T.; Kim, Y.-S. Investigation of the localized corrosion associated with phase transformation of tube-to-tube sheet welds of hyper duplex stainless steel in acidified chloride environments. *Corros. Sci.* **2012**, *64*, 164–173. [CrossRef]
9. Yang, Y.Z.; Jiang, Y.M.; Li, J. In situ investigation of crevice corrosion on UNS S32101 duplex stainless steel in sodium chloride solution. *Corros. Sci.* **2013**, *76*, 163–169. [CrossRef]
10. Zanotto, F.; Grassi, V.; Balbo, A.; Monticelli, C.; Zucchi, F. Stress corrosion cracking of LDX 2101 duplex stainless steel in chloride solutions in the presence of thiosulphate. *Corros. Sci.* **2014**, *80*, 205–212. [CrossRef]
11. Schmid, A.; Moria, G.; Hönig, S.; Weil, M.; Strobl, S.; Haubner, R. Comparison of the high-temperature chloride-induced corrosion between duplex steel and Ni based alloy in presence of H2S. *Corros. Sci.* **2018**, *139*, 76–82. [CrossRef]
12. Silva, R.; Kugelmeier, C.L.; Vacchi, G.S.; Martins Junior, C.B.; Dainezi, I.; Afonso, C.R.M.; Mendes Filho, A.A.; Rovere, C.A.D. A comprehensive study of the pitting corrosion mechanism of lean duplex stainless steel grade 2404 aged at 475 °C. *Corros. Sci.* **2021**, *191*, 109738. [CrossRef]
13. Sasikumar, C.; Sundaresan, R.; Merlin Medona, C.; Ramakrishnan, A. Corrosion study on AA-TIG welding of duplex stainless steel. *Mater. Today: Proc.* **2021**, *45*, 3383–3385. [CrossRef]
14. Shi, J.; Ming, J.; Sun, W. Passivation and chloride-induced corrosion of a duplex alloy steel in alkali-activated slag extract solutions. *Constr. Build. Mater.* **2017**, *155*, 992–1002. [CrossRef]

15. Briz, E.; Biezma, M.V.; Bastidas, D.M. Stress corrosion cracking of new 2001 lean–duplex stainless steel reinforcements in chloride contained concrete pore solution: An electrochemical study. *Constr. Build. Mater.* **2018**, *192*, 1–8. [CrossRef]
16. Alonso, M.C.; Luna, F.J.; Criado, M. Corrosion behavior of duplex stainless steel reinforcement in ternary binder concrete exposed to natural chloride penetration. *Constr. Build. Mater.* **2019**, *199*, 385–395. [CrossRef]
17. Welding procedures and welders. In *Guidelines of Approval Testing, WG 01*, 3rd ed.; Safety Assessment Federation (SAFed): London, UK, 2020.
18. ASTM G48-11; *Standard Test Method for Pitting and Crevice Corrosion Resistance of Stainless Steels and Related Alloys by Use of Ferric Chloride Solutions*; ASTM: West Conshohocken, PA, USA, 2011.
19. NORSOK M601; *Welding and Inspection of Piping*; NORSOK: Lysaker, Norway, 2008.
20. ASTM A923-03; *Standard Test Methods for Detecting Detrimental Intermetallic Phase in Duplex Austenitic/Ferritic Stainless Steels*; ASTM: West Conshohocken, PA, USA, 2003. [CrossRef]
21. Stankiewicz, M.; Slazak, B. Resistance to pitting corrosion of duplex steel and its welded joints. In *The Importance of Steel Selection, Modeling Methods and Experimental Tests in the Design of Building Structures*; Maslak, M., Ed.; Selected Issues; Publishing House of the Cracow University of Technology: Cracow, Poland, 2020; pp. 47–83. (In Polish)
22. Calderon-Uriszar-Aldaca, I.; Briz, E.; Garcia, H.; Matanza, A. The weldability of duplex stainless steel in structural components to withstand corrosive marine environments. *Metals* **2020**, *10*, 1475. [CrossRef]
23. Knyazeva, M.; Pohl, M. Duplex Steels. Part I: Genesis, Formation, Structure and Part II: Carbides and Nitrides. *Metallogr. Microstruct. Anal.* **2013**, *2*, 113–121, 343–351. [CrossRef]
24. Craig, B. Clarifying the applicability of PREN equations: A short focused review. *Corrosion* **2021**, *77*, 382–385. [CrossRef]
25. Peguet, L.; Gaugain, A. Localized corrosion resistance of duplex stainless steels: Methodology and properties: A review paper. *Revue de Métallurgie* **2011**, *108*, 231–243. [CrossRef]
26. Berns, H.; Theisen, W. *Ferrous Materials–Steel and Cast Iron*; Springer: Berlin, Germany, 2008; pp. 47, 322. [CrossRef]
27. Nilsson, J. Overview–super duplex stainless steels. *J. Mater. Sci. Technol.* **1992**, *8*, 685–700. [CrossRef]
28. Gunn, R.N. (Ed.) *Duplex Stainless Steels. Microstructure, Properties and Applications*; Abington Publishing: Nashville, TN, USA, 1997. [CrossRef]
29. Speidel, M.O. Corrosion science of stainless steels. In Proceedings of the International Conference on Stainless Steels, Chiba Japan, 10–13 June 1991; Volume 1, pp. 25–34.
30. Holloway, G. Effective welding of duplex & super duplex stainless steels. In Proceedings of the Welding Society Meeting, Singapore, 29 October 2003.
31. Charles, J. Super duplex stainless steels: Structure and properties. In Proceedings of the Conference on Duplex Stainless Steels, Beaune, Bourgogne, France, 28–30 October 1991; Volume 1, pp. 3–48.
32. Bernhardsson, S. The corrosion resistance of duplex stainless steels. In Proceedings of the Conference Duplex Stainless Steels, Beaune, Bourgogne, France, 28–30 October 1991; Volume 1, p. 185.
33. Elhoud, A.M.; Renton, N.C.; Deans, W.F. The effect of manufacturing variables on the corrosion resistance of a super duplex stainless steel. *Int. J. Adv. Manuf. Technol.* **2011**, *52*, 451–461. [CrossRef]
34. Chail, G.; Kangas, P. Super and hyper duplex stainless steels: Structures, properties and applications. In Proceedings of the 21st European Conference on Fracture (ECF21), Catania, Italy, 20–24 June 2016.
35. Hosseini, V.; Karlsson, L.; Engelberg, D.; Wessman, S. Time-temperature-precipitation and property diagrams for super duplex stainless weld metals. *Weld. World* **2018**, *62*, 517–533. [CrossRef]
36. Pohl, M.; Storz, O.; Glogowski, T. Effect of intermetallic precipitations on the properties of duplex stainless steel. *Mater. Charact.* **2007**, *58*, 67–71. [CrossRef]
37. Barbosa, C.A.; Sokolowski, A. Development of UNS S 32760 super-duplex stainless steel produced in large diameter rolled bars. *Revista Escola de Minas* **2013**, *66*, 201–208. [CrossRef]
38. Practical Guidelines for the Fabrication of Duplex Stainless Steel. Available online: http://www.edelstahl.de/fileadmin/user_upload/ess_duplex_steel.pdf (accessed on 2 July 2021).
39. Lippold, J.C.; Kotecki, D.J. *Welding Metallurgy and Weldability of Stainless Steels*; John Willey & Sons Inc.: New York, NY, USA, 2015.
40. Akpanyung, K.V.; Loto, R.T. Pitting corrosion evaluation: A review, International Conference on Engineering for Sustainable World. *J. Phys. Conf. Ser.* **2019**, *1378*, 022088. [CrossRef]
41. Örnek, C.; Davut, K.; Kocabaş, M.; Bayatli, A.; Ürgen, M. Understanding corrosion morphology of duplex stainless steel wire in chloride electrolyte. *Corros. Mater. Degrad.* **2021**, *2*, 397–411. [CrossRef]
42. Kotecki, D.J. Ferrite determination in stainless steel welds: Advances since 1974. *Weld. J.* **1997**, *76*, 24–37.
43. Olsson, C.-O.A.; Landolt, D. Passive films on stainless steels–chemistry, structure and growth. *Electrochim. Acta* **2003**, *48*, 1093–1104. [CrossRef]
44. Zhang, B.; Wei, X.X.; Wu, B.; Wang, J.; Shao, X.H.; Yang, L.X.; Zheng, S.J.; Zhou, Y.T.; Jin, Q.Q.; Oguzie, E.E.; et al. Chloride attack on the passive film of duplex alloy. *Corros. Sci.* **2019**, *154*, 123–128. [CrossRef]
45. Souier, T.; Martin, F.; Bataillon, C.; Cousty, J. Study of the passive film on duplex stainless steels and its breakdown by AFM and CS-AFM. In Proceedings of the International Conference EUROCORR, Nice, France, 6–10 September 2009; Volume 5, pp. 2869–2880.
46. Alvarez-Armas, I.; Degallaix-Moreuil, S. (Eds.) *Duplex Stainless Steels*; ISTE Ltd.: London, UK, 2009. [CrossRef]

47. Wiktorowicz, R.; Crouch, J. Shielding gas developments for TIG welding of duplex and super duplex stainless steels. *Weld. Metal. Fabr.* **1994**, *62*, 379–383.

48. Messer, B.; Oprea, V.; Wright, A. Duplex stainless steel welding: Best practices. *Stainless Steel World* **2007**, *53*, 53–63.

49. Babish, F. *Repair Welding of Stainless Steels*; Sandvik Materials Technology: Clarks Summit, PA, USA, 2014.

50. Stankiewicz, M.; Slazak, B. Arc welding of duplex steels in terms of maximizing the resistance of welded joints to pitting corrosion in the chloride environment. *Przegląd Spawalnictwa* **2018**, *90*, 142–150. (In Polish) [CrossRef]

51. Patrick, C.W.; Cox, M.A. Challenges welding duplex and super duplex stainless steel, American Fuels & Petrochemical Manufacturers. In Proceedings of the Reliability & Maintenance Conference, San Antonio, TX, USA, 20–23 May 2014. Available online: http://www.gowelding.com/met/duplex.html (accessed on 10 July 2021).

52. Sato, Y.S.; Kokawa, H.; Kuwana, T. Effect of nitrogen on sigma transformation in duplex stainless steel weld metal. *Sci. Technol. Weld. Join.* **1999**, *41*, 41–49. [CrossRef]

53. Van Nassau, L.; Meelker, H.; Neessen, F.; Hilkes, J. Welding Duplex and Superduplex Stainless Steel. An Update of the Guide for Industry. Available online: http://www.lasgroepzuid.com/documenten/Update_guide_Welding_duplex_and_superduplex_stainless_steel.pdf (accessed on 12 July 2021).

54. Baxter, C.; Young, M. Practical aspects for production welding and control of duplex stainless steel pressure and process plants. In Proceedings of the Conference Duplex America 2000; DA2 032; Stainless Steel World, KCI Publishing BV: Zuthphen, The Netherlands, 2000; pp. 129–139.

55. Faes, W.; Lecompte, S.; Zaaquib, Y.A.; Van Bael, J.; Salenbien, R. Corrosion and corrosion prevention in heat exchangers. *Corros. Rev.* **2019**, *2*, 131–155. [CrossRef]

56. Wegrzyn, T.; Piwnik, J.; Lazarz, B.; Wieszala, R.; Hadrys, D. Parameters of welding with micro-jet cooling. *Arch. Mater. Sci. Eng.* **2012**, *54*, 86–92.

57. Wegrzyn, T.; Piwnik, J.; Hadrys, D.; Wszolek, L. Low alloy steel structures after welding with micro-jet cooling. *Arch. Metall. Mater.* **2017**, *62*, 115–118. [CrossRef]

58. Ward, I. *Report on Weld Cleaning Methods, Document IW 150807*; Sandvik Australia Pty Ltd.: Mulgrave, VIC, Australia, 2007.

59. Alvarez, T.; Guttemberg, C.S.; Pardal, J.M.; Tavares, S. Influence of interpass temperature on the properties of duplex stainless steel during welding by submerged arc welding process. *Weld. Int.* **2016**, *30*, 348–358. [CrossRef]

60. Pan, J. Studying the passivity and breakdown of duplex stainless steel at micrometer and nanometer scales–the influence of microstructure. *Front. Mater.* **2020**, *7*, 133. [CrossRef]

61. Woollin, P. Ferric chloride testing for weld procedure qualification of duplex stainless steel weldments. In Proceedings of the International Conference UK Corrosion and EUROCORR, Bournemouth, UK, 31 October–3 November 1994.

62. Rajaguru, J.; Arunachalam, N. Effect of machined surface integrity on the stress corrosion cracking behaviour of super duplex stainless steel. *Eng. Fail. Anal.* **2021**, *125*, 105411. [CrossRef]

*materials*

Review

Electrochemical Deposition of Ni, NiCo Alloy and NiCo–Ceramic Composite Coatings—A Critical Review

Nyambura Samuel Mbugua, Min Kang *, Yin Zhang, Ndumia Joseph Ndiithi, Gbenontin V. Bertrand and Liang Yao

College of Engineering, Nanjing Agricultural University, Nanjing 210031, China; 2017112123@njau.edu.cn (N.S.M.); 2018212003@njau.edu.cn (Y.Z.); 2017112122@njau.edu.cn (N.J.N.); 2019212016@njau.edu.cn (G.V.B.); 2018112003@njau.edu.cn (L.Y.)
* Correspondence: kangmin@njau.edu.cn; Tel.: +86-25-5860-6578

Received: 1 July 2020; Accepted: 3 August 2020; Published: 6 August 2020

Abstract: In recent years, alloy and alloy-ceramic coatings have gained a considerable attention owing to their favorable physicochemical and technological properties. In this review, we investigate Ni, NiCo alloy and NiCo–ceramic composite coatings prepared by electrodeposition. Electrodeposition is a versatile tool and cost-effective electrochemical method used to produce high quality metal coatings. Surface finish and tribological properties of the coatings can be further improved by the addition of suitable agents and control of deposition operating conditions. In this review, Ni, NiCo alloy and NiCo–ceramic composite coatings prepared by electrodeposition are reviewed by critically evaluating previous researches. The use of the coatings and their potential for future research and development are discussed.

Keywords: nanostructured coatings; microhardness; corrosion resistance; electrochemical deposition

---

1. Introduction

Materials are a fundamental pillar in engineering technology. The materials' electrochemical, thermal and mechanical interaction begins on the surface. However, material surfaces face the constant threat of wear and corrosion resulting in massive losses in industry. The use of surface enhancement technology to prevent or mitigate the loss has therefore become inevitable. Over the last few decades, major scientific development in the fields of metallurgy and materials has occurred, giving rise to new engineering materials with superior properties. Developing suitable processing to produce desired materials is complex and requires altering the inherent properties of the materials. Consideration is made of the overall economic perspective and its environmental impact.

Electrodeposition is an electrochemical process that is used to modify the surface structure. Use of electrodeposition in surface engineering can be traced to nearly 200 years ago based on some hypotheses [1]. The galvanic cell, invented in early 1800's, paved way for use of electric current to produce coatings as a more cost-effective technique. The electrodeposition technique possesses an edge over other coating techniques due to several reasons [2]:

(i)   Low initial investment coupled with high rate of production.
(ii)   It can be used with a wide variety of shapes and sizes of substrates
(iii) Ease of producing economically viable quantities of nanocomposite materials, with grain sizes as small as 10 nm.
(iv) Products of electrodeposition require no further processing and can be used immediately after the process.

(v)   It is an easy concept that can be replicated in industry and laboratories with minimal technological barriers.

(vi)   Electrodeposited Ni coatings have exhibited superior density and lower porosity.

Electrodeposition is based on the principle that a layer of coating, either single or multilayer, forms as a result of the electrode–electrolyte electrochemical reactions occurring leading to electrodeposition of ions contained in electrolyte. It utilizes the properties of materials in their metastable state as a result of reduction of the grain size on the nano scale. In such a state, the grain boundaries contain a proportion of atoms that is higher or equivalent to those inside the grains [3]. These new types of materials consisting of nanoparticles are referred to as nanocomposite materials and they exhibit superior properties compared to traditional grain sized materials and sometimes offer completely new properties altogether. As a result, nanocomposite materials with grain sizes below 100 nm have received considerable attention from scientists and researchers all over the world. Several new electrodeposition synthesis methods have been devised over the years to increase production efficiency of new materials and minimize cost. This has ranged from using the basic direct current set ups to more ambitious procedures such as: pulse plating, jet electrodeposition, and the pulse reversal current technique [4–6].

Pure Ni is one of the most widely used alloying metals in the world owing to its superior corrosion and wear resistance [7]. Electrodeposited Ni coatings have uses in decoration, functional uses, as well as in engineering for surface protection [8]. Superplasticity in metals (materials exhibiting increased elongations to failure of >500%) is dependent on elevated testing temperatures and fine grain sizes of <10 μm [9]. Prasad and Chokshi [9] reported that electrodeposited nanocrystalline Ni is characterized by good superplasticity properties and is used in studying the phenomenon in electrodeposited metals owing to the ease of synthesizing coatings with small grain sizes. Ni coatings exhibit improved resistance to localized corrosion and this makes them perfect for use as anti-corrosive coatings [2]. Furthermore, research shows that thin films of different materials coated with epitaxial thin Ni coatings of a few nanometers have similar hardness to bulk nickel, such that the wear resistance of micro/nano electro mechanical system devices can be greatly improved if coated with a thin Ni coating [10]. Studies have shown that nano-sized nickel can aid the diamond yielding process. Ni atoms have a three dimension absent state which serves to attract electrons in the carbon fullerenes. This in turn causes the sp2 fullerene to be transformed into a diamondlike sp3 structure. This transformation can also be attributed to high reactivity which effectively aids the change at an impulse during shock wave loading [2].

Ni based alloy coatings can also be produced using electrodeposition. Choice of the alloying metal depends on the properties desired, which can range from good electrical conductivity, good wear and corrosion resistance, soft magnetic properties to special optical properties. Over the years, many different types of metals have been electrodeposited with Ni to form alloys: Co, Fe, Cu and W.

Ni–Co is one such alloy. There exist several different synthesis techniques for Ni–Co alloy coatings and they include: radio frequency magnetron sputtering [11], electrodeposition [12] and vacuum evaporation [13,14]. The electrodeposition technique has several advantages over the other two methods and they include: low cost, simplicity, scalability and manufacturability [15]. Furthermore, electrodeposition can be used to grow a wider range of materials.

Research shows that electrodeposited Ni–Co alloy coatings exhibit better properties compared to pure Ni coatings [12]. Ni–Co alloy exhibits higher hardness, better adhesion, excellent magnetic properties, high wear and corrosion resistance as well as good stability at high temperature [16–19]. Wang et al. [20] researched the effect of cobalt content on mechanical and microstructural properties on Ni–Co alloy coatings. It was reported that Ni–Co alloy coatings exhibited approximately double the microhardness when compared to pure Ni coatings. It was also reported that the Ni–49Co coatings exhibited a decreased rate of wear in comparison to pure Ni coatings. Hassani et al. [21] researched on low temperature superplasticity of nanocrystalline electrodeposited Ni–Co alloy with an average grain size of 20 nm. It was reported that a maximum elongation of 279% at a temperature of 773 K was obtained. Ni–Co alloys are considered to be the best suited materials for replacing hard chromium [22]. Research shows that microhardness in Ni–Co alloys increases gradually with increase in Co content up to an optimum level, after which the microhardness decreases with further increase in Co content [16].

To further improve the properties of Ni–Co alloy coatings, nanoparticles have been suspended in electrolyte and they become embedded into the electro-formed solid phase layer during electrodeposition [23]. In such materials, the inherent properties of the nanoparticles have been found to significantly influence the overall properties of the nanocomposite coatings. Many different types of nanoparticles have been electrodeposited with Ni–Co alloy including SiC, Al_2O_3, SiO_4, ZrO_2, Cr_2O_3, Si_3N_4 and TiO_2 [24]. These nano particles used in electrodeposition can be classified as either hard materials or soft materials depending on the desired properties. Soft materials such as graphite offer properties which include lower the friction coefficient to reduce wear and the coefficient of friction between shearing surfaces [25]. Hard materials such as Al_2O_3 improve the microhardness and wear resistance of surfaces [26]. The matrix phase microstructure coupled with nanoparticle content and distribution in the metallic matrix phase significantly influences the properties of nanoparticle reinforced Ni–Co matrix nanocomposite coatings. From past research, it is clear that addition of nanoparticles has served to improve the properties of Ni–Co coatings. In some instances, however, addition of Co has been observed to improve the overall properties of Ni-nanocomposite coatings such as hardness and residual stress [27]. Addition of Co^{2+} in Ni/diamond plating baths has been seen to greatly improve the deposition of diamond in the coatings resulting in higher diamond content, enhance bonding between matrix and particles, as well as more uniformly dispersed particles in the metal matrix, resulting in improved wear resistance and hardness properties [28]. The properties of Ni–Co alloys and their composite coatings have also been previously reported on by Karimzadeh A et al. [29]. This review aims takes a different approach to the wide research field of Ni–Co and Ni–Co-based composite coatings. Factors such as the effect of electrode orientation and forces existing in the electrolyte bath on the deposition process have been considered. Moreover, additives such as boric acid have been extensively explored, and the intricate working of the Watts solution has been more deeply discussed for easier understanding. Properties such as adhesion between substrate and coating have been extensively discussed, comparisons between techniques drawn, and recommendations for further adhesion improvement have been presented.

2. Electrodeposition Methods

2.1. Direct Current Electrodeposition

In Direct current (DC) electrodeposition, an electric current is continuously transferred through the system without any interruptions. DC electrodeposition is divided into two types depending on the orientation of the electrodes in electrolyte during the electrodeposition process. These are the conventional electrodeposition, and sediment codeposition (SCD) techniques. For conventional electrodeposition, the electrodes are placed vertically in the electrolyte, but for SCD they are placed horizontally. Adsorption of nanoparticles into the alloy matrix during electrodeposition is greatly influenced by forces acting on the suspended nanoparticles. The kinetics involved can be used to explain this difference. The two main forces at work during SCD electrodeposition are gravitational pull and the electrophoresis force, thereby giving more desirable properties compared to conventional deposition which solely relies on gravitational pull [30]. DC has several advantages over pulse electrodeposition and pulse reversal current (PRC) electrodeposition. These include simplicity, the availability of vast technical knowledge and affordability. Borkar T [31] reported that for all Ni and Ni nanocomposite coatings deposited, DC deposited coatings exhibited much stronger (less random) crystallographic textures compared to coatings deposited using PC and PRC deposition techniques. Furthermore, in DC electrodeposition, the Co content in the deposited Ni–Co coatings is dependent on the composition of the electrolyte, unlike in jet electrodeposition where Co electrodeposition is controlled by diffusion [4].

2.2. Pulse Current Electrodeposition

Pulse current electrodeposition (PC) has been used extensively over the years in Ni–Co electrodeposition [32]. The three most fundamental parameters that affect the properties of the coatings include in pulse current electrodeposition include: peak current (I_{peak}), pulse imposition time (ON-time, T_{ON}) and switch off time (OFF-time, T_{OFF}). These parameters relate mathematically to evaluate (1) pulse frequency, (2) duty cycle, and (3) average current density, as shown [32,33]:

$$f = \frac{1}{T_{OFF} + T_{ON}} \tag{1}$$

$$\gamma = \frac{T_{ON}}{T_{OFF} + T_{ON}} * 100 \tag{2}$$

$$I_{avg} \frac{T_{ON}}{T_{OFF} + T_{ON}} * I_{peak} = I_{peak} \tag{3}$$

where γ is the duty cycle, f the frequency, I_{peak} the peak current density, and I_{avg} the average current density.

In pulse electrodeposition, peak current density has a significant effect on microhardness, crystallite size, surface morphology, microstructure, composition, and tensile strength of PC deposited Ni–Co alloys and their nanocomposites [34]. Crystallization and growth in turn determine the microstructure of the nanocomposite. The texture of the coatings is determined by both peak current density and organic surfactants [35]. The quantity of adatoms located at the surface is higher due to the applied higher current density as compared to DC electrodeposition. This results in smaller grain sizes [5]. Padmanabhan [36] reported that pulse electrodeposition is used as an effective method for the reduction of grain sizes to nanoscale [5].

Although it is a relatively low-cost synthesis technique with simple implementation, pulse electrodeposition is ideal for production of full density nanocrystalline materials [33]. Yang and Cheng [32] reported that the morphology of the deposited coatings changed from nodular to acicular, and a finer grain size was observed with increasing pulse frequency and decreasing duty cycle. It has also been reported that at low current densities, smoother surface morphologies are observed [37], but an increase in peak current density produces distinct colony-like morphologies characterized by clearer colony boundaries [34].

Pulse electrodeposition exhibits several advantages over DC electrodeposition, such as improved wear resistance and hardness, particle distribution, structure, morphological structure and the ability to control the grain sizes of the deposits [38]. PC electrodeposition has higher instantaneous current density compared to DC electrodeposition, thereby increasing its effectiveness in agitating the adsorption–desorption processes that occur at the Ni electrolyte interface. This makes it possible to control the electrodeposited Ni coating microstructure [39,40]. It has also been reported that PC deposits contain a higher nanoparticle content compared to coatings deposited by the DC deposition technique [31]. Yang and Cheng [32] concluded that the high microhardness of deposited Ni–Co–SiC coatings was primarily associated with increasing SiC content in the coatings with increasing pulse frequency and decreasing duty cycle. This significantly increased the microhardness as a function of the dispersion strengthening mechanism. Watts type baths used for pulse plating of Ni based coatings have been reported to produce the best coating results [2].

The PC deposition technique suffers from a drawback called the double layer capacitance effect. The double charging layer of the electrodes is subjected to charging and discharging occurrences. In a case where the ON- and OFF-times are much shorter than the charging and discharging times, the PC would revert to DC current. As such, care is taken to select a high frequency thus ensuring the effect of the double layer capacitance effect is negligible.

2.3. Jet Electrodeposition

In this technique, a jet of plating solution is directed at the cathode surface directly. There exists an electrical field between the anode (located in the nozzle) and cathode (substrate). Figure 1 is a schematic representation of jet electrodeposition technique setup [4].

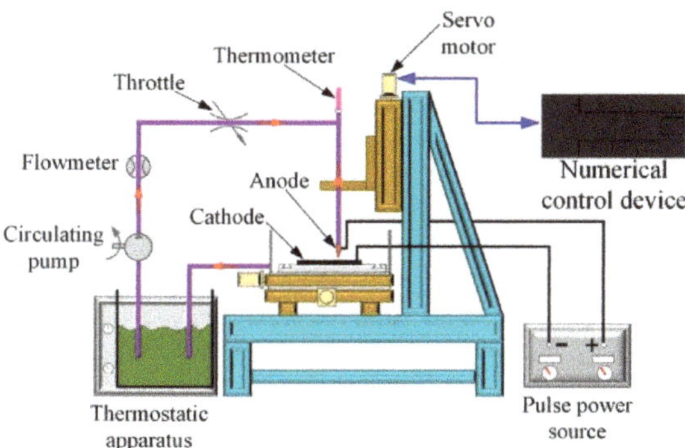

Figure 1. Schematic image of pulse jet electrodeposition.

As the plating solution flows, electric current is transferred along the stream of fluid to the substrate surface, thereby enabling deposition to occur on the cathode surface where the jet flows over [41]. Jet electrodeposition is a high-speed electroplating technique that offers a wide range of advantages [4]. These include: (i) a higher deposition rate compared to other conventional electrodeposition techniques, and (ii) a more efficient grain size refining effect. This is attributed to the cathode in the jet technique having a larger overpotential that can be used simultaneously with higher current densities. In this technique, Co content in the coatings increases with increases in electrolyte jet speed, Co^{2+} ion concentration in the electrolyte, and cathodic current density.

2.4. Pulse Reversal Current Electrodeposition

There are nine waveform parameters that can be used to represent the waveform characteristics according to the definition of bidirectional pulse parameters: (i) forward conduction time t_c, (ii) reverse conduction time t_d, (iii) forward peak current density I_{p+}, (iv) reverse peak current density I_{p-}, (v) average current density I_{av}, (vi) pulse period T, (vii) reverse current coefficient x, (viii) duty cycle λ, and (ix) frequency f [42]. The nine parameters are interrelated and they do not change independently. Where average current density I_{av}, duty cycle λ, reverse pulse coefficient x, and frequency f are considered as the independent variables, the parameters can be related mathematically, as shown in Equations (4)–(12) [43]:

$$I_{p+} = \frac{I_{av}}{x\lambda + \lambda - x} \tag{4}$$

$$I_{p-} = xI_{p+} \tag{5}$$

$$t_c = \lambda T = \frac{\lambda}{f} \tag{6}$$

$$t_d = T - t_c = \frac{1-\lambda}{f} \tag{7}$$

$$T = t_c + t_d \tag{8}$$

$$I_{av} = \frac{t_c\,I_{p+} - t_c\,I_{p-}}{T} \tag{9}$$

$$\lambda = \frac{t_c}{t_d} \tag{10}$$

$$x = \frac{I_{p-}}{I_{p+}} \tag{11}$$

$$f = \frac{1}{T} \tag{12}$$

Figure 2 shows the schematic diagram of a typical pulse-current wave form when $I_{av} = 15$ A·dm^2, $\lambda = 0.5$, $x = 0.5$, and $f = 10$ Hz.

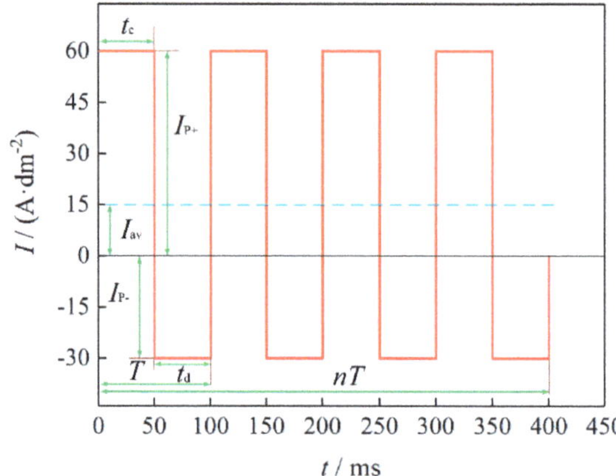

Figure 2. Schematic diagram of pulse waveform when $I_{av} = 15$ A·dm^2, $\lambda = 0.5$, $x = 0.5$, and $f = 10$ Hz.

Pulse reversal current (PRC) electrodeposition has been used over the years owing to its unique ability to enhance nanoparticle incorporation in electrodeposited composite coatings, with the highest reported content being 23 wt% [44–46]. Ni–Co coatings produced using the PRC synthesis technique have been reported to exhibit higher hardness, better anti-wear properties, more compact surfaces, lower residual macro stress and better compact surfaces than those produced using DC and PC electrodeposition techniques [6,40]. In particular, the hardness improvement has been attributed to increased distribution of nanoparticles in the deposited Ni–Co matrix. As for lower residual macro stress, PRC deposits are characterized by uniform coatings since the technique hinders thickening of the corners and edges, as is common with DC electrodeposition. It can also be postulated that adsorption of H atoms by the deposited coatings is suppressed by pulse intervals and reversal current in PRC technique [40]. PRC deposited Ni–Co nanocomposite coatings possess smoother surfaces, finer Ni matrix crystals, and smaller grain sizes compared to DC technique deposited coatings [40].

The smooth surfaces coupled with high hardness increase the load-carrying capacity of Ni–Co nanocomposite coatings and thereby improves their wear resistance properties. The lower macro-residual stress also plays a great part in wear resistance, where it causes lower brittleness and higher toughness in the coatings, and this decreases the rate of nanoparticle flaking and metal matrix peeling during wear testing [40]. PRC technique can be used in ultrasonic power conditions and this presents desirable effects on deposited Ni–Co coatings [47,48]. Overall, it presents several advantageous effects on the electrochemical process:

(i)   Hindering sundries adsorption thereby altering the reaction mechanism,
(ii)   An increase in exchange current density,
(iii)   Lowering cathodic polarization,
(iv)   Current efficiency and yield improvements, and
(v)   (Strengthening conversion and diffusion.

It has been reported that Ni–Co nanocomposite coatings electrodeposited using the PRC technique subject to ultrasonic conditions present finer grained, compact, and uniform coatings [6].

3. Electrodeposition Parameters for Ni–Co Alloys

3.1. Effect of Co Concentration in Electrolyte

Cobalt content has been reported to increase with increase in Co concentration in the electrolyte bath [12,49]. For Ni–Co deposition, Ni and Co combine to form a solid solution that engulfs the nanoparticles suspended in the electrolyte. As such, the higher the concentration of Co element in the electrolyte, the more formation of this solid solution matrix occurs on the cathode surface. Variation of Co^{2+} in the electrolyte has significant effects on grain sizes, and cobalt content of the electrodeposited Ni–Co alloys and nanocomposites. Grain refinement can be explained using two approaches. Firstly, a lattice strain is produced as a result of the difference in atomic sizes of Co and Ni. As the content of Co increases, the lattice distortion becomes aggravated causing vacancy and dislocation defects in the lattice. Grain size refinement is a direct consequence of these defects.

Secondly, as a result of anomalous deposition of Co, the deposited Co content is higher than the corresponding concentration of Co in the electrolyte [49,50]. In this process, the less noble element (Co in this case) is reduced preferentially, resulting in a much higher content. This increase in content reaches a critical value, beyond which the microstructure of the electrodeposited coating changes from single phase (a-phase) face-centered cubic (FCC) to a combination of a-phase and hexagonal close-packed (HCP) e-phase. This combination of two structural phases results in grain refinement of deposited coatings.

Addition of Co^{2+} into the electrolyte also enhances transport and deposition of suspended nanoparticles. Transfer rate of nanoparticles through the bulk of the electrolyte is a function of electrophoretic forces existing between charged nanoparticles and the cathode surface which in turn is influenced by the quantity of adsorbed cations on the nanoparticle surfaces (zeta potential). Therefore, the amount and type of adsorbed cations determines the magnitude of electrophoretic force [51]. It was reported that deposition of nanoparticles was enhanced by addition of Co^{2+} into the electrolyte, and it can be suggested that zeta potential in Co^{2+} containing baths is much more positive, owing to Co^{2+} being much more easily adsorbed onto the surfaces of nanoparticles than Ni^{2+}. As a result, the electrophoretic forces exerted on nanoparticles are more intense.

3.2. Current Density

Current density (CE) affects the current efficiency of the process. Current efficiency (%) describes the ratio of electrochemical current density for a specific reaction to total applied current density. It illustrates the transfer efficiency of electrons to the electrochemical system. Current efficiency (η) is calculated by factoring charge passed, weight of deposits given by the difference between the weight of samples before and after deposition, and the chemical composition of the coatings as shown in Equation (13) [52].

$$\eta = \frac{W}{It}\left(\sum \frac{(Fg_ie_i)}{KN_i}\right) \times 100 \tag{13}$$

where W is the weight of the deposit (g), I is the current passed (A), t is the deposition time (h), g_i is the weight fraction of the element in the binary alloy deposit, e_i is the number of electrons transferred in the reduction of 1 mol atoms of that element, N_i is the atomic weight of that element (g/mol), F is the Faraday constant (96,485.3 C/mol) and K is a unit conversion factor (3600 C/A h).

Grain size, microstructure, brightness, thickness distribution, composition, surface morphology, microhardness and tensile strength of PC electrodeposited Ni–Co alloys are significantly influenced by variation of peak current density J_p [34,53]. It has been reported that current density has a significant influence on rate of deposition, plating adherence and the quality of plating of Ni–Co coatings. The deposition rate increases with increases in current density [54]. Li et al. [34] used a pulse technique to research the effect of varying pulse frequency in electrodeposition of nanocrystalline Ni–Co deposits. It was found that increase in peak current density resulted in a lower cobalt content, smaller grain size, higher tensile strength and a colony like morphology. A recommendation for a peak current density range of 100–200 A·dm^{-2} was suggested for grain sizes ranging from 15–20 nm, a 7%–8% cobalt content, 590–600 kg mm^{-2} microhardness and 118–1200 MPa tensile strength. When using the PC technique to electrodeposit Ni–Co, it was reported that the lowest current densities achieved the most compact layer of alloys [5].

Extremely high current density J_p however has been reported to have a drastic effect on microhardness and tensile strength of electrodeposited Ni–Co coatings [34]. This has also held true for Ni–Co alloys electrodeposited using jet electrodeposition technique [4]. It could be suggested that this is because of the decrease in cobalt content with increase in the current density. Li et al. [34] researched the effects of peak current density on mechanical properties of Ni–Co alloys and reported that Co content decreased with peak current density increase. One of the key aspects in electrodeposited Ni–Co alloys and nanocomposite coatings is solid solution strengthening where the Ni and Co in electrolyte combine to form a solid solution. A decrease in Co content will therefore yield less solid solution strengthening for coatings electrodeposited at higher current densities. Figure 3 shows the relationship between peak current density and cobalt content [34].

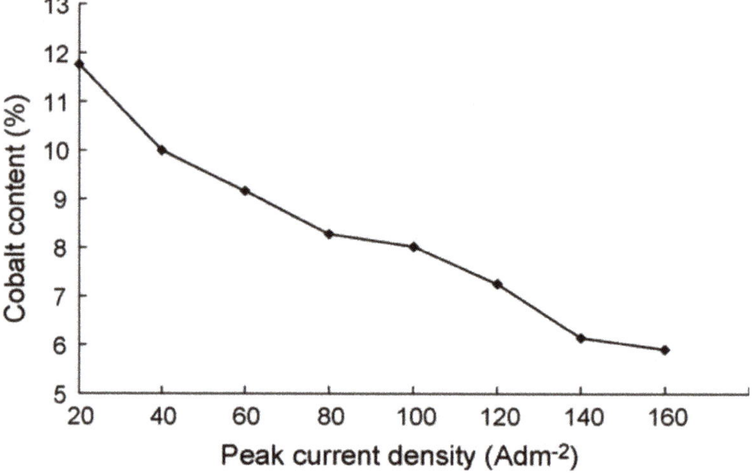

Figure 3. Cobalt content in Ni–Co alloy coatings with varying peak current density [34]. Reprinted from Applied Surface Science, 254, Yundong Li, Hui Jiang, Weihua Huang, Hui Tian, Effects of peak current density on the mechanical properties of nanocrystalline Ni–Co alloys produced by pulse electrodeposition/Pages No. 6865–6869, Copyright (2020), with permission from Elsevier.

In the case of Ni–Co, electrodeposition of the Co element is influenced and controlled by diffusion as compared to that of Ni which is predominantly controlled by activation. In light of this, an increase in cathodic current density results in an increase in cathodic overpotential. Concurrently, activation of the electrode reaction increases and, as a result, the rate of deposition of Ni element into the coating increases significantly [34].

In the case of Ni–Co nanocomposites, it has been reported that thinner coatings resulting from smaller amounts of electricity tend to have a higher content of nanoparticles compared to thicker coatings that are formed with larger amounts of electricity. This suggests that nanoparticles become adsorbed during the early phase of deposition and become unevenly distributed with increases in thickness of coatings [55].

3.3. Particle Content

The magnetic properties, corrosion resistance, and mechanical properties of electrodeposited Ni–Co nanocomposites are mostly governed by the amount and distribution of nanoparticles. There are several interrelated factors that influence incorporation and distribution of nanoparticles into the Ni–Co coatings. These factors can be broadly classified as (i) electrolyte composition (reagents, pH, additives); (ii) nanoparticle characteristics (size, shape, type); (iii) deposition parameters (current density, bath temperature, concentration of nanoparticles and rate of electrolyte agitation); (iv) cobalt content in the coating where the Co^{2+} cations that are adsorbed on the particle surface increase the incorporation of nano particles into the coating deposits, causing an increase in nanoparticle deposition with increase in cobalt content [28,51,56]; and (v) electrode orientation, which determines the incorporation efficiency of the nanoparticles. Increase in nano particle content on Ni–Co nano composite coatings can be improved by using the sediment deposition technique (SCD) as opposed to the conventional electrodeposition technique. In the SCD technique, the electrodes immersed in the electrolyte are placed horizontally and parallel to each other. As such, electrodeposition in this technique takes advantage of gravitational pull coupled with the electrophoresis force resulting in better incorporation of nano particles [30,49]. In conventional electrodeposition, only the electrophoresis force is utilized. Figure 4 shows the electrodeposition setups depending on electrode orientation in the electrolyte. Research shows that nanoparticle content in electrodeposited Ni–Co coatings increases steadily with increase in nanoparticle concentration to a given maximum value beyond which the nanoparticle content in the deposit decreases. This increase in nanoparticle content with increase in concentration can be attributed to increased transportation of the nanoparticles to the cathode surface where more and more nanoparticles can be engulfed in the growing Ni–Co matrix. At high concentrations of nanoparticles, the interaction equilibrium between the suspended nanoparticles and the embedded ones is exceeded, beyond which the surface of the cathode becomes covered such that more suspended nanoparticles cannot be embedded into the coatings. Moreover, there is increased mechanical collisions between the nanoparticles and this reduces their transportation efficiency across the electrolyte bulk. The content of nanoparticles in the coatings therefore decreases.

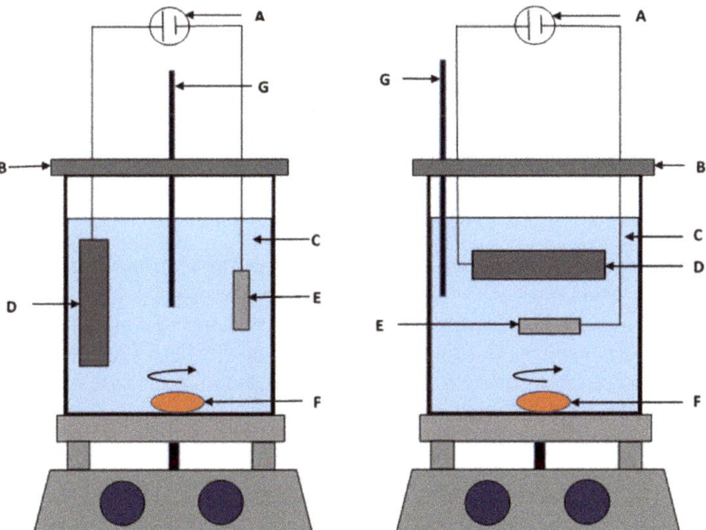

Figure 4. Schematic image of the deposition setups: (A) DC power supply, (B) Epoxy cover, (C) plating solution, (D) anode, (E) cathode, (F) magnetic bar and (G) external pH–temperature probe [30].

3.4. Electrolyte Agitation

High electrodeposition current density translates to high deposition rates, which almost always causes burrs on the cathode surface as well as increasing the coating roughness. Electrolyte agitation causes a distinct reduction of burrs that are formed on the edges of coated substrates and this reduces the coating roughness and improves uniformity [53]. Furthermore, increasing the stirring speed of the electrolyte during the deposition process increases the content of nanoparticles deposited in the Ni–Co coatings up to a given maximum level beyond which the content reduces [26,57]. At low agitation rates, the concentration of nanoparticles surrounding the cathode may reduce, resulting in the feed rate of the nanoparticles being lower than their adsorption into the Ni–Co matrix. Additionally, incomplete dispersion as a result of insufficient convection may cause agglomeration of nanoparticles and gravity settling. The surface energy of nanoparticles is greatly reduced when nanoparticles agglomerate, and this lowers the content in the deposited coatings. At higher agitation rates, the volume of nanoparticles reaching the cathode surface (mass transfer) increases thereby increasing the overall content of nanoparticles in the coatings. Goto et al. [55] reported that for the range of the experiment conducted, the nano diamond (ND) content in the coatings increased with increase in the stirring speed as shown in Figure 5. Where pulse electrodeposition is conducted using sediment deposition technique, agitation is a factor of T_{ON}/T_{OFF} ratio, where T_{ON} and T_{OFF} represent the time intervals when the agitation is on and off. In this technique, the nanoparticles settle on the horizontal cathode with aid from gravitational force. As such, the lower the T_{ON}/T_{OFF} ratio, the higher the rate of nanoparticle incorporation into the growing Ni–Co matrix [57].

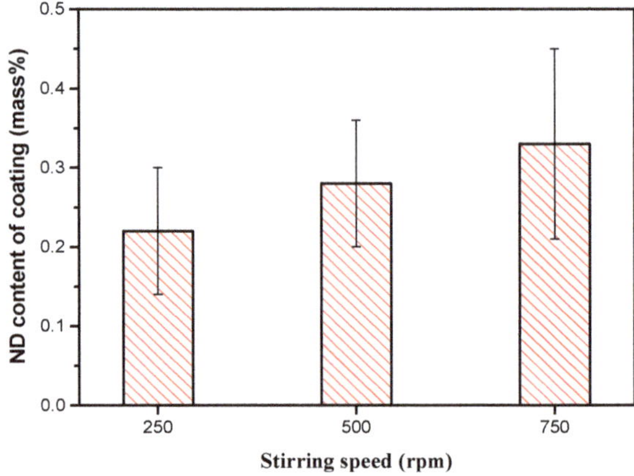

Figure 5. Relation between stirring speed and nano diamond (ND) content of the coatings [55].

Excessive rotation is however detrimental to coating quality. Vigorous hydrodynamic forces are generated at high agitation rates and these forces pluck out nanoparticles from the cathode surface before they become successfully embedded in the growing matrix surface, leading to lower nanoparticle content in the deposited coatings. This conclusion was reported by [26,58,59] who related the increase in corrosion resistance up to a maximum value to an increase in electrolyte agitation rate, beyond which it decreased with further increase in agitation rate.

3.5. Temperature

Shi et al. [60] postulated that there is a two-fold effect with respect to temperature. When electrolyte temperatures are low, there is an increase in nanoparticle kinetic activity with a rise in the temperature of the bath, and this boosts adsorption of nanoparticles into the metal matrix. Increasing the temperature decreases the density and viscosity of the bath, thereby improving the mobility of ions within the electrolyte. As such, coatings deposited at higher temperatures exhibit superior properties owing to thicker coatings and higher nanoparticle contents than those deposited at lower temperatures [53].

Temperature variation has also been reported to alter the size of crystals in the deposited coatings. Prabu and Wang [53] reported that increasing temperature from 20 to 60 °C resulted in larger and sharper crystals. This was attributed to increase in rate of reducible ion diffusion to the cathode with an increase in temperature which decreased the polarization resistance. Research on the effect of temperature on Ni–Co coatings shows that the deposition rate increases with increases in temperature of the electrolytic bath [4]. Idris et al. [54] reported that, for Ni–Co coatings electrodeposited using high speed jet electrodeposition, an increase in temperature from 55 to 65 °C (at 1 A/cm^2 current density) resulted in a subsequent increase in thickness of the electrodeposited coatings from 61.4 to 71.7 µm, respectively. Similar increasing trends of the thickness with rise in temperature were observed at current densities of 0.1, 0.3 and 0.5 A/cm^2. This can be attributed to grain growth as a result of a free growth mode of Ni resulting from the temperature rise.

When the temperature exceeds certain limits, thermodynamic ion movement is enhanced greatly, and the nanoparticle's kinetic energy increases. As a result, less nanoparticles become adsorbed into the metal matrix. This conforms to Langmuir's adsorption theory, where temperature increase beyond certain levels has a negative effect on nanoparticle absorbability. As a result, the electric field and the overpotential of the cathode are decreased, making it harder for nanoparticles to be embedded into the coating. As such, lower contents of nanoparticles are observed in the deposited coatings.

3.6. Electrolyte pH

According to past research, it can be seen that the composition and structures of Ni–Co alloys and their nanocomposites can significantly affect their physiochemical properties. The effect of pH on Ni–Co deposits is predominantly dominated by three factors [61]: (i) the acidic environment in the electrolyte dissolving newly deposited metal atoms on the cathode surface, (ii) metal hydroxide formation and adsorption on the surface of the electrode, and (iii) normal electrodeposition of metals. When the electrolyte pH is low, the newly deposited metal becomes dissolved at a faster rate and formation and adsorption of metal hydroxides becomes depressed. In this case, the electrodeposition process is mainly dominated by normal electrodeposition of metals and this results in lower Co^{2+} in the bath. At higher pH values however, the formation and adsorption of metallic hydroxides is promoted, and the newly deposited metal dissolution becomes suppressed. For higher pH values, the electrodeposition process is dominated by formation and adsorption of Co hydroxides on the cathode surface and this produces higher Co contents in deposited coatings [62,63]. Tian et al. [61] reported a gradual increase in Co content from 9.4% to 19.6% with an increase in the value of pH from 2.0 to 5.4. pH value in the electrolyte has also been observed to affect current efficiency. An increase in current efficiency from 52.1% to 81.2% was also reported with increase in pH from 2.0 to 5.4. Research linking pH value to hydrogen evolution has also been reported. Increase in pH in Ni–Co-based deposits has been associated with an increase the in hydrogen evolution rate, followed by creation of trace amounts of Ni and Co hydroxides which hinder the growth of crystals [64].

3.7. Pulse Frequency

Increased pulse frequency has been reported to achieve good Ni–Co films with smooth surface morphology and high microhardness, as well as better corrosion resistance of deposited coatings [5,37]. Pulse frequency has a significant influence on the morphology of the deposited coatings as well as the content of nanoparticles. At higher pulse frequencies, smaller grain sizes are obtained. Bigger grains tend to be more thermodynamically stable than smaller ones, and as such, an increase in OFF-time causes the grain size to increase [65]. Similar findings were reported by Yang and Cheng [32]. This was attributed to enhancement of the nucleation process by the SiC nanoparticles by providing electro crystallization nucleation sites and retarding crystal growth. Furthermore, the content of SiC in the Ni–Co/SiC nanocomposites increased with increasing pulse frequency. When pulse frequency levels are higher, the overpotential generated is much higher, and this provides more energy for nanoparticle adsorption into the coatings. As such, it can be concluded that higher pulse frequency offers better properties for deposited coatings as a factor of increased nanoparticle content.

3.8. Duty Cycle

When low duty cycles are used, the resulting coating is characterized by finer and more compact structures than if higher duty cycles had been used. It has been reported that a low duty cycle increases the microhardness and corrosion resistance of electrodeposited Ni–Co/SiC nanocomposite coatings. This can be related to the increase in SiC nanoparticle content in the coatings at lower duty cycles [32]. With the increase in duty cycle, there is a transformation of the surface morphology from a branched, acicular structure to a more nodular structure. It can therefore be concluded that lower duty cycles offer the best properties for deposited materials.

4. Mechanism of Ni–Co–Nanoparticle Electrodeposition

The process starts with complete combination of Ni and Co in the electrolyte to form a solid solution [66]. At the same time, the surfaces of nanoparticles suspended in the electrolyte adsorb positive and negative ions (particle charging). The nanoparticles are then transferred by electrophoresis force and gravitational force to the growing Ni–Co matrix where they become embedded into the deposit. Ni–Co deposition is believed to be anomalous deposition where the content of the less noble element (Cobalt in this case) is much higher than the concentration of the same element ions in the electrolyte [16,67,68]. This could be attributed to the relatively fast kinetics related to cobalt deposition. Moreover, the inhibition of nickel deposition in presence of cobalt ions is less likely as a result of evolution of metal hydroxides [67]. Figure 6 shows the cobalt content in deposited Ni–Co coatings as a function of Co concentration in the electrolyte [16].

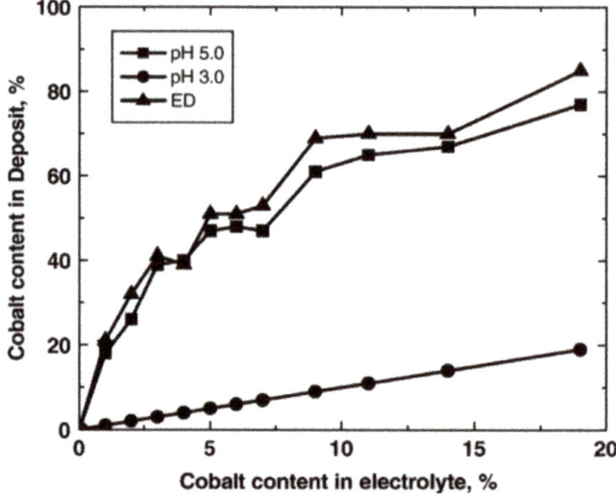

Figure 6. Alloy composition as a function of Co^{2+} concentration in the electrolyte [16]. Reprinted from Surface and Coatings Technology, 201, Meenu Srivastava, V. Ezhil Selvi, V.K. William Grips, K.S. Rajam, Corrosion resistance and microstructure of electrodeposited nickel-cobalt alloy coatings/Pages No. 3051–3060, Copyright (2020), with permission from Elsevier.

Many models have been proposed concerning co-deposition of nanoparticles. Celis [69] suggested that it was a five-step process which took into consideration the creation of an ionic cloud that engulfed the nanoparticles, movement of the nanoparticles through the electrolyte and the diffusion layer. These steps can be classified as: (i) formation of an ionic cloud on the surface of the nanoparticles from the adsorbed ions; (ii) the charged nanoparticles are then transferred through the electrolyte bulk until they reach the hydrodynamic boundary layer; (iii) through diffusion, the nanoparticles are transferred en-masse to electrode surface; (iv) adsorption of electroactive ions and the free ions occurs on the particles on the cathode; and (v) electroreduction of adsorbed ions occurs followed by incorporation of particles into the growing metal matrix [70]. Guglielmi's model proposed that the adsorption mechanism onto the cathode followed a two-step process for the charged nanoparticles. Firstly, the charged nanoparticles are loosely adsorbed while still being engulfed in a film of adsorbed cations [69].

5. Baths Used in Electrodeposition Process for Ni–Co Coating

Ni–Co alloys have been synthesized using many different electrodeposition techniques: jet, direct current and Pulse current electrodeposition techniques [5]. With the said techniques, researchers have also used different types of electrolyte baths during electrodeposition such as sulfate, sulfamate, and chloride solutions [21,22,71].

5.1. Chloride Baths

Many researchers over the years have used chloride baths to electrodeposit Ni–Co alloys and nanocomposites. Nickel chloride added to the chloride baths provides Cl^- ions which are essential for efficient dissolution of the nickel anode [72]. Fan and Piron [67] studied anomalous electrodeposition of Ni–Co alloy on Cu substrates using complex citrate and simple chloride baths. It was reported that the Ni–Co electrodeposited using the chloride baths exhibited anomalous deposition where the content of Co was higher than in the complex citrate bath. This was attributed to preferential deposition of Co which was more reactive (less noble) than Ni. It was also found that simple chloride baths were better suited at electrodepositing at current densities as low as 0.1 mA·cm^{-2}.

5.2. Sulfate Baths

Sulfate-based baths offer good electrodeposition prospects owing to their relative affordability compared to other baths and their ability to deposit high Co content coatings [64]. These baths contain both sulfate compounds as well as Ni chloride. The Ni chloride acts an additional source of Ni^{2+} ions in the electrolyte and this influences the thickness of the deposited coatings. The Cl^- ions from the Ni chloride increases conductivity of the electrolyte solution by causing dissolution of Ni anodes [72,73]. Internal stresses in deposited coatings have been observed to decrease where chlorides are absent in electrolytes. Absence of Cl^- ions is favorable in processes where the consumable anodes remain unused. However, they are used together with other reactive agents where the anodes are consumed during the electrodeposition process [74,75].

5.3. Sulphamate Baths

Sulphamate and sulphamate–sulfate baths have been reported to deposit qualified and coherent Ni–Co structures [68]. For the case of high-speed electrodeposition, nickel sulphamate baths containing boric acid have been preferred over watts baths because of the ease of obtaining relatively thicker coatings with less internal tensile stresses [76]. Ni–Co coatings deposited from sulfamate–sulfate baths have been reported to have smoother and finer surfaces and this can be attributed to lower stresses being generated in coatings deposited with these baths [68].

6. Additives

Additives are added into the electrolyte bath during electrodeposition to increase the range of current density, change physical and mechanical properties, reduce the size of crystallites by reducing growth, increase the coating's luster, reduce nanoparticle agglomeration, and reduce internal stresses generated during the deposition process [35].

6.1. Boric Acid

In 1916, Watts O.P. formulated and optimized an electrolyte comprising of nickel sulfate, nickel chloride and boric acid. This formula commonly known as Watts solution has been used extensively in nickel electrodeposition and its developmental impacts cannot be overstated [72]. The effect of boric acid in Nickel plating baths during electrodeposition has been researched over the years and the results can be summed up into six views:

(a) Boric acid suppresses oxygen evolution. Gadad and Harris [77] researched on oxygen incorporation in electrodeposited Mi, Fe and Ni–Fe alloy coatings. It was reported that an increase in applied current density resulted in an increase in the content of oxygen in Ni coatings and this posed a detrimental effect to the magnetic and electrical properties of the coatings. Addition of boric acid reduced the oxygen incorporation in all three electrodeposition systems, with less than 2 wt% oxygen observed in all cases. The buffering effect is not attributed directly to the boric acid but more to the complexing ability of boric acid with metal ions in the electrolyte.

(b) Boric acid promotes deposition of Nickel by acting as a catalyst. The adsorptive interaction of boric acid has also been observed in Ni–Zn alloy coatings where boric acid increased the current efficiency of the system at lower Zn (II) concentrations and increased the Ni content of the coatings at higher Zn (II) concentrations [78]. Significant change in primary nucleation rate coupled with suppressed secondary nucleation on coatings was also reported. Cyclic voltammetric deposition results have shown that the hydrogen evolution rate (HER) increases relative to the increase in boric acid concentration.

(c) Boric acid as a pH buffer. In electrodeposition, the practical buffer range is given at pKa ± 1, but this value is much higher in the case of boric acid (9.23 ± 1 at 25 °C). This is an anomaly considering the pH of the Ni electrolyte is 4.0. The anomaly can be attributed to formation of weak bound complexes between nickel ions and boric acid, such that the said complexes act as pH buffers [79,80]. The presence of these complexes however, has yet to be confirmed experimentally. This pH buffering phenomena has been found to be significantly influenced by the applied current density. Tsuru et al. [76] reported that at lower current densities (below 1.0 A dm^{-2}), the pH buffering properties of boric acid were exhibited.

(d) Suppression of hydrogen evolution by boric acid. During electrodeposition, electric current flowing through the system causes an increase in pH and as a result, hydrogen gas is produced at the cathode. Hydrogen evolution at the cathode is detrimental to the reduction of metal ions, and therefore boric acid is added into the plating bath solution to prevent electrode surface passivation as well as act as a surface agent which acts as a selective membrane to block passage of the reduction of Nickel, while permitting the reduction of iron in a retarded state. Improving the electrodeposition current density range thereby minimizes the effect [81]. Yin et al. [82] suggested that boric acid acted like a surfactant which was adsorbed onto the surface and hinders hydrogen evolution. The adsorbed boric acid interferes with the alloy nucleation process thereby reducing the hydrogen evolution rate in Ni-enriched phases [82]. It should be noted that the hydrogen evolution suppressing properties of boric acid have only been observed in presence of nickel ions and this suggests that there exists a mutual interaction between nickel and boric acid [76].

(e) Reduction of passive film formation by boric acid during Ni electrodeposition. Boric acid was found to significantly deter surface passivation on Ni reduction in Fe–Ni setups [82]. Tsuru et al. [76] suggested that, by acting as a surface agent, boric acid hindered passivation of the electrode surface during reduction of nickel.

(f) Accelerating growth rates of deposits. Boric acid improves the lateral as well as the outward growth rate during deposition of nickel [83].

It has also been reported that coatings electrodeposited in electrolytes containing boric acid exhibit better appearance coupled with reduced brittleness [81].

6.2. Surfactants

Nanoparticle stabilization in the electrolyte and non-agglomerated nanoparticle dispersion are key elements of producing Ni–Co nanocomposite coatings with excellent properties. Surfactants such as cetyltrimethylammonium bromide (CTAB) [84–86], sodium dodecyl sulfate (SDS) [87,88], and sodium lauryl sulfate (SLS) [89] are added into the Ni–Co nanocomposite coating baths to prevent nanoparticle agglomeration by reducing the electrolyte's surface tension to create smaller hydrogen bubbles thereby limiting pitting effect [35]. Surfactants of cationic and anionic nature are commonly used owing to their significant influence on ceramic-based nanoparticle dispersibility. Surfactants added into the bath work by changing the cathode polarization potentials, thereby changing the adhesion, smoothness, grain growth rate, and grain size of the deposited coatings [90]. Several researchers have used optimum quantities of different surfactants and reported improvement in corrosion resistance and mechanical properties [91].

Different surfactants have different effects on different nanoparticles. It was reported that CTAB surfactant offered a better alternative over SDS and triton X when they were compared in deposition of SiC nanoparticles using the PRC technique [92] and the PC technique [35]. However, the advantage offered by surfactant addition is not limitless. When the concentration of a surfactant exceeds a certain limit, it becomes counter-productive. Ger M [93] observed that excessive CTAB surfactant increased nanoparticle's adhesive force which resulted in deposition of coarser SiC nanoparticles.

6.3. Saccharin

Addition of saccharin results in an increase in alloy deposition overvoltage which promotes deposition of Ni while hindering that of Co, and as such the resulting Ni–Co coatings exhibit reduced Co content [4,94]. Research shows that surface morphologies of Ni–Co coatings exhibited colony-like morphologies which consist of grain morphologies where several grain colonies converge to form one larger colony. As such, grain size is greatly reduced owing to the grain refinement phenomena. This is achieved by inhibition of pyramidal growth by saccharin thereby leading to the production of shiny, smooth surfaces [94]. This concurs with Weil and Cook [95] who reported that addition of organic additives such as coumarin and thiourea into the Ni–Co electrolytes hindered the growth of pyramids, caused surface roughness reduction, grain size reduction and increased surface brightness.

Increase in saccharin content in the coatings up to a certain value also improves the microhardness of Ni–Co coatings beyond which the microhardness reduces with increase in saccharin content. Li et al. [94] reported that increase in saccharin content to 3 g/L, 4 g/L and 5 g/L resulted in increased microhardness values of 456 kg/mm^2, 507 kg/mm^2 and 554 kg/mm^2, respectively. This conclusion was also reached by Wang et al. [96]. The increase in microhardness with increase in saccharin content to a certain value can be attributed to grain refining effect, and the decrease in microhardness beyond that level of saccharin can be attributed to the inverse Hall–Petch relationship when refining of grain size reaches a certain level [97]. The drawback to adding saccharin to the electrolyte is the reduced ductility of the resulting Ni–Co coatings owing to the sulphur and carbon impurities that are usually present in saccharin laden nanocrystalline coatings. The said impurities separate into grain boundaries thereby preventing the efficient sliding of grain boundaries and hence the low ductility [98].

Saccharin has been reported to act as an internal stress reliever in electrodeposited Ni–Co coatings and this has been attributed to grain refinement. Internal stresses are developed within the coating layer during the electrodeposition process and they cause oriented resultant strain in deposited coatings. Hydrogen ion reduction occurs at the cathode, and the small sized H$^+$ promote favorable conditions for diffusion to the coating's active centers. As the H$^+$ become transformed into H molecules, internal stresses are developed as the volume changes. They are classified into three main categories [99–102]:

(i)   Macroscopic stresses. These are caused by inhomogeneity in the deposited coatings. These comprise of either compressive or tensile stresses and they occur in galvanic cells.
(ii) Microscopic stresses. These originate at grain boundaries and at locations where dislocations accumulate.
(iii) Sub-microscopic stresses.

While internal stresses have been known to improve hardness and abrasion resistance of deposited coatings, at high levels, these stresses increase the coating's brittleness. Brittle coatings develop extensive microcracks which expose the substrate surface to corrosive attack and degradation when exposed to a corrosive medium. The saccharin molecules are reversibly adsorbed on active sites thereby hindering growth of crystals and impeding surface diffusion of adatoms. As such the volume of grain boundaries increases and this dissipates the energy created by internal stresses [103].

Saccharin has been employed in tensile research of electrodeposited Ni–Co nanocomposite coatings. Wang et al. [96] reported that PC deposited Ni–Co/Al$_2$O$_3$ nanocomposites containing saccharin exhibited low-temperature superplasticity, where a maximum elongation of 632% was achieved at a strain of 1.67×10^{-3} s^{-1} and a temperature of 823 K. The dominant superplastic accommodation process was taken to be dislocation glide.

7. Properties of Electrodeposited Ni–Co Coatings

7.1. Microstructure

As stated earlier, Ni–Co coatings deposition is anomalous. This can be explained as a function of local pH increase resulting in creation and adsorption of metallic hydroxides, as well as a faster rate of cobalt hydroxide adsorption [104], deposition of Co cations (first step) followed by Ni deposition (second step) in a two-step process [105], and preferential Co element deposition which causes the diffusion layer to become depleted [4]. Ni–Co alloys with up to 58 wt% cobalt content exhibit the single phase of Ni matrix with FCC type of phase structure. When the cobalt content is in the range between 64 wt% to 80 wt% the phase structure becomes a combination of FCC and HCP as shown in Figure 7 [68].

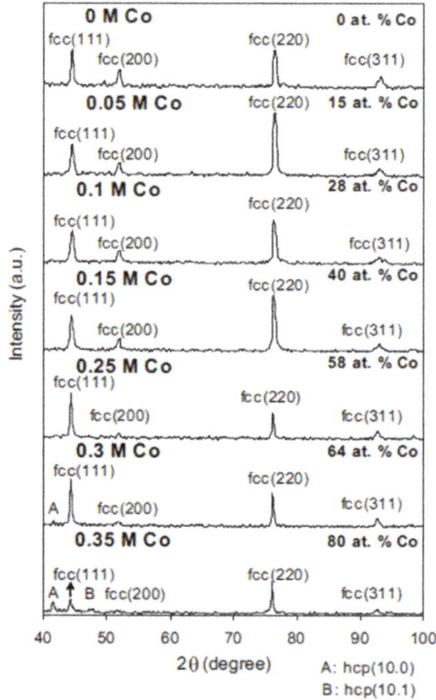

Figure 7. X-ray diffraction (XRD) patterns of Ni–Co films deposited from the electrolytes containing different Co concentrations [68]. Reprinted from Applied Surface Science, 258, Ali Karpuz, Hakan Kockar, Mursel Alper, Oznur Karaagac, Murside Haciismailoglu, Electrodeposited Ni–Co films from electrolytes with different Co contents/Pages No. 4005–4010, Copyright (2020), with permission from Elsevier.

Ni–Co alloys with Co content above 80 wt% exhibit complete HCP phase structure [16,40,68]. Ni–Co alloy surface morphologies are significantly influenced by the coating's chemical composition. Rafailovic et al. [106] researched the mechanical properties of Ni–Co alloys deposited on Cu substrates. It was reported that a platelet structured morphology was formed for coatings with a Ni^{2+}/Co^{2+} ratio of 0.25 at 65 mA cm^{-2}. The surface morphology exhibited enhanced dendritic growth when the ratios were 0.5 and 2. At the highest Ni^{2+}/Co^{2+} ratio of 4, the surface morphology exhibited cauliflower structure as shown in Figure 8.

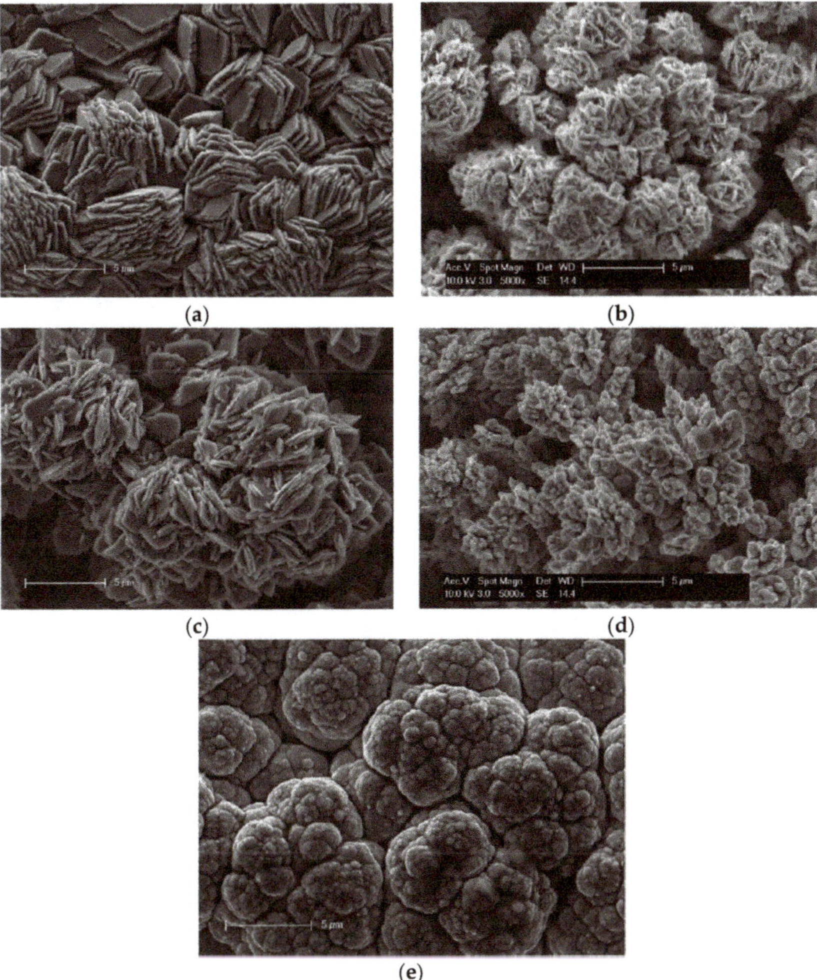

Figure 8. SEM micrographs of Ni–Co deposits obtained at a current density 65 mA cm^{-2} from an electrolyte with different Ni^{2+}/Co^{2+} concentration ratio: (**a**) 0.25, (**b**) 0.5, (**c**) 1, (**d**) 2 and (**e**) 4 [106]. Reprinted from Materials Chemistry and Physics, 120, L.D. Rafailovic, H.P. Karnthaler, T. Trisovic, D.M. Minic, Microstructure and mechanical properties of disperse Ni–Co alloys electrodeposited on Cu substrates/Pages No. 409–416, Copyright (2020), with permission from Elsevier.

The microstructure of electrodeposited Ni–Co-based coatings is affected by evolution of hydrogen gas at the cathode. Hydrogen is a by-product of the electrodeposition process owing to the breakdown of water molecules in the plating solution during the electrodeposition process. This holds both an advantageous and detrimental effect depending on the desired output of the process. Hydrogen offers a promising versatile, efficient, and clean candidate for use as an energy source to replace commonly used fossil fuels which cause CO_2 emissions that are harmful to the environment [107–109]. Hydrogen can be successfully generated using the less efficient (higher operating cost) alkaline water electrolysis [110,111], or using low hydrogen evolution reaction (HER) overpotential electrodes [112]. Ni-based alloys and compounds form such electrode materials owing to their low cost coupled with high catalytic activity and stability [113]. In the case of depositing quality coatings however, the evolution of hydrogen gas is detrimental to the structure, hence the critical need to control its production. The synthesized hydrogen gas attaches to the surface of the base metal creating a blanket of air that inhibits nucleation and deposition of the coatings and this causes poor adherence leading to non-uniform coatings [114]. The hydrogen evolution phenomenon has been reported to be more significant at higher current densities, owing to lower hydrogen overpotential where numerous gas pits are formed on the coating surface as a result of the hydrogen produced [112].

7.2. Mechanical Properties

Microhardness in electrodeposited Ni–Co coatings increases with an increase in Co content in the coatings. Baghal et al. [115] reported similar findings when the microhardness of Ni–SiC coatings was compared to Ni–Co/SiC coatings as a function of increasing current density, as shown in Figure 9.

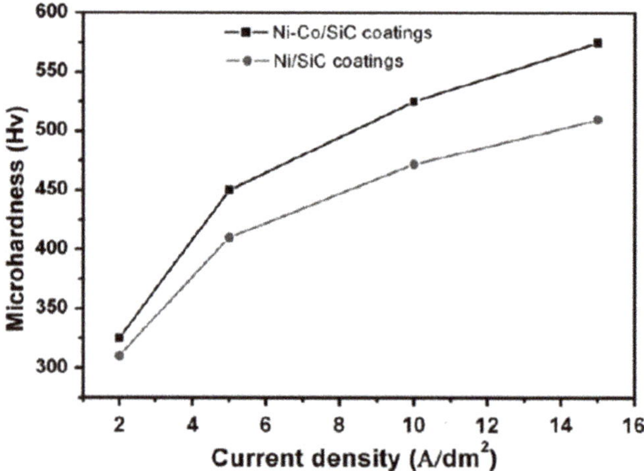

Figure 9. The effect of current density on microhardness of Ni–Co/SiC and Ni/SiC coatings [115]. Reprinted from Surface and Coatings Technology, 206, S.M. Lari Baghal, M. Heydarzadeh Sohi, A. Amadeh, A functionally gradient nano-Ni–Co/SiC composite coating on aluminum and its tribological properties/Pages No. 4032–4039, Copyright (2020), with permission from Elsevier.

Enhancement of micro-hardness in Ni–Co alloy coatings as a function of Co content can be linked to the (i) grain size reduction, (ii) phase composition, where two-phase structures are formed, and (iii) solid solution strengthening [16,20,116]. Babak [12] reported similar observations where coatings up to 45 wt·% Co were characterized solely by FCC lattices. No other phase was observed from the XRD pictographs and it was suggested, therefore, that the effect of phase composition on microhardness of the said coatings was not significant. This was held as evidence that solid solution hardening played a key role in the microhardness of the deposited Ni–Co coatings. The grain size of deposited Ni–Co alloy coatings decreased gradually with increase in content of Co element in the coatings [12].

The wear rate of deposited Ni–Co coatings has been observed to decrease with increase in Co content in the deposits. The phenomena can be associated with increase in microhardness with increase in Co content. A relationship exists between wear rate and microhardness of a coating, called Archard's law, which provides that the sliding wear volume loss is directly proportional to friction coefficient and inversely proportional to the hardness of the material [117]. This relationship can be seen in Figure 10 [60].

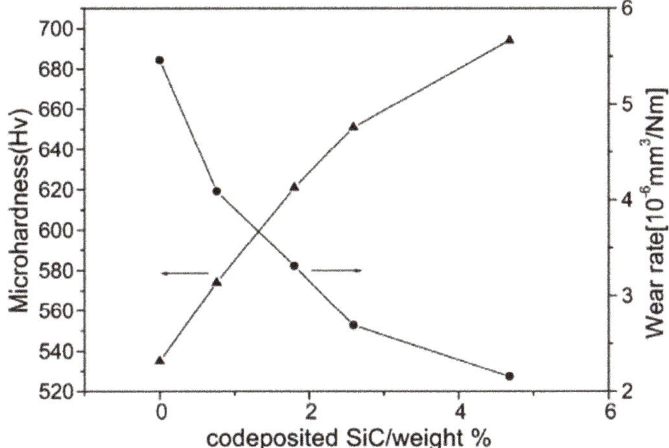

Figure 10. Microhardness and wear rate of the nanocomposite coating vs. weight percentage of co-deposited SiC particulates in the nanocomposite coating [60]. Reprinted from Applied Surface Science, 252, Lei Shi, Chufeng Sun, Ping Gao, Feng Zhou, Weimin Liu, Mechanical properties and wear and corrosion resistance of electrodeposited Ni–Co/SiC nanocomposite coating/Pages No. 3591–3599, Copyright (2020), with permission from Elsevier.

In some instances, however, the wear rate decreases with decrease in microhardness, a deviation from Archard's Law. This is concurrent with findings reported by Wang et al. [20] where the decrease in wear rate beyond 49 wt% with a concurrent decrease in microhardness from 462 HV to 298 HV was attributed to changes in the phase structure. As Co content increases, the phase structure of the deposited Ni–Co coatings changes from solely FCC structure, to FCC coupled with HCP structure, and when the Co content goes beyond 80 wt% (See Figure 7), the phase structure is transformed to a predominantly HCP structure. This transformation in phase structure to a higher ratio of HCP causes a decrease in the coefficient of friction (COF) of the deposited Co-rich coatings and therefore decreased wear loss [20]. As such, the decreasing wear rate was associated with the decreasing coefficient of friction (COF). This is shown in Figure 11.

Magnetic measurements done on electrodeposited Ni–Co coatings show that Co content has significant influence on the magnetic and structural properties of the coatings. Increase in Co content in the coatings results in a gradual increase in the saturation magnetization. This conclusion is concurrent with Karpuz et al. [68] where the highest in-plane saturation magnetization of 1000 emu/cm^3 was achieved at the highest contents of Co (80%). The same trend was also reported by [118].

Coatings improve the performance of the component by isolating the material's structure from the environment. Different substrate materials have been used for deposition of Ni, Ni–Co, and Ni–Co-nanocomposite based coatings ranging from steel, aluminum, to copper [106,115]. Failure of systems is usually associated with substrate–coating interface failure owing to the differences in mechanical and physical properties. Coating adhesion plays a key role in a surface's wear resistance and it is measured through friction testing where the friction abruptly changes when the coating breaks, also known as the critical load point. As such, a larger critical load indicates stronger coating adhesion [119]. Different mechanisms aimed at improving the critical load have been researched over the years. Wei et al. [120] reported that use of the magnetic jet electrodeposition technique (MJE) yielded a higher adhesion compared to traditional jet electrodeposition technique (TJE). At 4 g/L, TJE technique had a maximum adhesion of 23.58 N compared to 33.20 N under MJE technique as shown in Figure 12.

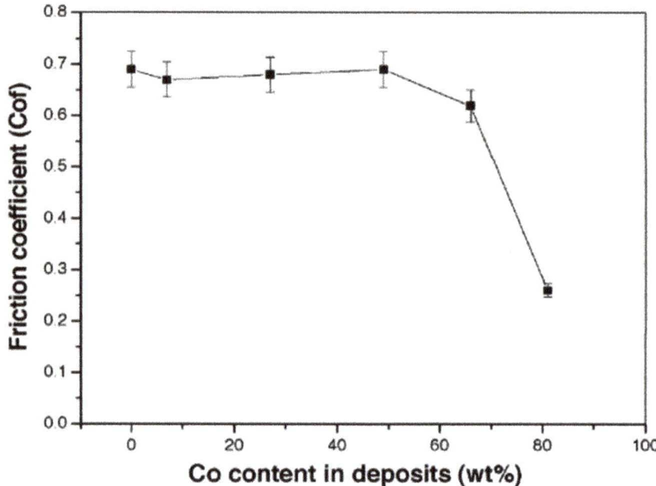

Figure 11. Friction coefficient as function of Co content in the Ni–Co alloy deposit [20]. Reprinted from Applied Surface Science, 242, Liping Wang, Yan Gao, Qunji Xue, Huiwen Liu, Tao Xu, Microstructure and tribological properties of electrodeposited Ni–Co alloy deposits/Pages No. 326–332, Copyright (2020), with permission from Elsevier.

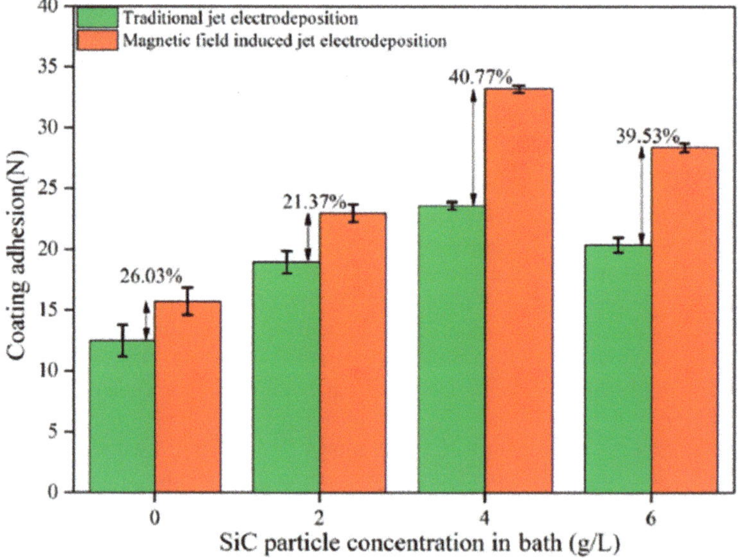

Figure 12. Adhesion of the composite coatings [120]. Reprinted from Journal of Alloys and Compounds, 791, Wei Jiang, Lida Shen, Mingyang Xu, Zhanwen Wang, Zongjun Tian, Mechanical properties and corrosion resistance of Ni–Co–SiC composite coatings by magnetic field-induced jet electrodeposition/Pages No. 847–855, Copyright (2020), with permission from Elsevier.

The adhesion of Ni–Co binary alloys can be further improved by incorporating nanoparticles into the matrix such that the effective area of contact between the substrate and coatings is increased, and the dispersion strengthening mechanism of the nanoparticles improves the coating's adhesion [121,122]. Another concept that has been considered for adhesion improvement is functionally graded materials (FGMs), whereby interfacial problems are mitigated by controlling progressive changes in structure and properties [123].

Ni–Co-nanocomposite coatings deposited at high current densities have exhibited higher surface roughness. This can be traced to adsorption of nanoparticles into the coating surface coupled with formation of pits and crevices as a result of an increase in hydrogen evolution rate at high current densities. Similar observations were made by Dheeraj et al. [35] where a surface roughness of 2.31 ± 1.78 μm was reported for sample S3 which represented coatings deposited at higher current densities.

7.3. Corrosion Behaviour

The corrosion behavior of electrodeposited Ni–Co alloy can be attributed to: chemical composition, grain size, preferred orientation, and phase structure. In the case of chemical composition, it has been reported that formation of Ni–Co alloy can change the nobility of the materials thereby affecting corrosion resistance [124]. Co is less noble that Ni, that is, it is more reactive compared to Ni. Therefore, increasing the Co content is bound to produce coatings with greater electrochemical activity than that of purely Ni coatings [125]. The polarization resistance of Ni–Co alloys has been reported to increase with increase in Co content up to a given limiting value, beyond which the corrosion resistance decreases with further increase in Co content. This is concurrent with findings reported by Babak et al. [126] where Ni–17Co coatings exhibited better corrosion resistance (10.08 kΩ cm^2) compared to Ni–42Co alloy coatings (3.32 kΩ·cm^2) as shown in Figure 13.

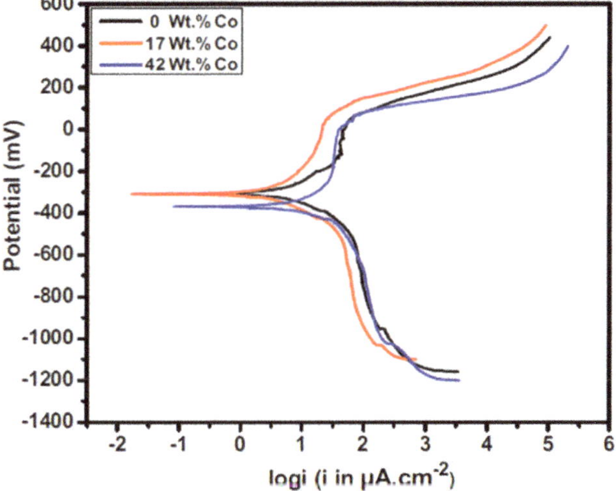

Figure 13. The potentiodynamic polarization curves of the Ni–Co alloy coating as a function of the cobalt content [126]. Reprinted from Applied Surface Science, 307, Babak Bakhit, Alireza Akbari, Farzad Nasirpouri, Mir Ghasem Hosseini, Corrosion resistance of Ni–Co alloy and Ni–Co/SiC nanocomposite coatings electrodeposited by sediment codeposition technique/Pages No. 351–359, Copyright (2020), with permission from Elsevier.

The phase structures of Ni–Co binary alloy coatings have been observed to consist of FCC single-phase solid solutions [12,126]. In the case of the coating's materials microstructure, single phase structures have proven more corrosion resistant than two-phase structures. Corrosion attacks usually occur along grain boundaries (between phases) owing to the galvanic cells that are created between the phases and their higher levels of energy compared to parts located within the crystal itself. Moreover, base centered cubic (bcc) phases have lower corrosion resistance compared to face centered cubic (FCC) phases as a result of their lower packing factor.

As a factor of preferred orientation, it has been reported that Zn–Ni alloys exhibit high corrosion resistance owing to crystallographic planes being predominantly present with higher packing densities [127]. Babak et al. reported that Ni–17Co alloy coatings exhibited high corrosion resistance as a function of predominantly (111) preferred orientation. Lupi et al. [66] found that electrodeposited Ni–Co alloys containing 40%–50% Co content provide the best catalytic properties of the alloy for the reaction leading to evolution of hydrogen in alkaline media.

8. NiCo–Ceramic Composites

Nanoparticles have been added to the Ni–Co matrix to further improve the properties. Ni–Co nanocomposite coatings exhibit superior hardness and wear resistance to alloy coatings. This can be attributed to inherent hardness of nanoparticles used, coupled with nanoparticle strengthening through Orowan mechanism [30]. According to this mechanism, the adsorbed nanoparticles hinder the formation and propagation of dislocations as the metallic matrix carries the load.

In some instances, the increase in Co concentration in the bath has been reported to improve nanoparticle incorporation thereby improving coating properties such as corrosion resistance [30]. This effect of nanoparticles can be linked to several factors:

(i)   Reduction of exposed area open to corrosive media. Nanoparticles used in Ni–Co nanocomposite electrodeposition are usually ceramics. When these nanoparticles are uniformly distributed in the Ni–Co matrix, they minimize the metallic area that is exposed to corrosive attacks and, as a result, the corrosion potential is shifted to nobler values [128].

(ii) SiC nanoparticles acting as physical barriers that hinder creation and propagation of corrosive pits.

(iii) Nanosized SiC particles which offer greater corrosion resistance than micro-sized particles when used to deposit Ni–Co/SiC nanocomposites. Owing to their smaller sizes, such nanoparticles can access structural defects such as porosities and cracks thereby mitigating the corrosive effect at such locations.

(iv) Formation of micro-galvanic cells. The metallic matrix acts as an anode while the nanoparticles act as cathodes when the Ni–Co nanocomposite coatings are exposed to corrosive media. Where the metallic matrix's electrochemical potential is less positive than that of the nanoparticles, the corrosion mechanism of the micro–galvanic cells is transformed to uniform corrosion from pitting and localized corrosion [126].

Other factors that may be associated with metallic alloy nanocomposite coatings include: reinforcing phase induced hardening, texture evolution, grain refinement of the matrix, and solid solution strengthening depending on selected matrix [129–131]. Wear resistance of Ni–Co alloy coating is greatly improved by incorporation of nanoparticles in the matrix. This increase can be attributed to the dispersion strengthening effect of adsorbed hard nanoparticles coupled with grain-refining tendencies. The uniform dispersion in the matrix and to a small extent particle agglomeration may be linked to improved wear resistance in Ni–Co nanocomposites. Similar conclusions were reached by Shi et al. [60].

8.1. Effect of Al$_2$O$_3$ Nanoparticles

Al$_2$O$_3$ nanoparticles have inherent high corrosion resistance, and in the nano-scale range, can efficiently contribute to porosity reduction, thereby further boosting the corrosion resistance of deposited coatings [132,133]. Incorporation of Al$_2$O$_3$ nanoparticles into the Ni–Co matrix produces nanocomposite coatings characterized by nodular, uniform, and compact morphology [26]. The content of Al$_2$O$_3$ in the coatings increases with increases in Al$_2$O$_3$ concentration in the electrolyte. The same trend was reported by Borkar [31], with a maximum Al_2O_3 content being achieved with 40 g L^{-1} nanoparticle concentration, beyond which the content decreased. Ni–Co alloy coatings exhibit a face centered cubic (FCC) crystal structure and the same has been reported for Ni–Co/Al_2O_3. However, the crystal orientation of the resulting Ni–Co/Al_2O_3 nanocomposite coating undergoes transformation from crystal face (200) lattice to (111) lattice [40].

Like with most Ni–Co nanocomposite coatings, adsorption of Al_2O_3 nanoparticles into the Ni–Co matrix results in a subsequent increase in the coating's hardness and wear to a certain maximum, beyond which the coating becomes brittle and spalls off. Tian [26] concluded that an increase in Al_2O_3 nanoparticle concentration in the bath caused a subsequent increase in the corrosion resistance of the Ni–Co/Al_2O_3 nanocomposite coatings up to a certain limiting value, suggesting that optimal operating conditions and parameters are key to corrosion resistance maximization. This may be attributed to uniform dispersion of Al_2O_3 nanoparticles in the Ni–Co matrix.

It has been suggested that Al_2O_3 nanoparticles may also improve deposition of Ni and Co elements in the coatings [26]. Wear resistance is also higher for Ni–Co/Al_2O_3 compared to their alloy counterparts and this too increases with increases in nanoparticle content.

8.2. Effect of SiC Nanoparticles

Ni–Co/SiC nanocomposite coatings exhibit a porous free and dense microstructure characterized by uniformly distributed SiC nanoparticles throughout the deposited coating surface [126]. The phase structure of Ni–Co/SiC nanocomposites is predominantly face centered cubic. Silicon carbide (SiC) is a chemically inert semi-conductor material [134]. Embedding of SiC nanoparticles causes a significant improvement in the corrosion resistance, wear resistance and microhardness of the deposited coatings [31]. Babak [126] reported the same findings, with the highest corrosion resistance achieved with coatings containing 8.1 vol.% SiC nanoparticles.

In the case of Ni–Co/SiC nanocomposite coatings, it has been observed that the concentration of Co element in the electrolyte has a significant effect on SiC content, whereby SiC content increases considerably with increasing Co element concentration. Babak [135] reported that SiC content increased from 2.0 vol.% to 8.1 vol.% with increases in the concentration of Co in the electrolyte. This phenomenon was attributed to ease of Co^{2+} cations adsorbing on nanoparticle surfaces compared to Ni^{2+} cations. Therefore, there was an increase in the adsorbed positive charge on the surface of the SiC nanoparticles as the Co element concentration increased. Generally, mass transfer of positively charged SiC nanoparticles towards the cathode surface is enhanced since there is an increase in the electrophoresis force exerted on them [28]. However, when the metallic cations become saturated on the SiC nanoparticle surface, a decrease in SiC content with increasing Co element concentration is observed.

Wear rate has been reported to increase with increases in SiC nanoparticle content in Ni–Co matrices [60]. As more content of SiC nanoparticles is added into the coatings, the grain refining and dispersion strengthening effects become magnified thereby improving the wear resistance of the nanocomposite coatings.

8.3. Effect of ZrO_2 Nanoparticles

ZrO_2 nanoparticles are known for their fracture toughness, stress induced transformation, and strength [136–138]. As such, they are an important consideration in nanocomposite coating deposition for wear protection in environments at high pressure and high temperature. Ni–Co/ZrO_2 nanocomposite coatings also exhibit high hardness and excellent corrosion resistance.

9. Applications

Electrodeposited Ni–Co alloys and nanocomposites exhibit unique properties and, as such, they are used for a wide variety of industrial applications. Their combined reduced localized corrosion and microhardness greatly improves protective coating performance. These coatings can therefore be used to protect less wear resistant and softer substrate surfaces for use in industry.

Karpuz et al. [68] reported that Ni–Co coatings electrodeposited from baths containing nickel sulfamate, boric acid and cobalt sulfate have potential for application in magnetic sensors. These magnetic properties of deposited Ni–Co coatings and their nanocomposites offers attractive potential to serve as soft magnets for motors, power supplies and high-efficiency transformers [3].

Ni–Co hydroxide nanosheets have been identified as candidates for pseudocapacitor application to meet the ever-growing demand for new energy storage devices. Pu J, et al. [139] researched on Ni–Co layered double hydroxides (LDHs) nanosheets and reported that the nanosheets exhibited excellent specific capacitance of 1734 F g^{-1} at 6 A g^{-1}. The nanosheets also exhibited better stability with a capacitance retention of 86% in the galvanostatic charge–discharge test after 1000 cycles.

10. Future Scope and Recommendations

Key areas that have been identified for future additional research include [140,141]:

(i)   Corrosion resistance behavior in varying environments, such as steady and dynamic conditions.
(ii)   Tribological properties such as dry and wet abrasive behavior under controlled loads.
(iii)   Electroless deposition of Ni–Co coatings, which offers a more competitive and specialized option.
(iv)   Thermal oxidation resistance of Ni–Co alloy matrices.
(v)   More efficient electrolyte agitation techniques such as submerged jet impingement and flow cells, jet eductors, and ultrasound.
(vi)   Extensive research on hydrogen evolution mitigation in electrodeposited Ni-based coatings by using pulse electrodeposition and additives.
(vii)   Use of response surface methodology to optimize the Ni–Co electrodeposition process and increase accuracy of the desired properties and also predict tested properties.

Hydrogen evolution affects the coating structure of deposited coatings. Several approaches have been used with different materials to great success. Kannan and Wallipa [114] coated a magnesium alloy with calcium phosphate using constant-potential and pulse-potential methods and analyzed the in vitro corrosion resistance properties. It was reported that the polarization resistance of pulse-potential deposited coatings was three times higher than that exhibited by constant-potential deposited coatings, and this was associated with the calcium phosphate particles being closely packed for pulse-potential coatings. This provides an interesting approach that can be researched on using Ni and Ni–Co based coatings.

Several other hypotheses have been postulated for the mitigation of hydrogen evolution in the electrodeposition process in different coatings. In past research, an organic solvent (ethanol) was added to the electrolyte bath to slow down hydrogen evolution on a magnesium alloy by decreasing conductivity of the plating solution, resulting in decreased hydrogen bubble bursting rate and hence a highly dense coating was deposited [142]. This approach of slowing down hydrogen evolution by decreasing the conductivity, however, results in lower deposition rates. Metal deposition utilizes the OH^- ions generated during H_2O breakdown whereby the metallic ions are reduced to form the metal. The mechanisms for hydrogen generation and metal-hydroxyl ion adsorption are shown in Equations (14)–(18) [143,144].

$$2H_2O + 2e^- = H_2 + 2OH^- \tag{14}$$

$$2H + 2e^- = H_2 \tag{15}$$

$$M^{2+} + OH^- = M(OH)^+ \tag{16}$$

$$M(OH)^+ \rightarrow M(OH)^+_{ads} \tag{17}$$

$$M(OH)^+_{ads} + 2e^- = M + OH^- \tag{18}$$

where M can be Ni or Co ions. As such, a balance must be struck between the rate of hydrogen evolution and the deposition rate, and this offers an interesting area for research and application in Ni and Ni–Co based electrodeposited coatings. Other additives used include polyethylene glycol and di-sodium ethylenediamine tetraacetic acid (EDTA) in pulse copper deposition, and it has been reported that this improves the throwing power, current efficiency, and thickness of deposited coatings [145]. These additives can also be considered for electrodeposited Ni and Ni–Co based coatings.

It is advisable to use larger sample sizes because they hinder the manifestation of edge effects that are common in smaller sample sizes. This is especially common where the current density used is high with respect to the substrate's surface area.

The adhesive force that exists between the substrate and the coatings plays a major role in the wear resistance of the material. As such, ensuring a good bond exists between the deposit and substrate surface is key. Copper electrodes tend to exhibit better adhesive properties compared to their steel counterparts, but they are also more costly. Using a pre-treatment step offers the chance to improve this bond, especially where the substrate is made of steel. From personal experience, a triple immersion procedure of electronic cleaning can be used for better adhesion. This comprises of degreasing using electro-hydrostatic fluid, then removal of oxide layer by passing the substrates through a strong activating solution, and finally removal of carbon-black by passing the substrates through a weak activating solution [49]. A similar pre-treatment process has been used in other coatings like Ni–W nanocomposite coatings with good adhesion translating to superb wear resistance achieved [146]. This pre-treatment process provides interesting possibilities for future use in Ni–Co alloys and nanocomposites. De-ionized water should be used to clean the substrate surface after each pre-treatment step.

Research shows that orientation of electrodes in the electrolyte plays a major role in the deposition process. Results obtained from Ni–Co deposition suggest that the sediment deposition technique (SCD) is more favorable compared to conventional deposition technique. Ni–Co alloy and nanocomposite coatings deposited from the SCD technique have exhibited superior properties of higher Co content and higher nanoparticle content which translate to better microhardness, and improved wear and corrosion resistance of the deposited coatings. As such, selection of SCD in DC electrodeposition of Ni–Co composites should be considered for further research. Based on current trends, it can be seen that owing to their exceptional wear resistance and corrosion properties, deposited Ni–Co alloys and their nanocomposites are strong contenders for further application in the aviation industry, for use in jet engine fabrication, automotive engineering, textiles and general engineering. In recent years, a keen interest has developed in specialized engineering where deposited coatings consist of mixed functional properties, as well as deposition of superhydrophobic surface coatings which exhibit excellent wear and corrosion resistance, better self-cleaning and good tribological properties.

In essence, nanoparticles can be selected to match the desired properties of any coating, and with such capacity for discovery coupled with the ever-growing need of better material properties, the possibilities for future applications are endless.

Author Contributions: Conceptualization, N.S.M. and M.K.; methodology, N.S.M., and Y.Z.; validation, N.S.M. and M.K.; formal analysis, N.S.M. and L.Y.; investigation, N.S.M., Y.Z., and N.J.N.; resources, M.K.; writing—original draft preparation, N.S.M.; G.V.B., and N.J.N.; writing—review and editing, N.S.M., Y.Z., and M.K.; visualization, N.S.M.; G.V.B., and L.Y.; supervision, M.K.; project administration, M.K.; funding acquisition, M.K. All authors have read and agreed to the published version of the manuscript.

Funding: This research was funded by the Technology development programmer for the Northern Jiangsu area, grant number BN2014019.

Conflicts of Interest: The authors declare no conflict of interest. The funders had no role in the design of the study; in the collection, analyses, or interpretation of data; in the writing of the manuscript, or in the decision to publish the results.

References

1. Martin, P. *Introduction to Surface Engineering and Functionally Engineered Materials*; John Wiley & Sons: New York, NY, USA, 2011.
2. Shriram, S.; Mohan, S.; Renganathan, N.G.; Venkatachalam, R. Electrodeposition of nanocrystalline nickel—A brief review. *Int. J. Surf. Eng. Coat.* **2000**, *78*, 194–197. [CrossRef]
3. Gurrappa, I.; Binder, L. Electrodeposition of nanostructured coatings and their characterization—A review. *Sci. Technol. Adv. Mat.* **2008**, *9*, 1–11. [CrossRef] [PubMed]
4. Qiao, G.; Jing, T.; Wang, N.; Gao, Y.; Zhao, X.; Zhou, J.; Wang, W. High-speed jet electrodeposition and microstructure of nanocrystalline Ni–Co alloys. *Electrochim. Acta* **2005**, *51*, 85–92. [CrossRef]
5. Tury, B.; Lakatos-Varsányi, M.; Roy, S. Ni–Co alloys plated by pulse currents. *Surf. Coat. Technol.* **2006**, *200*, 6713–6717. [CrossRef]
6. Chang, L.; Guo, H.; An, M. Electrodeposition of Ni–Co/Al$_2$O$_3$ composite coating by pulse reverse method under ultrasonic condition. *Mater. Lett.* **2008**, *62*, 3313–3315. [CrossRef]
7. Wielage, B.; Lampke, T.; Zacher, M.; Dietrich, D. Electroplated nickel composites with micron-to nano-sized particles. *Key Eng. Mater.* **2008**, *384*, 283–309. [CrossRef]
8. Schlesinger, M.; Paunovic, M. *Modern Electroplating*; John Wiley & Sons: New York, USA, 2011.
9. Prasad, M.; Chokshi, A. Superplasticity in electrodeposited nanocrystalline nickel. *Acta Mater.* **2010**, *58*, 5724–5736. [CrossRef]
10. Saraev, D.; Miller, R.E. Atomic-scale simulations of nanoindentation-induced plasticity in copper crystals with nanometer-sized nickel coatings. *Acta Mater.* **2006**, *54*, 33–45. [CrossRef]
11. Kim, D.; Kim, M.K.; Son, J.T.; Kim, H.G. Effect of target properties on deposition of lithium nickel cobalt oxide thin-films using RF magnetron sputtering. *J. Power Sources.* **2002**, *108*, 239–244. [CrossRef]
12. Bakhit, B.; Akbari, A. Nanocrystalline Ni–Co alloy coatings: Electrodeposition using horizontal electrodes and corrosion resistance. *J. Coat. Technol. Res.* **2013**, *10*, 285–295. [CrossRef]
13. Rashkova, V.; Kitova, S.; Konstantinov, I.; Vitanov, T. Vacuum evaporated thin films of mixed cobalt and nickel oxides as electrocatalyst for oxygen evolution and reduction. *Electrochim. Acta* **2002**, *47*, 1555–1560. [CrossRef]
14. Koikeda, T.; Fujiwara, S.; Chikazumi, S. Perpendicular anisotropy of evaporated magnetic iron-nickel and cobalt-nickel thin films. *J. Phys. Soc. Jpn.* **1966**, *21*, 1914–1921. [CrossRef]
15. Dharmadasa, I.; Haigh, J. Strengths and advantages of electrodeposition as a semiconductor growth technique for applications in macroelectronic devices. *J. Electrochem. Soc.* **2006**, *153*, G47–G52. [CrossRef]
16. Srivastava, M.; Selvi, V.E.; Grips, V.K.W.; Rajam, K.S. Corrosion resistance and microstructure of electrodeposited nickel–cobalt alloy coatings. *Surf. Coat. Technol.* **2006**, *201*, 3051–3060. [CrossRef]
17. Yang, X.; Li, Q.; Zhang, S.; Gao, H.; Luo, F.; Dai, Y. Electrochemical corrosion behaviors and corrosion protection properties of Ni–Co alloy coating prepared on sintered NdFeB permanent magnet. *J. Solid. State. Electr.* **2010**, *14*, 1601–1608. [CrossRef]
18. Ranjith, B.; Kalaignan, G.P. Ni–Co–TiO$_2$ nanocomposite coating prepared by pulse and pulse reversal methods using acetate bath. *Appl. Surf. Sci.* **2010**, *257*, 42–47. [CrossRef]
19. Wang, G.; Chan, K.; Zhang, K. Low temperature superplasticity of nanocrystalline electrodeposited Ni–Co alloy. *Scr. Mater.* **2006**, *54*, 765–770. [CrossRef]
20. Wang, L.; Gao, Y.; Xue, Q.; Liu, H.; Xu, T. Microstructure and tribological properties of electrodeposited Ni–Co alloy deposits. *Appl. Surf. Sci.* **2005**, *242*, 326–332. [CrossRef]
21. Hassani, S.; Raeissi, K.; Golozar, M. Effects of saccharin on the electrodeposition of Ni–Co nanocrystalline coatings. *J. Appl. Electrochem.* **2008**, *38*, 689–694. [CrossRef]
22. Hansal, W.E.G.; Tury, B.; Halmdienst, M.; Varsányi, M.L.; Kautek, W. Pulse reverse plating of Ni–Co alloys: Deposition kinetics of Watts, sulfamate and chloride electrolytes. *Electrochim. Acta* **2006**, *52*, 1145–1151. [CrossRef]
23. Gogotsi, Y. *Nanomaterials Handbook*; CRC Press: Florida, FL, USA, 2006.
24. Wang, C.; Chan, K. Enhanced low-temperature superplasticity of Ni–Co alloy by addition of nano-Si$_3$Ni$_4$ particles. *Mat. Sci. Eng. A-Struct.* **2008**, *491*, 266–269. [CrossRef]
25. Afshar, A.; Ghorbani, M.; Mazaheri, M. Electrodeposition of graphite-bronze composite coatings and study of electroplating characteristics. *Surf. Coat. Technol.* **2004**, *187*, 293–299. [CrossRef]

26. Tian, B.; Cheng, Y. Electrolytic deposition of Ni–Co–Al$_2$O$_3$ composite coating on pipe steel for corrosion/erosion resistance in oil sand slurry. *Electrochim. Acta* **2007**, *53*, 511–517. [CrossRef]

27. Cai, F.; Jiang, C.; Fu, P.; Ji, V. Effects of Co contents on the microstructures and properties of electrodeposited NiCo–Al composite coatings. *Appl. Surf. Sci.* **2015**, *324*, 482–489. [CrossRef]

28. Wang, L.; Gao, Y.; Liu, H.; Xue, Q.; Xu, T. Effects of bivalent Co ion on the co-deposition of nickel and nano-diamond particles. *Surf. Coat. Technol.* **2005**, *191*, 1–6. [CrossRef]

29. Karimzadeh, A.; Aliofkhazraei, M.; Walsh, F.C. A review of electrodeposited Ni-Co alloy and composite coatings: Microstructure, properties and applications. *Surf. Coat. Technol.* **2019**, *372*, 463–498. [CrossRef]

30. Bakhit, B.; Akbari, A. Effect of particle size and co-deposition technique on hardness and corrosion properties of Ni–Co/SiC composite coatings. *Surf. Coat. Technol.* **2012**, *206*, 4964–4975. [CrossRef]

31. Borkar, T. *Electrodeposition of Nickel Composite Coatings*; Oklahoma State University: Oklahoma, OK, USA, 2010.

32. Yang, Y.; Cheng, Y.F. Fabrication of Ni–Co–SiC composite coatings by pulse electrodeposition—Effects of duty cycle and pulse frequency. *Surf. Coat. Technol.* **2013**, *216*, 282–288. [CrossRef]

33. Qu, N.S.; Zhu, D.; Chan, K.C.; Lei, W.N. Pulse electrodeposition of nanocrystalline nickel using ultra narrow pulse width and high peak current density. *Surf. Coat. Technol.* **2003**, *168*, 123–128. [CrossRef]

34. Li, Y.; Jiang, H.; Tian, H. Effects of peak current density on the mechanical properties of nanocrystalline Ni–Co alloys produced by pulse electrodeposition. *Appl. Surf. Sci.* **2008**, *254*, 6865–6869. [CrossRef]

35. Dheeraj, P.R.; Patra, A.; Sengupta, S.; Das, S.; Das, K. Synergistic effect of peak current density and nature of surfactant on microstructure, mechanical and electrochemical properties of pulsed electrodeposited Ni-Co-SiC nanocomposites. *J. Alloy. Compd.* **2017**, *729*, 1093–1107. [CrossRef]

36. Padmanabhan, K. Grain boundary sliding controlled flow and its relevance to superplasticity in metals, alloys, ceramics and intermetallics and strain-rate dependent flow in nanostructured materials. *J. Mater. Sci.* **2009**, *44*, 2226–2238. [CrossRef]

37. Chung, C.K.; Chang, W. Effect of pulse frequency and current density on anomalous composition and nanomechanical property of electrodeposited Ni–Co films. *Thin Solid Film.* **2009**, *517*, 4800–4804. [CrossRef]

38. Gyftou, P.; Pavlatou, E.; Spyrellis, N. Effect of pulse electrodeposition parameters on the properties of Ni/nano-SiC composites. *Appl. Surf. Sci.* **2008**, *254*, 5910–5916. [CrossRef]

39. Tury, B.; Lakatos-Varsányi, M.; Roy, S. Effect of pulse parameters on the passive layer formation on pulse plated Ni–Co alloys. *Appl. Surf. Sci.* **2007**, *253*, 3103–3108. [CrossRef]

40. Chang, L.M.; An, M.Z.; Guo, H.F.; Shi, S.Y. Microstructure and properties of Ni–Co/nano-Al$_2$O$_3$ composite coatings by pulse reversal current electrodeposition. *Appl. Surf. Sci.* **2006**, *253*, 2132–2137. [CrossRef]

41. Karakus, C.; Chin, D.T. Metal distribution in jet plating. *J. Electrochem. Soc.* **1994**, *141*, 691. [CrossRef]

42. Wang, W.; Hou, F.Y.; Wang, H.; Guo, H.T. Fabrication and characterization of Ni–ZrO$_2$ composite nano-coatings by pulse electrodeposition. *Scr. Mater.* **2005**, *53*, 613–618. [CrossRef]

43. Chandrasekar, M.S.; Pushpavanam, M. Pulse and pulse reverse plating–Conceptual, advantages and applications. *Electrochim. Acta* **2008**, *53*, 3313–3322. [CrossRef]

44. Podlaha, E.; Landolt, D. Pulse-Reverse Plating of Nanocomposite Thin Films. *J. Electrochem. Soc.* **1997**, *144*, L200. [CrossRef]

45. Vidrine, A.; Podlaha, E. Composite Electrodeposition of Ultrafine γ-Alumina Particles in Nickel Matrices; Part I: Citrate and chloride electrolytes. *J. Appl. Electrochem.* **2001**, *31*, 461–468. [CrossRef]

46. Xiong-Skiba, P.; Engelhaupt, D.; Hulguin, R.; Ramsey, B. Effect of pulse plating parameters on the composition of alumina/nickel composite. *J. Electrochem. Soc.* **2005**, *152*, C571–C576. [CrossRef]

47. Sáez, V.; González, V.; Iniesta, J.; Frías, A.; Aldaz, A. Electrodeposition of PbO$_2$ on glassy carbon electrodes: Influence of ultrasound frequency. *Electrochem. Commun.* **2004**, *6*, 757–761. [CrossRef]

48. Touyeras, F.; Hihn, J.Y.; Bourgoin, X.; Jacques, B.; Hallez, L.; Branger, V. Effects of ultrasonic irradiation on the properties of coatings obtained by electroless plating and electro plating. *Ultrason. Sonochem.* **2005**, *12*, 13–19. [CrossRef] [PubMed]

49. Mbugua, N.S.; Kang, M.; Li, H.; Liu, Y.; Joseph, N.; Zhang, Y. The Influence of Co Concentration on the Properties of Conventionally Electrodeposited Ni–Co–Al$_2$O$_3$–SiC Nanocomposite Coatings. *Prot. Met. Phys. Chem.* **2020**, *56*, 94–102. [CrossRef]

50. Popov, K.; Grgur, B.; Djokić, S.S. *Fundamental Aspects of Electrometallurgy*; Kluwer Academic Publishers: Moscow, Russia, 2007.

51. Wu, G.; Li, N.; Wang, D.L.; Zhou, D.R.; Xu, B.Q.; Mitsuo, K. Effect of α-Al_2O_3 particles on the electrochemical codeposition of Co–Ni alloys from sulfamate electrolytes. *Mater. Chem. Phys.* **2004**, *87*, 411–419. [CrossRef]

52. Yari, S.; Dehghanian, C. Deposition and characterization of nanocrystalline and amorphous Ni–W coatings with embedded alumina nanoparticles. *Ceram. Int.* **2013**, *39*, 7759–7766. [CrossRef]

53. Prabu, S.; Wang, H.W. Factors Affecting the Electrodeposition of Aluminum Metal in an Aluminum Chloride–Urea Electrolyte Solution. *J. Chin. Chem. Soc-Taip.* **2017**, *64*, 1467–1477. [CrossRef]

54. Idris, J.; Christian, C.; Gaius, E. Nanocrystalline Ni-Co alloy synthesis by high speed electrodeposition. *J. Nanomater.* **2013**, *2013*, 1–8. [CrossRef]

55. Goto, Y.; Kamebuchi, Y.; Hagio, T.; Kamimoto, Y.; Ichino, R.; Bessho, T. Electrodeposition of copper/carbonous nanomaterial composite coatings for heat-dissipation materials. *Coatings* **2018**, *8*, 5. [CrossRef]

56. Srivastava, M.; Grips, V.W.; Rajam, K. Influence of Co on Si_3N_4 incorporation in electrodeposited Ni. *J. Alloy. Compd.* **2009**, *469*, 362–365. [CrossRef]

57. Bakhit, B.; Akbari, A. Synthesis and characterization of Ni–Co/SiC nanocomposite coatings using sediment co-deposition technique. *J. Alloy. Compd.* **2013**, *560*, 92–104. [CrossRef]

58. Bercot, P.; Pena-Munoz, E.; Pagetti, J. Electrolytic composite Ni–PTFE coatings: An adaptation of Guglielmi's model for the phenomena of incorporation. *Surf. Coat. Technol.* **2002**, *157*, 282–289. [CrossRef]

59. Lozano-Morales, A.; Podlaha, E. The Effect of Al_2O_3 Nanopowder on Cu Electrodeposition. *J. Electrochem. Soc.* **2004**, *151*, C478–C483. [CrossRef]

60. Shi, L.; Sun, C.; Gao, P.; Zhou, F.; Liu, W. Mechanical properties and wear and corrosion resistance of electrodeposited Ni–Co/SiC nanocomposite coating. *Appl. Surf. Sci.* **2006**, *252*, 3591–3599. [CrossRef]

61. Tian, L.; Xu, J.; Xiao, S. The influence of pH and bath composition on the properties of Ni–Co coatings synthesized by electrodeposition. *Vacuum.* **2011**, *86*, 27–33. [CrossRef]

62. Gomez, E.; Pane, S.; Valles, E. Electrodeposition of Co–Ni and Co–Ni–Cu systems in sulphate–citrate medium. *Electrochim. Acta* **2005**, *51*, 146–153. [CrossRef]

63. Oriňáková, R.; Orinak, A.; Vering, G.; Talian, I.; Smith, M.R.; Arlinghaus, H. Influence of pH on the electrolytic depositon of Ni–Co Film. *Thin Solid Film.* **2008**, *516*, 3045–3050. [CrossRef]

64. Ma, C.; Wang, S.C.; Low, C.T.J.; Wang, L.P.; Walsh, F.C. Effects of additives on microstructure and properties of electrodeposited nanocrystalline Ni–Co alloy coatings of high cobalt content. *Int. J. Surf. Eng. Coat.* **2014**, *92*, 189–195. [CrossRef]

65. Puippe, J.C.; Leaman, F. *Theory and Practice of Pulse Plating*; Amer Electroplaters Soc.: Orlando, FL, USA, 1986.

66. Lupi, C.; Dell'Era, A.; Pasquali, M.; Imperatori, P. Composition, morphology, structural aspects and electrochemical properties of Ni–Co alloy coatings. *Surf. Coat. Technol.* **2011**, *205*, 5394–5399. [CrossRef]

67. Fan, C.; Piron, D. Study of anomalous nickel-cobalt electrodeposition with different electrolytes and current densities. *Electrochim. Acta* **1996**, *41*, 1713–1719. [CrossRef]

68. Karpuz, A.; Kockar, H.; Alper, M.; Karaagac, O.; Haciismailoglu, M. Electrodeposited Ni–Co films from electrolytes with different Co contents. *Appl. Surf. Sci.* **2012**, *258*, 4005–4010. [CrossRef]

69. Celis, J.P.; Roos, J. Kinetics of the deposition of alumina particles from copper sulfate plating baths. *J. Electrochem. Soc.* **1977**, *124*, 1508–1511. [CrossRef]

70. Ahmad, Y.H.; Mohamed, A. Electrodeposition of nanostructured nickel-ceramic composite coatings: A review. *Int. J. Electrochem. Sci.* **2014**, *9*, 1942–1963.

71. Gomez, E.; Pane, S.; Alcobe, X.; Vallés, E. Influence of a cationic surfactant in the properties of cobalt–nickel electrodeposits. *Electrochim. Acta* **2006**, *51*, 5703–5709. [CrossRef]

72. Di Bari, G.A. Electrodeposition of nickel. In *Modern Electroplating*, 5th ed.; Schlesinger, M., Paunovic, M., Eds.; John Wiley & Sons, Inc.: Hoboken, NJ, USA, 2000; Volume 3, pp. 79–114.

73. Gezerman, A.O.; Corbacioglu, B.D. Analysis of the characteristics of nickel-plating baths. *Int. J. Chem.* **2010**, *2*, 124–137. [CrossRef]

74. Ispas, A.; Matsushima, H.; Bund, A.; Bozzini, B. A study of external magnetic-field effects on nickel–iron alloy electrodeposition, based on linear and non-linear differential AC electrochemical response measurements. *Electroanal. Chem.* **2011**, *651*, 197–203. [CrossRef]

75. Ramazani, A.; Asgari, V.; Montazer, A.H.; Kashi, M.A. Tuning magnetic fingerprints of FeNi nanowire arrays by varying length and diameter. *Curr. Appl. Phys.* **2015**, *15*, 819–828. [CrossRef]

76. Tsuru, Y.; Nomura, M.; Foulkes, F. Effects of boric acid on hydrogen evolution and internal stress in films deposited from a nickel sulfamate bath. *J. Appl. Electrochem.* **2002**, *32*, 629–634. [CrossRef]

77. Gadad, S.; Harris, T.M. Oxygen Incorporation during the Electrodeposition of Ni, Fe, and Ni-Fe Alloys. *J. Electrochem. Soc.* **1998**, *145*, 3699. [CrossRef]

78. Karwas, C.; Hepel, T. Influence of Boric Acid on Electrodeposition and Stripping of Ni-Zn Alloys. *J. Electrochem. Soc.* **1988**, *135*, 839. [CrossRef]

79. Saubestre, E.B. The Chemistry of Watts Nickel Plating Solutions. *Plating* **1958**, *45*, 927–936.

80. Tilak, B.; Gendron, A.; Mosoiu, M. Borate buffer equilibria in nickel refining electrolytes. *J. Appl. Electrochem.* **1977**, *7*, 495–500. [CrossRef]

81. Šupicová, M.; Rozik, R.; Tmkova, L.; Oriňáková, R.; Gálová, M. Influence of boric acid on the electrochemical deposition of Ni. *J. Solid State Electr.* **2006**, *10*, 61–68. [CrossRef]

82. Yin, K.M.; Lin, B.T. Effects of boric acid on the electrodeposition of iron, nickel and iron-nickel. *Surf. Coat. Technol.* **1996**, *78*, 205–210. [CrossRef]

83. Abyaneh, M.; Hashemi-Pour, M. The effect of the concentration of boric acid on the kinetics of electrocrystallization of nickel. *Int. J. Surf. Eng. Coat.* **1994**, *72*, 23–26. [CrossRef]

84. Narasimman, P.; Pushpavanama, M.; Periasamyb, V. Effect of surfactants on the electrodeposition of Ni-SiC composites. *Port. Electrochim. Acta* **2012**, *30*, 1–14. [CrossRef]

85. Li, Q.; Fu, W.; Mu, Y.; Zhang, W.; Lv, P.; Zhou, L.; Yang, H.; Chi, K.; Yang, L. The effects of CTAB concentration on the properties of electrodeposited cadmium telluride films. *CrystEngComm* **2014**, *16*, 5227–5233. [CrossRef]

86. Kılıc, F.; Gul, H.; Aslan, S.; Alp, A.; Akbulut, H. Effect of CTAB concentration in the electrolyte on the tribological properties of nanoparticle SiC reinforced Ni metal matrix composite (MMC) coatings produced by electrodeposition. *Colloid Surf. A.* **2013**, *419*, 53–60. [CrossRef]

87. Sabri, M.; Sarabi, A.A.; Kondelo, S.M.N. The effect of sodium dodecyl sulfate surfactant on the electrodeposition of Ni-alumina composite coatings. *Mater. Chem. Phys.* **2012**, *136*, 566–569. [CrossRef]

88. Baghal, S.M.L.; Amadeh, A.; Sohi, M.H.; Hadavi, S.M.M. The effect of SDS surfactant on tensile properties of electrodeposited Ni–Co/SiC nanocomposites. *Mat. Sci. Eng. A-Struct.* **2013**, *559*, 583–590. [CrossRef]

89. Sen, R.; Bhattacharya, S.; Das, S.; Das, K. Effect of surfactant on the co-electrodeposition of the nano-sized ceria particle in the nickel matrix. *J. Alloy. Compd.* **2010**, *489*, 650–658. [CrossRef]

90. Gamburg, Y.D.; Zangari, G. *Theory and Practice of Metal Electrodeposition*; Springer Science & Business Media: Berlin, Germany, 2011.

91. Guo, C.; Zuo, Y.; Zhao, X.; Zhao, J.; Xiong, J. Effects of surfactants on electrodeposition of nickel-carbon nanotubes composite coatings. *Surf. Coat. Technol.* **2008**, *202*, 3385–3390. [CrossRef]

92. Pradhan, A.K.; Das, S. Pulse reverse electrodeposition of Cu-SiC nanocomposite coating: Effects of surfactants and deposition parameters. *Met. Mater. Trans. A* **2014**, *45*, 5708–5720. [CrossRef]

93. Ger, M.D. Electrochemical deposition of nickel/SiC composites in the presence of surfactants. *Mater. Chem. Phys.* **2004**, *87*, 67–74. [CrossRef]

94. Li, Y.; Jiang, H.; Wang, D.; Ge, H. Effects of saccharin and cobalt concentration in electrolytic solution on microhardness of nanocrystalline Ni–Co alloys. *Surf. Coat. Technol.* **2008**, *202*, 4952–4956. [CrossRef]

95. Weil, R.; Cook, H. Electron-Microscopic Observations of the Structure of Electroplated Nickel. *J. Electrochem. Soc.* **1962**, *109*, 295. [CrossRef]

96. Wang, G.; Jiang, S.; Zhen, L.U.; Zhang, K. Preparation and tensile properties of Al_2O_3/Ni-Co nanocomposites. *T Nonferr. Met. Soc.* **2011**, *21*, s374–s379. [CrossRef]

97. Koch, C. Optimization of strength and ductility in nanocrystalline and ultrafine grained metals. *Scr. Mater.* **2003**, *49*, 657–662. [CrossRef]

98. Yin, W.M.; Whang, S.H.; Mirshams, R. Effect of interstitials on tensile strength and creep in nanostructured Ni. *Acta Mater.* **2005**, *53*, 383–392. [CrossRef]

99. Hadian, S.; Gabe, D. Residual stresses in electrodeposits of nickel and nickel–iron alloys. *Surf. Coat. Technol.* **1999**, *122*, 118–135. [CrossRef]

100. Dzedzina, R.; Hagarova, M. Effect of Additive on the Internal Stress in Galvanic Coatings. *Int. J. Electrochem. Sci.* **2013**, *8*, 8291–8298.

101. Kubart, T.; Mala, Z.; Novak, R.; Novakova, D. Effect of coated edge geometry on internal stress distribution in multilayered coatings. *Surf. Coat. Technol.* **2001**, *142*, 610–614. [CrossRef]

102. El-Sherik, A.; Shirokoff, J.; Erb, U. Stress measurements in nanocrystalline Ni electrodeposits. *J. Alloy. Compd.* **2005**, *389*, 140–143. [CrossRef]

103. Chen, C.J.; Lin, K.L. Internal stress and adhesion of amorphous Ni–Cu–P alloy on aluminum. *Thin Solid Film.* **2000**, *370*, 106–113. [CrossRef]

104. Bai, A.; Hu, C.C. Effects of electroplating variables on the composition and morphology of nickel–cobalt deposits plated through means of cyclic voltammetry. *Electrochim. Acta* **2002**, *47*, 3447–3456. [CrossRef]

105. Go, E.; Ramirez, J.; Valle, E. Electrodeposition of Co–Ni alloys. *J. Appl. Electrochem.* **1998**, *28*, 71–79.

106. Rafailović, L.D.; Karnthaler, H.P.; Trisovic, T.; Minić, D.M. Microstructure and mechanical properties of disperse Ni–Co alloys electrodeposited on Cu substrates. *Mater. Chem. Phys.* **2010**, *120*, 409–416. [CrossRef]

107. Barbir, F. Transition to renewable energy systems with hydrogen as an energy carrier. *Energy* **2009**, *34*, 308–312. [CrossRef]

108. Veziro, T.N.; Barbir, L.F. Hydrogen: The wonder fuel. *Int. J. Hydrog. Energy* **1992**, *17*, 391–404. [CrossRef]

109. Bockris, J.N.; Veziroğlu, T.M. A solar-hydrogen economy for USA. *Int. J. Hydrog. Energy* **1983**, *8*, 323–340. [CrossRef]

110. Mazloomi, K.; Gomes, C. Hydrogen as an energy carrier: Prospects and challenges. *Renew. Sust. Energy Rev.* **2012**, *16*, 3024–3033. [CrossRef]

111. Ganley, J.C. High temperature and pressure alkaline electrolysis. *Int. J. Hydrog. Energy* **2009**, *34*, 3604–3611. [CrossRef]

112. Vijayakumar, J.; Mohan, S.; Kumar, S.A.; Suseendiran, S.R.; Pavithra, S. Electrodeposition of Ni–Co–Sn alloy from choline chloride-based deep eutectic solvent and characterization as cathode for hydrogen evolution in alkaline solution. *Int. J. Hydrog. Energy* **2013**, *38*, 10208–10214. [CrossRef]

113. Lasia, A.; Rami, A. Kinetics of hydrogen evolution on nickel electrodes. *J. Electroanal Chem.* **1990**, *294*, 123–141. [CrossRef]

114. Kannan, M.B.; Wallipa, O. Potentiostatic pulse-deposition of calcium phosphate on magnesium alloy for temporary implant applications—An in vitro corrosion study. *Mat. Sci. Eng. C* **2013**, *33*, 675–679. [CrossRef]

115. Baghal, S.M.L.; Sohi, M.H.; Amadeh, A. A functionally gradient nano-Ni–Co/SiC composite coating on aluminum and its tribological properties. *Surf. Coat. Technol.* **2012**, *206*, 4032–4039. [CrossRef]

116. Dieter, P.P. *Mechanical Metallurgy*; McGraw-Hill Book Co.: New York, NY, USA, 1986.

117. Jeong, D.H.; Gonzalez, F.; Palumbo, G.; Aust, K.T.; Erb, U. The effect of grain size on the wear properties of electrodeposited nanocrystalline nickel coatings. *Scripta. Mater.* **2001**, *44*, 493–499. [CrossRef]

118. Kim, D.; Park, D.Y.; Yoo, B.Y.; Sumodjo, P.T.A.; Myung, N.V. Magnetic properties of nanocrystalline iron group thin film alloys electrodeposited from sulfate and chloride baths. *Electrochim. Acta* **2003**, *48*, 819–830. [CrossRef]

119. Kusakabe, S.; Rawls, H.R.; Hotta, M. Relationship between thin-film bond strength as measured by a scratch test, and indentation hardness for bonding agents. *Dent. Mater.* **2016**, *32*, e55–e62. [CrossRef]

120. Jiang, W.; Shen, L.; Xu, M.; Wang, Z.; Tian, Z. Mechanical properties and corrosion resistance of Ni-Co-SiC composite coatings by magnetic field-induced jet electrodeposition. *J. Alloy. Compd.* **2019**, *791*, 847–855. [CrossRef]

121. Chen, X.H.; Chen, C.S.; Xiao, H.N.; Cheng, F.Q.; Zhang, G.; Yi, J.G. Corrosion behavior of carbon nanotubes–Ni composite coating. *Surf. Coat. Technol.* **2005**, *191*, 351–356. [CrossRef]

122. Rogal, Ł.; Kalita, D.; Tarasek, A.; Bobrowski, P.; Czerwinski, F. Effect of SiC nano-particles on microstructure and mechanical properties of the CoCrFeMnNi high entropy alloy. *J. Alloy. Compd.* **2017**, *708*, 344–352. [CrossRef]

123. Jasim, K.M.; Rawlings, R.D.; West, D.R.F. Metal-ceramic functionally gradient material produced by laser processing. *J. Mater. Sci.* **1993**, *28*, 2820–2826. [CrossRef]

124. Cramer, S.D.; Covino, B.S. *ASM Handbook*; ASM international Materials Park: Ohio, OH, USA, 2003.

125. Edward, J. *Coating and Surface Treatment Systems for Metals: A Comprehensive Guide to Selection*; ASM International: Michigan, MI, USA, 1997.

126. Bakhit, B.; Akbari, A.; Nasirpouri, F.; Hosseini, M.G. Corrosion resistance of Ni–Co alloy and Ni–Co/SiC nanocomposite coatings electrodeposited by sediment codeposition technique. *Appl. Surf. Sci.* **2014**, *307*, 351–359. [CrossRef]

127. Ramanauskas, R.; Quintana, P.; Maldonado, L.; Pomés, R.; Pech, M.A. Corrosion resistance and microstructure of electrodeposited Zn and Zn alloy coatings. *Surf. Coat. Technol.* **1997**, *92*, 16–21. [CrossRef]

128. García, I.; Conde, A.; Langelaan, G.; Fransaer, J.; Celis, J.P. Improved corrosion resistance through microstructural modifications induced by codepositing SiC-particles with electrolytic nickel. *Corros. Sci.* **2003**, *45*, 1173–1189. [CrossRef]

129. Dieter, G.E.; Bacon, D. *Mechanical Metallurgy*; McGraw Hill: New York, NY, USA, 1986.

130. Pavlatou, E.A.; Stroumbouli, P.; Gyftou, P.; Spyrellis, N. Hardening effect induced by incorporation of SiC particles in nickel electrodeposits. *J. Appl. Electrochem.* **2006**, *36*, 385–394. [CrossRef]

131. Zimmerman, A.F.; Palumbo, G.; Aust, K.T.; Erb, U. Mechanical properties of nickel silicon carbide nanocomposites. *Mat. Sci. Eng. A-Struct.* **2002**, *328*, 137–146. [CrossRef]

132. Zhou, Y.B.; Ding, Y.Z. Oxidation resistance of co-deposited Ni-SiC nanocomposite coating. *Trans. Nonferr. Met. Soc.* **2007**, *17*, 925–928. [CrossRef]

133. Shi, L.; Sun, C.; Liu, W. Electrodeposited nickel–cobalt composite coating containing MoS2. *Appl. Surf. Sci.* **2008**, *254*, 6880–6885. [CrossRef]

134. Bhatnagar, M.; Baliga, B.J. Comparison of 6H-SiC, 3C-SiC, and Si for power devices. *IEEE Trans. Electron. Dev.* **1993**, *40*, 645–655. [CrossRef]

135. Bakhit, B. The influence of electrolyte composition on the properties of Ni–Co alloy coatings reinforced by SiC nano-particles. *Surf. Coat. Technol.* **2015**, *275*, 324–331. [CrossRef]

136. Yang, G.; Yin, L.; Fang, X.; Fang, M.; Liu, Y.; Huang, Z.; Liu, B. Fabrication and liquid–solid, two-phase erosion wear behaviour of β-Sialon ceramic from pyrophyllite by carbothermal reduction and nitridation. *Ceram. Int.* **2014**, *40*, 10737–10741. [CrossRef]

137. Basu, B.; Vleugels, J.; Van Der Biest, O. Microstructure–toughness–wear relationship of tetragonal zirconia ceramics. *J. Eur. Ceram. Soc.* **2004**, *24*, 2031–2040. [CrossRef]

138. Beiyue, M.; Jingkun, Y. Phase composition of SiC-ZrO$_2$ composite materials synthesized from zircon doped with La$_2$O$_3$. *J. Rare Earths* **2009**, *27*, 806–810.

139. Pu, J.; Tong, Y.; Wang, S.; Sheng, E.; Wang, Z. Nickel–cobalt hydroxide nanosheets arrays on Ni foam for pseudocapacitor applications. *J. Power Sources* **2014**, *250*, 250–256. [CrossRef]

140. Ma, C.; Wang, S.C.; Walsh, F.C. Electrodeposition of nanocrystalline nickel and cobalt coatings. *Int. J. Surf. Eng. Coat.* **2015**, *93*, 8–17. [CrossRef]

141. Walsh, F.C.; Wang, S.; Zhou, N. The electrodeposition of composite coatings: Diversity, applications and challenges. *Curr. Opin. Electrochem.* **2020**, *20*, 8–19. [CrossRef]

142. Kannan, M.B. Improving the packing density of calcium phosphate coating on a magnesium alloy for enhanced degradation resistance. *J. Biomed. Mater. Res. Part. A* **2013**, *101*, 1248–1254. [CrossRef]

143. Tury, B.; Radnoczi, G.Z.; Radnoczi, G.; Varsányi, M.L. Microstructure properties of pulse plated Ni–Co alloy. *Surf. Coat. Technol.* **2007**, *202*, 331–335. [CrossRef]

144. Darband, G.B.; Aliofkhazraei, M.; Rouhaghdam, A.S.; Kiani, M.A. Three-dimensional Ni-Co alloy hierarchical nanostructure as efficient non-noble-metal electrocatalyst for hydrogen evolution reaction. *Appl. Surf. Sci.* **2019**, *465*, 846–862. [CrossRef]

145. Mohan, S.; Raj, V. The effect of additives on the pulsed electrodeposition of copper. *Trans. IMF* **2005**, *83*, 194–198. [CrossRef]

146. Nyambura, S.M.; Kang, M.; Zhu, J.; Liu, Y.; Zhang, Y.; Ndiithi, N.J. Synthesis and Characterization of Ni–W/Cr$_2$O$_3$ Nanocomposite Coatings Using Electrochemical Deposition Technique. *Coatings* **2019**, *9*, 815. [CrossRef]

© 2020 by the authors. Licensee MDPI, Basel, Switzerland. This article is an open access article distributed under the terms and conditions of the Creative Commons Attribution (CC BY) license (http://creativecommons.org/licenses/by/4.0/).

Review

Aqueous Corrosion of Aluminum-Transition Metal Alloys Composed of Structurally Complex Phases: A Review

Libor Ďuriška, Ivona Černičková, Pavol Priputen and Marián Palcut *

Faculty of Materials Science and Technology in Trnava, Institute of Materials Science,
Slovak University of Technology in Bratislava, 917 24 Trnava, Slovakia; libor.duriska@stuba.sk (L.Ď.);
ivona.cernickova@stuba.sk (I.Č.); pavol.priputen@stuba.sk (P.P.)
* Correspondence: marian.palcut@stuba.sk

Abstract: Complex metallic alloys (CMAs) are materials composed of structurally complex intermetallic phases (SCIPs). The SCIPs consist of large unit cells containing hundreds or even thousands of atoms. Well-defined atomic clusters are found in their structure, typically of icosahedral point group symmetry. In SCIPs, a long-range order is observed. Aluminum-based CMAs contain approximately 70 at.% Al. In this paper, the corrosion behavior of bulk Al-based CMAs is reviewed. The Al–TM alloys (TM = transition metal) have been sorted according to their chemical composition. The alloys tend to passivate because of high Al concentration. The Al–Cr alloys, for example, can form protective passive layers of considerable thickness in different electrolytes. In halide-containing solutions, however, the alloys are prone to pitting corrosion. The electrochemical activity of aluminum-transition metal SCIPs is primarily determined by electrode potential of the alloying element(s). Galvanic microcells form between different SCIPs which may further accelerate the localized corrosion attack. The electrochemical nobility of individual SCIPs increases with increasing concentration of noble elements. The SCIPs with electrochemically active elements tend to dissolve in contact with nobler particles. The SCIPs with noble metals are prone to selective de-alloying (de–aluminification) and their electrochemical activity may change over time as a result of de-alloying. The metal composition of the SCIPs has a primary influence on their corrosion properties. The structural complexity is secondary and becomes important when phases with similar chemical composition, but different crystal structure, come into close physical contact.

Keywords: aluminum; transition metal; corrosion; quasicrystal; approximant

Citation: Ďuriška, L.; Černičková, I.; Priputen, P.; Palcut, M. Aqueous Corrosion of Aluminum-Transition Metal Alloys Composed of Structurally Complex Phases: A Review. *Materials* **2021**, *14*, 5418. https://doi.org/10.3390/ma14185418

Academic Editor: Daniel de la Fuente

Received: 15 July 2021
Accepted: 14 September 2021
Published: 19 September 2021

Publisher's Note: MDPI stays neutral with regard to jurisdictional claims in published maps and institutional affiliations.

Copyright: © 2021 by the authors. Licensee MDPI, Basel, Switzerland. This article is an open access article distributed under the terms and conditions of the Creative Commons Attribution (CC BY) license (https://creativecommons.org/licenses/by/4.0/).

1. Introduction

Aluminum alloys are frequently used materials in many applications due to their low specific mass, good mechanical properties, formability, high recycling potential and superior corrosion resistance [1,2]. Two principal classifications of Al alloys exist: wrought and casting alloys, both of which are heat-treatable and non-heat-treatable. Alloying elements, including transition metals (TM), are often added to Al alloys to improve their physical and mechanical properties [3]. Complex metallic alloys (CMAs) are materials composed of structurally complex intermetallic phases (SCIPs). In SCIPs a long-range order is observed. Well-defined atomic clusters are found in the structure, typically of icosahedral symmetry. Such phases also include quasicrystals (QCs) characterized by an infinitely large unit cell (often with icosahedral symmetry) and quasicrystalline approximants, whose unit cell can be described by "classical" crystallography but contains hundreds to thousands of atoms [4–8]. Aluminum-based CMAs contain approximately 70 at.% Al. Most Al-based CMAs are alloyed with TM. However, non-transition metals like Mg, can also be used. Depending on their chemical composition, the structural complexity may vary from dozens of atoms per unit cell up to thousands of atoms. This group of materials has been receiving a significant attention since the discovery of quasicrystals in melt-spun Al–Mn

59

alloy by Shechtman et al. [4]. The discovery of quasicrystals was awarded by Nobel prize in Chemistry in 2011.

In 1982, D. Shechtman discovered a quasiperiodic arrangement of atoms in an Al–Mn alloy showing an icosahedral symmetry but no unit cell [4]. Many such compounds correspond to stable or metastable states in various phase diagrams [5]. The icosahedral phase, as the first quasicrystalline structure discovered in the Al–Mn system, has a long-distance arrangement without translational symmetry [4–7]. Many elements in thermodynamically stable QCs observed yet belong to alkali, alkali-earth, transition, or rare-earth metals (Figure 1). Typical examples of QCs and quasicrystalline approximants are Al–TM alloys, where the TM is formed by one or more transition metals (TM = Cu, Co, Ni, Fe, etc.) [9]. Thus, the quasicrystalline materials are often low-cost materials that are easy to produce in large amounts [5–7,9]. Examples of quasicrystalline phases are summarized in Table 1 [7,8,10–24].

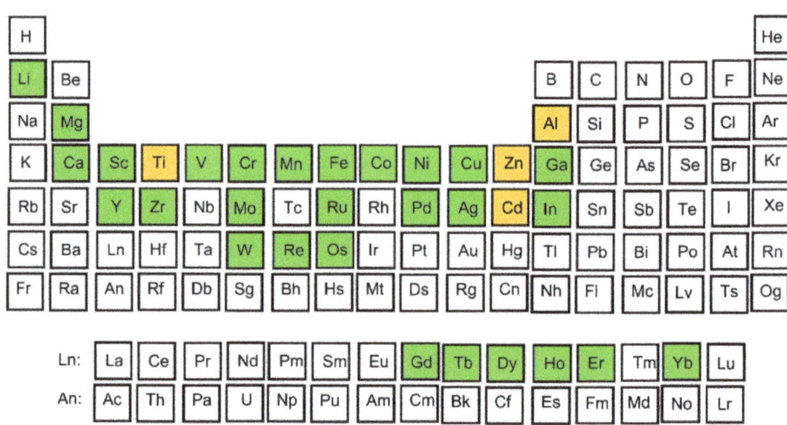

Figure 1. Chemical elements forming thermodynamically stable QCs. Main forming elements (Al, Ti, Zn, Cd) are marked by orange color, alloying elements are denoted by green.

Table 1. Examples of quasicrystalline phases, compiled from references [7,8,10–23].

Quasiperiodicity	Alloy Systems
one-dimensional quasiperiodic arrangement	Al–Ni–Si, Al–Cu–Co, Al–Cu–Mn, Mo–V
two-dimensional octagonal quasiperiodic arrangement	Cr–Ni–Si, V–Ni–Si, Mn–Si–Al, Mn–Si, Mn–Fe–Si
two-dimensional decagonal quasiperiodic arrangement	Al–Pd, Al–Ir, Al–Os, Al–Pt, Al–Rh, Al–Ru, Al–Fe, Al–Mn, Al–Ni, Al–Ni(Si), Al–Cr(Si), Al–Co, Al–Cu–Mn, Al–Mn–Fe, Al–Cu–Ni, Al–Cu–Co, Al–Co–Ni, V–Ni–Si, Al–Pd–Mn, Al–Pd–Co, Al–Pd–Fe, Al–Pd–Cr, Al–Pd–Os, Al–Pd–Ru, Ga–Fe–Cu–Si, Al–Mn–Fe–Ge, Zn–Ge–Dy
two-dimensional dodecagonal quasiperiodic arrangement	V–Ni, Cr–Ni, V–Ni–Si
three-dimensional icosahedral quasiperiodic arrangement	Al–Fe, Al–Mn, Al–Re, Al–Ru, Al–W, Al–Mo, Ti–Ni, Al–Cr–Ru, Mn–Ni–Si, Ni–Nb, Al–Cu–Mn, Al–Cu–Fe, Al–Pd–Mn, Zn–Mg–Y

The atomic structure of QCs offers an interesting combination of properties, such as low thermal conductivity combined with high electrical resistivity [25], low coefficient of friction [26,27] and/or high hardness [28–30]. The CMAs have a large potential for technical applications as the combination of the properties is unique and not observed in conventional materials. Potential applications in thermoelectrics, coatings, composites, and catalysts have been reported [5,9,31]. It has been shown that quasicrystalline materials may be more efficient catalysts compared to their crystalline counterparts. In 1994, a superior catalytic activity of quasicrystalline Al–Pd alloy relative to crystalline Al–Pd, pure Pd, and pure Cu was described [32]. Surface energies of icosahedral $Al_{70}Pd_{21}Mn_9$ and $Al_{65}Cu_{23}Fe_{12}$ phases are comparable to polytetrafluoroethylene, and yet quite different from pure aluminum and related crystalline materials. Studies revealed that surface energy is decreasing by increasing structural perfection [33]. Considering the above fact, a direct application of the quasicrystalline coatings as scratch resistant films is already on the market, offering a lowered adhesion to some polymers or food [5].

Other possible applications of QCs are found in energy saving, namely thermal insulation, light absorption, power generation, and hydrogen storage [34]. Thermal barrier demonstrators could be assessed in real conditions during the aircraft engine test on the ground. The design of selective quasicrystalline light absorbers takes advantage of the specific optical properties, e.g., the $Al_{65}Cu_{23}Fe_{12}$ alloy has solar absorbance of 90% [9,34].

The corrosion of Al-based CMAs is a relatively new field, with first investigations emerging 28 years ago. Initial studies were focused on a small family of alloys. First, corrosion parameters were reported for quasicrystalline Al–Cr–Cu–Fe and Al–Cu–Fe alloys in aqueous Na_2SO_4 (0.5 mol dm^{-3}) and in solutions of different pH [35]. The corrosion behavior of the quasicrystalline Al–Pd–Mn alloy was later studied in aqueous NaCl [36]. Nevertheless, the experimental polarization curves have not been analyzed in terms of electrochemical reactions. In later years, a more systematic approach to corrosion of Al–based CMAs has been adapted. Alloys with carefully chosen chemical composition and phase constitution have been prepared and investigated [37]. Furthermore, different electrolytes were studied [37,38]. A good thermodynamic stability of the materials at pH between 4 and pH 9 has been observed [37]. Several authors found that the electrochemical properties of the materials are determined by their chemical composition rather than by their complex crystal structure [37,39]. A high temperature oxidation of several Al-based CMAs was also studied [40–42]. The presence of high Al concentration improves the corrosion resistance of SCIPs [43,44]. Despite their practical potential, however, the corrosion behavior of Al–TM SCIPs has not been systematically reviewed yet. In the present work, we aim to provide a systematic review of aqueous corrosion behavior of bulk Al-based CMAs and compare them with traditional alloys.

2. Crystal Structure of Quasicrystals and Their Approximants

The discovery of QCs changed the view of the composition of solids and prompted a change in the definition of crystals. QCs contain such types of atomic arrangements whose symmetry does not correspond to the "classical" rules of filling the three-dimensional space with building units. The quasicrystalline arrangement contains a five-fold axial symmetry, which was originally considered to be forbidden in classical crystallography. The original definition, which specified the crystal as a material with a regular periodic arrangement of building units, was changed after the discovery of QCs. At present, the crystal is defined as a material with discrete diffraction peaks or as a material with a point diffraction pattern, respectively [4,6–8,45].

The building blocks of QCs are mostly arranged in clusters. The clusters are composed of several layers that are formed by individual atoms. Three main types of clusters are known: Mackay, Bergman, and Tsai (Figure 2). The clusters differ from one another in the different layering [4–9,46]. Clusters are composed of several shells. In the following lines, three shells will be considered to describe both Mackay and Bergman clusters for the sake of simplicity. The Mackay cluster (Figure 2a) with 54 atoms consists of an inner icosahedron (12 atoms) followed by an icosidodecahedron with 30 atoms. The third shell is an icosahedron with 12 atoms. The Bergman cluster (Figure 2b) differs from the Mackay cluster in the second shell that is formed by a dodecahedron with 20 atoms. Thus, three shells of the Bergman cluster consist of 44 atoms. The Tsai cluster (Figure 2c) has five shells with 158 atoms. The first tetrahedron shell with four atoms sits inside a dodedahedron (20 atoms). The dodecahedron is encased in an icosahedron (12 atoms). The icosahedron sits inside an icosidodecahedron with 30 atoms. The fifth shell is a rhombic triacontahedron (32 atoms + 60 atoms in its edges) [46].

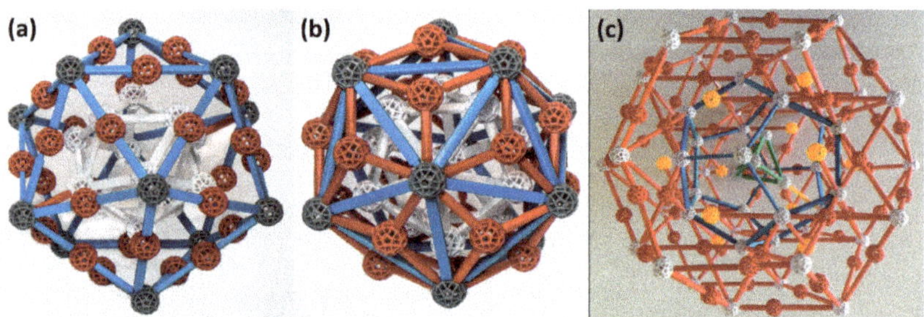

Figure 2. Models of Mackay (**a**), Bergman (**b**) and Tsai (**c**) clusters, reproduced from reference [46].

A one-dimensional Fibonacci sequence constitutes the simplest case of quasicrystalline arrangement [47]. In the Fibonacci sequence the sum of the two preceding numbers forms the next number. The ratio of two consecutive numbers is close to $\tau = (1 + \sqrt{5})/2 = 1.61803398875$ and is called the "golden ratio". If two segments are selected, with L representing a longer segment and S a shorter segment, then to form the quasicrystalline arrangement, the shorter segment S is replaced by a longer segment L and the combination SL replaces the longer segment L. This results in a following sequence

$$S \rightarrow L \rightarrow SL \rightarrow LSL \rightarrow SLLSL \rightarrow LSLSLLSL \rightarrow SLLSLLSLSLLSL \tag{1}$$

The sequence (1) has features of order, but it is not periodically ordered. There are sections SL and SLL, which alternate but do not repeat with regular periodicity. In fact, there is no periodically repeating segment, a so-called unit cell, in the Fibonacci sequence [6,47]. Based on the Fibonacci sequence, it is possible to draw a one-dimensional quasicrystal graphically (Figure 3). In Figure 3a, letters S and L correspond to shorter and longer segments used in the Fibonacci sequence, respectively. Figure 3b shows the arrangement of S and L segments according to the Fibonacci sequence. If the points dividing the line into segments are atoms, a lattice with the quasicrystalline arrangement in one dimension can be obtained. By adding the second and third dimensions, a simple example of a quasicrystalline lattice can be drawn (Figure 3c).

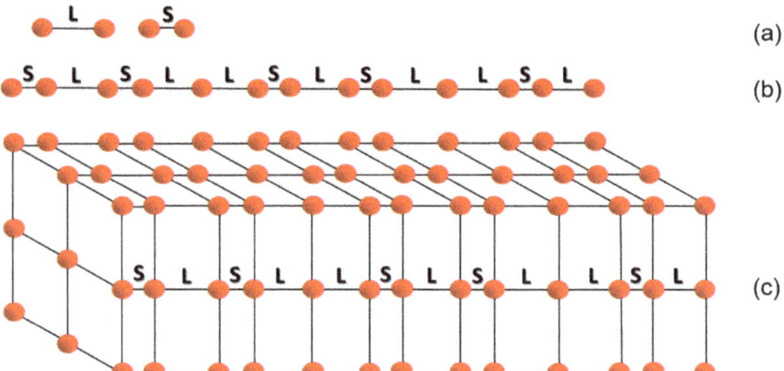

Figure 3. One-dimensional quasicrystalline arrangement: structural segments (**a**), one-dimensional quasiperiodicity (**b**), and simple quasicrystal in three-dimensional space (**c**).

The quasiperiodic arrangement in two-dimensional space can be represented by Penrose tiling [9,45]. This arrangement consists of two tiles that do not repeat periodically in two-dimensional space. The construction is given by two basic units: a wider rhombus with acute angle $\alpha = 2\pi/5$ and a narrower rhombus with acute angle $\alpha = \pi/5$. These tiles are arranged in shapes with a five-fold axis of symmetry. As with the Fibonacci sequence, the Penrose tiling is not arranged completely randomly. Basic repeating motifs, such as a star with a 5-axis axis of symmetry, are found in the arrangement of tiles, but are not repeated periodically. Such a star in the Penrose tiling may correspond to a cluster in a real quasicrystalline arrangement.

In three-dimensional space, two types of rhombohedra form the three-dimensional quasicrystalline arrangement containing elements with the icosahedral symmetry. This arrangement is called icosahedral. In the structure of QCs, no unit cell repeating regularly can be found. Thus, its size is theoretically infinite. In fact, the size of the whole grown crystal is the size of the cell. These large unit cells are in contrast to other metallic materials, whose unit cells are built from small number of atoms [48].

Two-dimensional QCs containing a quasicrystalline arrangement along two axes (in one plane) can be further divided based on crystallographic rules of symmetry, which are based on their diffraction patterns. There are octagonal QCs (O-type) with eight-fold rotational symmetry, decagonal (D-type) with ten-fold rotational symmetry, and dodecagonal (DD-type) with twelve-fold rotational symmetry [12]. The three-dimensional QC, also called icosahedral QC (i-QC), is quasiperiodic along all three axes [49]. The clusters usually comprise one of the icosahedral–shaped layers, or the entire clusters can be arranged in the icosahedral shape. The presence of five-fold axes of symmetry in the icosahedral structure is related to the point diffraction spectrum of the i-QC showing ten-fold symmetry.

In addition to QCs, there are also arrangements with many atoms in the lattice along with the presence of a cluster-based structure. The arrangements are called quasicrystalline approximants. The QCs and the quasicrystalline approximants (schematic representation given in Figure 4, [50,51]) may consist of equally formed clusters. While QCs have the clusters arranged quasiperiodically in space (i.e., non-periodically), the quasicrystalline approximants have clusters arranged with regular periodicity. Therefore, a unit cell is present in the structure of the quasicrystalline approximant, which is periodically repeated in three-dimensional space. However, a cluster in a quasicrystalline approximant structure may also comprise icosahedron-shaped layers or other shapes with the presence of five- or ten-fold axis of symmetry, and thus their structure may exhibit a diffraction pattern with a hint of a ten-fold axis of symmetry despite the regular arrangement.

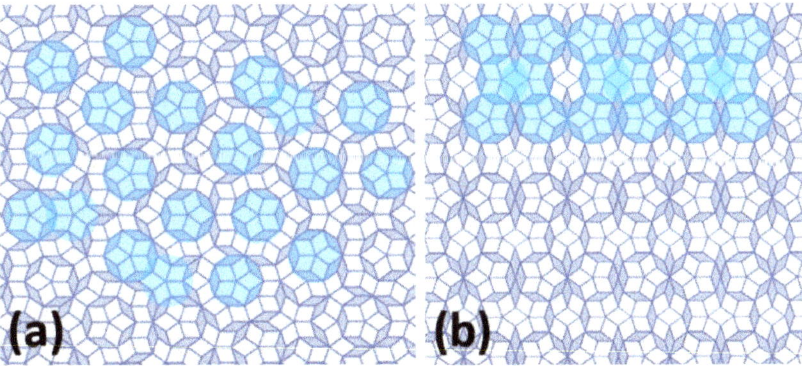

Figure 4. Schematic structure of quasicrystal (**a**) and quasicrystalline approximant (**b**) with denoted star-shape corresponding to clusters in real structure. Quasicrystalline arrangement is represented by Penrose tiling.

In Figure 5, the comparison of electron diffraction patterns of both QC and quasicrystalline approximant is shown. The electron diffraction pattern of the QC (Figure 5a) has a perfect five-fold symmetry. The electron diffraction pattern of a quasicrystalline approximant (Figure 5b) with an orthorhombic unit cell has either two-fold or four–fold axis of symmetry, but there are indications of five-fold symmetry related to the icosahedral arrangement of atoms in clusters [10,11,52–59].

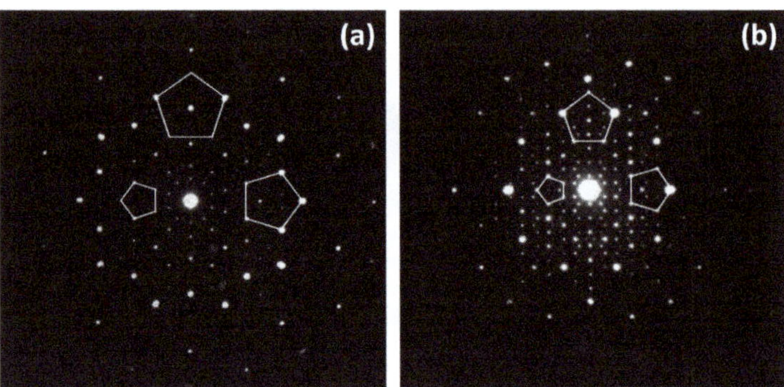

Figure 5. Comparison of electron diffraction patterns in [010] zone axis: decagonal quasicrystal in Al–Co-Cu system (**a**), ε_{16} decagonal quasicrystalline approximant in Al–Pd–Co system (**b**).

The surface structure of SCIPs is significantly less understood compared to bulk [60,61]. Preliminary results show that the adsorption of small, covalently bonding molecules on icosahedral quasicrystals is very similar to that of pure Al substrate. This is consistent with other studies, which indicate that the surface termination of most SCIPs is Al-rich [60]. A scanning tunneling microscopy (STM) has been utilized to study the surfaces of Al–Pd–Mn quasicrystals [62–64]. The STM permits a visualization of the local atomistic surface structure. Specific planes of the bulk structure have been observed as surface terminations [63]. The termination planes are characterized by high atomic density and include elements with the lowest surface energy. Nevertheless, the interpretation of individual STM images is challenging and often needs to be accompanied by theoretical models of the surface [61]. Therefore, ab initio density functional theory (DFT) calculations have been utilized to model quasicrystalline surfaces [65,66]. To perform the calculations, a Vienna ab initio simulation package (VASP) has been used [66]. The atomic structure model of the five-fold Al–Pd–Mn surface is derived from the icosahedral approximant model. In the model, the surface was cut perpendicular to one of its pseudo-five-fold axes. The cleavage position was selected to create high density surface layers consistent with experimental findings. The resulting surface structure is characterized by Penrose tiling [65]. Most tiling vertices coincide with the center of Bergman clusters.

3. Overview of Electrochemical Corrosion

The corrosion is a natural process that occurs when metallic materials are exposed to aqueous environments [67,68]. When a metal is immersed in aqueous solution, its cations spontaneously evolve on the metal–electrolyte interface and pass into the solution. During reaction, the material microscopically dissolves. Along with the cations, electrons are also released, and an electrical double layer is created at the metal–electrolyte interface [67]. The release of electrons causes the metal to become electrically charged. As a result of reaction, an electrode potential is established on the metal–electrolyte interface. After some time of immersion, an equilibrium is restored at the electrolyte–metal interface.

An overview of the metal corrosion is presented in Figure 6. The corrosion leads to an oxidation of metal and transfer of metal cations to the electrolyte. The oxidation occurs on a metal surface at a specific site known as an anode (anodic reaction site, [67]). Electrons that are released by metal are subsequently consumed by either dissolved oxygen or hydrogen cations in the electrolyte. The reduction reaction takes place at cathode (cathodic reaction site). The relative sizes and locations of cathodic and anodic sites are important variables influencing the overall corrosion rate. The sizes of cathodic and anodic areas may vary greatly; from atomic scales to macroscopically large dimensions.

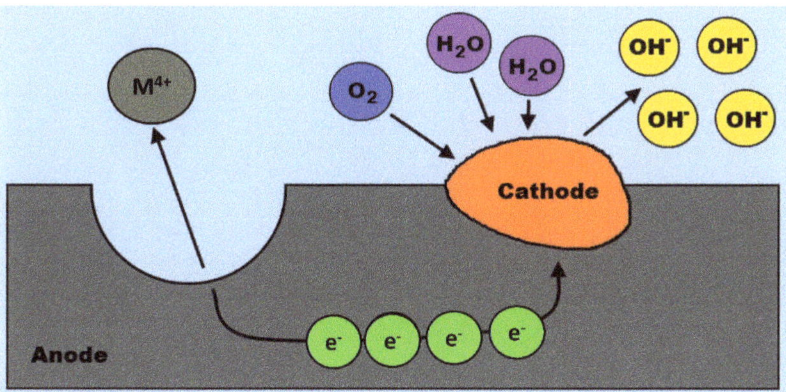

Figure 6. An elementary electrochemical corrosion cell on metal surface.

The metal oxidation is given by the following reaction

$$M \rightarrow M^{z+} + ze^- \tag{2}$$

The Gibbs free energy change (ΔG_r) of the reaction is given as

$$\Delta G_r = -zFE \tag{3}$$

In this equation, z is the number of electrons involved in the reaction, F is a Faraday's constant (96 481 C mol^{-1}), and E is the electrode potential. At standard conditions (T = 298.15 K, p = 101 325 Pa), the standard Gibbs free energy of the reaction (ΔG_r^0) is related to the standard electrode potential (E^0)

$$\Delta G_r^0 = -zFE^0 \tag{4}$$

Standard electrode potentials of metals are compared in Table 2. Since the Gibbs energy is related to electrode potential (Equation (3)), the tendency of a metal to corrode in given environment may be evaluated using E–pH plots. The diagrams have been calculated for most metals by Pourbaix and are available in ref. [69]. Figure 7 displays the E–pH diagram for Al–H$_2$O system [69,70]. The plot indicates the stability regions of different phases in aqueous solutions. The E–pH diagram shows four different regions where metallic aluminum, aluminum cations (Al^{3+}), aluminum hydroxide and complex anion [Al(OH)$_4$]$^-$ are stable. The region, where the metallic Al is stable, is labelled as immunity region. The areas with aluminum cations and anions as stable species are marked as corrosive regions. In these areas the corrosion occurs. The passivity region is where the solid hydroxide exists. In this region, Al is protected by a passive layer. The E–pH diagram demonstrates that corrosion takes place in both alkaline and acidic environments. The protective layer is formed at pH 4–9 [71]. The diagram also shows that the equilibrium electrode potential between [Al(OH)$_4$]$^-$ and Al, shifts to less noble values with increasing pH.

Table 2. Standard potentials, E^0, for metal electrodes, compiled from reference [68].

Electrode	$E^0[V_{SHE}]$ [a]	Electrode	$E^0[V_{SHE}]$	Electrode	$E^0[V_{SHE}]$
Au/Au^{3+}	+1.498	Ni/Ni^{2+}	−0.250	Zn/Zn^{2+}	−0.763
Pt/Pt^{2+}	+1.200	Co/Co^{2+}	−0.277	Ti/Ti^{2+}	−1.630
Pd/Pd^{2+}	+0.978	Cd/Cd^{2+}	−0.403	Al/Al^{3+}	−1.662
Ag/Ag^+	+0.799	Fe/Fe^{2+}	−0.440	Mg/Mg^{2+}	−2.363
Cu/Cu^{2+}	+0.337	Cr/Cr^{3+}	−0.744	Li/Li^+	−3.045

[a] Volts versus standard hydrogen electrode.

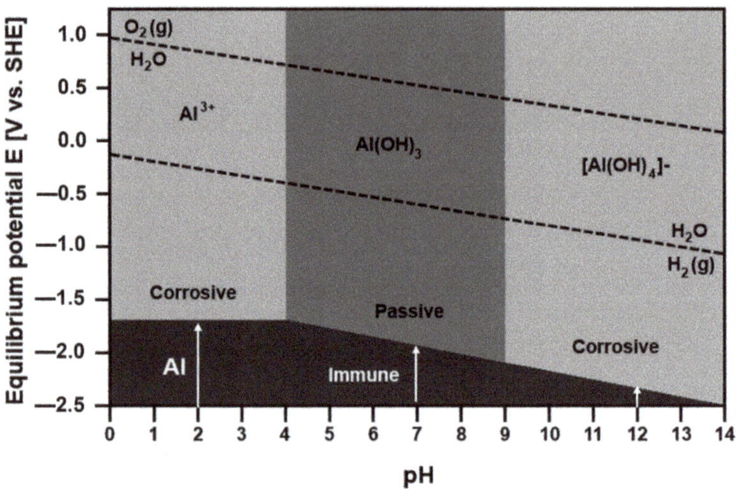

Figure 7. Equilibrium E–pH diagram of Al, replotted from [70].

Although E–pH plots are useful in determining the metal's tendency to corrode in the given environment, they do not provide a kinetic information. The rate of corrosion therefore needs to be determined separately by experimental methods. Corrosion rates are obtained either by weight loss measurements or electrochemical methods [67,68]. The weight loss measurement is a simple experiment to determine corrosion rates. In the experiment, a clean weighed piece of material with well-defined dimensions is exposed to the corrosive environment for a sufficient period. The corrosion rate (v_{corr}) is then calculated based on the recorded weight loss according to the following equation [19,20]

$$v_{corr} = \frac{\Delta w}{St} \tag{5}$$

In this equation, Δw is the weight loss, t is the reaction time and S is the exposed surface area.

Weight loss measurements, although useful, can be time-consuming and may not provide a complete information about reaction mechanism. Electrochemical techniques are therefore widely used to study the corrosion mechanisms of metals in different electrolytes. A potentiodynamic polarization is an electrochemical technique where the progress of reaction is controlled by potentiostat [67,68,70]. It brings in a variety of parameters and provides valuable information about reaction mechanism. In the experiment, three electrodes are assembled in a corrosion cell [72]. The corrosion cell includes a working electrode (sample), counter (auxiliary) electrode and reference electrode. During the experiment, the potential of the working electrode is systematically varied with respect to reference electrode. The resulting current is measured by counter electrode. The potential of reference electrode is constant and serves as reference value. Silver chloride (Ag/AgCl) and calomel electrodes (Hg/Hg_2Cl_2) immersed in a saturated KCl solution are most frequently used reference electrodes. Platinum mesh is used as counter electrode as this metal is corrosion resistant in most environments.

A schematic potential versus current density curve recorded during the polarization experiment is given in Figure 8. Several different regions can be distinguished on the curve. The first region is immune region. In this region, observed at low potentials, the metal is thermodynamically stable. In immune region, cathodic reactions prevail at the metal surface. The second region is labelled as active region. It is observed once a corrosion potential, E_{corr}, has been reached. In the active region, the metal actively corrodes according to Equation (2). The active corrosion means that the anodic dissolution of the metal takes place in the studied solution. Some metals can passivate. Therefore, a passivation region can be also observed on the polarization curve. The passive region corresponds to passive layer formation on the metal surface. The passivation is reflected by rapid current density increase or stabilization at potentials higher than E_p (passivation potential) on the polarization curve. At very high potentials, the current may start to abruptly increase. The increase is a result of passive film breakdown and happens at potential higher than transpassive potential, E_{tr}. The passive film breakdown may be initiated by aggressive halide anions and lead to localized corrosion (pitting). A given alloy system may contain either some or all regions shown in Figure 8a.

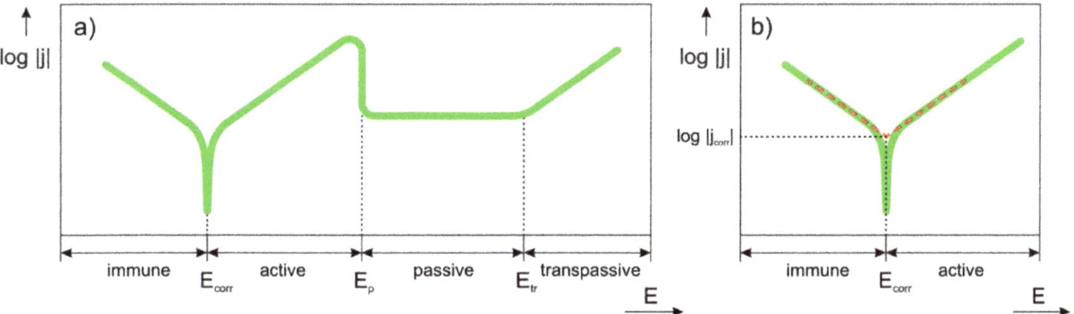

Figure 8. Schematic polarization curve of a passivating metal: (**a**) full curve, (**b**) Tafel extrapolation of cathodic and anodic regions.

The polarization curve provides a variety of electrochemical parameters. The corrosion potential and corrosion current density are important parameters that can be determined by Tafel extrapolation of the polarization curve. The procedure is shown in Figure 8b. In Tafel extrapolation, tangents to the polarization curve measured in immune and active regions are plotted. The intersection of the tangents determines the corrosion potential and corrosion current density. The corrosion potential reflects the tendency of the metal to corrode in given environment. The corrosion current corresponds to the rate of corrosion. In corrosion cell, a Faraday's law is valid [67,68]. The weight loss of the metal at the working electrode, Δw, is calculated by the following equation

$$\Delta w = \frac{A}{zF} I_{corr} t \tag{6}$$

In this equation, A is the atomic weight of the metal, I_{corr} is the corrosion current, t is the reaction time, F is Faraday's constant and z is the number of electrons involved in the electrochemical reaction. The corrosion rate, v_{corr}, is calculated from the corrosion current as

$$v_{corr} = \frac{\Delta w}{St} = \frac{A}{zF} j_{corr} \tag{7}$$

In this equation S is the sample surface area and j_{corr} is the corrosion current density ($j_{corr} = I_{corr}/S$). Equation (6) is valid for pure metals. For alloys, however, an equivalent weight, E_w must be introduced to account for different molar masses of constituent metals and different valence states. The following equation defines the equivalent weight of an alloy [73]

$$E_w = \frac{1}{\sum \frac{z_i f_i}{A_i}} \tag{8}$$

In the equation, z_i is the valence state, f_i is the weight fraction and A_i is the atomic weight of metal i in the alloy. The corrosion rate than becomes

$$v_{corr} = \frac{E_w}{F} j_{corr} \tag{9}$$

The assignment of valence states for TMs is often ambiguous as these elements have multiple stable valences. An independent experimental technique is therefore required, in addition to corrosion experiments, to establish the proper valence state. Another approach is to consult equilibrium Pourbaix diagrams [69]. The equilibrium E–pH diagrams can be used estimate the stable valence state of TM at the experimental conditions (electrode potential and pH of the electrolyte during corrosion test).

Metals become anodic and corrode only if their equilibrium half-cell potentials are smaller than the half-cell potential of the corresponding cathodic reaction [67,68]. When metals are combined into alloys it is no longer possible to define a unique half-cell potential. In multiphase alloys, different phases may act as local anodes and cathodes. The physical condition of the material may also be important. Constitutional variables such as the type and amount of structural defects (dislocations, grain boundaries) and crystal orientation are also important factors influencing the overall corrosion behavior.

Aluminum has a low standard electrode potential (Table 2, [68]). Therefore, aluminum and aluminum alloys are prone to corrosion. Nevertheless, the materials are also easily passivated. The passivation is related to spontaneous aluminum oxide/hydroxide film formation at the interface [74]. The passive film protects the material and impedes further reaction with the environment. Oxide layers grown on aluminum alloys at ambient temperatures are generally non-crystalline, although short-range cubic ordered structure has also been observed. In humid atmospheres, hydroxyl-oxides such as AlOOH or $Al(OH)_3$ may also form on aluminum surface. The passive film is generally self-renewing and self-healing. Therefore, an accidental loss of the passive film due to, for example, abrasion is rapidly restored.

Aluminum and its alloys are prone to pitting corrosion [75,76]. This type of local corrosion is often observed in seawater as it is initiated by chlorides and other halide anions in the electrolyte. The process may lead to passivity breakdown. Secondary phase particles are important constitutional variables affecting the corrosion rate. They can be classified into three different groups based on their electrochemical potential [76]: particles with active elements, noble elements, and particles with both active and noble elements. Reactive particles with active metals (such as Li, Mg or Zn) have low electrode potentials. These particles behave as anodes and subsequently dissolve when embedded in aluminum matrix. Particles with more noble elements (such as Fe or Cu) have higher electrode potentials and constitute local cathodes. They initiate anodic dissolution of Al matrix. The matrix adjacent to local cathode is preferentially attacked due to galvanic microcell created at the matrix/particle interface (Figure 9, [76]).

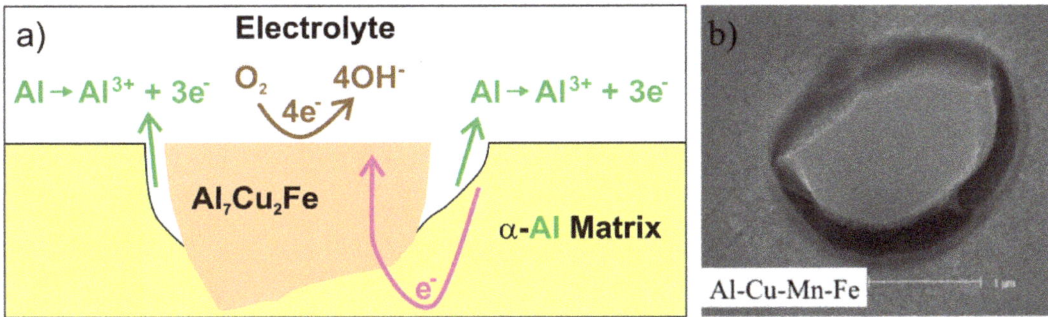

Figure 9. Schematic of the de-alloying and subsequent trenching of Al_2CuMg intermetallic in AA2024 aluminum alloy (**a**); microstructure image of corroded alloy surface after 1-h exposure in aerated H_2O at 30 °C (**b**), redrawn (**a**) and reproduced (**b**) from ref. [76].

If intermetallic particles contain both noble and active elements, their electrochemical behavior changes over time. The active elements may preferentially dissolve, leaving behind the noble metals. This process is known as de–alloying. It is schematically shown in Figure 10 for Al_2CuMg [76]. The galvanic interactions at the matrix-particle interface change because of de–alloying. The de–alloyed particle becomes nobler over time and may initiate an anodic dissolution of the surrounding matrix. Experimental conditions may also influence the particle dissolution behavior. For example, $Al_{20}Cu_2Mn_3$ is a noble particle with respect to matrix at room temperature. Nevertheless, it may become anodic at temperatures higher than 50 °C. At 50 °C a dealloying behavior of $Al_{20}Cu_2Mn_3$ has been observed, with de–alloying features much the same as Al_2CuMg [77,78].

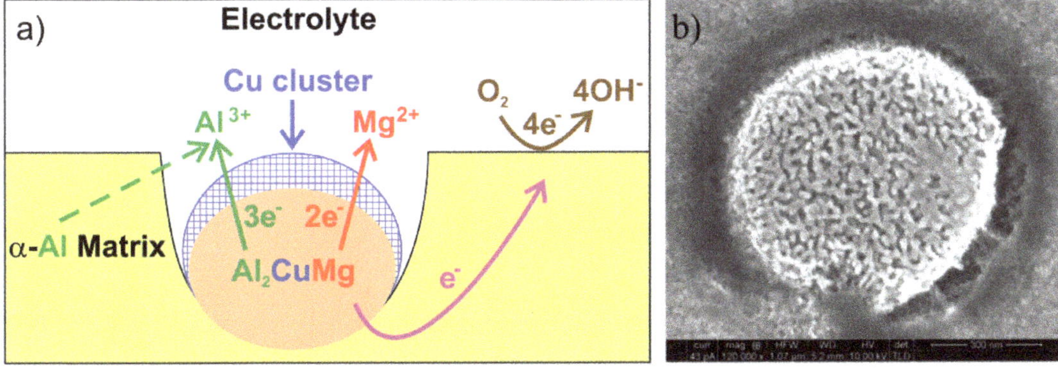

Figure 10. Schematic of trenching of Al_7Cu_2Fe intermetallic in AA2024-T3 aluminum alloy (**a**); microstructure image of corroded alloy surface after 1-h exposure in aerated H_2O at 30 °C, redrawn (**a**) and reproduced (**b**) from ref. [76].

In this review we aim to address the following fundamental questions:

- Which SCIP of the alloy has the highest tendency to corrode?
- Which factors influence the positions of anodic and cathodic sites on the metal surface?
- Which factors affect the corrosion rate?
- In this paper the complex Al–TM alloys have been sorted according to their chemical composition.

4. Al–Co Alloys

The Al–Co alloys composed of SCIPs were investigated by Lekatou et al. [79–82] The authors prepared a series of novel Al–Co alloys with 3.3–10.3 at.% Co by arc-melting. The microstructures obtained were ranging from fully eutectic to hypereutectic microstructures with primary precipitation of structurally complex Al_9Co_2. Relatively uniform and directional microstructures were obtained (Figure 11, [81]). The fraction of directionally solidified Al_9Co_2 was increasing with increasing Co concentration. Microstructures of the materials before and after corrosion are compared in Figure 11. The alloys displayed a similar corrosion behavior in 3.5 wt.% NaCl. The corrosion attack resulted in a preferential dissolution of Al solid solution (ss).

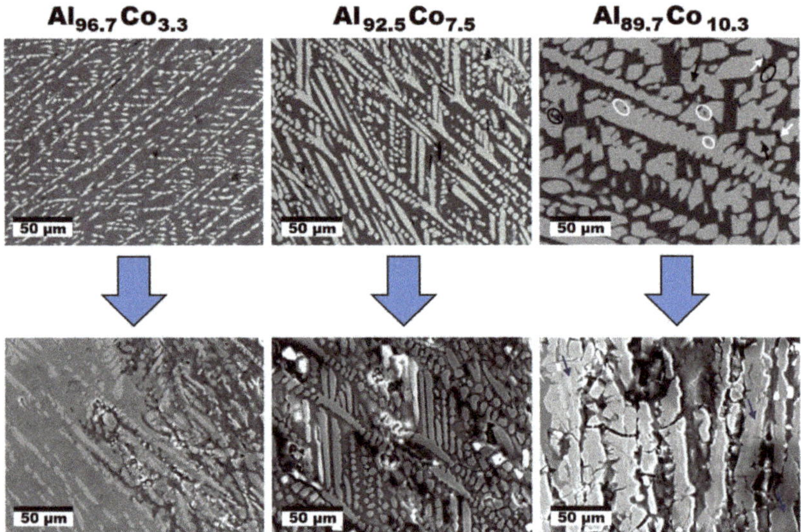

Figure 11. Microstructure of the rapidly solidified Al–Co alloys before and after corrosion in aqueous NaCl, adapted from reference [81].

A potentiodynamic polarization behavior of the Al–Co alloys in aqueous NaCl has been studied in refs. [81,82]. The $Al_{96.7}Co_{3.3}$ alloy showed a slightly superior corrosion performance compared to the remainder of the alloys as it had a relatively low corrosion current. Nevertheless, all Al–Co had a substantially higher corrosion resistance compared to Al. The anodic dissolution behavior was found to consist of four stages [81,82]. The individual stages were assigned to the following processes:

1. Active dissolution of Al(ss)
2. Passivation of Al(ss)
3. Breakdown of the passive film at the Al_9Co_2/Al(ss) interface and dissolution of the Al(ss) due to galvanic interaction with the nobler intermetallic (IMC)
4. Passivation of Al_9Co_2

At higher Co concentrations, a fragmentation of Al_9Co_2 in the corroded alloys occurred. The fragments piling in the gaps resulting from Al(ss) dissolution retarded the corrosion attack of the electrolyte. Al_9Co_2 had a higher electrochemical potential compared to Al(ss).

The corrosion behavior of Al–Co complex metallic alloys with 24–29 at.% Co was studied by Palcut et al. [83,84]. The following relative nobility of Al–Co CMAs has been found:

$$Al(ss) < Al_9Co_2 < Al_{13}Co_4 < Al_5Co_2 < \beta(AlCo) \tag{10}$$

The nobility of IMCs increases with increasing Co concentration. The volume fractions of the phases and physical contacts between them play an important role in the alloy corrosion behavior. Results indicate that a galvanic mechanism is involved. Moreover, it should be mentioned that Al–Co IMCs are brittle [85]. Therefore, a piling of noble but brittle particles, such as β(AlCo), in pores resulting from massive dissolution of surrounding less-noble phases may significantly influence the alloy stability [82]. The structural defects in the alloy may act as rapid diffusion paths leading to a significant material degradation over time. The galvanic coupling of noble IMCs with more active phases may be critical to the alloy corrosion stability in halide-containing environments.

The parallel occurrence of SCIPs with similar chemical compositions has a positive effect on the corrosion susceptibility of the alloy [84,85]. The $Al_{74}Co_{24}$ alloy was composed of three phases with close chemical compositions (Z–Al_3Co, Al_5Co_2, and $Al_{13}Co_4$, [84]). The $Al_{74}Co_{24}$ alloy had a higher corrosion potential compared to the remaining alloys which is an indicator of a superior corrosion resistance. The inspection of the alloy after corrosion testing revealed a relatively uniform phase dissolution [84]. The potential differences between constituent phases were probably small enough to initiate galvanic corrosion. The alloy corrosion could only be initiated at high electrode potentials. A polarization at high potentials resulted into a massive degradation of this alloy.

To further investigate the corrosion susceptibility of individual SCIPs with close chemical composition, an annealing of the $Al_{74}Co_{26}$ alloy at 1000 °C for 330 h has been carried out [86]. The annealing resulted in equilibrium microstructure of the alloy composed of Z–Al_3Co and Al_5Co_2. The Z–Al_3Co phase in the as-annealed $Al_{74}Co_{26}$ alloy was significantly less attacked. Although the bulk of this phase comprises less aluminum, it appears to be nobler and less susceptible to pitting corrosion compared to Al_5Co_2. The reason for this behavior could stem in a different structure of the phase surface. The Al_5Co_2 surface is terminated in puckered layers [87]. The surface of Z–Al_3Co, on the other hand, is more densely populated compared to Al_5Co_2 [88]. Therefore, Z–Al_3Co was less prone to corrosion attack.

The corrosion behavior of the as-annealed $Al_{74}Co_{26}$ alloy was investigated in neutral (NaCl, 0.6 mol dm^{-3}), alkaline (NaOH, 10^{-2} mol dm^{-3}) and acidic electrolytes (HCl, 10^{-2} mol dm^{-3}) by cyclic potentiodynamic polarization [86]. The potentiodynamic curves are shown in Figure 12. Anodic parts of the curves measured in HCl and NaCl solutions displayed a passive region which was followed by an abrupt current density increase. When the polarization scan was reversed, a positive hysteresis was found. These features indicate pitting corrosion. The polarization behavior in NaOH, on the contrary, corresponds to uniform alloy corrosion.

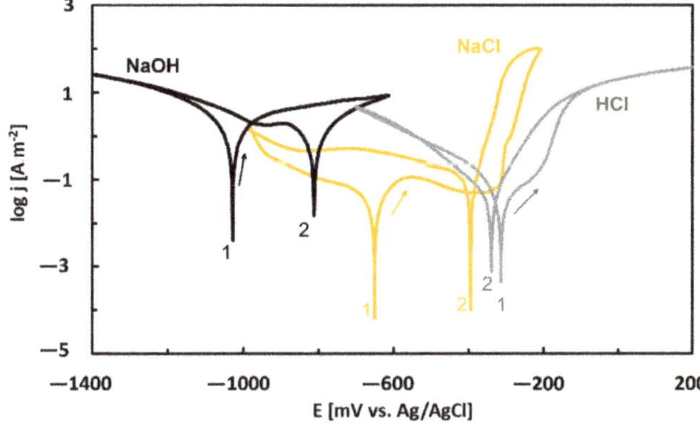

Figure 12. Potentiodynamic cyclic polarization curves of near-equilibrium $Al_{74}Co_{26}$ alloy in different electrolytes, re-plotted from reference [86].

The forward curves were evaluated by Tafel extrapolation and corrosion currents and corrosion potentials were obtained [86]. The lowest corrosion potential and highest corrosion current were found for NaOH. The highest corrosion potential and lowest corrosion current, on the other hand, were observed for the HCl solution. This behavior is in accordance with equilibrium E–pH diagram of Al (Figure 7).

5. Al–Cr Alloys

The Al–Cr alloys are expected to demonstrate a good corrosion resistance due to high concentrations of Al and Cr [89]. Both are passivating elements producing protective scales. The corrosion resistance of an $Al_{70}Cr_{20}Fe_{10}$ alloy was studied by Li et al. [90]. The authors used commercial gas-atomized $Al_{70}Cr_{20}Fe_{10}$ powders that were consolidated by spark plasma sintering. The phases present in the sintered Al–Cr–Fe pellets were the following: an icosahedral phase (i–Al–Cr–Fe), decagonal phase (d–Al–Cr–Fe) and crystalline $Al_8(Cr,Fe)_5$ and $Al_9(Cr,Fe)_4$ phases. Authors measured an open circuit potential (OCP) of the alloy in 3.5 wt. % NaCl and found that the OCP was nobler compared to Al. The OCP of the alloy was stable over time, indicating that an equilibrium has been rapidly established on the alloy surface. The $Al_{70}Cr_{20}Fe_{10}$ alloy had a nobler corrosion potential and hence a lower susceptibility to corrosion compared to Al. It passivated in saline solution spontaneously due to significant amount of Cr. The alloy had a higher corrosion rate compared to pure Al [90]. Nevertheless, the corrosion rate was close to that of 316 stainless steel and smaller than AISI 440C stainless steel or AISI H13 tool steel.

The passivation behavior of Al–Cr–Fe alloys was studied by Ott et al. [91] The authors used a flow microcapillary plasma mass spectrometry. The schematic of the experimental set up is shown in Figure 13 [91]. In the experiment, a tiny microcapillary was positioned on the alloy surface and continuously filled with the desired solution. The flow injection was operated in loops by switching the valve. The loop volume was continuously transferred to the inductively coupled plasma mass spectrometer (ICP MC) for element analysis. The microcapillary was refilled with fresh electrolyte from the reservoir. The circulation was ensured by a peristaltic pump. A microscope was included to control precise positioning of the capillary on the alloy surface. The technique provided time-resolved information about transient electrochemical processes and element-specific dissolution at the metal–electrolyte interface.

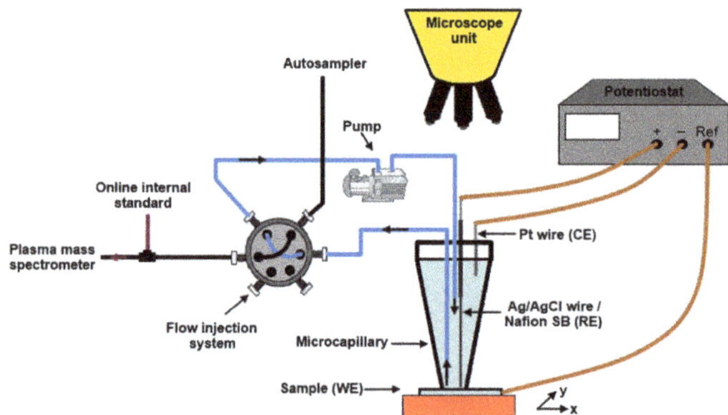

Figure 13. Schematic of the microcapillary flow ICP MS setup, re-drawn from reference [91].

The authors prepared and studied a polycrystalline γ-phase $Al_{64.2}Cr_{27.2}Fe_{8.1}$ alloy (composition given in at.%, [91]). The corrosion behavior was studied in two acidic solutions: H_2SO_4 (pH 0) and HCl (pH 2). In sulfuric acid, very low element dissolution rates were found. Neither Fe nor Al is stable at low pH [69]. Therefore, Cr is an essential element in the passive film stability. It helps to stabilize the Al cations within the passive film, as evidenced by a low Al release over 2 h. A possibly mixed oxy-hydroxide of Al and Cr was suggested to have been formed on the alloy surface (Figure 14, [91]). The dynamic passivation mechanism is related to the fact that the cation dissolution occurring at the oxyhydroxide–solution interface (②) is compensated by additional film growth at the metal–oxyhydroxide interface (①). Longer air-aging was found to be beneficial for stabilizing the passive film.

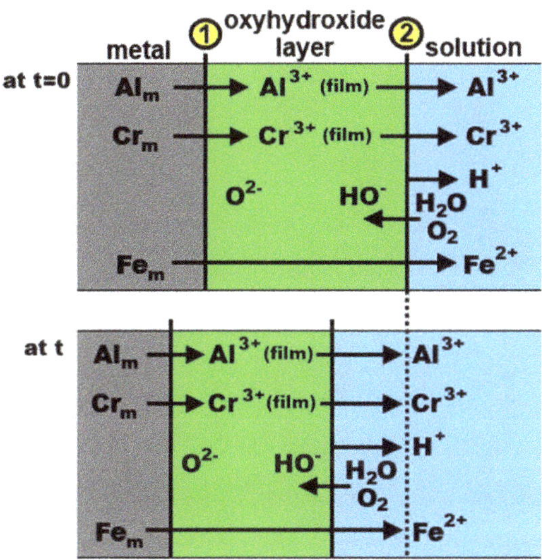

Figure 14. Schematic of the passive film evolution on γ-phase Al$_{64.2}$Cr$_{27.2}$Fe$_{8.1}$ alloy, re-drawn from reference [91].

In chloride-containing hydrochloric acid, ten times higher Al dissolution rates were found at the OCP, suggesting a decreasing stability of the spontaneously formed passive film [91]. The thickness of the dissolved passive film was much higher compared to H$_2$SO$_4$ (Table 3). But even in HCl, a potentiostatic polarization at 0.18V$_{SCE}$ slowed down the dissolution processes at the oxyhydroxide–solution interface by a factor of 6. The electrochemical polarization at low passive potentials induces electrical field generated oxide film modification, thereby increasing the chemical stability at the oxyhydroxide–solution interface. In the high potential passive region, a localized attack was initiated with subsequent active metal dissolution.

Table 3. Estimated equivalent passive film thickness [91].

Electrolyte	pH	Ageing Time [h]	Applied Potential [V$_{SCE}$] [a]	Dissolved Passive Film Thickness [nm]	Formed Passive Film Thickness [nm]
H$_2$SO$_4$	0	0.5	OCP [b]	106	
H$_2$SO$_4$	0	0.5	0.18	55.6	8.4
H$_2$SO$_4$	0	0.5	0.68	153	6.9
H$_2$SO$_4$	0	3	OCP	116	
HCl	2	0.5	OCP	880	
HCl	2	0.5	0.18	147	24.6

[a] Volts versus saturated calomel electrode, [b] Open circuit potential.

The passivation behavior of Al–Cr–Fe complex metallic alloys in NaCl+HCl mixtures was further investigated by Beni et al. [92] The authors prepared three alloys: polycrystalline single phase Al$_{79.5}$Cr$_{12.5}$Fe$_{8.0}$ (composed of orthorhombic phase), Al$_{64.2}$Cr$_{27.2}$Fe$_{8.1}$ (single phase, composed of cubic γ phase) and single crystalline orthorhombic $Al_{79.0}Cr_{15.0}Fe_{6.0}$. The corrosion behavior of the different alloys could be explained considering the passivating role of Cr combined with Fe oxyhydroxide precipitation. The anticipated reaction mechanism is presented in Figure 15 [92]. The $Al_{79.5}Cr_{12.5}Fe_{8.0}$ alloy was found to undergo an active dissolution in the electrolyte, as proven by the high element concentrations in solution measured by ICP MS. The chromium concentration (12.5 at.%) was small but sufficient to stabilize the initially air-formed oxyhydroxide for 22 days, as evidenced by the constant low pH of the solution and low dissolution compared to Al. The concentration of Cr was, however, too low to provide a long-term protection. A thick and non-protective layer has been formed on the surface. With increasing Cr concentration, a protective layer on the alloys started to form. The Cr concentration of 15.0 at.% was sufficient to stabilize the passive film up to 78 days. A complete and long-lasting protective scale was finally achieved at 27.2 at.% Cr [92,93].

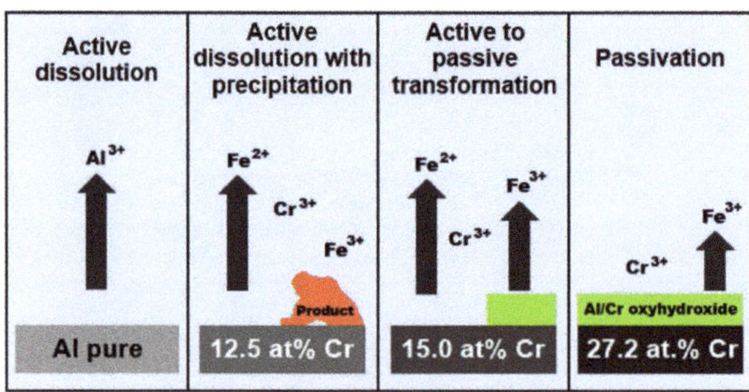

Figure 15. Schematic of passivation/dissolution processes on pure Al and Al–Cr–Fe alloys in aqueous HCl + NaCl mixture with initial pH 2, re-drawn from reference [92].

6. Al–Noble-Metal Alloys

Massiani et al. [35] investigated the corrosion behavior of crystalline and quasicrystalline phases in the Al–Cu–Fe(–Cr) alloys by potentiodynamic polarization in strongly acidic and alkaline solutions. They found that the corrosion resistance was determined by the alloy chemical composition. The complex crystal structure had only a minor influence. Rüdiger and Köster [94] found that the corrosion behavior of quasicrystals and their approximants in the Al–Cu–Fe alloy system could be explained based on the electrochemical behavior of the component elements. The surface of the icosahedral $Al_{63}Cu_{25}Fe_{12}$ was covered by a thick non-protective layer composed of Cu_2O, $Al(OH)_3$ and metallic Cu. The scale chemical composition was comparable to crystalline Al_7Cu_2Fe. The complex crystal structure thus did not have a substantial influence on the corrosion resistance [94]. Furthermore, the authors observed a formation of porous Cu layer in i–$Al_{63}Cu_{25}Fe_{12}$ phase at pH 0.

While Rüdiger and Köster studied single phase quasicrystalline alloys, Huttunen et al. investigated Al–Cu–Fe alloys composed of several different phases (Table 4, [95]). The corrosion behavior was determined by anodic polarization. The microstructural features and phase constitution of the alloys before and after the polarization were studied by scanning electron microscopy and X-ray diffraction.

Table 4. Chemical composition of SCIPs in Al–Cu–Fe alloys (in at.%) studied in reference [95].

Alloy	θ–Al$_2$Cu			ψ–Al$_{65}$Cu$_{20}$Fe$_{15}$			β–AlFe			λ–Al$_{13}$Fe$_4$		
	Al	Cu	Fe	Al	Cu	Fe	Al	Cu	Fe	Al	Cu	Fe
Al$_{60}$Cu$_{27.5}$Fe$_{12.5}$	49.8	49.5	0.7	65.2	22.5	12.3	67.6	32.2	0.2	73.4	4.5	22.1
Al$_{62.5}$Cu$_{25}$Fe$_{12.5}$	47.2	52.3	0.5	62.0	27.4	10.6	65.7	33.9	0.4	71.8	7.5	20.7
Al$_{65}$Cu$_{20}$Fe$_{15}$	51.5	47.7	0.8	65.3	22.9	11.8	68.3	31.4	0.3	73.7	4.3	22.0
Al$_{67.5}$Cu$_{20}$Fe$_{12.5}$	66.4	33.0	0.6	70.3	20.1	9.6	93.8	5.8	0.4	73.7	3.4	22.9

The study was focused on four different Al–Cu–Fe alloys: Al$_{67.5}$Cu$_{20}$Fe$_{12.5}$, Al$_{65}$Cu$_{20}$Fe$_{15}$, Al$_{62.5}$Cu$_{25}$Fe$_{12.5}$ and Al$_{60}$Cu$_{27.5}$Fe$_{12.5}$ [95]. The authors found that the presence of structurally complex phases did not improve the alloys corrosion resistance [95,96]. The chemical composition of the phases, however, was of great importance. The corrosion potentials of Al–Cu–Fe alloys with Cu-rich phases were nobler and had lower corrosion rates compared to Cu-lean alloys [95]. Relative amounts of the phases and their electrical contacts were also significant factors influencing the overall corrosion behavior. Phases with high Cu concentration remained virtually unaffected by corrosion. The phases with low Cu atomic fractions were susceptible to corrosion attack. This behavior could be explained by the higher electrode potential of Cu compared to Al and Fe (Table 2). The corrosion was found to occur by galvanic mechanism near phase boundaries. The corrosion behavior of Al–Cu–Fe alloys was studied in alkaline, neutral, and acidic solutions. In alkaline and neutral electrolytes, an oxidation of Al and Cr occurred on the surface of the alloys. The oxidation was accompanied with Cu deposition on the alloy surface. The Cu deposition interfered with passive layer formation and introduced pores into the oxide film. The icosahedral ψ–Al$_{65}$Cu$_{20}$Fe$_{15}$ was the only phase capable of forming a stable passive layer on the surface [95].

The Al–noble-metal alloys are interesting materials from electrochemical point of view. The alloys are prone to selective dissolution of less noble elements (leaching) because of markedly different electrode potentials of the constituent metals [97]. The less noble elements tend to dissolve in the electrolyte, leaving behind their vacant positions. As a result of leaching, a porous de-alloyed structure forms on the alloy surface [97]. The leaching can be either uniform or localized. Examples of leaching include preferential dissolution of Zn from brass (de-zincification) or Fe removal from gray cast iron (a so-called graphitic corrosion) [98]. Other examples include de-aluminification, de-nickelification and de-cobaltification [99].

Mishra et al. studied a chemical leaching of Al–Cu–Co decagonal quasicrystals [100]. The authors prepared two alloys with Al$_{65}$Cu$_{15}$Co$_{20}$ and Al$_{65}$Cu$_{20}$Co$_{15}$ chemical compositions and studied their corrosion behavior in aqueous NaOH (10 mol L^{-1}). The alloys were immersed in the alkaline solution at room temperature for 8 h. Most Al atoms were removed (Figure 16, [100]). A nearly uniformly distributed metallic Cu, Co and Co$_3$O$_4$ nanoparticles were found on the alloy surface after leaching. The crystallite size was calculated from the XRD reflections' broadening and further confirmed by TEM [100]. The nanostructure formation of the leached layer was controlled by Al dissolution rate during leaching. The dispersed Cu and Co nanoparticles were stable in the leached layer and Cu agglomeration was suppressed.

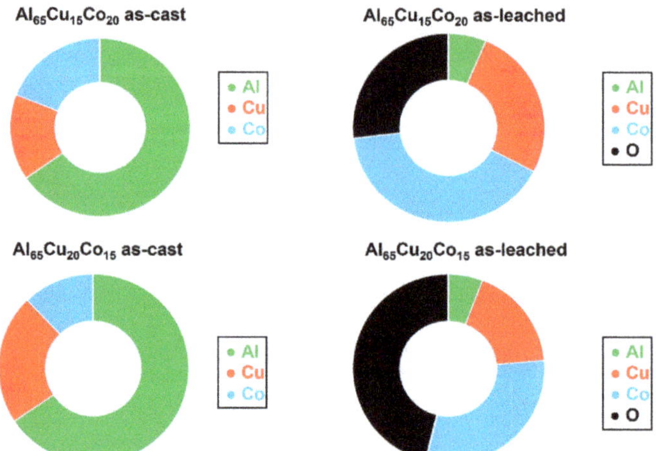

Figure 16. Chemical composition (in at.%) of as-cast and as-leached Al–Cu–Co alloys, plotted from data in reference [100].

Porous nanostructures composed of noble metals are important catalysts. The formation of Cu-rich porous nanostructure from decagonal $Al_{65}Co_xCu_{35-x}$ alloys (x = 12.5, 15, 17.5 at.%) was studied by Kalai Vani et al. [101] A selective dissolution of Al and Co was achieved by combined immersion of the alloys in both NaOH (5 mol L^{-1}) and Na_2CO_3 (0.5 mol L^{-1}) electrolytes. A high specific surface of 30 m^2 g^{-1} of the porous Cu structure was achieved.

The electrochemical de-alloying of binary Al–noble metal alloys was also studied [102]. It has been shown that nano-porous Pd, Ag and Au with various structures can be produced through electrochemical leaching of the Al–based alloys in NaCl aqueous solution. Galvanic interactions between coexisting phases dominate during corrosion of double phase alloys. The level of de-alloying depends on the critical de–alloying potential [103], diffusion of the noble element and reactivity of the noble element and chloride anion. The porosity evolution is a dynamic process. It is not a simple excavation of the less noble phase from two phase material. The formation of the porous nanostructure involves selective leaching of Al and is accompanied with coarsening of the noble element due to surface diffusion.

The corrosion behavior of Al–Pd alloys composed of SCIPs was studied in references [104,105]. The open circuit potentials are given in Figure 17. The OCPs decrease in the following order:

$$Al_{67}Pd_{33}, Al_{72}Pd_{28} \text{ (group I)} > Al_{77}Pd_{23}, Al_{88}Pd_{12} \text{ (group II)} \tag{11}$$

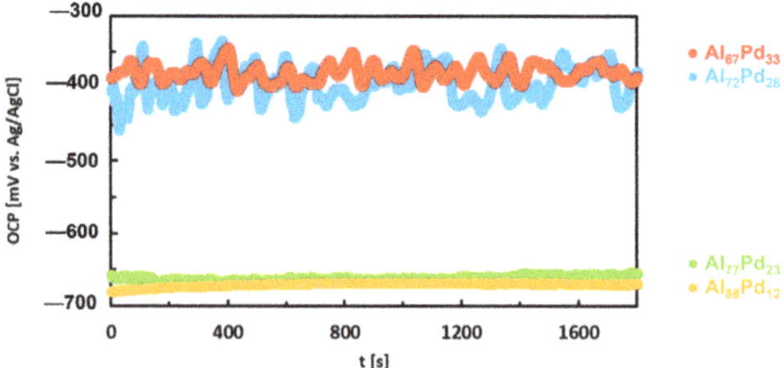

Figure 17. Open circuit potentials of Al–Pd alloys in aqueous NaCl, re-plotted from data in reference [105].

The OCPs of the alloys decrease with decreasing Pd concentration. This observation is in accordance with expectations since Al is electrochemically more active compared to Pd (Table 2). The corrosion resistance of both as-annealed and as-solidified alloys was comparable. A large difference, however, between OCP and E_{corr} has been found for group I alloys ($Al_{67}Pd_{33}$ and $Al_{72}Pd_{28}$). The OCPs of these alloys were comparable to their pitting potentials obtained by potentiodynamic polarization. The $Al_{67}Pd_{33}$ and $Al_{72}Pd_{28}$ alloys were probably in a pitting corrosion stage during the OCP measurement. This suggestion was manifested by oscillations of OCP resulting from a possible pitting behavior (Figure 17). The anodic dissolution of the alloy at pits requires a generation of cathodic current from the surrounding surface. The electric current bursts are transient and cause a temporary decrease in the OCP value. The pitting corrosion sites are usually very small. However, the current densities during transient bursts inside the pits can be up to 1 A/m^2 [106]. The significant corrosion rates of the alloy are due to aggressive environments developed inside the pits. Although the pits are small, they may affect the electrochemical response of much larger surface areas. Therefore, the differences in current densities on separated anodic and cathodic sites are reflected in potential oscillations (so-called electrochemical noise associated with localized corrosion).

Interactions between phases with different chemical composition play a significant role in alloy corrosion. The $Al_{67}Pd_{33}$ and $Al_{72}Pd_{28}$ alloys were found to be composed of structurally complex ε_n (Al_3Pd) and δ(Al_3Pd_2). The electrochemical nobility of Al–Pd phases in aqueous NaCl (0.6 mol L^{-1}) increases in the following order

$$(Al) < \varepsilon_n(Al_3Pd) < \delta(Al_3Pd_2) \tag{12}$$

The δ phase has a higher concentration of Pd. It serves as a cathode, and thereby further accelerates the anodic dissolution of the surrounding ε_n phase. The corrosion mechanism of Al–Pd alloys in aqueous NaCl involves a rapid passivation stage on the alloy surface [105,106]. However, once a breakdown potential is reached during potentiodynamic polarization, the passive layer becomes unstable and susceptible to local attack by chloride anions. Consequently, chloro–aluminum complex cations are formed and released into the solution. The local disruption of the passive layer reveals a naked alloy surface which becomes more susceptible to further corrosion attack.

The microstructures of as-annealed and as-solidified $Al_{72}Pd_{28}$ and $Al_{67}Pd_{33}$ alloys had similar features after corrosion testing. In the alloys a high number of inter-penetrating channels have been found [105,106]. Pits were also observed in the inter-connection between the channels. The formation of channels was driven by pitting and de–alloying. The pits were probably initiation sites of the channels. A preferential de-alloying of Al (de-aluminification) has also been observed. The preferential leaching of Al led to initiation of microcracks. During rapid solidification residual stresses have been accumulated in the alloys. The stresses were released during leaching, resulting in continuous tunnels inter-penetrating the surfaces of de-alloyed materials. A similar corrosion behavior was also found for the Al–Pd–Co alloys (Figure 18, [107]). The de-alloying of Al was more pronounced in the as-solidified alloys. This is probably a consequence of their higher defect concentrations compared to as-annealed alloys. The de–alloying behavior was significantly reduced in as-annealed alloys [105].

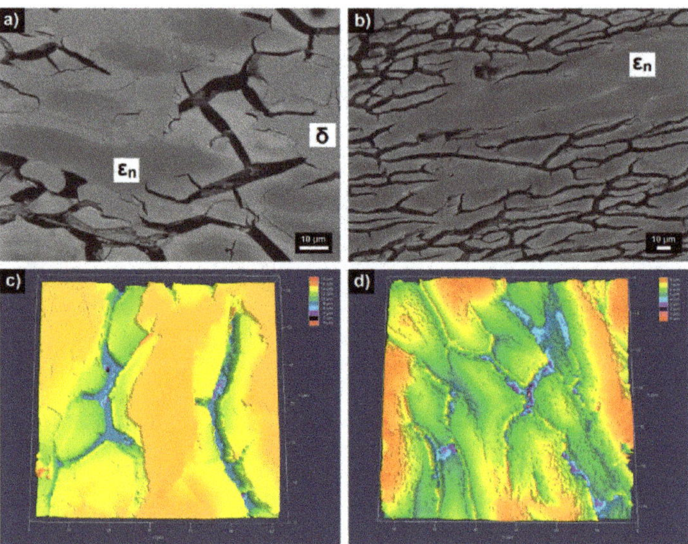

Figure 18. Microstructure of as-corroded Al–Pd–Co alloys: backscatter scanning electron microscopy images of $Al_{70}Pd_{25}Co_5$ (**a**) and $Al_{74}Pd_{12}Co_{14}$ (**b**) and confocal laser scanning images of $Al_{70}Pd_{25}Co_5$ (**c**) and $Al_{74}Pd_{12}Co_{14}$ (**d**). Reproduced from reference [107].

7. Comparison of Al–TM Alloys with Different Chemical Composition

Corrosion parameters of previously studied Al–based CMAs are given in Table 5. The corrosion behavior of Al–Co, Al–Pd and Al–Pd–Co alloys in aqueous NaCl (0.6 mol L^{-1}) is compared in Figure 19. Corrosion potentials of Al–Co alloys decrease with increasing Al concentration. The Al–Pd alloys have lower corrosion potentials. The values are smaller than the corrosion potentials of the remaining two alloy groups. Furthermore, the corrosion currents of Al–Pd alloys are higher compared to the Al–Pd–Co and Al–Co alloys (Figure 19b). These observations indicate that Al–Pd alloys are more susceptible to corrosion attack compared to the remaining two alloy groups.

Table 5. Electrochemical corrosion parameters of Al-based complex metallic alloys.

Alloy	Condition	Electrolyte	E_{corr} [mV vs. Ag/AgCl]	j_{corr} [A m^{-2}]	Reference
$Al_{96.7}Co_{3.3}$	Cast	Aerated NaCl (0.6 mol dm^{-3})	-838 ± 20	0.7 ± 0.1	[79]
$Al_{96.7}Co_{3.3}$	Arc-melted	Aerated NaCl (0.6 mol dm^{-3})	-820 ± 36	0.3 ± 0.1	[79]
$Al_{96.7}Co_{3.3}$	Powder-metallurgy sintered	Aerated NaCl (0.6 mol dm^{-3})	-890 ± 50	0.9 ± 0.2	[79]
$Al_{96.7}Co_{3.3}$	Cast	Aerated NaCl (0.6 mol dm^{-3})	-820 ± 36	0.3 ± 0.1	[81]
$Al_{95.1}Co_{4.9}$	Cast	Aerated NaCl (0.6 mol dm^{-3})	-823 ± 23	0.6 ± 0.1	[81]
$Al_{92.5}Co_{7.5}$	Cast	Aerated NaCl (0.6 mol dm^{-3})	-799 ± 23	0.7 ± 0.1	[81]
$Al_{89.7}Co_{10.3}$	Cast	Aerated NaCl (0.6 mol dm^{-3})	-816 ± 23	0.9 ± 0.1	[81]
$Al_{82.3}Co_{17.7}$	Cast	Aerated NaCl (0.6 mol dm^{-3})	-843 ± 16	1.6 ± 0.1	[82]
$Al_{82.3}Co_{17.7}$	Arc-melted	Aerated NaCl (0.6 mol dm^{-3})	-825 ± 18	0.8 ± 0.1	[82]
$Al_{82.3}Co_{17.7}$	Powder-metallurgy sintered	Aerated NaCl (0.6 mol dm^{-3})	-877 ± 23	5.8 ± 0.6	[82]

Table 5. *Cont.*

Alloy	Condition	Electrolyte	E_{corr} [mV vs. Ag/AgCl]	j_{corr} [A m^{-2}]	Reference
$Al_{99.1}Co_{0.9}$	Arc-melted	Aerated H_2SO_4 (1 mol dm^{-3})	-400 ± 7	1.9 ± 0.3	[80]
$Al_{97.6}Co_{2.4}$	Arc-melted	Aerated H_2SO_4 (1 mol dm^{-3})	-406 ± 2	3.6 ± 0.6	[80]
$Al_{96.7}Co_{3.3}$	Arc-melted	Aerated H_2SO_4 (1 mol dm^{-3})	-388 ± 10	2.6 ± 0.6	[80]
$Al_{95.1}Co_{4.9}$	Arc-melted	Aerated H_2SO_4 (1 mol dm^{-3})	-390 ± 5	1.9 ± 0.6	[80]
$Al_{92.5}Co_{7.5}$	Arc-melted	Aerated H_2SO_4 (1 mol dm^{-3})	-381 ± 18	3.1 ± 0.8	[80]
$Al_{89.3}Co_{10.3}$	Arc-melted	Aerated H_2SO_4 (1 mol dm^{-3})	-372 ± 7	2.9 ± 0.4	[80]
$Al_{76}Co_{24}$	Cast	Aerated NaCl (0.6 mol dm^{-3})	-706	0.13	[84]
$Al_{75}Co_{25}$	Cast	Aerated NaCl (0.6 mol dm^{-3})	-729	0.039	[84]
$Al_{74}Co_{26}$	Cast	Aerated NaCl (0.6 mol dm^{-3})	-515	0.58	[84]
$Al_{73}Co_{27}$	Cast	Aerated NaCl (0.6 mol dm^{-3})	-646	0.05	[84]
$Al_{72}Co_{28}$	Cast	Aerated NaCl (0.6 mol dm^{-3})	-672	0.04	[84]
$Al_{71}Co_{29}$	Cast	Aerated NaCl (0.6 mol dm^{-3})	-530	0.10	[83]
$Al_{74}Co_{26}$	Annealed in Ar 1050 °C 330 h	Aerated NaCl (0.6 mol dm^{-3})	-651	0.051	[86]
$Al_{74}Co_{26}$	Annealed in Ar 1050 °C 330 h	Aerated HCl (0.01 mol dm^{-3})	-314	0.032	[86]
$Al_{74}Co_{26}$	Annealed in Ar 1050 °C 330 h	Aerated NaOH (0.01 mol dm^{-3})	-1026	2.6	[86]
$Al_{72}Fe_{15}Ni_{13}$	Cast	Aerated NaCl (0.87 mol dm^{-3})	-	1.4	[108]
$Al_{69}Co_{21}Ni_{10}$	Cast	Aerated NaCl (0.87 mol dm^{-3})	-	1.2	[108]
$Al_{70}Cr_{20}Fe_{10}$	Powder metallurgy sintered	Aerated NaCl (0.6 mol dm^{-3})	-938	0.018	[90]
$Al_{65}Cu_{20}Fe_{15}$	Cast	NaCl (0.6 mol dm^{-3})	-638 ± 100	0.37	[109]
$Al_{78}Cu_7Fe_{15}$	Cast	NaCl (0.6 mol dm^{-3})	-586 ± 100	0.056	[109]
$Al_{80}Cu_5Fe_{14}Si_1$	Cast	NaCl (0.6 mol dm^{-3})	-570 ± 100	0.14	[109]
$Al_{70}Cu_9Fe_{10.5}Cr_{10.5}$	Cast	Na_2SO_4 (0.5 mol dm^{-3})	-556	1.6×10^{-2}	[35]
$Al_{64}Cu_{24}Fe_{12}$	Cast	Na_2SO_4 0.5 mol dm^{-3})	-555	7.3×10^{-2}	[35]
$Al_{63}Cu_{20}Co_{15}Si_2$	Cast	Na_2SO_4 (0.5 mol dm^{-3})	-635	2.2×10^{-2}	[35]
$Al_{70}Cu_9Fe_{10.5}Cr_{10.5}$	Cast	Na_2SO_4 (0.5 mol dm^{-3}) + H_2SO_4 (pH 2)	-496	1.4×10^{-2}	[35]
$Al_{64}Cu_{24}Fe_{12}$	Cast	Na_2SO_4 (0.5 mol dm^{-3}) + H_2SO_4 (pH 2)	-512	0.8×10^{-2}	[35]
$Al_{63}Cu_{20}Co_{15}Si_2$	Cast	Na_2SO_4 (0.5 mol dm^{-3}) + H_2SO_4 (pH 2)	-186	0.6×10^{-2}	[35]

Table 5. *Cont.*

Alloy	Condition	Electrolyte	E_{corr} [mV vs. Ag/AgCl]	j_{corr} [A m^{-2}]	Reference
$Al_{70}Cu_9Fe_{10.5}Cr_{10.5}$	Cast	NaOH (0.1 mol dm^{-3})	−921	1.6×10^{-2}	[35]
$Al_{64}Cu_{24}Fe_{12}$	Cast	NaOH (0.1 mol dm^{-3})	−1508	336×10^{-2}	[35]
$Al_{63}Cu_{20}Co_{15}Si_2$	Cast	NaOH (0.1 mol dm^{-3})	−1441	462×10^{-2}	[35]
$Al_{72}Pd_{20}Mn_8$	Annealed in Ar 800 °C 12 h	Deaerated NaCl (0.5 mol dm^{-3})	−355	0.5	[36]
$Al_{88}Pd_{12}$	Cast	Aerated NaCl (0.6 mol dm^{-3})	−794	0.89	[105]
$Al_{77}Pd_{23}$	Cast	Aerated NaCl (0.6 mol dm^{-3})	−809	0.82	[105]
$Al_{72}Pd_{28}$	Cast	Aerated NaCl (0.6 mol dm^{-3})	−797	0.63	[105]
$Al_{67}Pd_{33}$	Cast	Aerated NaCl (0.6 mol dm^{-3})	−798	0.62	[105]
$Al_{77}Pd_{23}$	Annealed in Ar 700 °C 500 h	Aerated NaCl (0.6 mol dm^{-3})	−763	0.75	[105]
$Al_{72}Pd_{28}$	Annealed in Ar 700 °C 500 h	Aerated NaCl (0.6 mol dm^{-3})	−841	0.68	[105]
$Al_{67}Pd_{33}$	Annealed in Ar 700 °C 500 h	Aerated NaCl (0.6 mol dm^{-3})	−783	0.72	[105]
$Al_{88}Pd_{12}$	Cast	Aerated HCl (0.01 mol dm^{-3})	−478	0.26	[104]
$Al_{77}Pd_{23}$	Cast	Aerated HCl (0.01 mol dm^{-3})	−450	0.27	[104]
$Al_{72}Pd_{28}$	Cast	Aerated HCl (0.01 mol dm^{-3})	−253	0.03	[104]
$Al_{67}Pd_{33}$	Cast	Aerated HCl (0.01 mol dm^{-3})	−200	0.17	[104]
$Al_{88}Pd_{12}$	Cast	Aerated NaOH (0.01 mol dm^{-3})	−1019	0.42	[104]
$Al_{77}Pd_{23}$	Cast	Aerated NaOH (0.01 mol dm^{-3})	−1033	0.25	[104]
$Al_{72}Pd_{28}$	Cast	Aerated NaOH (0.01 mol dm^{-3})	−879	0.25	[104]
$Al_{67}Pd_{33}$	Cast	Aerated NaOH (0.01 mol dm^{-3})	−892	0.34	[104]
$Al_{70}Pd_{25}Co_5$	Cast	Aerated NaCl (0.6 mol dm^{-3})	−677	0.10	[107]
$Al_{74}Pd_{12}Co_{14}$	Cast	Aerated NaCl (0.6 mol dm^{-3})	−758	0.18	[107]
$Al_{93}Co_5Ti_2$	Powder metallurgy sintered	Aerated NaCl (0.6 mol dm^{-3})	−707	1.1	[110]
$Al_{88}Co_{10}Ti_2$	Powder metallurgy sintered	Aerated NaCl (0.6 mol dm^{-3})	−676	14	[110]
$Al_{83}Co_{15}Ti_2$	Powder metallurgy sintered	Aerated NaCl (0.6 mol dm^{-3})	−490	4.0×10^{-4}	[110]
$Al_{78}Co_{20}Ti_2$	Powder metallurgy sintered	Aerated NaCl (0.6 mol dm^{-3})	−669	2.39	[110]
$Al_{73}Co_{25}Ti_2$	Powder metallurgy sintered	Aerated NaCl (0.6 mol dm^{-3})	−661	0.64	[110]
$Al_{68}Co_{30}Ti_2$	Powder metallurgy sintered	Aerated NaCl (0.6 mol dm^{-3})	−649	7.0	[110]

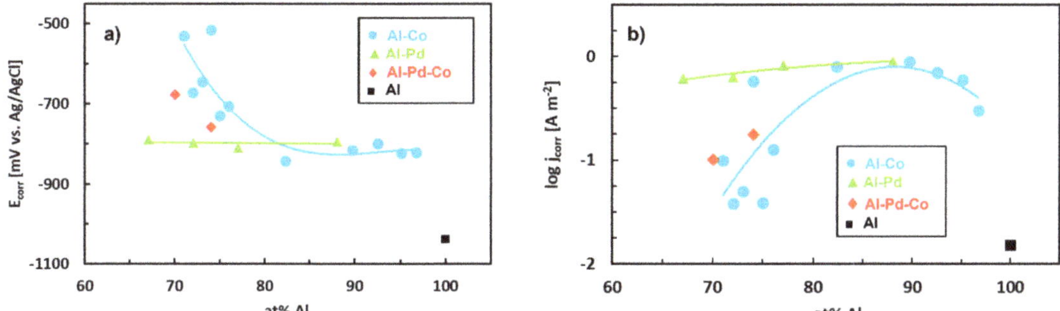

Figure 19. Corrosion parameters of Al–Pd, Al–Co and Al–Pd–Co alloys in aqueous NaCl (0.6 mol L^{-1}). Data is compiled from references [81,82,84,105,107].

The corrosion behavior of the Al–Pd–Co alloys is closer to Al–Co alloys (Figure 19). This observation is unexpected, since Al–Co–Pd and Al–Co alloys have different phase constitutions. Moreover, the preferentially corroding phase is ε_n in the Al–Pd–Co alloys. ε_n is absent in the Al–Co alloys. It can be noted that Co substitution for Pd significantly improves the corrosion resistance of ε_n. The positive influence of Co on the corrosion resistance of Al–TM alloys has also been observed by Sukhova and Polonskyy [108]. It is therefore the chemical composition and not the crystal structure of the phase that plays a dominant role in the corrosion resistance.

To further probe the role of chemical composition, we have compared the corrosion parameters of the previously discussed Al–TM alloys. The data compilation is plotted in Figure 20. The parameters are relatively scattered due to large differences in the overall alloy chemical compositions (Table 5). Nevertheless, some general trends can be noted. The as-solidified Al–Pd–Co alloys have corrosion current densities comparable to Al–Cu–Fe alloys. The corrosion potentials of the Al–Pd–Co and Al–Cu–Fe alloys are close to −650 mV (vs. Ag/AgCl)). The Al–Cr–Fe alloy is also included in Figure 20. This alloy has a lower corrosion potential compared to the remainder of the alloys. This is related to the absence of noble metals, such as Pd, in the alloy. Furthermore, the Al–Cr–Fe alloy has a low corrosion current due to the presence of Cr [90]. This element is responsible for a rapid passive layer formation on the alloy surface.

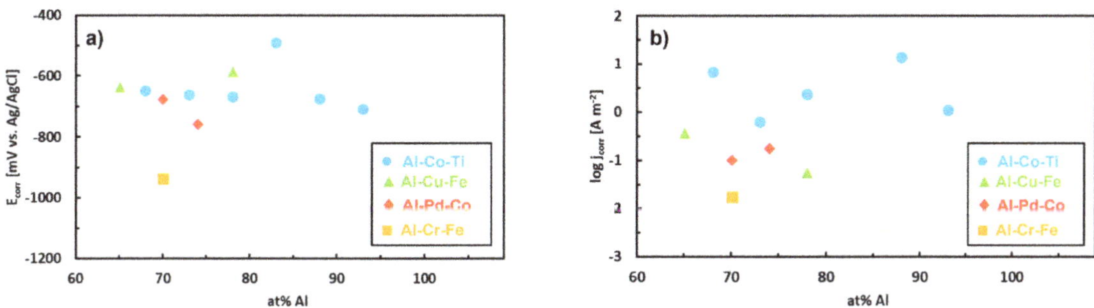

Figure 20. Corrosion parameters of ternary Al–TM alloys in aqueous NaCl (0.6 mol L^{-1}). Data is compiled from references [90,107,109,110].

The corrosion parameters of Al–Co–Ti alloys [110] are also included in the same figure. The corrosion potentials of these alloys are comparable to Al–Pd–Co alloys (Figure 20). The concentration of Ti in the Al–Co–Ti alloys was constant (2 at.%). The atomic fraction of Co was varied between 5–30 at.%. Due to small and constant Ti concentration, the microstructural features of the Al–Co–Ti alloys were comparable to Al–Co alloys [81,84]. The corrosion current densities of Al–Pd–Co alloys, however, were smaller compared to Al–Co–Ti alloys. The Al–15Co–2Ti alloy was an exception as the alloy demonstrated a lower corrosion current compared to the remainder of the alloys. The difference was related to different intermetallic particles contained in the alloy ($Al_{13}Co_4$, Al_9Co_2, and Al_3Ti). They had different volume fractions and morphologies compared to the remaining Al–Ti–Co alloys [110]. The observations show that specific Co atomic fractions may significantly increase the corrosion resistance of the bulk Al–TM alloys. The ε_n phase of the Al–Pd–Co alloys had a high concentration of Co. The Co additions significantly contributed to the superior corrosion performance of the bi-phasic $Al_{70}Pd_{25}Co_5$ alloy.

The corrosion parameters of the structurally complex Al–TM phases are comparable to previously studied Al–TM intermetallic phases with simpler structures [111]. Therefore, it is the chemical composition of the SCIP and not the crystal structure that influences the corrosion behavior. The electrochemical activity of the SCIPs may also vary with time. Zhu et al. investigated the corrosion performance of Al–TM intermetallic phases over time [78]. At early stages of exposure, a de–alloying was the primary corrosion mechanism. The de–alloying led to an ennoblement of intermetallic particles over time due to preferential Al leaching [78]. The ennoblement speeded up an anodic dissolution of the adjacent matrix and worsened the corrosion behavior. A long-term annealing may also influence the corrosion performance of the alloy constituent phases. It reduces internal stresses generated during casting and contributes to a more uniform element redistribution in the SCIPs.

8. Conclusions

In this paper the electrochemical corrosion behavior of Al–TM alloys composed of SCIPs has been reviewed. The following conclusions can be drawn:

1. The Al–TM alloys have a capability of forming passive layers because of high Al concentration. The Al–Cr alloys, for example, can form protective passive layers of considerable thickness in different electrolytes.

2. In halide-containing solutions the Al–TM alloys are prone to pitting corrosion. Galvanic microcells between different SCIPs form which may further accelerate the localized corrosion attack.

3. The electrochemical activity of aluminum–transition-metal SCIPs is primarily determined by electrode potential of the alloying element(s). The electrochemical nobility of individual SCIPs increases with increasing concentration of noble elements. The SCIPs with less noble elements tend to dissolve in contact with nobler particles. The SCIPs with noble metals are prone to selective de-alloying (de-aluminification). The electrochemical activity of SCIPs may change over time.

4. The chemical composition of the SCIPs has a primary influence on their corrosion properties. The structural complexity is secondary. It becomes important when phases with similar chemical composition come into close physical contact. The phase with higher structural complexity tends to be cathodic and can be retained during corrosion.

Author Contributions: Conceptualization, M.P. and L.Ď.; methodology, M.P., L.Ď. and I.Č.; formal analysis, M.P., L.Ď., I.Č. and P.P.; investigation, M.P., L.Ď., I.Č. and P.P.; resources, M.P. and P.P.; data curation, M.P.; writing—original draft preparation, M.P., L.Ď. and I.Č.; writing—review and editing, M.P., L.Ď., I.Č. and P.P.; visualization, M.P., L.Ď., I.Č. and P.P.; supervision, M.P.; project administration, M.P., and P.P.; funding acquisition, M.P. and P.P. All authors have read and agreed to the published version of the manuscript.

Funding: This research was funded by the Grant Agency VEGA of the Ministry of Education, Science, Research and Sport of the Slovak Republic and Slovak Academy of Sciences, grant number 1/0330/18, the Slovak Research and Development Agency project number APVV-20-0124 and the European Regional Development Fund, project No. ITMS2014+: 313011W085.

Institutional Review Board Statement: Not applicable.

Informed Consent Statement: Not applicable.

Data Availability Statement: This is a review article. The data used in this paper are publicly available in cited references.

Conflicts of Interest: The authors declare no conflict of interest. The funders had no role in the design of the study; in the collection, analyses, or interpretation of data; in the writing of the manuscript, or in the decision to publish the results.

References

1. Georgantzia, E.; Gkantou, M.; Kamaris, G.S. Aluminium alloys as structural material: A review of research. *Eng. Struct.* **2021**, *227*, 111372. [CrossRef]
2. Starke, E.A., Jr.; Staley, J.T. Application of modern aluminium alloys to aircraft. In *Fundamentals of Aluminum Metallurgy*; Lumley, R., Ed.; Woodhead Publishing: Cambridge, UK, 2011; pp. 747–783. [CrossRef]
3. Davis, J.R. Aluminum and aluminum alloys. In *Alloying: Understanding the Basics*; ASM International: Materials Park, OH, USA, 2001; pp. 351–416. [CrossRef]
4. Shechtman, D.; Blech, I.; Gratias, D.; Cahn, J.W. Metallic phase with long-range orientational order and no translational symmetry. *Phys. Rev. Lett.* **1984**, *53*, 1951–1953. [CrossRef]
5. Dubois, J.M. Quasicrystals. *J. Phys. Condens. Matter* **2001**, *13*, 7753–7762. [CrossRef]
6. Dubois, J.M. *Useful Quasicrystals*; World Scientific Publishing: Singapore, 2005; pp. 30–34. ISBN 978-9812561886.
7. Steurer, W.; Deloudi, S. *Crystallography of Quasicrystals—Concepts, Methods, and Structures*; Springer: Berlin/Heidelberg, Germany, 2009; pp. 258–293. ISBN 978-3-642-01899-2. [CrossRef]
8. Janot, C. *Quasicrystals*, 2nd ed.; Clarendon Press: Oxford, UK, 1994; ISBN 0-19-851778-5. [CrossRef]
9. Dubois, J.M.; Ferré, E.B.; Feuerbacher, M. Introduction to the Science of Complex Metallic Alloys. In *Complex Metallic Alloys: Fundamentals and Applications*; Dubois, J.-M., Belin-Ferré, E., Eds.; Wiley-VCH: Hoboken, NJ, USA, 2011; pp. 1–39. [CrossRef]
10. Priputen, P.; Liu, T.; Černičková, I.; Janičkovič, D.; Kolesár, V.; Janovec, J. Experimental study of Al-Co-Cu phase diagram in temperature range of 800–1050 °C. *J. Phase Equilib. Diff.* **2013**, *34*, 425–429. [CrossRef]
11. Priputen, P.; Černičková, I.; Lejček, P.; Janičkovič, D.; Janovec, J. A partial isothermal section at 1000 °C of Al-Mn-Fe phase diagram in vicinity of Taylor phase and decagonal quasicrystal. *J. Phase Equilib. Diff.* **2016**, *37*, 130–134. [CrossRef]
12. Socolar, J.E.S. Simple octagonal and dodecagonal quasicrystals. *Phys. Rev. B* **1989**, *39*, 10519–10551. [CrossRef] [PubMed]
13. Wang, N.; Cen, H.; Kuo, K.H. Two-dimensional quasicrystal with eightfold rotational symmetry. *Phys. Rev. Lett.* **1987**, *59*, 1010–1013. [CrossRef]
14. Cao, W.; Ye, H.Q.; Kuo, K.H. A new octagonal quasicrystal and related crystalline phases in rapidly solidified Mn_4Si. *Phys. Stat. Solidi* **1988**, *A107*, 511–519. [CrossRef]
15. Wang, N.; Fung, K.K.; Kuo, K.H. Symmetry study of the Mn-Si-Al octagonal quasicrystal by convergent beam electron-diffraction. *Appl. Phys. Lett.* **1988**, *52*, 2120–2121. [CrossRef]
16. Wang, Z.M.; Kuo, K.H. The octagonal quasilattice and electron-diffraction patterns of the octagonal phase. *Acta Crystallogr.* **1988**, *A44*, 857–863. [CrossRef]
17. He, L.X.; Zhang, Z.; Wu, Y.K.; Kuo, K.H. Stable decagonal quasicrystals with different periodicities along the tenfold axis in $Al_{65}Cu_{20}Co_{15}$. *Inst. Phys. Conf. Ser.* **1988**, *93*, 501–502.
18. Tsai, A.P.; Inoue, A.; Masumoto, T. Stable decagonal quasicrystals with a periodicity of 1.6 nm in Al-Pd-(Fe, Ru or Os) alloys. *Philos. Mag. Lett.* **1991**, *64*, 163–167. [CrossRef]
19. Ge, S.P.; Kuo, K.H. Icosahedral and stable decagonal quasicrystals in $Ga_{46}Fe_{23}Cu_{23}Si_8$, $Ga_{50}Co_{25}Cu_{25}$ and $Ga_{46}V_{23}Ni_{23}Si_8$. *Phil. Mag. Lett.* **1997**, *75*, 245–253. [CrossRef]
20. Yokoyama, Y.; Yamada, Y.; Fukaura, K.; Sunada, H.; Inoue, A.; Note, R. Stable decagonal quasicrystal in an Al-Mn-Fe-Ge system. *Jpn. J. Appl. Phys.* **1997**, *36*, 6470–6474. [CrossRef]
21. Sato, T.J.; Abe, E.; Tsai, A.P. A novel decagonal quasicrystal in Zn-Mg-Dy system. *Jpn. J. Appl. Phys.* **1997**, *36*, L1038–L1039. [CrossRef]
22. Yang, Q.B.; Wei, W.D. Description of the dodecagonal quasicrystal by a projection method. *Phys. Rev. Lett.* **1987**, *58*, 1020–1023. [CrossRef] [PubMed]
23. Chen, H.; Li, D.X.; Kuo, K.H. New type of two-dimensional quasicrystal with twelvefold rotational symmetry. *Phys. Rev. Lett.* **1988**, *60*, 1645–1648. [CrossRef] [PubMed]
24. Ďuriška, L. Phase-Constitutional, Thermodynamic, and Electrochemical Studies of Al-Base Complex Metallic Alloys. Ph.D. Thesis, Slovak University of Technology, Trnava, Slovakia, 2017. Available online: https://opac.crzp.sk/?fn=detailBiblioForm&sid=7F4700268CB7D7D1FF59AFA53547 (accessed on 6 July 2021).
25. Yokoyama, Y.; Miura, T.; Tsai, A.P.; Inoue, A.; Masumoto, T. Preparation of a large $Al_{70}Pd_{20}Mn_{10}$ single-quasicrystal by the Czochralski method and its electrical resistivity. *Mater. Trans. JIM* **1992**, *33*, 97–101. [CrossRef]
26. Dubois, J.M.; Kang, S.S.; Stebut, J. Quasicrystalline low-friction coatings. *J. Mater. Sci. Lett.* **1991**, *10*, 537–541. [CrossRef]
27. Rabson, D.A. Toward theories of friction and adhesion on quasicrystals. *Prog. Surf. Sci.* **2012**, *87*, 253–271. [CrossRef]
28. Tsai, A.P.; Suenaga, H.; Ohmori, M.; Yokoyama, Y.; Ioue, A.; Masumoto, T. Temperature dependence of hardness and expansion in an icosahedral Al-Pd-Mn alloy. *Jpn. J. Appl. Phys.* **1992**, *31*, 2530–2531. [CrossRef]

29. Köster, U.; Liu, W.; Liebertz, H.; Michel, M. Mechanical properties of quasicrystalline and crystalline phases in Al-Cu-Fe alloys. *J. Non-Cryst. Sollids* **1993**, *153–154*, 446–452. [CrossRef]

30. Yokoyama, Y.; Inoue, A.; Masumoto, T. Mechanical properties, fracture mode and deformation behavior of $Al_{70}Pd_{20}Mn_{10}$ single-quasicrystal. *Mater. Trans. JIM* **1993**, *34*, 135–145. [CrossRef]

31. Dubois, J.M.; Kang, S.S.; Perrot, A. Towards applications of quasicrystals. *Mater. Sci. Eng. A* **1994**, *A179–A180*, 122–126. [CrossRef]

32. Masumoto, T.; Inoue, A. YKK Corporation, Honda Giken Kogyo Kabushiki Kaisha. European Patent 94115137.5.

33. Jenks, C.J.; Thiel, P.A. Comments on quasicrystals and their potential use as catalysts. *J. Mol. Catal. A—Chem.* **1998**, *131*, 301–306. [CrossRef]

34. Dubois, J.M. New prospects from potential applications of quasicrystalline materials. *Mater. Sci. Eng. A* **2000**, *294–296*, 4–9. [CrossRef]

35. Massiani, Y.; Yaazza, S.A.; Croussier, J.P.; Dubois, J.M. Electrochemical behaviour of quasicrystalline alloys in corrosive solutions. *J. Non-Cryst. Solids* **1993**, *159*, 92–100. [CrossRef]

36. Asami, K.; Tsai, A.-P.; Hashimoto, K. Electrochemical behavior of a quasicrystalline Al-Pd-Mn alloy in a chloride-containing solution. *Mater. Sci. Eng. A* **1994**, *181*, 1141. [CrossRef]

37. Rüdiger, A.; Köster, U. Corrosion behavior of Al-Cu-Fe quasicrystals. *Mater. Sci. Eng. A* **2000**, *294–296*, 890–893. [CrossRef]

38. Torres, A.; Serna, S.; Patiño, C.; Rosas, G. Corrosion Behavior of W and b Quasicrystalline Al–Cu–Fe Alloy. *Acta Metall. Sin.* **2015**, *28*, 1117–1122. [CrossRef]

39. Veys, D.; Rapin, C.; Li, X.; Aranda, L.; Fournee, V.; Dubois, J.-M. Electrochemical behavior of approximant phases in the Al–(Cu)–Fe–Cr system. *J. Non-Cryst. Solids* **2004**, *347*, 1–10. [CrossRef]

40. Wehner, B.I.; Köster, U.; Rüdiger, A.; Pieper, C.; Sordelet, D.J. Oxidation of Al–Cu–Fe and Al–Pd–Mn quasicrystals. *Mater. Sci. Eng. A* **2000**, *294–296*, 830–833. [CrossRef]

41. Rampulla, D.M.; Mancinelli, C.M.; Brunell, I.F.; Gellman, A.J. Oxidative and Tribological Properties of Amorphous and Quasicrystalline Approximant Al-Cu-Fe Thin Films. *Langmuir* **2005**, *21*, 4547–4553. [CrossRef]

42. Yamasaki, M.; Pang Tsai, A. Oxidation behavior of quasicrystalline $Al_{63}Cu_{25}Fe_{12}$ alloys with additional elements. *Mater. Sci. Eng. A* **2000**, *294-296*, 890–893. [CrossRef]

43. Pinhero, P.J.; Anderegg, J.W.; Sordelet, D.J.; Besser, M.F.; Thiel, P.A. Surface oxidation of Al-Cu-Fe alloys: A comparison of quasicrystalline and crystalline phases. *Philos. Mag. B* **1999**, *79*, 91–110. [CrossRef]

44. Šulhánek, P.; Drienovský, M.; Černičková, I.; Ďuriška, L.; Skaudžius, R.; Gerhátová, Ž.; Palcut, M. Oxidation of Al-Co alloys at high temperatures. *Materials* **2020**, *13*, 3152. [CrossRef] [PubMed]

45. Graef, M.; McHenry, M. *Structure of Materials: An Introduction to Crystallography, Diffraction, and Symmetry*, 3rd ed.; Cambridge University Press: New York, NY, USA, 2008; ISBN 978-0-521-65151-6. [CrossRef]

46. Taylor, J.E.; Teich, E.G.; Damasceno, P.F.; Kallus, Y.; Senechal, M. On the form and growth of complex crystals: The case of Tsai-type clusters. *Symmetry* **2017**, *9*, 188. [CrossRef]

47. Livio, M. *Zlatý Řez*, 1st ed.; Dokořán: Prague, Czech Republic, 2006; ISBN 80-7363-064-8.

48. Smontara, A.; Smiljanić, I.; Bilušić, A.; Grushko, B.; Balanetskyy, S.; Jagličić, Z.; Vrtnik, S.; Dolinšek, J. Complex ε-phases in the Al-Pd-transition-metal systems. *J. Alloys Compd.* **2008**, *450*, 92–102. [CrossRef]

49. Bancel, P.A.; Heiney, P.A. Icosahedral aluminum-transition-metal alloys. *Phys. Rev. B* **1986**, *33*, 7917–7922. [CrossRef]

50. Create and Colour a Penrose Tiling. Available online: https://craftdesignonline.com/penrose/ (accessed on 6 July 2021).

51. Periodic Rhombus Tiling with Penrose Tiles. Available online: https://sk.pinterest.com/pin/268245721535341551/?d=t&mt=login (accessed on 6 July 2021).

52. Černičková, I.; Švec, P.; Watanabe, S.; Čaplovič, Ľ.; Mihalkovič, M.; Kolesár, V.; Priputen, P.; Bednarčík, J.; Janičkovič, D.; Janovec, J. Fine structure of phases of epsilon-family in Al73.8Pd11.9Co14.3 alloy. *J. Alloys Compd.* **2014**, *609*, 73–79. [CrossRef]

53. Černičková, I.; Priputen, P.; Liu, T.; Zemanová, A.; Illeková, E.; Janičkovič, D.; Švec, P.; Kusý, M.; Čaplovič, Ľ.; Janovec, J. Evolution of phases in Al-Pd-Co alloys. *Intermetallics* **2011**, *19*, 1586–1593. [CrossRef]

54. Černičková, I.; Ďuriška, L.; Priputen, P.; Janičkovič, D.; Janovec, J. Isothermal section of the Al-Pd-Co phase diagram at 850 °C delimited by homogeneity ranges of phases epsilon, U, and F. *J. Phase Equilib. Diff.* **2016**, *37*, 301–307. [CrossRef]

55. Černičková, I.; Ďuriška, L.; Drienovský, M.; Janičkovič, D.; Janovec, J. Phase transitions in selected Al-Pd-Co alloys during continuous cooling. *Kovové Mater.* **2017**, *55*, 403–411. [CrossRef]

56. Ďuriška, L.; Černičková, I.; Čička, R.; Janovec, J. Contribution to thermodynamic description of Al-Pd system. *J. Phys. Conf. Ser.* **2017**, *809*, 012008. [CrossRef]

57. Adamech, M.; Černičková, I.; Ďuriška, L.; Kolesár, V.; Drienovský, M.; Bednarčík, J.; Svoboda, M.; Janovec, J. Formation of less-know structurally complex zéta b and orthorhombic quasicrystalline approximant epsilon n on solidification of selected Al-Pd-Cr alloys. *Mater. Charact.* **2014**, *97*, 189–198. [CrossRef]

58. Priputen, P.; Kusý, M.; Drienovský, M.; Janičkovič, D.; Čička, R.; Černičková, I.; Janovec, J. Experimental reinvestigation of Al-Co phase diagram in vicinity of $Al_{13}Co_4$ family of phases. *J. Alloys Compd.* **2015**, *647*, 486–497. [CrossRef]

59. Kolesár, V.; Priputen, P.; Bednarčík, J.; Černičková, I.; Svoboda, M.; Drienovský, M.; Janovec, J. Evolution of phases in $Al_{55}Ni_{30}Pd_{15}$ alloy at temperatures up to 600 °C. *Intermetallics* **2014**, *46*, 141–146. [CrossRef]

60. Jenks, C.J.; Thiel, P.A. Quasicrystals: A Short Review from a Surface Science Perspective. *Langmuir* **1998**, *14*, 1392–1397. [CrossRef]

61. Fournée, V.; Ledieu, J.; Park, J.Y. Surface Science of Complex Metallic Alloys. In *Complex Metallic Alloys: Fundamentals and Applications*; Dubois, J.-M., Belin-Ferré, E., Eds.; Wiley-VCH: Hoboken, NJ, USA, 2011; pp. 155–206. [CrossRef]

62. Ebert, P.H.; Feuerbacher, M.; Tamura, N.; Wollgarten, M.; Urban, K. Evidence for a Cluster-Based Structure of AlPdMn Single Quasicrystals. *Phys. Rev. Lett.* **1996**, *77*, 3827–3830. [CrossRef]

63. Papadopolos, Z.; Kasner, G.; Ledieu, J.; Cox, E.J.; Richardson, N.V.; Chen, Q.; Diehl, R.D.; Lograsso, T.A.; Ross, A.R.; McGrath, R. Bulk termination of the quasicrystalline fivefold surface of $Al_{70}Pd_{21}Mn_9$. *Phys. Rev. B* **2002**, *66*, 184207. [CrossRef]

64. Krajčí, M.; Hafner, J. Structure, stability, and electronic properties of the i-AlPdMn quasicrystalline surface. *Phys. Rev. B* **2005**, *71*, 054202. [CrossRef]

65. Hafner, J. Ab initio density-functional calculations in materials science: From quasicrystals over microporous catalysts to spintronics. *J. Phys. Condens. Matter* **2010**, *22*, 384205. [CrossRef] [PubMed]

66. Krajčí, M.; Hafner, J. Surfaces of Complex Intermetallic Compounds: Insights from Density Functional Calculations. *Acc. Chem. Res.* **2014**, *47*, 3378–3384. [CrossRef]

67. Stansbury, E.E.; Buchanan, R.A. Introduction and overview of electrochemical corrosion. In *Fundamentals of Electrochemical Corrosion*; ASM International: Materials Park, OH, USA, 2000; pp. 1–21. [CrossRef]

68. Perez, N. Electrochemistry. In *Electrochemistry and Corrosion Science*; Kluwer Academic Publishers: New York, NY, USA, 2004; pp. 27–70. [CrossRef]

69. Pourbaix, M. *Atlas of Electrochemical Equilibria in Aqueous Solution*; National Association of Corrosion Engineers: Houston, TX, USA, 1974.

70. Hasannaeimi, V.; Sadeghilaridjani, M.; Mukherjee, S. *Electrochemical and Corrosion Behavior of Metallic Glasses*; MDPI: Basel, Switzerland, 2021. [CrossRef]

71. Burgleigh, T.D. Corrosion of aluminum and its alloys. In *Handbook of Aluminum. Alloys Production and Materials Manufacturing*; Totten, G.E., Mackenzie, D.S., Eds.; ASM International: Materials Park, OH, USA, 2003; Volume 2, pp. 421–463. [CrossRef]

72. ISO 17475:2005 Corrosion of Metals and Alloys—Electrochemical Test Methods—Guidelines for Conducting Potentiostatic and Potentiodynamic Polarization Measurements. Available online: https://www.iso.org/standard/31392.html (accessed on 6 August 2021).

73. *ASTM G102–89 Standard Practice for Calculation of Corrosion Rates and Related Information from Electrochemical Measurements*; ASTM International: West Conshohocken, PA, USA, 2015. Available online: https://www.astm.org/Standards/G102.htm (accessed on 6 August 2021).

74. Sukiman, N.L.; Zhou, X.; Birbilis, N.; Hughes, A.E.; Mol, J.M.C.; Garcia, S.J.; Zhou, X.; Thompson, G.E. Durability and Corrosion of Aluminium and Its Alloys: Overview, Property Space, Techniques and Developments. In *Aluminium Alloys—New Trends in Fabrication and Applications*; Ahmad, Z., Ed.; Intechopen: London, UK, 2012; pp. 47–97.

75. Szklarska-Smialowska, Z. Pitting Corrosion of aluminum. *Corros. Sci.* **1999**, *41*, 1743–1767. [CrossRef]

76. Li, J.; Dang, J. A Summary of Corrosion Properties of Al-Rich Solid Solution and Secondary Phase Particles in Al Alloys. *Metals* **2017**, *7*, 84. [CrossRef]

77. Li, J.; Hurley, B.; Buchheit, R. Effect of temperature on the localized corrosion of AA2024-t3 and the electrochemistry of intermetallic compounds during exposure to a dilute NaCl solution. *Corrosion* **2016**, *72*, 1281–1291. [CrossRef]

78. Zhu, Y.; Sun, K.; Frankel, G.S. Intermetallic Phases in Aluminum Alloys and Their Roles in Localized Corrosion. *J. Electrochem. Soc.* **2018**, *165*, C807–C820. [CrossRef]

79. Lekatou, A.G.; Sfikas, A.K.; Karantzalis, A.E. The influence of the fabrication route on the microstructure and surface degradation properties of Al reinforced by Al_9Co_2. *Mater. Chem. Phys.* **2017**, *200*, 33–49. [CrossRef]

80. Sfikas, A.K.; Lekatou, A.G. Electrochemical Behavior of Al–Al_9Co_2 Alloys in Sulfuric Acid. *Corros. Mater. Degrad.* **2020**, *1*, 249–272. [CrossRef]

81. Lekatou, A.; Sfikas, A.K.; Petsa, C.; Karantzalis, A.E. Al-Co alloys prepared by vacuum arc melting: Correlating microstructure evolution and aqueous corrosion behavior with Co content. *Metals* **2016**, *6*, 46. [CrossRef]

82. Lekatou, A.; Sfikas, A.K.; Karantzalis, A.E.; Sioulas, D. Microstructure and corrosion performance of Al-32%Co alloys. *Corros. Sci.* **2012**, *63*, 193–209. [CrossRef]

83. Palcut, M.; Priputen, P.; Kusý, M.; Janovec, J. Corrosion behaviour of Al-29at%Co alloy in aqueous NaCl. *Corros. Sci.* **2013**, *75*, 461–466. [CrossRef]

84. Palcut, M.; Priputen, P.; Šalgó, K.; Janovec, J. Phase constitution and corrosion resistance of Al-Co alloys. *Mater. Chem. Phys.* **2015**, *166*, 95–104. [CrossRef]

85. Eckert, J.; Scudino, S.; Stoica, M.; Kenzari, S.; Sales, M. Mechanical engineering properties of CMAs. In *Complex Metallic Alloys: Fundamentals and Applications*; Dubois, J.-M., Belin-Ferré, E., Eds.; Wiley-VCH: Hoboken, NJ, USA, 2011; pp. 273–315. [CrossRef]

86. Priputen, P.; Palcut, M.; Babinec, M.; Mišík, J.; Černičková, I.; Janovec, J. Correlation between microstructure and corrosion behavior of near-equilibrium Al-Co alloys in various environments. *J. Mater. Eng. Perform.* **2017**, *26*, 3970–3976. [CrossRef]

87. Meier, M.; Ledieu, J.; De Weerd, M.-C.; Huang, Y.-T.; Abreu, G.J.P.; Pussi, K.; Diehl, R.; Mazet, T.; Fournée, V.; Gaudry, E. Interplay between bulk atomic clusters and surface structure in complex intermetallic compounds: The case study of the $Al_5Co_2(001)$ surface. *Phys. Rev. B* **2015**, *91*, 085414. [CrossRef]

88. Anand, K.; Fournée, V.; Prevot, G.; Ledieu, J.; Gaudry, E. Nonwetting Behavior of Al-Co Quasicrystalline Approximants Owing to Their Unique Electronic Structures. *ACS Appl. Mater. Interfaces* **2020**, *12*, 15793–15801. [CrossRef] [PubMed]

89. Demange, V.; Machizaud, F.; Dubois, J.M.; Anderegg, J.W.; Thiel, P.A.; Sordelet, D.J. New approximants in the Al–Cr–Fe system and their oxidation resistance. *J. Alloys Compd.* **2002**, *342*, 24–29. [CrossRef]

90. Li, R.T.; Murugan, V.K.; Dong, Z.L.; Khor, K.A. Comparative Study on the Corrosion Resistance of Al–Cr–Fe Alloy Containing Quasicrystals and Pure Al. *J. Mater. Sci. Technol.* **2016**, *32*, 1054–1058. [CrossRef]

91. Ott, N.; Beni, A.; Ulrich, A.; Ludwig, C.; Schmutz, P. Flow microcapillary plasma mass spectrometry-based investigation of new Al–Cr–Fe complex metallic alloy passivation. *Talanta* **2014**, *120*, 230–238. [CrossRef]

92. Beni, A.; Ott, N.; Caporali, S.; Guseva, O.; Schmutz, P. Passivation/precipitation mechanisms of Al-Cr-Fe complex metallic alloys in acidic chloride containing electrolyte. *Electrochim. Acta* **2015**, *179*, 411–422. [CrossRef]

93. Ura-Bińczyk, E.; Homazava, N.; Ulrich, A.; Hauert, R.; Lewandowska, M.; Kurzydlowski, K.J.; Schmutz, P. Passivation of Al-Cr-Fe and Al-Cu-Fe-Cr complex metallic alloys in 1 M H_2SO_4 and 1 M NaOH solutions. *Corros. Sci.* **2011**, *53*, 1825–1837. [CrossRef]

94. Rüdiger, A.; Köster, U. Corrosion of Al-Cu-Fe quasicrystals and related crystalline phases. *J. Non-Cryst. Solids* **1999**, *250–252*, 898–902. [CrossRef]

95. Huttunen-Saarivirta, E.; Tiainen, T. Corrosion behaviour of Al-Cu-Fe alloys containing a quasicrystalline phase. *Mater. Chem. Phys.* **2004**, *85*, 383–395. [CrossRef]

96. Huttunen-Saarivirta, E. Microstructure, fabrication and properties of quasicrystalline Al–Cu–Fe alloys: A review. *J. Alloys Compd.* **2004**, *363*, 150–174. [CrossRef]

97. Erlebacher, J.; Aziz, M.J.; Karma, A.; Dimitrov, N.; Sieradzki, K. Evolution of nanoporosity in dealloying. *Nature* **2001**, *410*, 450–453. [CrossRef] [PubMed]

98. Battezzati, L.; Scaglione, F. De-alloying of rapidly solidified amorphous and crystalline alloys. *J. Alloys Compd.* **2011**, *509S*, S8–S12. [CrossRef]

99. Yu, J.; Ding, Y.; Xu, C.; Inoue, A.; Sakurai, T.; Chen, M. Nanoporous Metals by Dealloying Multicomponent Metallic Glasses. *Chem. Mater.* **2008**, *20*, 4548–4550. [CrossRef]

100. Mishra, S.S.; Pandey, S.K.; Yadav, T.P.; Srivastava, O.N. Influence of chemical leaching on Al-Cu-Co decagonal quasicrystals. *Mater. Chem. Phys.* **2017**, *200*, 23–32. [CrossRef]

101. Kalai Vani, V.; Kwon, O.J.; Hong, S.M.; Fleury, S.M. Synthesis of porous Cu from Al-Cu-Co decagonal quasicrystalline alloys. *Philos. Mag.* **2011**, *91*, 2920–2928. [CrossRef]

102. Zhang, Q.; Zhang, Z. On the electrochemical dealloying of Al-based alloys in a NaCl aqueous solution. *Phys. Chem. Chem. Phys.* **2010**, *12*, 1453–1472. [CrossRef]

103. Sieradzki, K.; Dimitrov, N.; Movrin, D.; McCall, C.; Vasiljevic, N.; Erlebacher, J. The Dealloying Critical Potential. *J. Electrochem. Soc.* **2002**, *149*, B370–B377. [CrossRef]

104. Palcut, M.; Ďuriška, L.; Špoták, M.; Vrbovský, M.; Gerhátová, Ž.; Černičková, I.; Janovec, J. Electrochemical corrosion of Al-Pd alloys in HCl and NaOH solutions. *J. Min. Metall. B* **2017**, *53*, 333–340. [CrossRef]

105. Ďuriška, L.; Palcut, M.; Špoták, M.; Černičková, I.; Gondek, J.; Priputen, P.; Čička, R.; Janičkovič, D.; Janovec, J. Microstructure, phase occurrence and corrosion behavior of as-solidified and as-annealed Al–Pd alloys. *J. Mater. Eng. Perform.* **2018**, *27*, 1601–1613. [CrossRef]

106. Kelly, R.G.; Scully, J.R.; Shoesmith, D.W.; Buchheit, R.G. *Electrochemical Techniques in Corrosion Science and Engineering*; Marcel Dekker Inc.: New York, NY, USA, 2003. [CrossRef]

107. Palcut, M.; Ďuriška, L.; Černičková, I.; Brunovská, S.; Gerhátová, Ž.; Sahul, M.; Čaplovič, Ľ.; Janovec, J. Relationship between Phase Occurrence, Chemical Composition, and Corrosion Behavior of as-Solidified Al–Pd–Co Alloys. *Materials* **2019**, *12*, 1661. [CrossRef]

108. Sukhova, O.V.; Polonskyy, V.A. Structure and corrosion of quasicrystalline cast Al-Co-Ni and Al-Fe-Ni alloys in aqueous NaCl solution. *East Eur. J. Phys.* **2020**, *3*, 5–10. [CrossRef]

109. Babilas, R.; Bajorek, A.; Spilka, M.; Radoń, A.; Łoński, W. Structure and corrosion resistance of Al–Cu–Fe alloys. *Prog. Nat. Sci. Mater.* **2020**, *30*, 393–401. [CrossRef]

110. Debili, M.Y.; Sassane, N.; Boukhris, N. Structure and corrosion behavior of Al-Co-Ti alloy system. *Anti Corros. Methods Mater.* **2017**, *64*, 443–451. [CrossRef]

111. Birbilis, N.; Buchheit, R.G. Electrochemical Characteristics of Intermetallic Phases in Aluminum Alloys: An experimental survey and discussion. *J. Electrochem. Soc.* **2005**, *152*, B140–B151. [CrossRef]

Article

Intergranular Corrosion and Microstructural Evolution in a Newly Designed Al-6Mg Alloy

Kweon-Hoon Choi [1,2], Bong-Hwan Kim [2,*], Da-Bin Lee [2], Seung-Yoon Yang [2], Nam-Seok Kim [2], Seong-Ho Ha [2], Young-Ok Yoon [2], Hyun-Kyu Lim [2] and Shae-Kwang Kim [2]

[1]   Department of Industrial Materials and Smart Manufacturing Engineering, Korea University of Science and Technology, Daejeon 34113, Korea; kchoi74@kitech.re.kr

[2]   Advanced Process and Materials R&BD Group, Korea Institute of Industrial Technology (KITECH), Cheonan 31056, Korea; dabin25@kitech.re.kr (D.-B.L.); sy8357@kitech.re.kr (S.-Y.Y.); kimns@kitech.re.kr (N.-S.K.); shha@kitech.re.kr (S.-H.H.); veryoon@kitech.re.kr (Y.-O.Y.); hklim@kitech.re.kr (H.-K.L.); shae@kitech.re.kr (S.-K.K.)

*   Correspondence: bonghk75@kitech.re.kr; Tel.: +82-32-850-0440

Abstract: In this work, the microstructure and corrosion behavior of a novel Al-6Mg alloy were investigated. The alloy was prepared by casting from pure Al and Mg+Al$_2$Ca master alloy. The ingots were homogenized at 420 °C for 8 h, hot-extruded and cold-rolled with 20% reduction (CR20 alloy) and 50% reduction (CR50 alloy). The CR50 alloy exhibited a higher value of intergranular misorientation due to a higher cold rolling reduction ratio. The average grain sizes were 19 ± 7 µm and 17 ± 9 µm for the CR20 and CR50 alloys, respectively. An intergranular corrosion (IGC) behavior was investigated after sensitization by a nitric acid mass-loss test (ASTM G67). The mass losses of both the CR20 and CR50 alloys were similar at early periods of sensitization, however, the CR20 alloy became more susceptible to IGC as the sensitization time increased. Grain size and β phase precipitation were two critical factors influencing the IGC behavior of this alloy system.

Keywords: light alloys; Al-Mg; high-strength; mechanical properties; intergranular corrosion; precipitation

Citation: Choi, K.-H.; Kim, B.-H.; Lee, D.-B.; Yang, S.-Y.; Kim, N.-S.; Ha, S.-H.; Yoon, Y.-O.; Lim, H.-K.; Kim, S.-K. Intergranular Corrosion and Microstructural Evolution in a Newly Designed Al-6Mg Alloy. *Materials* **2021**, *14*, 3314. https://doi.org/10.3390/ma14123314

Academic Editor: Marián Palcut

Received: 11 May 2021
Accepted: 9 June 2021
Published: 15 June 2021

Publisher's Note: MDPI stays neutral with regard to jurisdictional claims in published maps and institutional affiliations.

Copyright: © 2021 by the authors. Licensee MDPI, Basel, Switzerland. This article is an open access article distributed under the terms and conditions of the Creative Commons Attribution (CC BY) license (https://creativecommons.org/licenses/by/4.0/).

1. Introduction

The 5xxx series Al-Mg alloys are widely used materials in the automotive industry due to their high strength-to-weight ratios, weldability, and good corrosion resistance [1–3]. Strength and ductility can be improved by adding solute Mg to the alloys because of the solid solution strengthening mechanism [4,5]. Nevertheless, as the amount of Mg increases, a selective oxidation prevails and causes a formation of oxide inclusions at elevated temperatures [6]. The rapid oxidation of magnesium alloys can be suppressed by additions of Ca [7]. With this background, a new alloying strategy has been developed. The strategy uses an Mg + Al$_2$Ca master alloy instead of pure Mg during casting. The Mg + Al$_2$Ca master alloy improves the oxidation resistance of Al-Mg alloys by forming a protective CaO/MgO mixed layer on the surface [8,9]

Previous studies have shown that anodic β-phase (β-Mg$_2$Al$_3$) precipitation along grain boundaries is an important factor affecting the intergranular corrosion (IGC) susceptibility of Al-Mg alloy [6–8]. The Mg segregation leads to anodic β-phase formation at 50–200 °C. The β-phase precipitation is referred to as sensitization [10–13]. The β-phase formation is usually observed at grain boundaries (GB), sub-grains, intermetallics, dislocations and other defects. Generally, the sequences of β precipitation have been reported as follows [14–16]:

Solid solution α → Guinier–Peston zones → β″ → β′ → β precipitation

The β phases are electrochemically active compared to the Al matrix which leads to galvanic corrosion. The distribution and morphology of β precipitates affect the susceptibility of the alloy to IGC. To evaluate the IGC susceptibility, a nitric acid mass-loss test (NAMLT, ASTM G67) can be used to measure the corrosion rate.

In this research, the IGC susceptibility of sensitized Al-6Mg alloy with two different cold rolling conditions was studied. It is not yet clearly understood how the cold rolling process affects the corrosion behavior [17]. D'Antuono et al. observed that an increased rolling reduction increased the growth rate of β precipitation due to lowering nucleation temperature [18]. Additionally, an initial β precipitation was observed preferentially at low-angle grain boundaries rather than high-angle grain boundaries [19]. On the contrary, Wang et al. found that the maximum corrosion depth decreases with increasing cold rolling reduction ratio. They also stated that larger thickness reductions are attributable to an increased number of small-sized grains formed at the grain boundary, which can eventually break off the continuity of corrosion.

This paper aims to figure out the effect of cold rolling and sensitization treatment on the IGC susceptibility of a newly designed Al-6Mg alloy. In this study, a microstructure evolution of the alloy was analyzed by electron backscattered diffraction (EBSD). The continuity of β-phase precipitation at grain boundary was studied by scanning electron microscopy (SEM) and transmission electron microscopy (TEM). The effects of the cold rolling reduction ratio and sensitization heat treatment on the IGC susceptibility are discussed.

2. Materials and Methods

2.1. Sample Preparation

The ingots of Al-6Mg alloy were prepared in an electric resistance furnace by the casting of pure Al and Mg + Al_2Ca master alloy. The ingots were homogenized at 420 °C into a plate with an extrusion ratio of 21.7:1. The initial extrusion thickness was 12 mm. The alloy was cold-rolled to 3 mm by using a 2-high mill rolling machine (FENN, East Berlin, CT, USA). The cold-rolled specimen was further annealed at 420 °C for 1 h in a box furnace and air-cooled. Finally, the alloy was cold-rolled with 20% reduction (CR20 alloy) and 50% reduction (CR50 alloy), respectively. The chemical composition of the Al-6Mg alloy is shown in Table 1. The chemical composition was measured by inductively coupled plasma-atomic emission spectroscopy (ICP-AES, SPECTRO).

Table 1. Chemical composition of Al-6Mg alloy (wt.%, measured by ICP-AES).

Alloy	Si	Fe	Cu	Mg	Mn	Ca	Al
Al-6Mg	0.05	0.07	<0.01	6.04	0.03	~0.02	Bal.

2.2. Intergranular Corrosion Test (NAMLT)

The specimens were sensitized at 100 °C for 0, 3, 7, 48, 144, and 207 h prior to corrosion testing. A nitric acid mass-loss test (NAMLT, ASTM G67) was used to determine the IGC susceptibility [20]. According to the standard, the test specimen was machined into blocks. The surfaces of the blocks were polished with 320 grit abrasive paper, and the specimen dimensions were measured to the nearest 0.02 mm. Before the test, the samples were etched in 5% NaOH (1 min at 80 °C). The initial mass of each specimen was measured on a digital scale. The samples were then immersed in 70% HNO_3 for 24 h. After the test, all specimens were carefully rinsed with water and brush with a stiff plastic brush. The final weight of the specimens was measured and used to calculate the mass-loss rate [20].

Microstructural evolutions of the alloys were investigated by optical microscopy (OM, Nikon MA200, Nikon, Tokyo, Japan), scanning electron microscopy (FE-SEM/EDSFEI Quanta200F, Hitachi, Japan), electron backscatter diffraction (EBSD, EDAX Hikari EBSD detector, Mahwah, NJ, USA), and transmission electron microscopy (TEM, JEOL JEM-2100F, Akishima, Japan). The obtained EBSD data were analyzed using the MTEX software (open-source MATLAB toolbox) [21]. Image J software (open-source Java image program, NIH Image) was used to calculate the maximum intergranular corrosion depth and area.

3. Results

3.1. Microstructure

Inverse pole figure (IPF) maps of newly developed Al-6Mg with two different cold rolling conditions are shown in Figure 1. The results reveal that the CR20 alloy has more equiaxed grains compared to the CR50 alloy. The CR50 alloy has a deformed microstructure due to a higher cold rolling reduction ratio. Figure 2 shows the orientation distribution functions (ODF) of the CR20 and the CR50 alloy. Typical deformation textures of the cold-rolled Al-6Mg alloys are also shown in Figure 2. As the cold rolling reduction ratio increased, more deformed textures tended to be obtained.

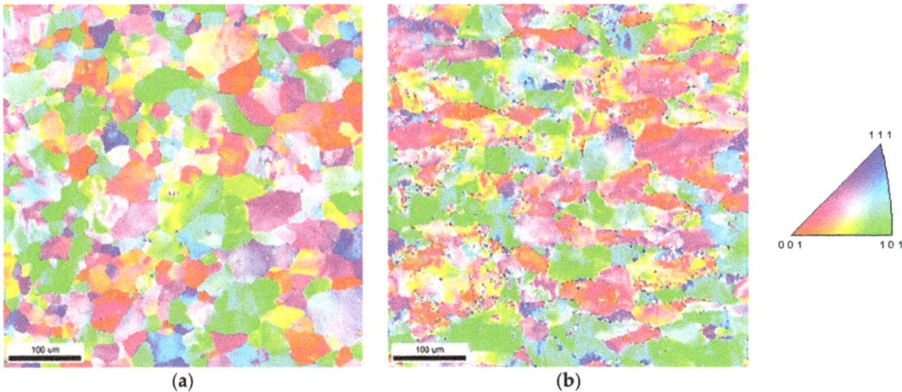

Figure 1. Inverse pole figure (IPF) maps of the new Al-6Mg CR20 (**a**) and CR50 alloys (**b**).

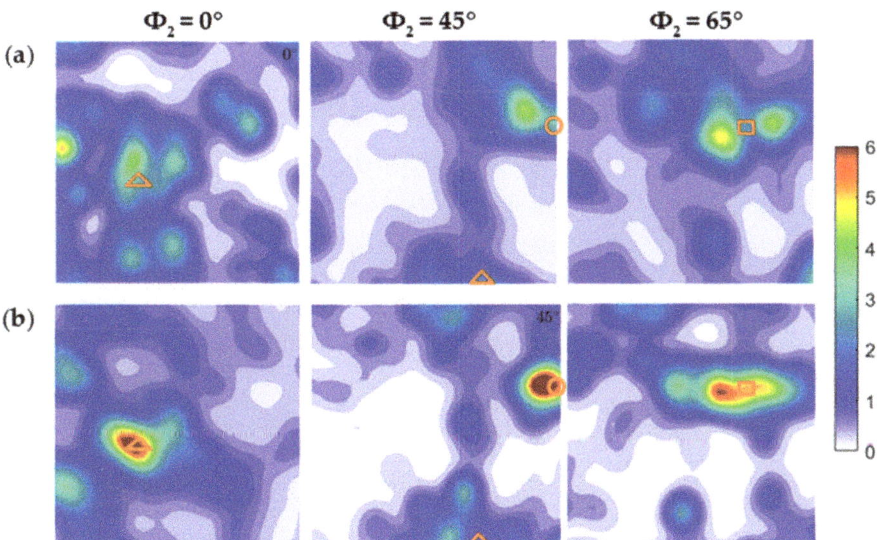

Figure 2. ODF sections of the new Al-6Mg CR20 (**a**) and CR50 alloys (**b**). The triangle represents Brass texture, circle is Copper texture, and the square is used for S texture.

Miller indices of the texture components for rolled samples are listed in Table 2. In the previous studies it was reported that the fraction of deformation texture components (Brass {110} <112>, Copper {112} <111> and S {123} <634>) increased while the recrystallization texture (Cube {100} <010> and Goss {011} <100>) did not change with increasing cold rolling reduction ratio in the Al-Mg alloy [22,23]. Figure 2a indicates that the CR20 alloy has an evolution of copper texture in $\Phi2 = 45$ section. The CR50 alloy shows a strong copper texture (Figure 2b). In addition, the CR50 alloy shows a stronger brass and S texture compared to the CR20 alloy. The results are in accordance with previous research [22,23]. However, the recrystallization texture was not found in the newly designed Al-6Mg alloy, which shows a partial disagreement with previous studies [22,23]. Therefore, the texture of the newly developed Al-6Mg alloy is yet to be completely understood.

Table 2. List of texture components for cold-rolled samples.

Type	Components	Orientation
Recrystallization textures	Cube	{100} <010>
	Goss	{011} <100>
Deformation textures	Brass	{011} <211>
	S	{123} <634>
	Copper	{112} <111>

Kernel average misorientation (KAM) is a measure of local grain misorientation based on the set of all neighboring misorientation. This map can be used to explain the effect of the rolling reduction ratio on the intergranular misorientation and evaluate the stored strain energy for a given point [24]. Figure 3 shows that the KAM map of the CR20 and CR50 alloys. The CR50 alloy exhibits a higher value of intergranular misorientation. As the reduction ratio of cold rolling increases, the dislocation density and volume fraction of low-angle grain boundary (LAGB) increase, which results in increasing KAM value. Figure 4 illustrates the grain size areas of CR20 and CR50 alloys. The average grain size areas of the CR20 and CR50 alloys were calculated as 365.8 ± 51.21 μm^2 and 283.6 ± 87.92 μm^2, respectively. These values correspond to average grain sizes of 19 ± 7 μm for the CR20 alloy and 17 ± 9 μm for the CR50 alloy, respectively.

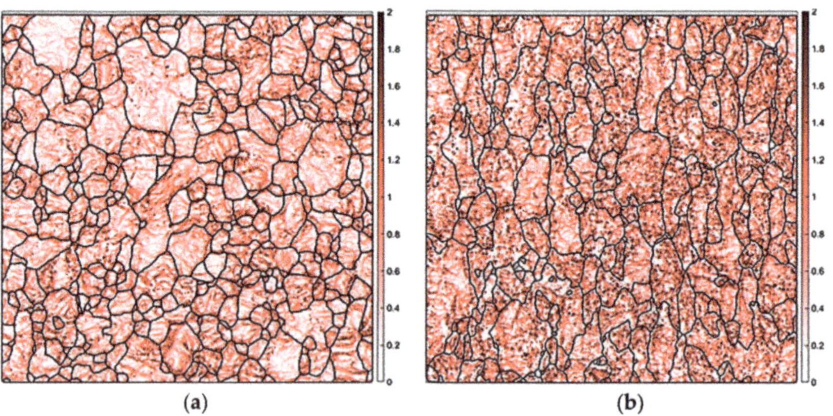

(a)   (b)

Figure 3. KAM maps of the CR20 (a) and CR50 alloys (b).

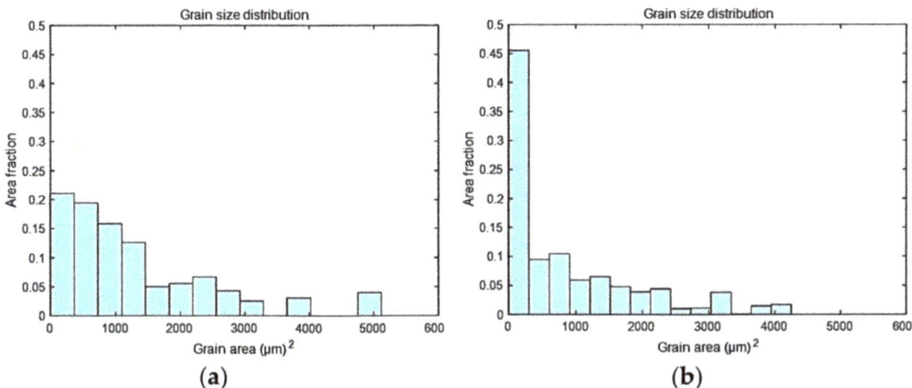

Figure 4. Grain size area histograms of the CR20 (**a**) and CR50 alloys (**b**).

3.2. Mass-Loss Test Results

Figure 5 shows the NAMLT results of the CR20 and CR50 alloys. The specimens were sensitized at 100 °C for 0, 3, 7, 48, 144, and 207 h prior to corrosion testing. The figure shows that both the CR20 and CR50 alloys had similar mass losses in concentrated nitric acid at sensitization for less than 7 h. As the sensitization time increased, the CR50 showed a slightly smaller mass-loss rate compared to the CR20 alloy. A previous study indicates that the IGC susceptibility of the Al alloys gradually decreases as the cold rolling reduction ratio increases [25].

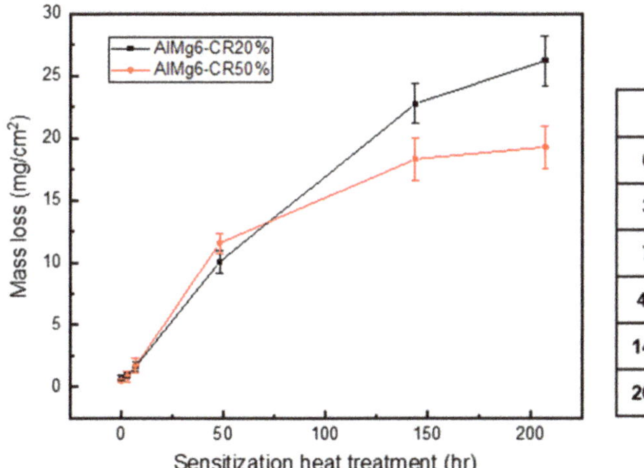

	CR20	CR50
0hr	0.71	0.53
3hr	0.96	0.85
7hr	1.66	1.75
48hr	10.05	11.58
144hr	22.75	18.33
207hr	26.22	19.30

Figure 5. NAMLT results of the CR20 alloy and CR50 alloy with different aging times at 100 °C.

The mass-loss of 25 to 75 mg/cm^2 indicates that the specimen is susceptible to IGC. Smaller values correspond to better corrosion resistance. The effect of the manufacturing process on the IGC of Al-Mg alloys has been previously studied [25–27]. Zhang et al. studied how the grain size modification by various manufacturing processes affected the intergranular corrosion [26]. Previous researchers reported that the NAMLT value of AA5083 alloy (~4.5 wt% of solute Mg) was 18 mg/cm^2 after 8 days (192 h) of sensitization exposure at 100 °C [27]. Similar values (19 mg/cm^2 after 200 h of sensitization at 100 °C) were obtained in the current study. Therefore, the IGC susceptibility of high Mg-containing Al-Mg alloys appears not to be negatively affected by the cold rolling process.

3.3. Microstructure of Corrosion Attacked Surface

Figures 6 and 7 show the metallographic cross sections of the CR20 and CR50 alloys after the IGC test. To explore the IGC behavior, the specimens were etched using Keller's reagent for 10 s. Figures 6a and 7a indicate the estimated corrosion depths after 0 h of sensitization. The depths are 16.75 μm and 17.71 μm for the CR20 and CR50 alloys, respectively. The penetration depth increases with increasing annealing time. The precipitation of β at grain boundaries increases with increasing sensitization time. It is assumed that the anodic precipitation of the CR50 alloy is deeper than in the CR20 alloy, which results in the IGC rate being higher in this alloy. As shown in Figures 6c and 7c, the grains start to be detached from the specimen after 48 h at 100 °C.

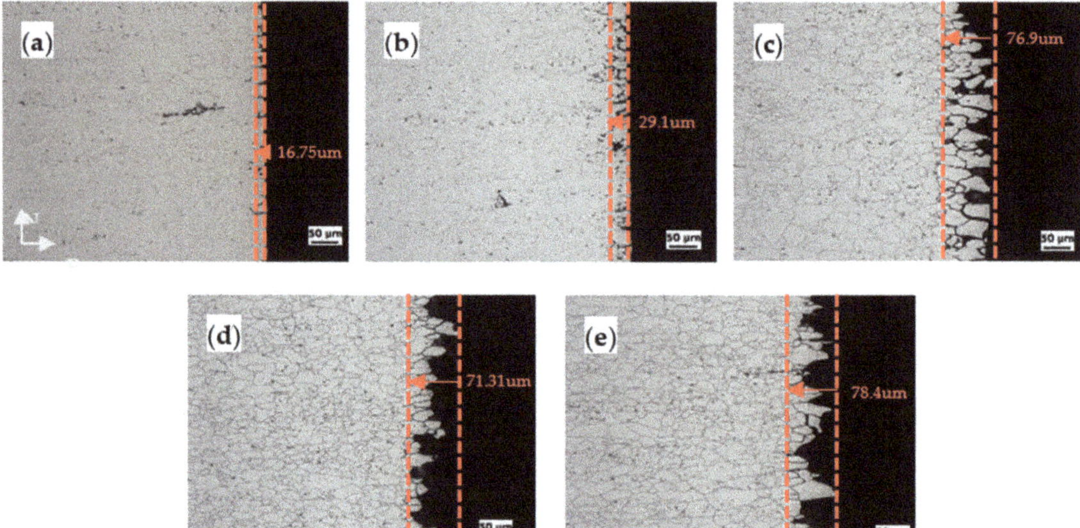

Figure 6. The surface of the CR20 alloy after NAMLT at different sensitization times: (**a**) 0 h, (**b**) 7 h, (**c**) 48 h, (**d**) 144 h, and (**e**) 207 h.

Figure 8 shows the maximum corrosion depth of both CR20 and CR50 alloys. The values are similar at the maximum sensitization time. On the other hand, the CR20 alloy shows a higher mass-loss rate compared to the CR50 alloy (Figure 5).

The continuity of β precipitation at grain boundary is critical for the IGC depth. The grain size, on the other hand, is a more important factor in the mass-loss rate at the early periods of sensitization. While there is not a significant difference between the alloys in maximum IGC depth, the NAMLT results show a higher mass-loss rate in the CR20 alloy after the long heat treatment time. This means that the CR20 alloy is more susceptible to IGC.

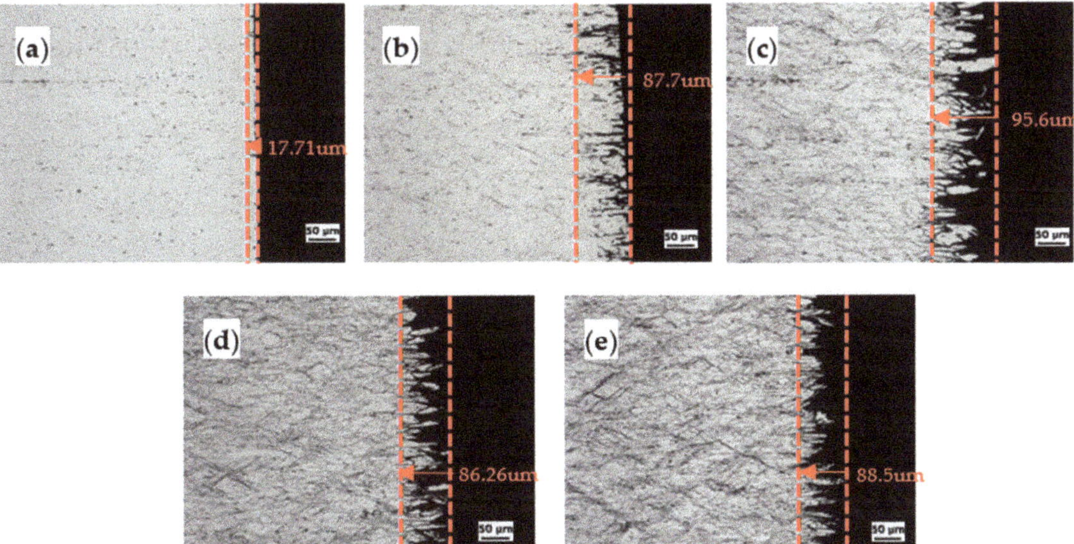

Figure 7. The surface of the CR50 alloy after NAMLT at different sensitization times: (**a**) 0 h, (**b**) 7 h, (**c**) 48 h, (**d**) 144 h, and (**e**) 207 h.

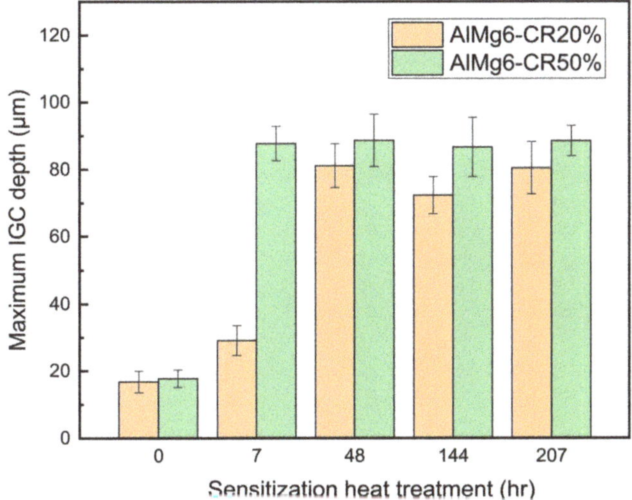

Figure 8. Maximum IGC depths of CR20 and CR50 alloys.

3.4. Schematic of β Phase Distribution in Grain Boundary

Figure 9 shows the SEM images of the CR20 and CR50 alloy surface after sensitization. The grain boundary is covered with β precipitation. The specimens were etched using H_3PO_4 etchant to selectively reveal the grain boundary covered by the anodic β precipitation. The continuity of β precipitation can be indirectly observed by the microstructure of the etched surface. At sensitization time of 7 h, the grain boundaries were more discernible in the CR50 alloy. This indicates that the CR50 alloy is more susceptible to IGC at 7 h sensitization. Figures 10 and 11 show a close look-up at the microstructure of the etched surface. The grain boundary is covered by β precipitation. The β precipitation is more discontinuous in the CR20 alloy.

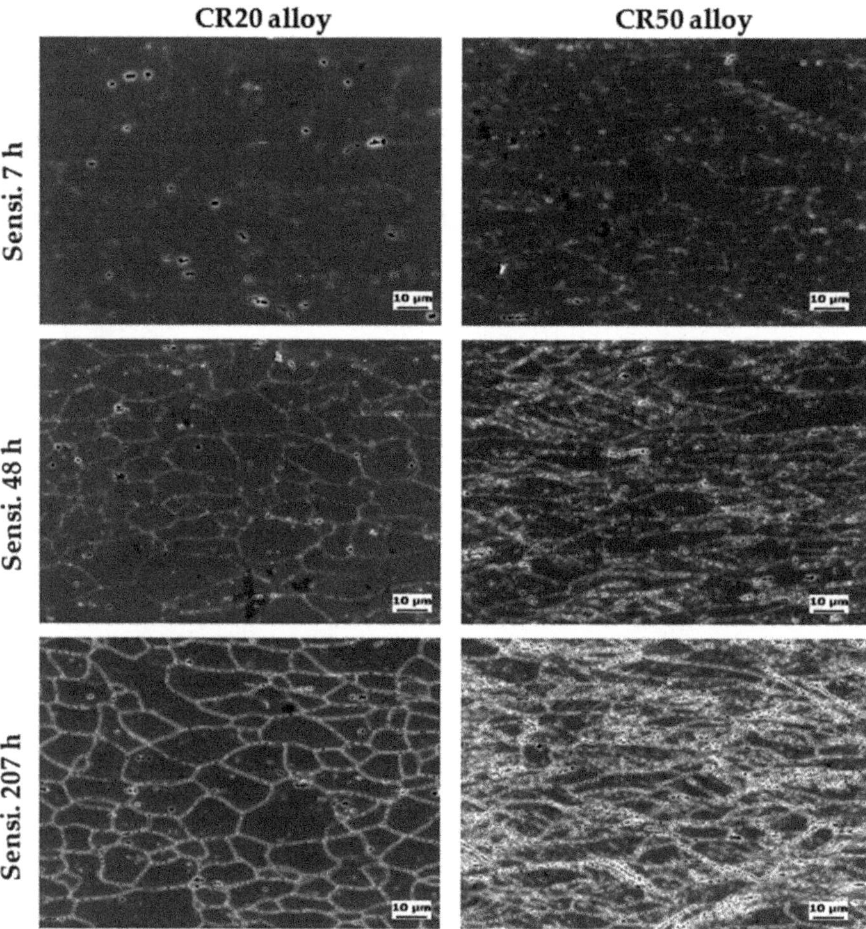

Figure 9. Microstructure of sensitized CR20 and CR50 alloys.

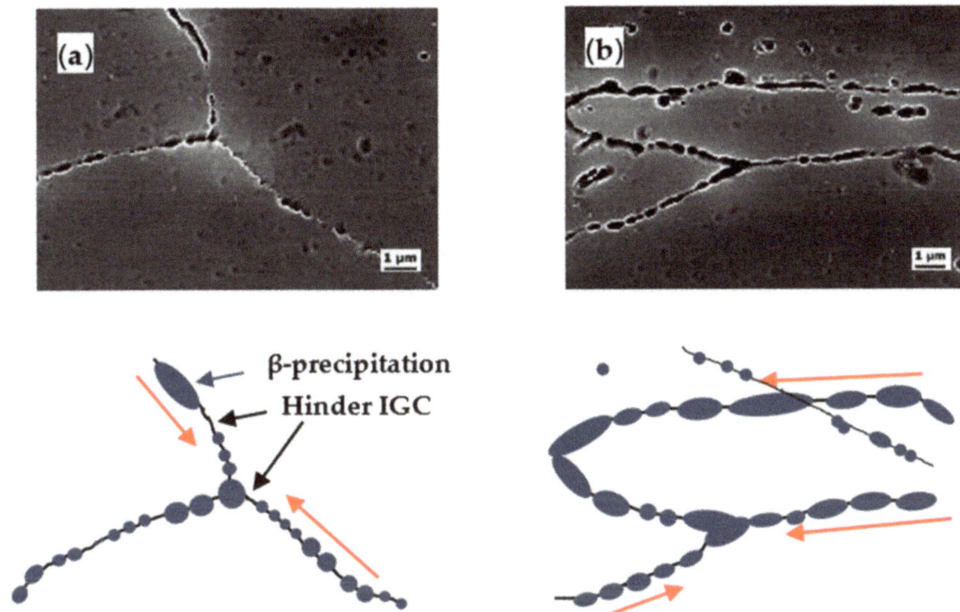

Figure 10. Close inspection of β precipitation covered grain boundary: (**a**) CR20 alloy, Sensi. 48 h, (**b**) CR50 alloy, Sensi. 48 h.

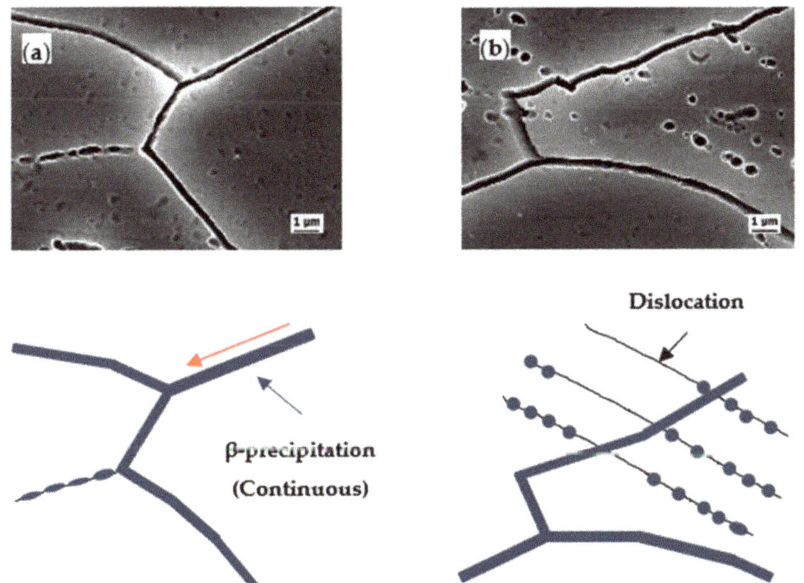

Figure 11. Close inspection of β precipitation covered grain boundary: (**a**) CR20 alloy, Sensi. 207 h, (**b**) CR50 alloy, Sensi. 207 h.

Previous studies found that the precipitate growth rates increased with rolling reduction [18,19]. A high density of dislocations can lower the activation energy, which most likely initiates the precipitation in the rolled specimen [18]. Increasing the dislocation density results in enhancing the diffusivity of Mg atoms at sensitization treatment due to pipeline diffusion [19]. On the other hand, there is no difference in the continuity of the precipitation at grain boundary at long sensitization times (Figure 11) by precipitation [18]. Considering both mass-loss results and maximum IGC depths (Figures 5 and 9), the grain size thus becomes a crucial factor in IGC at longer sensitization times.

3.5. Direct Observation of β Precipitation Distribution

Figure 12 shows the thickness of the precipitates for the CR20 and CR50 alloys. The size of the precipitates was calculated from the crossline thickness of the precipitates by TEM image (Figure 13). Figure 13e represents the TEM image of the CR20 alloy with EDS mapping of magnesium at 48 h of sensitization. The thickness was found to be 6.1 ± 1.7 nm for the CR20 alloy and 7.5 ± 3.0 nm for the CR50 alloy at 48 h of sensitization time. The precipitate thickness at 207 h is higher. The thickness of β-precipitates is crucial for the IGC rate as it affects the continuity of the precipitation at the grain boundary. Figure 13a reveals that the β precipitation of the CR20 alloy in the early period of sensitization is discontinuous. This results in a superior corrosion resistance compared to the CR50 alloy at the early sensitization time. The size of β precipitates at longer sensitization times becomes larger for the CR20 alloy. Both Figure 13c,d show that β precipitation was almost continuously distributed at the grain boundary with almost the same thickness as shown in Figure 12. Zhang et al. also agree that the kinetics of precipitation growth is reduced with sensitization time [28]. The results of this study show that the IGC significantly depended on the grain size for long-term sensitization, as compared to the size of precipitates.

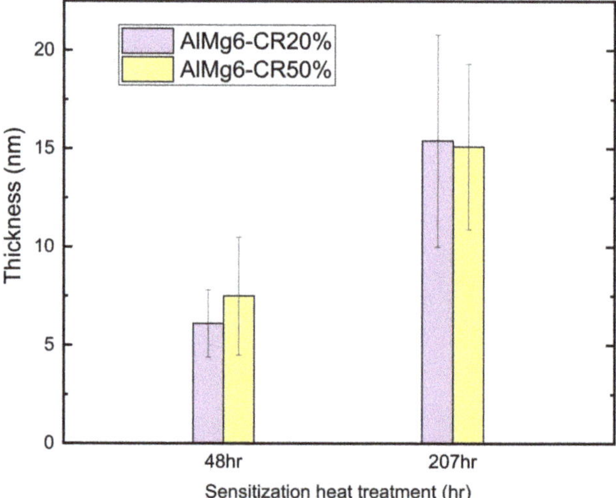

Figure 12. Thickness of β-phase precipitates in the CR20, and CR50 alloys sensitized for 48 and 207 h, respectively.

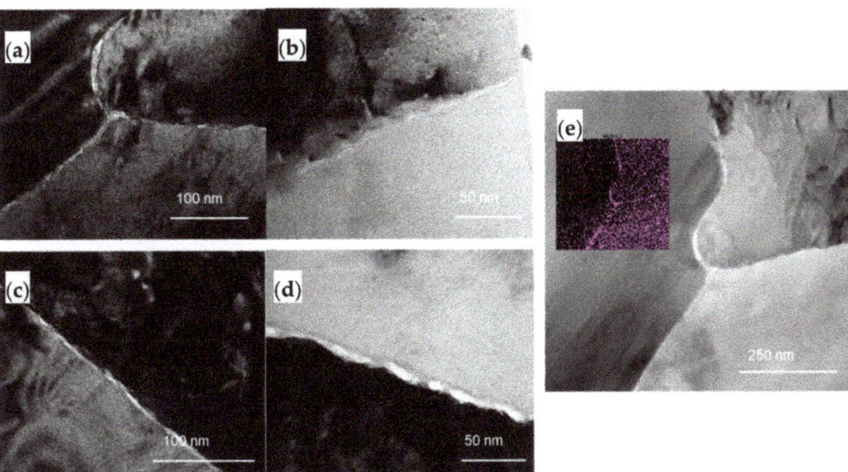

Figure 13. TEM of β precipitates in: (**a**) CR20 alloy, 48 h; (**b**) CR50 alloy, 48 h; (**c**) CR20 alloy, 207 h; (**d**) CR50 alloy, 207 h; and (**e**) TEM-EDS (Mg element) CR20 alloy, 48 h.

4. Discussion

In this research, we found that both grain size and continuity of β precipitation at grain boundaries are important factors affecting the Al-Mg IGC susceptibility.

The TEM image in Figure 13 shows that the β precipitates are much thicker in the CR50 alloy at an early period of sensitization. The β precipitates thickness, however, is almost the same at long-term sensitization. Some researchers suggested that grain boundary misorientation is a crucial factor for the growth rate and the final size of β precipitation. These factors affect the continuity of β precipitation at grain boundaries [10,12,26,29–34]. Wang et al. concluded that some grain boundaries, e.g., low-angle grain boundaries generated by plastic deformation, are not susceptible to IGC [28]. On the other hand, D'Antuono reported that although the β precipitation was preferentially formed at low-angle grain boundary, the final size of precipitation was larger at high-angle grain boundary [18]. The influence of grain boundary plane orientation was reported to affect the continuity of precipitation. It was found that grain boundary (GB) planes close to {110} direction facilitate the β precipitation while the GB plane near {100} direction may be resistant to β precipitation [31,32].

Previous studies showed that the rolled specimen had a high resistance to IGC coming from the confluence of refined grain size and the fraction of low-angle grain boundaries [17,25,26]. In this study, it was revealed that the effect of grain size on IGC needs to be considered depending on the sensitization heat treatment which affects the formation of anodic β-Mg_2Al_3 precipitation at the grain boundary. High dislocation density induced by cold rolling facilitates the precipitate growth rates. The formation of anodic β-Mg_2Al_3 is affected by temperature and the presence of prior strain [35,36]. The increased dislocation density tends to lower the nucleation temperature and reduce Mg diffusion at a lower temperature [18,19]. These factors are reflected in increasing the susceptibility of the CR50 alloy in the early period of sensitization. In this situation, the dislocation density and grain boundary type affect the IGC susceptibility more significantly compared to the grain size. On the other hand, the grain size affects the IGC susceptibility of cold-rolled Al-6Mg alloy more dramatically than the grain boundary type. It was found that the large-grained material tends to be more susceptible to IGC when the precipitation is continuously formed at the grain boundary due to sufficient sensitization time.

5. Conclusions

This study has explored the IGC behavior of a newly designed Al-6Mg alloy with two different cold rolling conditions. It was revealed that the grain size and the continuity of β precipitation play an important role in IGC. The precipitation growth rate and final size of precipitates affect how the grain boundary is covered by β precipitation. At the early period of sensitization, the precipitation growth rate is a crucial factor in IGC. The dislocation density and grain boundary orientation affect the precipitation growth rate. The CR50 alloy has a slightly higher precipitation growth compared to the CR20 alloy because of the high density of dislocations. This results in a higher maximum IGC depth. However, the grain size effect is more dominant when the sensitization time is long enough to cover the grain boundary by anodic β precipitation.

Author Contributions: Conceptualization, K.-H.C.; methodology, K.-H.C. and H.-K.L.; software, K.-H.C.; validation, D.-B.L., S.-Y.Y. and N.-S.K.; formal analysis, S.-H.H.; investigation, D.-B.L.; resources, S.-K.K.; data curation, Y.-O.Y.; writing—original draft preparation, K.-H.C.; writing—review and editing, B.-H.K. and H.-K.L.; visualization, K.-H.C.; supervision, B.-H.K. and H.-K.L.; project administration, S.-K.K.; funding acquisition, S.-K.K. All authors have read and agreed to the published version of the manuscript.

Funding: This research received no external funding.

Institutional Review Board Statement: Not applicable.

Informed Consent Statement: Not applicable.

Data Availability Statement: Data sharing not applicable. No new data were created or analyzed in this study. Data sharing is not applicable to this article.

Conflicts of Interest: The authors declare no conflict of interest.

References

1. Polmear, I.J. *Light Alloys*, 4th ed.; Butterworth-Heinemann: Oxford, UK, 2005.
2. Sanders, R.E., Jr.; Hollinshead, P.A.; Simielli, E.A. Industrial Development of Non-Heat Treatable Aluminum Alloys. In Proceedings of the 9th International Conference on Aluminium Alloys, Brisbane, Australia, 2–5 August 2004.
3. Kubota, M.; Nie, J.F.; Muddle, B.C. Characterisation of Precipitation Hardening Response and As-Quenched Microstructures in Al-Mg(-Ag) Alloys. *Mater. Trans.* **2004**, *45*, 3256–3263. [CrossRef]
4. Lee, B.-H.; Kim, S.-H.; Park, J.-H.; Kim, H.-W.; Lee, J.-C. Role of Mg in simultaneously improving the strength and ductility of Al–Mg alloys. *Mater. Sci. Eng. A* **2016**, *657*, 115–122. [CrossRef]
5. Huskins, E.L.; Cao, B.; Ramesh, K.T. Strengthening mechanisms in an Al–Mg alloy. *Mater. Sci. Eng. A* **2010**, *527*, 1292–1298. [CrossRef]
6. Czerwinski, F. The oxidation behaviour of an AZ91D magnesium alloy at high temperatures. *Acta Mater.* **2002**, *50*, 2639–2654. [CrossRef]
7. You, B.-S.; Park, W.-W.; Chung, I.-S. The effect of calcium additions on the oxidation behavior in magnesium alloys. *Scr. Mater.* **2000**, *42*, 1089–1094. [CrossRef]
8. Kim, B.-H.; Ha, S.-H.; Yoon, Y.-O.; Lim, H.-K.; Kim, S.K.; Kim, D.-H. Effect of Ca addition on selective oxidation of Al3Mg2 phase in Al-5 mass% Mg alloy. *Mater. Lett.* **2018**, *228*, 108–111. [CrossRef]
9. Ha, S.-H.; Yoon, Y.-O.; Kim, B.-H.; Lim, H.-K.; Lee, T.-W.; Lim, S.-H.; Kim, S.K. Pilling-Bedworth Ratio Approach to Surface Oxidation of Al–Mg Alloys Containing Al 2 Ca and Its Experimental Verification. *Sci. Adv. Mater.* **2018**, *10*, 697–700. [CrossRef]
10. Steiner, M.A.; Agnew, S.R. Modeling sensitization of Al–Mg alloys via β-phase precipitation kinetics. *Scr. Mater.* **2015**, *102*, 55–58. [CrossRef]
11. Tan, L.; Allen, T.R. Effect of thermomechanical treatment on the corrosion of AA5083. *Corros. Sci.* **2010**, *52*, 548–554. [CrossRef]
12. Zhang, R.; Zhang, Y.; Yan, Y.; Thomas, S.; Davies, C.H.J.; Birbilis, N. The effect of reversion heat treatment on the degree of sensitisation for aluminium alloy AA5083. *Corros. Sci.* **2017**, *126*, 324–333. [CrossRef]
13. Jain, S.; Hudson, J.L.; Scully, J.R. Effects of constituent particles and sensitization on surface spreading of intergranular corrosion on a sensitized AA5083 alloy. *Electrochim. Acta* **2013**, *108*, 253–264. [CrossRef]
14. Jain, S.; Lim, M.L.C.; Hudson, J.L.; Scully, J.R. Spreading of intergranular corrosion on the surface of sensitized Al-4.4Mg alloys: A general finding. *Corros. Sci.* **2012**, *59*, 136–147. [CrossRef]
15. Yan, J.; Hodge, A.M. Study of β precipitation and layer structure formation in Al 5083: The role of dispersoids and grain boundaries. *J. Alloy. Compd.* **2017**, *703*, 242–250. [CrossRef]

16. Lim, M.L.C.; Matthews, R.; Oja, M.; Tryon, R.; Kelly, R.G.; Scully, J.R. Model to predict intergranular corrosion propagation in three dimensions in AA5083-H131. *Mater. Des.* **2016**, *96*, 131–142. [CrossRef]

17. Lin, L.; Liu, Z.; Li, Y.; Han, X.; Chen, X. Effects of Severe Cold Rolling on Exfoliation Corrosion Behavior of Al-Zn-Mg-Cu-Cr Alloy. *J. Mater. Eng. Perform.* **2012**, *21*, 1070–1075. [CrossRef]

18. Scotto D'Antuono, D.; Gaies, J.; Golumbfskie, W.; Taheri, M.L. Direct measurement of the effect of cold rolling on β phase precipitation kinetics in 5xxx series aluminum alloys. *Acta Mater.* **2017**, *123*, 264–271. [CrossRef]

19. Scotto D'Antuono, D.; Gaies, J.; Golumbfskie, W.; Taheri, M.L. Grain boundary misorientation dependence of β phase precipitation in an Al–Mg alloy. *Scr. Mater.* **2014**, *76*, 81–84. [CrossRef]

20. ASTM G67-04. *Standard Test Method for Determining the Susceptibility to Intergranular Corrosion of 5XXX Series Aluminum Alloys by Mass Loss After Exposure to Nitric Acid (NAMLT Test)*; ASTM International: West Conshohocken, PA, USA, 2004.

21. Bachmann, F.; Hielscher, R.; Schaeben, H. Grain detection from 2d and 3d EBSD data—Specification of the MTEX algorithm. *Ultramicroscopy* **2011**, *111*, 1720–1733. [CrossRef] [PubMed]

22. Duan, X.; Jiang, H.; Mi, Z.; Cheng, L.; Wang, J. Reduce the Planar Anisotropy of AA6016 Aluminum Sheets by Texture and Microstructure Control. *Crystals* **2020**, *10*, 1027. [CrossRef]

23. Choi, S.-H.; Choi, J.-K.; Kim, H.-W.; Kang, S.-B. Effect of reduction ratio on annealing texture and r-value directionality for a cold-rolled Al–5% Mg alloy. *Mater. Sci. Eng. A* **2009**, *519*, 77–87. [CrossRef]

24. Takayama, Y.; Szpunar, J.A. Stored Energy and Taylor Factor Relation in an Al-Mg-Mn Alloy Sheet Worked by Continuous Cyclic Bending. *Mater. Trans.* **2004**, *45*, 2316–2325. [CrossRef]

25. Wang, Z.; Zhu, F.; Zheng, K.; Jia, J.; Wei, Y.; Li, H.; Huang, L.; Zheng, Z. Effect of the thickness reduction on intergranular corrosion in an under–aged Al–Mg–Si–Cu alloy during cold–rolling. *Corros. Sci.* **2018**, *142*, 201–212. [CrossRef]

26. Zhao, Y.; Polyakov, M.N.; Mecklenburg, M.; Kassner, M.E.; Hodge, A.M. The role of grain boundary plane orientation in the β phase precipitation of an Al–Mg alloy. *Scr. Mater.* **2014**, *89*, 49–52. [CrossRef]

27. Zhang, R.; Knight, S.P.; Holtz, R.L.; Goswami, R.; Davies, C.H.J.; Birbilis, N. A Survey of Sensitization in 5xxx Series Aluminum Alloys. *Corrosion* **2016**, *72*, 144–159. [CrossRef]

28. Gupta, R.K.; Zhang, R.; Davies, C.H.J.; Birbilis, N. Influence of Mg Content on the Sensitization and Corrosion of Al-xMg(-Mn) Alloys. *Corrosion* **2013**, *69*, 1081–1087. [CrossRef]

29. Zhang, R.; Qiu, Y.; Qi, Y.; Birbilis, N. A closer inspection of a grain boundary immune to intergranular corrosion in a sensitised Al-Mg alloy. *Corros. Sci.* **2018**, *133*, 1–5. [CrossRef]

30. Ding, Q.; Zhang, D.; Zuo, J.; Hou, S.; Zhuang, L.; Zhang, J. The effect of grain boundary character evolution on the intergranular corrosion behavior of advanced Al-Mg-3wt%Zn alloy with Mg variation. *Mater. Charact.* **2018**, *146*, 47–54. [CrossRef]

31. Guo, C.; Zhang, H.; Wu, Z.; Wang, D.; Li, B.; Cui, J. Effects of Ag on the age hardening response and intergranular corrosion resistance of Al-Mg alloys. *Mater. Charact.* **2019**, *147*, 84–92. [CrossRef]

32. Ramachandran, D.C.; Murugan, S.P.; Kim, Y.-M.; Kim, D.; Kim, G.-G.; Nam, D.-G.; Jeong, C.; Do Park, Y. Effect of Microstructural Constituents on Fusion Zone Corrosion Properties of GMA Welded AA 5083 with Novel Al–Mg Welding Wires of High Mg Contents. *Met. Mater. Int.* **2020**, *26*, 1341–1353. [CrossRef]

33. Yan, J.; Heckman, N.M.; Velasco, L.; Hodge, A.M. Improve sensitization and corrosion resistance of an Al-Mg alloy by optimization of grain boundaries. *Sci. Rep.* **2016**, *6*, 26870. [CrossRef]

34. Holroyd NJ, H.; Burnett, T.L.; Seifi, M.; Lewandowski, J.J. Improved understanding of environment-induced cracking (EIC) of sensitized 5XXX series aluminium alloys. *Mater. Sci. Eng. A* **2017**, *682*, 613–621. [CrossRef]

35. Nebtp, S.; Hamanai, D.; Cizeron, G. Calorimetric study of pre-precipitation and precipitation in Al-Mg alloy. *Acta Metall. Mater.* **1994**, *43*, 3583–3588. [CrossRef]

36. Starink, M.J.; Zahra, A.-M. β′ and β precipitation in an Al–Mg alloy studied by DSC and TEM. *Acta Mater.* **1998**, *46*, 3381–3397. [CrossRef]

 *materials*

Article

Effect of Natural Aging on the Stress Corrosion Cracking Behavior of A201-T7 Aluminum Alloy

Mien-Chung Chen [1], Ming-Che Wen [2], Yang-Chun Chiu [1], Tse-An Pan [1], Yu-Chih Tzeng [3] and Sheng-Long Lee [1,*]

[1] Institute of Material Science and Engineering, National Central University, Taoyuan 320, Taiwan; mianzhongchen@gmail.com (M.-C.C.); albert77918@yahoo.com.tw (Y.-C.C.); peterpan.ck@gmail.com (T.-A.P.)

[2] Department of Mechanical Engineering, National Central University, Taoyuan 320, Taiwan; j0e9rr3y0@gmail.com

[3] Department of Power Vehicle and Systems Engineering, Chung-Cheng Institute of Technology, National Defense University, Taoyuan 334, Taiwan; a0932467761@gmail.com

* Correspondence: shenglon@cc.ncu.edu.tw; Tel.: +886-3-4267-325

Received: 2 November 2020; Accepted: 6 December 2020; Published: 10 December 2020

Abstract: The effect of natural aging on the stress corrosion cracking (SCC) of A201-T7 alloy was investigated by the slow strain rate testing (SSRT), transmission electron microscopy (TEM), scanning electron microscopy (SEM), differential scanning calorimetry (DSC), conductivity, and polarization testing. The results indicated that natural aging could significantly improve the resistance of the alloys to SCC. The ductility loss rate of the unaged alloy was 28%, while the rates for the 24 h and 96 h aged alloys were both 5%. The conductivity of the as-quenched alloy was 30.54 (%IACS), and the conductivity of the 24 h and 96 h aged alloys were decreased to 28.85 and 28.65. After T7 tempering, the conductivity of the unaged, 24 h, and 96 h aged alloys were increased to 32.54 (%IACS), 32.52 and 32.45. Besides, the enthalpy change of the 24 h and 96 h aged alloys increased by 36% and 37% compared to the unaged alloy. The clustering of the solute atoms would evidently be enhanced with the increasing time of natural aging. Natural aging after quenching is essential to improve the alloy's resistance to SCC. It might be due to the prevention of the formation of the precipitation free zone (PFZ) after T7 tempering.

Keywords: Al-Cu-Mg-Ag alloy; natural aging; stress corrosion cracking; SSRT; PFZ

1. Introduction

A201 (Al-4.5Cu-0.3Mg-0.7Ag) is a heat treatable aluminum alloy, which has the highest strength among the casting aluminum alloys, so it has been used in the aerospace and military industries for many years [1] The primary strengthening phases of A201 are θ′ and Ω, both having a similar composition to that of $CuAl_2$ [2,3]. The crystal structure of θ′ is tetragonal and with a = b = 0.414 nm and c = 0.580 nm, forming large rectangular or octagonal plates parallel to the {100}$_\alpha$ plane of the matrix α phase [4]. The Ω phase has a face-centered orthorhombic structure, with a = 0.496 nm, b = 0.859 nm and c = 0.848 nm, which forms hexagonal plate-like precipitates on the {111}$_\alpha$ plane of the matrix α phase [5–8].

To enhance the mechanical properties, especially the tensile strength, a T6 temper treatment (solution heat treated then artificially aged) is usually applied in heat treatable alloys [1]. However, for high strength Al-Cu-Mg (2XX series) or Al-Zn-Mg-Cu (7XXX series) alloys, T6 temper treatment is not recommended because it will increase the susceptibility of the alloy to stress corrosion cracking (SCC) [9–14]. SSC can occur when alloys are simultaneously subjected to stress and corrosive environments. Burleigh [15] specified three SCC mechanisms for aluminum alloys, including anodic dissolution, hydrogen-induced cracking, and the brittle passive film's rupturing. Speidel [16] have indicated that anodic dissolution is the primary mechanism of SCC in Al-Cu alloys. Misra [17] showed that Al-Cu alloys formed a precipitate free zone (PFZ) along the grain boundary following artificial aging, and this zone acts as an anode relative to the base of alloy. Eventually, under a corrosive environment, the grain boundary corrodes quickly and resulting in grain boundary cracking of the alloy.

T7 tempering (solution heat treatment then overaging) is recommended to lower the susceptibility of high strength Al-Cu-Mg (2XX series) or Al-Zn-Mg-Cu (7XXX series) alloys to SCC [1]. In the AA7075 (Al-Zn-Mg-Cu) alloy, for example, the primary strengthening phase is η ($MgZn_2$), which has the lowest potential compared to the α matrix and PFZ [18]. The T7 temper coarsens the precipitation, resulting in a discontinuous structure along the grain boundary, thereby decreasing the alloy's susceptibility to SCC [19]. However, for a high strength Al-Cu-Mg alloy, PFZ has the lowest potential than the $CuAl_2$ and α matrix. The inhibition of the formation of PFZ is the primary way to improve the resistance of the Al-Cu-Mg alloy to SCC [20].

Although the effect of aging on the SCC behavior of high strength aluminum alloys had been investigated for decades [13–21], and these works mainly focused on the effect of single artificial aging on the SCC, such as T6 (peak aging) or T7 (over aging) treatment. However, the lack of research on multiple heat treatments (combined natural aging with artificial aging) is the primary purpose of this work. Hence, the influence of natural aging on the SCC behavior, microstructure, and mechanical properties of A201-T7 alloy were investigated in this work to find the feasibility of multiple heat treatments. The results can provide a reference for the development of high-performance alloys with lower SCC suspicious while the mechanical properties could be maintained.

2. Materials and Methods

2.1. Melting and Casting

The specimens for testing were prepared as follows. High purity aluminum ingots (99.9 wt.%) were first melted in an electric resistance furnace using a graphite crucible. Suitable amounts of pure Cu, Mg, Ag, Al-75Mn, and Al-60Ti master alloys were then added. Pure argon was used for degassing the melt for 40 min. After being held for 10 min at 700 °C, it was poured into a 125 mm × 100 mm × 25 mm steel mold preheated to 300 °C. Table 1 shows the experimental alloy's chemical composition as determined by inductively coupled plasma optical emission spectrometry.

Table 1. Chemical composition of the experimental alloy (wt.%).

Alloy	Cu	Mg	Ag	Ti	Fe	Si	Al
A201	4.5 (0.1) *	0.3 (0.05)	0.7 (0.05)	0.3 (0.05)	0.05 (0.01)	0.03 (0.01)	Balance

* Standard deviations are listed in parentheses.

2.2. Heat Treatment of As-Cast Alloys

Table 2 shows the heat treatment cycles used to produce the as-cast alloy. First, they were solution treated at 510 °C for 2 h and then 530 °C for 20 h. Subsequently, water quenched (WQ) and immediately subjected to natural aging for different lengths of time (unaged, 24 h, 96 h). Finally, artificial aging was performed with T7 tempering at 190 °C for 5 h.

Table 2. Heat treatment cycles for preparation of as-cast alloys.

Alloy Notation	Solution Treatment	Natural Aging	Artificial Aging
NA0d	510 °C/2 h + 530 °C/20 h + WQ	none	190 °C/5 h
NA1d	510 °C/2 h + 530 °C/20 h + WQ	24 h	190 °C/5 h
NA4d	510 °C/2 h + 530 °C/20 h + WQ	96 h	190 °C/5 h

2.3. Slow Strain Rate Test

Slow strain rate testing (SSRT) was carried out in a 3.5% NaCl solution at pH = 7 using a tensile testing rate of 1.25×10^{-7} s^{-1}. The test specimens prepared according to ASTM B557M [22] had dimensions of 4mm in diameter and 20 mm gauge length. The occurrence of suspected stress corrosion cracking was determined from the ductility loss rate $[(E_{scc} - E_{air})/E_{air})]$ and ultimate tensile strength loss rate $[(UTS_{scc} - UTS_{air})/UTS_{air})]$. This is the ratio of elongation and tensile strength in 3.5% NaCl compared to that in air.

2.4. Materials Characterization

A Sigmascope SMP10 electrical conductivity meter was used to measure the alloys' electrical conductivity under different heat treatment conditions. The unit of electrical conductivity were percentages of the international annealed copper standard (%IACS). Sample disks 3 mm in diameter and 1mm thick were prepared for the differential scanning calorimetry (DSC) experiments. A SEIKO-DSC6200 differential scanning calorimeter (Chiba, Japan) at a heating rate of 10 °C/min was used to capture the DSC traces. The T7 state specimens were prepared for transmission electron microscopy (FEI-TEM, Tecnai, G2-F20, Hillsboro, OR, USA) by twin-jet electro-polishing in a solution of 30% HNO$_3$ methanol at −30 °C and 12 V. SSRT fractography was observed by scanning electron microscopy (FEI-SEM, JEOL, JSM-7800F Prime, Akishima, Japan). The Rockwell hardness B scale was used to measure the hardness of the experimental alloys.

2.5. Polarization Testing

Anodic potentiodynamic polarization experiments were performed in a 3.5% NaCl solution to determine the breakdown potentials of A201-T7 alloys prepared with various natural aging times. Silver-Silver chloride was used as the reference electrode and platinum as the auxiliary electrode. Each 15 mm diameter sample (working electrode) was exposed to the solution for 48 h before the start of the measurement and then potentiodynamically polarized from −630 mV below the open circuit potential (OCP) to a potential above the breakdown potential at 0.5 mV·s^{-1}. The corrosion rates and polarization resistance were determined by the polarization resistance method and Stern-Geary equation. Besides, the polarization test was repeated three times in order to check the reproducibility.

3. Results and Discussion

3.1. Electrical Conductivity

As shown in Table 3, the alloys' electrical conductivity decreased with an increase in the natural aging time. It was because the clustering of solute atoms led to a lattice distortion of the α matrix [23]. The longer the natural aging time was, the worse the lattice distortion became. As a result, the electrical conductivity of the alloys became lower. The changes in conductivity $[(C_{AN}-C_0)/C_0 \times 100\%]$ of the NA1d and NA4d alloys of 5.5% and 6.2%, indicated the flattening of clustering with an extension of the aging time from 24 h to 96 h.

Table 3. Electrical conductivity of alloys under different heat treatment conditions.

Alloy Notation	IACS (%)			Percentage Change (%)	
	As-Quenched (C_0)	Natural Aging (C_{NA})	T7 (C_{T7})	$\frac{C_{NA}-C_0}{C_0} \times 100\%$	$\frac{C_{T7}-C_{NA}}{C_{NA}} \times 100\%$
NA0d	30.54 (0.12) *	30.54 (0.12)	32.54 (0.24)	-	6.5
NA1d	30.54 (0.12)	28.85 (0.11)	32.52 (0.11)	−5.5	12.7
NA4d	30.54 (0.12)	28.65 (0.09)	32.45 (0.09)	−6.2	13.4

* Standard deviations are listed in parentheses.

3.2. DSC Analysis

An enlarged view of the DSC traces for the experimental alloys in the range of 150 °C to 350 °C is presented in Figure 1. The exothermic peak at 250 °C represents the precipitation of θ′ and Ω strengthening phase [24]. The enthalpy changes of the NA0d, NA1d, and NA4d alloys were 1.35, 1.83, and 1.85 W/g, respectively, showing that natural aging could significantly increase the exothermic heating. However, there was no obvious increase in exothermic heat while the increased period of natural aging time from 24 h to 96 h.

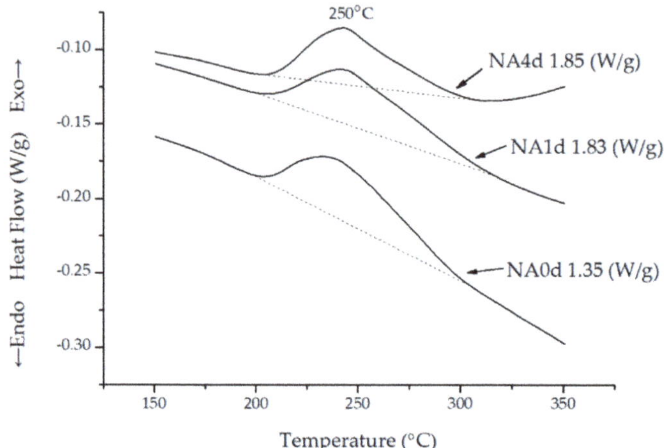

Figure 1. Differential scanning calorimetry (DSC) profile of different naturally aged A201 alloys.

3.3. Mechanical Properties

The mechanical properties of the alloys produced under different heat treatment conditions are shown in Table 4. The hardness of the alloy after solution treatment then water quenching was 33.5HRB, which increased gradually following natural aging. The percentage changes [($H_{NA}-H_0$)/H_0 × 100%] of the alloys, NA1d and NA4d, were 40% and 54%, indicating that there was no obvious promotion of the clustering of solute atoms with an increase in the natural aging time from 24 h to 96 h, which was consistent with the electrical conductivity and DSC analysis results.

Table 4. Mechanical properties of alloys under different heat treatment conditions.

Alloy Notation	Hardness		Tensile Test			Percentage Change (%)	
	As-Quenched	Natural Aging	T7			$\frac{H_{NA}-H_0}{H_0}$ × 100%	$\frac{H_{T7}-H_{NA}}{H_{NA}}$ × 100%
	H_0 (HRB)	H_{NA} (HRB)	H_{T7} (HRB)	YS (MPa)	UTS (MPa)	EL (%)	
NA0d	33.5 (2.2) *	33.5 (2.2)	71.1 (2.1)	320 (2.6)	398 (2.7)	3.5 (0.2)	-	112
NA1d	33.5 (2.2)	46.8 (1.5)	71.6 (1.6)	325 (2.0)	396 (2.8)	3.6 (0.1)	40	53
NA4d	33.5 (2.2)	54.5 (1.7)	70.6 (2.5)	318 (3.2)	397 (3.0)	3.5 (0.2)	54	30

* Standard deviations are listed in parentheses.

After T7 tempering, with the precipitation of the strengthening phases θ′ and Ω, there was an increase in the hardness of all the alloys, NA0d, NA1d, and NA4d, to 71HRB. Evidenced that the clustering of the solute atoms remained at the same level regardless of whether natural aging was adopted or not, and that lattice distortion would be eliminated after T7 tempering. The percentage changes [($H_{T7}-H_{NA}$)/H_{NA} × 100%] of the alloys, NA0d, NA1d, and NA4d, were 112%, 53%, and 30%. Obviously, the hardness did not increase as much with the extension of the natural aging time from 24 h to 96 h. In addition, there was no difference in the yielding stress (YS), ultimate tensile stress (UTS), or elongation (EL) whether natural aging was adopted or not. The YS, UTS, and EL of the three alloys were approximately 320 MPa, 397 MPa, and 3.5%, respectively, after T7 tempering.

3.4. Polarization Test

The polarization curves of A201-T7 alloys after different natural aging (unaged, 24 h, 96 h) are shown in Figure 2. No passivation areas could be found after the samples were immersed in a 3.5% NaCl solution for 48 h before the test. The electrochemical analysis results were shown in Table 5. The corrosion potential, rates, and polarization resistance were represented as E_{corr} (V), I_{corr} (A/cm^2), and R_p (Ω/cm^2), separately. The unaged alloy exhibited a lower corrosion potential and lower corrosion rates than the aged alloys. The corrosion potential of the NA0d alloy was −0.71 V, lower than for NA1d (−0.60 V), and NA4d (−0.59 V). The corrosion rates of the alloys, NA0d, NA1d, and NA4d, were 3.91×10^{-5}, 5.94×10^{-5}, and 6.77×10^{-5} (A/cm^2), respectively, as determined by the polarization resistance method. Correspondingly, the polarization resistance of the naturally aged alloys was lower than that of the unaged alloy due to the clearly seen of active region in Figure 2. It is worth noting that the SCC resistance may be decreased with the long-term experiment. The polarization resistance of the unaged allot was 828 (Ω/cm^2), while the 24 h and 96 h aged alloy were 633 and 604 (Ω/cm^2). However, there was no obvious decrease in the polarization resistance of the A201-T7 alloys with an increase in the natural aging time from 24 h to 96 h.

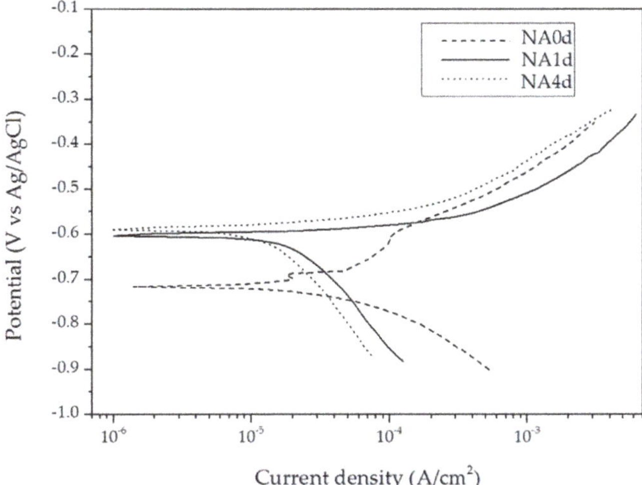

Figure 2. Polarization curves of different naturally aged A201-T7 alloys in a 3.5% NaCl solution.

Table 5. Parameters of polarization test of different naturally aged A201-T7 alloys in a 3.5% NaCl solution.

Alloy Notation	E_{corr} (V)	I_{corr} (A/cm^2)	R_p (Ω/cm^2)
NA0d	−0.71 (0.05) *	3.91×10^{-5} (1.2×10^{-5})	828 (61)
NA1d	−0.60 (0.04)	5.94×10^{-5} (3.3×10^{-6})	633 (35)
NA4d	−0.59 (0.05)	6.77×10^{-5} (2.8×10^{-6})	604 (28)

* Standard deviations are listed in parentheses.

3.5. Slow Strain Rate Testing

The results of slow strain rate testing of the A201-T7 alloys are presented in Table 6. Natural aging did not affect their elongation in air, approximately 3.5% for all three alloys. However, when it came to the saltwater environment, the elongation of the NA0d alloy decreased to 2.6%. In comparison, that of the NA1d and NA4d alloys remained about 3.5%, indicating that natural aging could significantly improve the alloy's resistance to SCC. The losses of ductility of NA1d and NA4d were 5.4% and 5.7%, respectively, while the loss in the unaged alloy could be as much as to 27.8%. Moreover, the loss of strength also showed the same tendency. The loss of strength of the unaged alloy was 15%, while the 24 h and 96 h aged alloys showed losses of 2.5% and 3.0%, respectively.

Table 6. Slow strain rate testing results of A201-T7 alloys.

Alloy Notation	EL in Air E_{air} (%)	EL in Salt Water E_{scc} (%)	UTS in Air UTS_{air} (MPa)	UTS in Salt Water UTS_{scc} (MPa)	$\frac{E_{scc}-E_{air}}{E_{air}} \times 100\%$	$\frac{UTS_{scc}-UTS_{air}}{UTS_{air}} \times 100\%$
NA0d	3.6 (0.1) *	2.6 (0.3)	399 (2.9)	339 (2.1)	−27.8	−15.0
NA1d	3.7 (0.2)	3.5 (0.2)	395 (2.6)	385 (2.8)	−5.4	−2.5
NA4d	3.5 (0.1)	3.3 (0.1)	397 (2.8)	385 (2.3)	−5.7	−3.0

* Standard deviations are listed in parentheses.

The SSRT results indicated that natural aging before T7 tempering was essential, for it had the great benefit of increased resistance of the A201-T7 alloy to SCC. It also showed that aging for 24 h was sufficient. Extending the aging time further had no additional benefit. We would also like to remind the reader that the slow strain rate testing (SSRT) might not be suitable to determine the SCC behavior of alloys in the latest research due to the sub-critical cracking and to invalidate the SSRT results [25,26].

An examination of Figure 3 shows the fracture surface of the A201-T7 alloys after slow strain rate testing. During SSRT in air, the fracture surfaces of the alloys, NA0d, NA1d, and NA4d, were similar with many dimples of different shapes and sizes observed, implying that the fracture mechanism was ductile fracturing. As a result, natural aging did not affect the elongation in the air. However, the fracture surfaces of the alloys were quite different when SSRT was conducted in the 3.5% NaCl solution. For the unaged alloy, nearly no dimples were observed, and the fracture mechanism was brittle fracturing. For the 24 h and 96 h aged alloys, cleavages and dimples were observed in the sample, and the fracture mechanism was a combination of ductile and brittle fracturing. The results were consistent with the mechanical properties presented in Table 4.

3.6. Transmission Electron Microscopy (TEM) Characterization

The Schematic diagram of the calculated diffraction pattern of Ω and θ' strengthening phases along $[011]_\alpha$ zone axis were shown in Figure 4 [27]. Strengthening phases Ω and θ' have a similar composition to $CuAl_2$, but with a different crystal structure. As shown in Figure 5d, the selected area electron diffraction patterns (SAED) along $[011]_\alpha$ zone axis evidenced Ω and θ' precipitation strengthening phases exist in A201 alloys after T7 tempered. The bright streaks and $(10\bar{1})$ diffraction spot of Ω phase and (100) diffraction spot of θ' phase were found at $(200)_\alpha$ position. The TEM bright field images of the grain boundaries of A201-T7 alloys shown in Figure 5 reveal the presence of precipitation at the grain boundaries. The needle-like strengthening phases Ω and θ' inside the grains can also be seen. The primary precipitation phase of A201-T7 alloys was Ω phases, and minor θ' phases with different orientation can also be observed. It seems that natural aging did not affect the size of precipitation phases, Ω and θ', after T7 heat treatment, just as shown in Figure 5d–f. The maximum length of Ω and θ' phase of three different natural aged alloys (unaged, 24 h, 96 h) were around 100 nm and 70 nm, separately. The NA0d alloy had significant prefer-precipitation during T7 tempering [28]. The oversaturation of solute atoms near the grain boundaries would diffuse to the grain boundaries, forming coarse and discontinuous precipitation phases along with a 200 nm wide precipitation free zone (PFZ) as shown in Figure 5a. However, for NA1d and NA4d, no obvious PFZ could be found at the grain boundaries, as shown in Figure 5b,c. This might be due to the clustering of solute atoms (Cu, Mg), which occurred during natural aging, which would directly transfer to the precipitation phases (Ω and θ') during T7 tempering, thus inhibiting the formation of PFZ.

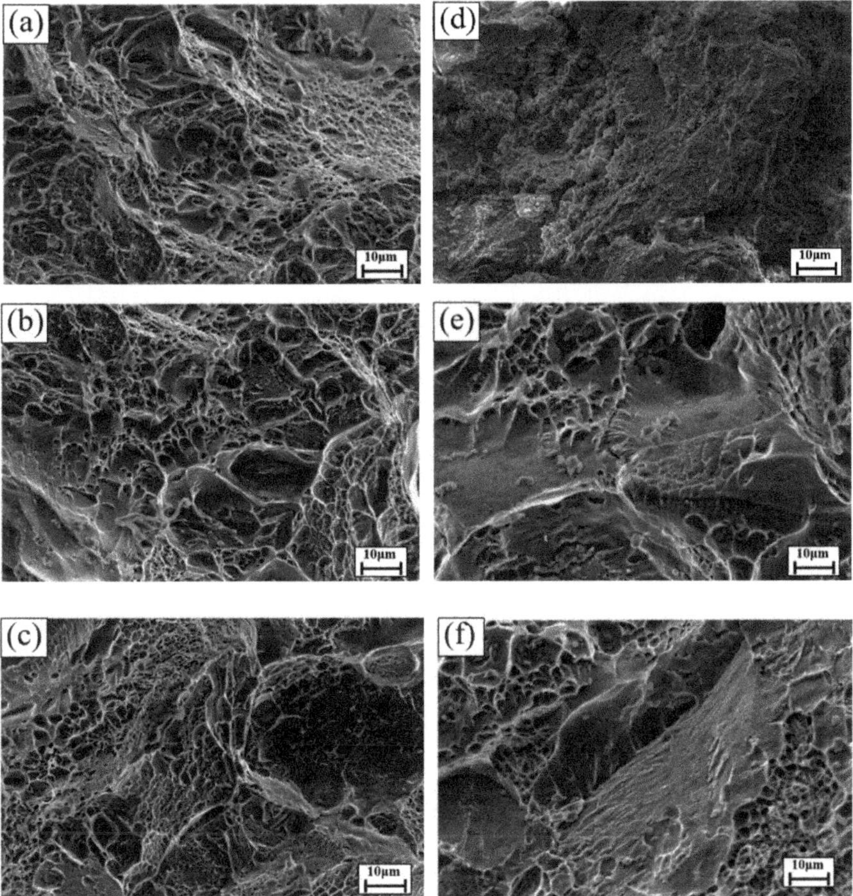

Figure 3. Fracture surface of A201-T7 alloys after slow strain rate testing: (**a**) NA0d sample in air; (**b**) NA1d sample in air; (**c**) NA4d sample in air; (**d**) NA0d sample in salt water; (**e**) NA1d sample in salt water; (**f**) NA4d sample in salt water.

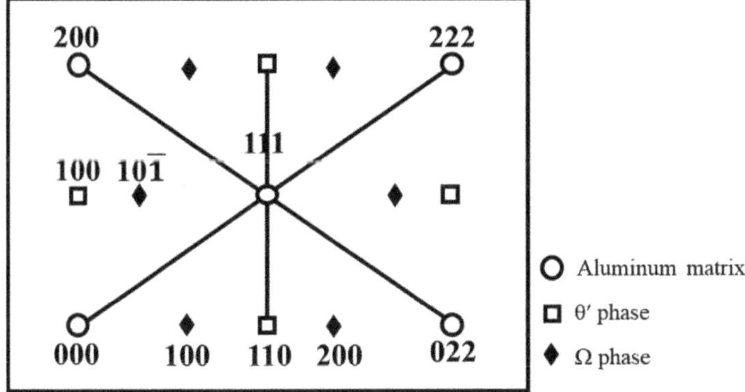

Figure 4. Schematic diagram of the diffraction pattern of θ′ and Ω strengthening phases along $[011]_\alpha$ zone in A201-T7 alloy.

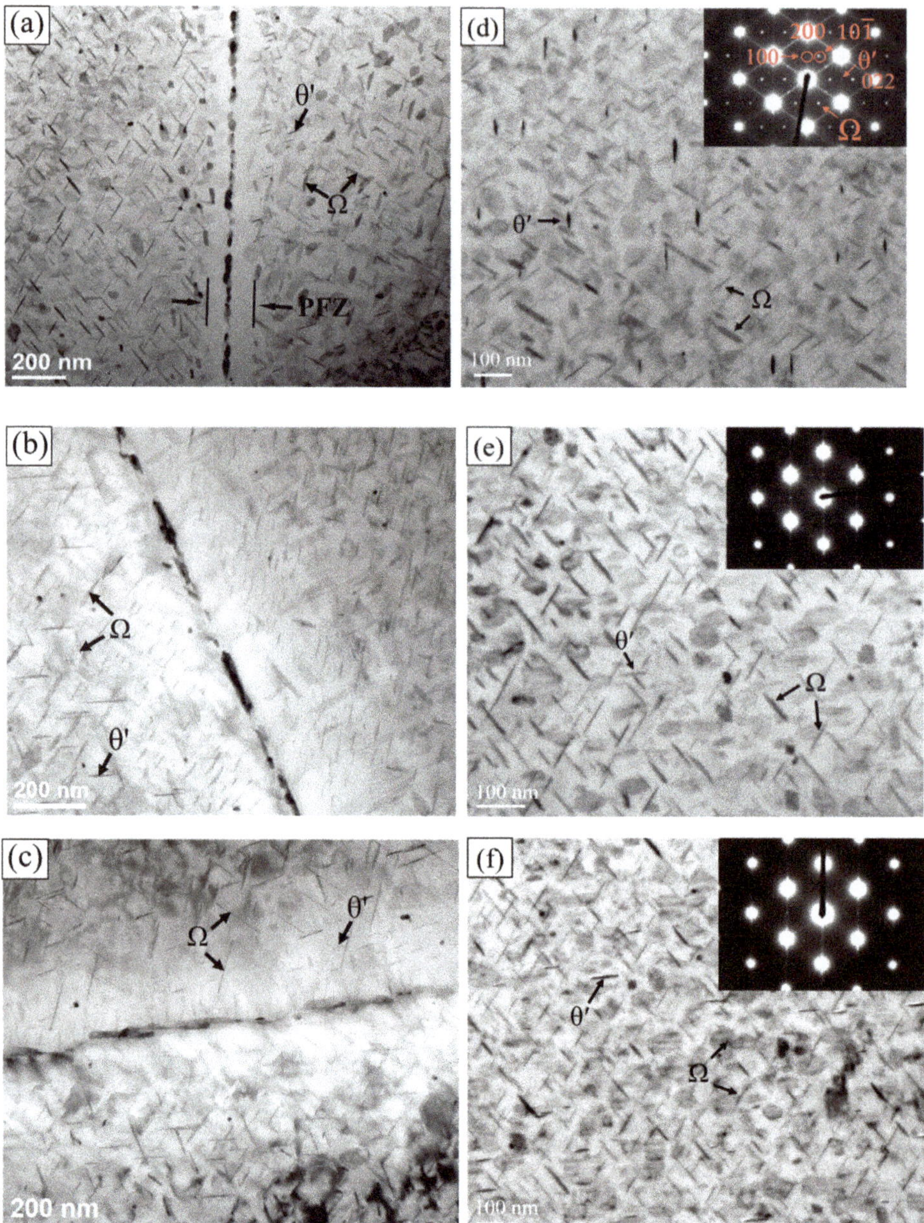

Figure 5. TEM bright field images and the selected area electron diffraction pattern (SAED) of: (**a,d**) Alloy NA0d; (**b,e**) Alloy NA1d; (**c,f**) Alloy NA4d after T7 tempering.

It is worth noting that the PFZ has the lowest potential compared to CuAl$_2$ phases (θ' and Ω) and α matrix in the 2XX series and 2XXX series Al-Cu-Mg alloys [29]. To lower the galvanic corrosion effect, the inhibition of the formation of PFZ can help prevent SCC [30]. The discussion above is supported by the polarization test and slow strain rate test results, indicating that naturally aged alloys have better SCC resistance than unaged alloys.

4. Conclusions

The effects of natural aging on stress corrosion cracking in A201-T7 alloy were investigated in this study and the following conclusions can be drawn:

(1)   For the as-quenched alloy, the conductivity decrease and the hardness increase during natural aging. However, the conductivity and mechanical properties (hardness, strength, and elongation) were unaffected by natural aging after T7 tempering.

(2)   Natural aging improved the resistance of A201-T7 alloys to SCC. 24 h aging was sufficient. Extending the aging time provided no additional benefit.

(3)   In the unaged alloys, PFZ existed and brittle fractures could be found on the SCC fracture surface; for the aged alloys, no PFZ existed, but a combination of fracture types with cleavages and dimples could be observed on the fracture surface.

Author Contributions: Data curation, Writing—original draft preparation, M.-C.C.; Conceptualization; Methodology, M.-C.W.; Performed experiment, Formal analysis, T.-A.P.; Investigation; Validation, Y.-C.C.; Project administration, Resources, Y.-C.T.; Supervision, Funding acquisition, Review and editing the draft, S.-L.L. All authors have read and agreed to the published version of the manuscript.

Funding: This research was funded by the National Chung-Shan Institute of Science & Technology (Contract: XV07116P418PE-CS).

Acknowledgments: The author would like to thank Tsai-Fu Chung from National Taiwan University for the TEM observation.

Conflicts of Interest: The authors declare no conflict of interest.

References

1.   Davis, J.R. *ASM Specialty Handbook: Aluminum and Aluminum Alloys*; ASM International: Materials Park, OH, USA, 1994.

2.   Chester, R.J.; Polmear, I.J. TEM Investigation of Precipitates in Al-Cu-Mg-Ag and Al-Cu-Mg Alloys. *Micron* **1980**, *11*, 311–312. [CrossRef]

3.   Kim, K.D.; Zhou, B.C.; Wolverton, C. Interfacial Stability of θ'/Al in Al-Cu Alloys. *Scr. Mater.* **2019**, *159*, 99–103. [CrossRef]

4.   Phillips, V.A. High Resolution Electron Microscope Observations on Precipitations in Al-3.0% Cu Alloy. *Acta Mater.* **1975**, *23*, 751–767. [CrossRef]

5.   Purnendu, K.M. Influence of Micro-alloying with Silver on Microstructure and Mechanical Properties of Al-Cu Alloy. *Mater. Sci. Eng. A* **2018**, *722*, 99–111.

6.   Knowles, K.M.; Stobbs, W.M. The Structure of {111} Age-Hardening Precipitates in Al-Cu-Mg-Ag Alloys. *Acta Cryst.* **1988**, *44*, 207–227. [CrossRef]

7.   Ivan, Z.; Rustam, K. Aging Behavior of an Al-Cu-Mg Alloy. *J. Alloys Compd.* **2018**, *759*, 108–119.

8.   Saeed, K.M.; Mohsen, K.; Roland, L. Mechanical Behavior and Texture Development of Over-aged and Solution Treated Al-Cu-Mg Alloy during Multi-directional Forging. *Mater. Charact.* **2018**, *135*, 221–227.

9.   Muddle, B.C.; Polmear, I.J. The Precipctate Ω Phase in Al-Cu-Mg-Ag Alloys. *Acta Metal.* **1989**, *37*, 777–789. [CrossRef]

10.   Garg, A.; Chang, Y.C.; Howe, J.M. Precipitation of the Ω Phase in an Al-4.0Cu-0.5Mg Alloy. *Scr. Mater.* **1990**, *24*, 677–680. [CrossRef]

11.   Hu, Y.C.; Liu, Z.Y.; Zhao, Q.; Bai, S.; Liu, F. P-Texture Effect on the Fatigue Crack Propagation Resistance in an Al-Cu-Mg Alloy Bearing a Small Amount of Silver. *Materials* **2018**, *11*, 2481. [CrossRef]

12.   ASTM B917. *Standard Practice for Heat Treatment of Aluminum-Alloy Castings form All Processes*; ASTM International: West Conshohocken, PA, USA, 2020.

13.   Rajan, K.; Wallace, W.; Beddoes, J.C. Microstructure Study of a High Strength Stress-Corrosion Resistant 7075 Aluminum Alloy. *J. Mater. Sci.* **1982**, *17*, 2817–2848. [CrossRef]

14.   Islam, M.U.; Wallace, W. Retrogression and Reaging Response of 7475 Aluminum Alloy. *Met. Technol.* **1983**, *10*, 386–392. [CrossRef]

15. Burleigh, T.D. The Postulated Mechanisms for Stress Corrosion Cracking of Aluminum Alloys. *Corrosion* **1991**, *47*, 89–98. [CrossRef]

16. Speidel, M.O.; Hyatt, M.V. *Advances in Corrosion Science and Technology*; Springer: Boston, MA, USA, 1972.

17. Misra, M.S.; Oswalt, K.J. Corrosion Behavior of Al-Cu-Mg-Ag (201) Alloy. *Met. Eng. Q.* **1976**, *16*, 39–44.

18. Alexander, I.I.; Zhang, B.; Wang, J.Q.; Han, E.H.; Ke, W.; Peter, C.O. SVET and SIET Study of Galvanic Corrosion of Al/MgZn$_2$ in Aqueous Solutions at Different pH. *J. Electrochem. Soc.* **2018**, *165*, 180–194.

19. Shi, Y.J.; Pan, Q.L. Influence of Alloyed Sc and Zr, and Heat Treatment on Microstructures and Stress Corrosion Cracking of Al-Zn-Mg-Cu Alloys. *Mater. Sci. Eng. A* **2015**, *621*, 173–181. [CrossRef]

20. Lee, H.J.; Kim, Y.J.; Jeong, Y.; Kim, S.S. Effects of Testing Variables on Stress Corrosion Cracking Susceptibility of Al 2024-T351. *Corros. Sci.* **2012**, *55*, 10–19. [CrossRef]

21. Cabrini, M.; Bocchi, S.; D'Urso, G.; Giardini, C.; Lorenzi, S.; Testa, C.; Pastore, T. Stress Corrosion Cracking of Friction Stir-Welded AA-2024 T3 Alloy. *Materials* **2020**, *13*, 2610. [CrossRef]

22. ASTM B557. *Standard Test Methods for Tension Testing Wrought and Cast Aluminum- and Magnesium-Alloy Products*; ASTM International: West Conshohocken, PA, USA, 2020.

23. Ivanov, R.; Deschamps, A.; Geuser, F.D. Clustering Kinetics during Natural Ageing of Al-Cu Based Alloys with (Mg, Li) Additions. *Acta Mater.* **2018**, *157*, 186–195. [CrossRef]

24. Chang, C.H.; Lee, S.L.; Lin, J.C.; Yeh, M.S.; Jeng, R.R. Effect of Ag Content and Heat Treatment on The Stress Corrosion Cracking of Al-4.6Cu-0.3Mg alloy. *Mater. Chem. Phys.* **2005**, *91*, 454–462. [CrossRef]

25. Martínez-Pañeda, E.; Harris, Z.D.; Fuentes-Alonso, S.; Scully, J.R.; Burns, J.T. On the suitability of slow strain rate tensile testing for assessing hydrogen embrittlement susceptibility. *Corros. Sci.* **2019**, *163*, 108291. [CrossRef]

26. Nikolaos, D.A.; Christina, C.; Panagiotis, S.; Stavros, K.K. Synergy of Corrosion-induced Micro-cracking and Hydrogen Embrittlement on The Structural Integrity of Aluminum Alloy (Al-Cu-Mg) 2024. *Corros. Sci.* **2017**, *121*, 32–42.

27. Beffort, O.; Solenthaler, C.; Uggowitzer, P.J.; Speidel, M.O. High toughness and high strength spray-deposited AlCuMgAg-base alloys for use at moderately elevated temperatures. *Mater. Sci. Eng. A* **1995**, *191*, 121–134. [CrossRef]

28. Wang, H.S.; Jiang, B.; Zhang, J.Y.; Wang, N.H.; Yi, D.Q.; Wang, B.; Liu, H.Q. The Precipitation Behavior and Mechanical Properties of Cast Al-4.5Cu-3.5Zn-0.5Mg Alloy. *J. Alloys Compd.* **2018**, *768*, 707–713. [CrossRef]

29. Qi, H.; Liu, X.Y.; Liang, S.X.; Zhang, X.L.; Cui, H.X.; Zheng, L.Y.; Gao, F.; Chen, Q.H. Mechanical Properties and Corrosion Resistance of Al-Cu-Mg-Ag Heat-resistant Alloy Modified by Interrupted Aging. *J. Alloys Compd.* **2016**, *657*, 318–324. [CrossRef]

30. Liu, X.Y.; Li, M.J.; Gao, F.; Liang, S.H.; Zhang, X.L.; Cui, H.X. Effects of Aging Treatment on The Intergranular Corrosion Behavior of Al-Cu-Mg-Ag alloy. *J. Alloys Compd.* **2015**, *639*, 263–267. [CrossRef]

Publisher's Note: MDPI stays neutral with regard to jurisdictional claims in published maps and institutional affiliations.

© 2020 by the authors. Licensee MDPI, Basel, Switzerland. This article is an open access article distributed under the terms and conditions of the Creative Commons Attribution (CC BY) license (http://creativecommons.org/licenses/by/4.0/).

 *materials*

Article

Corrosion Behavior of Cold-Formed AA5754 Alloy Sheets

Anna Dobkowska *, Agata Sotniczuk, Piotr Bazarnik, Jarosław Mizera and Halina Garbacz

Faculty of Materials Science and Engineering, Warsaw University of Technology, 02-507 Warsaw, Poland; agata.sotniczuk.dokt@pw.edu.pl (A.S.); piotr.bazarnik@pw.edu.pl (P.B.); jaroslaw.mizera@pw.edu.pl (J.M.); halina.garbacz@pw.edu.pl (H.G.)

* Correspondence: anna.dobkowska@pw.edu.pl; Tel.: +48-22-234-8399

Abstract: In this work, the influence of bending an AA5457 alloy sheet and the resulting microstructural changes on its corrosion behavior was investigated. Scanning electron microscopy (SEM) and transmission electron microscopy (TEM) were used to perform detailed microstructural analyses of the alloy in its original form and after bending. After immersion in naturally-aged NaCl under open-circuit conditions (0.5 M, adjusted to 3 by HCl), post-corrosion observations were made, and electrochemical polarization measurements were performed to investigate the corrosion mechanisms occurring on both surfaces. The results showed that the corrosion of AA5457 is a complex process that mainly involves trenching around coarse Si-rich particles, crystallographically-grown large pits, and the formation of multiple tiny pits around Si-rich nanoparticles. The experimental data showed that bending AA5457 changed the shape and distribution of Si-rich coarse particles, cumulated a higher dislocation density in the material, especially around Si-rich nanoparticles, and all of these factors caused that corrosion behavior of the AA5754 in the bending area was lowered.

Keywords: AA5754; corrosion; microstructure

1. Introduction

Due to their high mechanical and anti-corrosion properties, low density, and good formability, the 5xxx series of Al alloys is widely used in many branches of industry, particularly in the automotive industry [1–6]. Their chemical composition mainly includes Al and Mg (up to 7 wt.% Mg), along with secondary additions of Mn and trace amounts of Cr [7]. Their corrosion resistance arises from a complex combination of factors, but it strongly depends on the microstructural features of the alloy [8–10]. Several studies have reported the role of intermetallic particles in the corrosion resistance of Al alloys, and these particles can be organized by their size and electrochemical behavior with respect to the Al matrix [11–16]. If the particles are relatively large (a few µm), they are classified as coarse particles, but nanoprecipitates also often form in Al-Mg alloys. Due to their various electrochemical properties (anodic/cathodic), they can either suppress matrix dissolution or have only a small influence on the corrosion of Al alloys [17,18]. Particles rich in Zn or Si (such as $MgZn_2$ or Mg_2Si) are prone to dissolution [19]. Generally, Fe-rich particles, i.e., Al_n (Fe, Mn), are nobler with respect to the Al matrix and decrease the localized corrosion resistance of Al alloys depending on the content of other impurities [20–22]. The local corrosion attack around intermetallic particles has been explained in terms of various particle behaviors in corrosive media: particle fall-out, selective particle dissolution or particle dealloying [19,23]. Recent works on the microstructure of alloys have revealed different types of coarse particles in Al-Mg alloys, depending on the additional alloying elements.

The second type of local attack also occurs at small precipitates formed in 5xxx series Al alloys. When the Mg content exceeds 3.0 wt.% and alloys are exposed to temperatures of 50–225 °C, Mg atoms preferentially diffuse from the supersaturated solid solution (α) to grain boundaries (GBs). Then, the variety of Mg-rich phases changes from metastable

Citation: Dobkowska, A.; Sotniczuk, A.; Bazarnik, P.; Mizera, J.; Garbacz, H. Corrosion Behavior of Cold-Formed AA5754 Alloy Sheets. *Materials* **2021**, *14*, 394. https://doi.org/10.3390/ma14020394

Received: 29 December 2020
Accepted: 11 January 2021
Published: 14 January 2021

Publisher's Note: MDPI stays neutral with regard to jurisdictional claims in published maps and institutional affiliations.

Copyright: © 2021 by the authors. Licensee MDPI, Basel, Switzerland. This article is an open access article distributed under the terms and conditions of the Creative Commons Attribution (CC BY) license (https://creativecommons.org/licenses/by/4.0/).

(β′ and β″) to equilibrium β (Al$_3$Mg$_2$) precipitates at GBs (sensitization) [24–26]. Al$_3$Mg$_2$, which behaves as an anode with respect to the Al matrix, plays a predominant role in the corrosion resistance of Al-Mg alloys [19,27–29]; however, due to its small size (~100 nm or smaller), it is difficult to characterize this phase in situ [30,31]. The quantitative detection of the dissolution of β precipitates regarding the corrosion behavior of 5xxx alloys has been thoroughly studied by Guan et al. [25]. Vignesh et al. [32] analyzed the susceptibility to intergranular corrosion (IGC) of AA5083 and showed that the alloy's susceptibility to IGC may be reduced by grain structure refinement, dispersion, and the partial dissolution of secondary Mg$_2$Al$_3$ in the matrix. This has been confirmed in another study [33], which showed that the pitting, IGC and exfoliation all depended on the precipitate distribution. The study suggested that grain refinement can effectively enhance the IGC resistance of the commercial AA5083-H111. Moreover, the selective dissolution of Al$_3$Mg$_2$ phase in AA5083 also determines its resistance to stress corrosion cracking [34]. In 5xxx Al alloys, pitting is explained in terms of the formation of a defective oxide film [35] and is strongly dependent on the NaCl concentration and pH of the solution [36–38].

Since the 5xxx alloys do not respond to age hardening, previous reports regarding the electrochemical behavior of 5xxx alloys have also focused on post-processing heat-treatment and/or plastic deformation [39–41]. As stated by Neetu et al. [42], the subsequent annealing of the AA5080 produced by multi-axial forging at room temperature drastically improved its corrosion resistance. The homogenization heat treatment of AA5083-O plates after casting had the optimal effect on the overall corrosion resistance of the material [40]. Various plastically-deformed (i.e., friction stir processing, accumulative roll-bonding) Al-Mg alloys have shown better corrosion resistance compared with their parent material [43,44]. In contrast, extruded AA5083 showed similar overall corrosion resistance compared with conventional Al-Mg alloys [45].

Clearly, the corrosion behavior of Al-Mg alloys is complex and requires more studies to distinguish the dependence between the formation of various kinds of localized corrosion and their mutual interactions. Although Al-Mg alloys display fairly good corrosion resistance [46,47], subjecting them to additional cold-forming to produce semi-products or final elements may change the material's properties and compromise their electrochemical behavior. Therefore, the optimization of the corrosion performance of commercially applicable Al-Mg alloys is of critical importance in terms of using these alloys for structurally lightweight devices and when plastic deformation is necessary to obtain the desired shape of a design element (i.e., bending). For the needs of this project, we produced part of a construct using AA5457, which has an optimal combination of mechanical properties and is lightweight, commercially available, and prone to plastic deformation. In this work, Al-Mg type AA5754 alloy was used to produce the outer components of an experimental prototype of a capsule that allows for teleperformance assessment tests and telerehabilitation of sensory diseases (i.e., hearing, sight, smell, balance). The main goal of this study was to analyze how cold-working can change corrosion resistance of AA5754 (O/H111 according to [48]).

2. Materials

A schematic of the construction design is given in Figure 1a. To prepare a final element in the form of an adjustable angle plate, a 4 mm-thick sheet made of AA5754 (chemical composition given in Table 1) was cut into strips and subsequently bent to 90° (Figure 1). The main focus was to determine the influence of microstructural changes induced by bending on the corrosion behavior of the original AA5754 sheet. To accomplish this, the corrosion behavior in two specified areas of the final element was measured: in the undeformed area located at the arm of the bent Sheet (**A**) and at the bent area (**B**), as shown in Figure 1b.

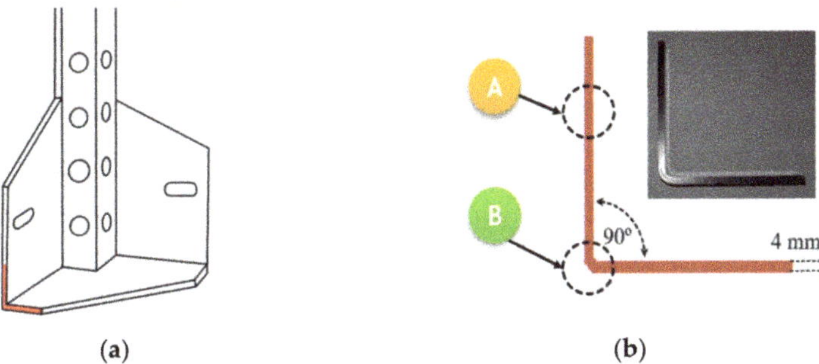

(a)   (b)

Figure 1. The outer design of the transferable capsule used for teleperformance assessment tests and telerehabilitation: (**a**), schematic of the outer elements, and (**b**), locations of corrosion tests.

Table 1. The chemical composition of 5754 alloy in mass% [48].

Si	Fe	Cu	Mn	Mg	Cr	Zn	Ti	Al
max 0.40	max 0.40	max 0.10	max 0.50	2.6–3.6	max 0.30	max 0.20	max 0.15	- Bal.

3. Methodology

3.1. Microstructure Characterization

Microstructure characterization was performed in two areas (undeformed and bent areas) using scanning electron microscopy (SEM, Hitachi SU8000, Tokyo, Japan) in back-scattered electron mode (BSE) with an accelerating voltage of 5 kV. The microstructure was observed on the cross-section of the Sheet in areas A and B shown in Figure 1b. Specimens for SEM observations were prepared by grinding samples with up to 1200-grit SiC paper and ion polishing with a Hitachi IM-4000 ion milling system. The ion milling is a damage-free process that does not cause surface deformation, eliminates stresses and oxide layers; therefore, the surface quality is good enough to observe the structure under channeling contrast using SEM. Structural investigations were combined with an energy dispersive X-ray (EDX) analysis of precipitates and intermetallic particles. To more thoroughly describe the role of intermetallic particles, the density of coarse particles was calculated by dividing the number of counted particles (N) by the studied area (A) of the specimen. Moreover, their size (A—area) and standard deviation (S_D) were calculated using computer software as per references [49,50].

Detailed microstructural observations of selected areas were performed using a transmission electron microscope (TEM, JEOL 1200, Tokyo, Japan) operating at an accelerating voltage of 120 kV, and the EDX spectra of the nanoparticles were obtained using a high-resolution scanning transmission electron microscope (STEM equipped with an EDX spectrometer, Hitachi S5500). The TEM samples were prepared using twin-jet electropolishing at 3 °C/20 V with an electrolyte composed of perchloric acid ($HClO_4$), ethanol and 2-butyxoethanol (type A2 purchased from Struers, Ltd., Ballerup, Denmark) Based on the observations, the dislocation density (ρ) was calculated using the formula [51]:

$$\rho = \frac{N}{L_r h} \tag{1}$$

where N is the number of dislocation-line intersections, L_r is the total length of all lines, and h is TEM sample thickness (h = 150 nm).

3.2. Corrosion Tests and Characterization of Corroded Surfaces

Electrochemical tests were performed in an aerated 0.5 M NaCl acidified to pH 3 by the addition of hydrochloric acid (HCl) as per reference [52]. The electrolyte was made from analytical-grade reagents and distilled water. A FAS1 Gamry potentiostat (Gamry Instruments, Warminster, PA, USA) was equipped with three electrodes: Pt as the counter electrode, Ag/AgCl as the reference electrode (Ag/AgCl wire immersed in a saturated KCl solution), and the measured sample as the working electrode. The corrosion potential (E_{corr}) was recorded during 3 or 6 h of immersion under open-circuit conditions. After immersion, the corrosion rate of samples was determined using linear polarization resistance (LPR), and potentiodynamic tests were conducted. LPR data were recorded over a potential range of $E_{corr} \pm 10$ mV at a scan rate of 5 mV/min. After LPR, the samples were conditioned for 3 min and potentiodynamic tests were performed at a scan rate of 5 mV/s, starting from -0.5 V/OCP to 2 V/REF at room temperature. To prevent noise, all measurements were conducted in a Faraday cage. The polarization curves were fitted using Gamry Elchem software (Gamry Instruments, Warminster, PA, USA) in Tafel mode. To ensure the reproducibility of the results, all measurements were repeated at least three times on the same samples. The surfaces of the samples after immersion under open-circuit potential were observed using a Hitachi SU8000 SEM.

4. Results

4.1. Microstructure Characterization

Figures 2 and 3 show BSE images of the AA5754 alloy in an undeformed area and a deformed area, respectively. The BSE image shown in Figure 2a at a lower magnification shows a typical granular microstructure with many large, irregularly-shaped intermetallic particles randomly distributed in the alloy. The SEM-EDS analysis reveals that there are two kinds of particles: the white particles marked as P_1 are Al, Mg, Mn, and Fe-rich particles (Figure 2a,c), and the dark grey precipitates P_2 are composed of Mg, Al, and Si (Figure 2b,c). In the deformed region (Figure 3), the subsequent bending altered the microstructure of the material, and the distribution of the intermetallic particles also changed; more coarse white precipitates were observed in the bending area (Figure 3a). Additionally, during deformation, a significant strain was introduced into the material and resulting from this, higher dislocation density occurred [42]. Therefore, a granular structure is not clearly distinguishable in the BSE contrast image; however, many dislocations, especially multiple slip bands, are visible (Figure 3b). Similar to the undeformed area, the coarse intermetallic particles shown in Figure 3c marked as P_1 are mainly enriched in Mg, Al, Mn, and Fe (Si was detected in the bending area), while those marked as P_2 are composed of Al, Mg, and Si (Figure 3d).

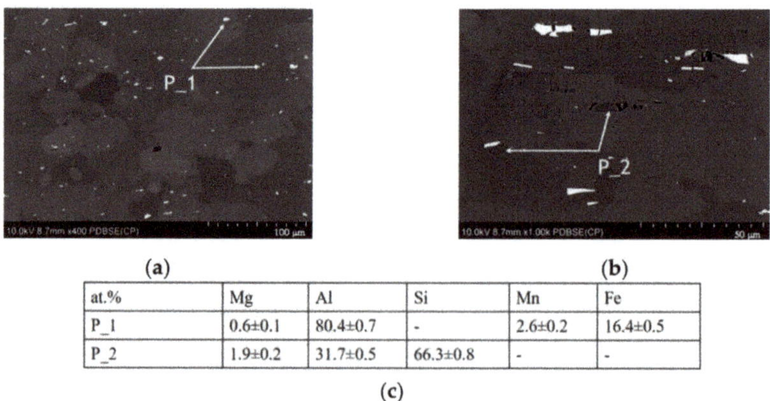

(a)

(b)

at.%	Mg	Al	Si	Mn	Fe
P_1	0.6±0.1	80.4±0.7	-	2.6±0.2	16.4±0.5
P_2	1.9±0.2	31.7±0.5	66.3±0.8	-	-

(c)

Figure 2. SEM images of the microstructure of the AA5457 sheet in the undeformed area (A): (**a**), an image showing the overall distribution of intermetallic particles, (**b**), image showing the morphology and surrounding area of intermetallic particles, and (**c**), chemical analysis of the intermetallic particles P_1 and P_2 marked in panels a and b.

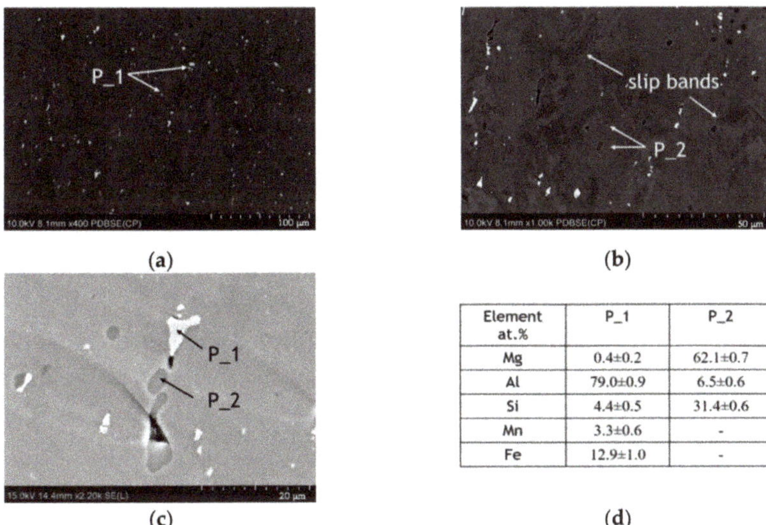

Element at.%	P_1	P_2
Mg	0.4±0.2	62.1±0.7
Al	79.0±0.9	6.5±0.6
Si	4.4±0.5	31.4±0.6
Mn	3.3±0.6	-
Fe	12.9±1.0	-

(a)   (b)   (c)   (d)

Figure 3. SEM images of the microstructure of the AA5457 sheet in the deformed area (B): (**a**) and (**b**), images showing the overall distribution of intermetallic particles, (**c**), image showing the morphology and area surrounding the intermetallic particles P_1 and P_2, and (**d**), the chemical analysis of the intermetallic particles P_1 and P_2 marked in panel c.

The density of P_1 particles in the undeformed and deformed areas was comparable, $N_A = 0.01$ and $N_A = 0.02$ (given in $1 \times \mu m^{-2}$), for the undeformed and deformed areas, respectively; however, after deformation, they showed smaller sizes (Table 2). The SEM images indicate that there was an abrupt change in the density of P_2 particles before and after deformation (compare Figures 2a and 3a). The calculations demonstrate that the density of the P_2 particles changed to $N_A = 0.001$ and $N_A = 0.01$ (given in $1 \times \mu m^{-2}$) in the undeformed and deformed areas, respectively (Table 2). As a result of bending, the observed particles were slightly smaller in size (Table 2).

Table 2. The mean surface area (A), standard deviation (S_D), and number per 1 μm^2 (N_A) of coarse intermetallic particles P_1 and P_2.

AA5754	P_1			P_2		
Area	A (μm^2)	S_D	N_A ($1 \times \mu m^{-2}$)	A (μm^2)	S_D	N_A ($1 \times \mu m^{-2}$)
undeformed (A)	1.09	2.92	0.01	2.74	2.82	0.001
deformed (B)	0.59	1.68	0.02	2.33	3.89	0.01

SEM structural analysis was complemented by TEM observations. Figure 4 presents TEM images of the undeformed (Figure 4a–c) and deformed regions (Figure 4d–f). The coarse grains (Figure 4a) contained only a few dislocations in the grain interior (Figure 4b). TEM investigations also revealed the presence of small uniformly-distributed nanoparticles, as illustrated in Figure 4c. The observed particles have sizes from 50–200 nm and are rich in Si and Mn (the chemical composition is presented below Figure 4f).

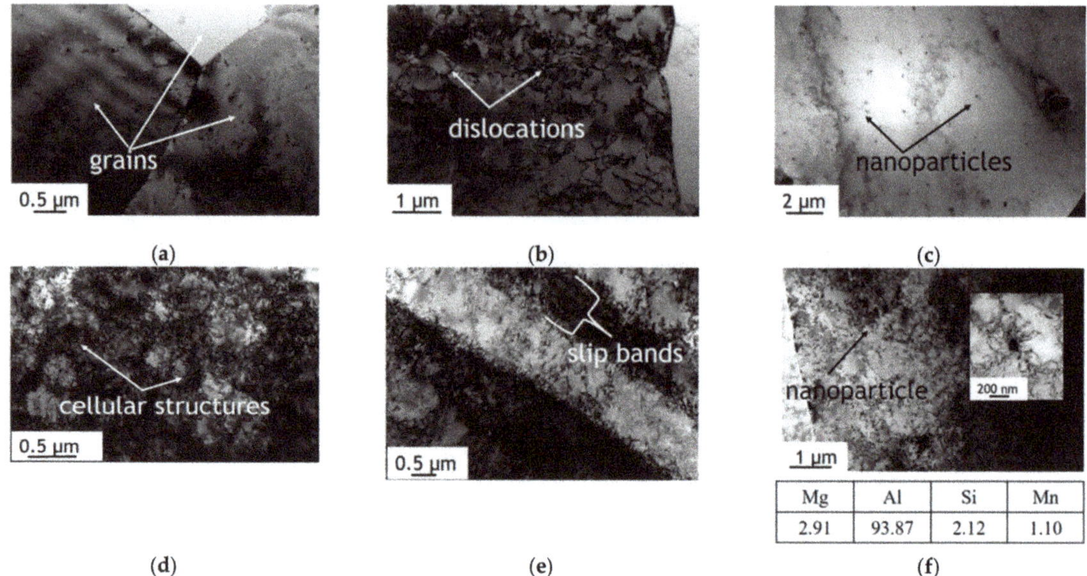

Mg	Al	Si	Mn
2.91	93.87	2.12	1.10

Figure 4. TEM images of the microstructure of the AA5457 sheet in the undeformed area (A): (**a**,**b**), images showing the overall grain boundaries and dislocation structure, and (**c**), image showing the overall distribution of nanosized intermetallic particles. TEM images of the microstructure of the AA5457 sheet in the deformed area (B): (**d**,**e**), images showing the increase in the dislocation density and the formation of cellular dislocation structures and slip bands, and (**f**), image showing an increase in the dislocation density around nanosized intermetallic particles. The representative chemical composition of a nanoparticle is presented below panel (**f**).

The results from the deformed region clearly show an increase in the number of dislocations compared with the undeformed section. The dislocation density increased by nearly an order of magnitude, from 4.5×10^{12} m^{-2} in the undeformed region (Figure 4b) to 4.4×10^{13} m^{-2} in the deformed region (Figure 4d,e). This is a typical phenomenon for deformed metallic materials. It should be noted that dislocations are distributed rather uniformly; however, in some regions, the formation of characteristic cellular structures (Figure 4d) and slip bands (Figure 4e) was observed. Moreover, an increase in the dislocation density was also observed around Si and Mn-rich nanoparticles (Figure 4f). Hard intermetallic particles pin dislocations during deformation, resulting in the formation of many dislocations around such particles.

4.2. Corrosion Testing

4.2.1. Electrochemical Testing

E_{corr} measurements recorded in naturally-aerated 0.5 M NaCl adjusted to pH 3 are shown in Figure 5. The evolution of E_{corr} in both samples exhibits a similar trend, with a rapid initial increase followed by a longer-term stabilization. The E_{corr} of the undeformed sample started at -0.78 V/REF, and increased to around -0.70 V/REF after the first 5 min of immersion. It maintained a near steady-state value of -0.70 V/REF for the remainder of the experiment. The initial increase of E_{corr} in the deformed area began from -0.94 V/REF and reached -0.72 V/REF after 10 min immersion. Then, a small but stable decline in E_{corr} to -0.78 V/REF was observed at the end of the experiment. The initial rapid increase in E_{corr} for both alloys may be related to the extensive corrosion and passivation processes [53], while the slight fluctuations observed in the E_{corr} curves were attributed to the pitting process caused by the constituents formed in the alloy [54–56].

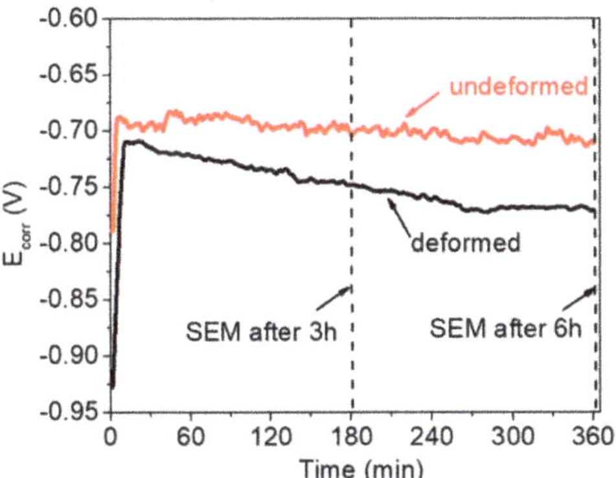

Figure 5. The E_{corr} measurements recorded during 6 h of immersion in 0.5 M NaCl (pH 3) in the undeformed and deformed areas of AA5754. SEM observations were made after 3 and 6 h of immersion, as marked by the dashed lines.

The potentiodynamic polarization curves recorded for the undeformed and deformed samples of AA554 after 3 and 6 h of immersion under open-circuit conditions are given in Figure 6. All curves exhibit similar trends with passive regions, which suggests an anodic control of the corrosion processes. All curves exhibit a wide passive region when the current density plateaued, extended by 0.04 V in the undeformed area (both 3 and 6 h immersion), to around 0.05 V for the deformed area after 3 h of immersion, and 0.07 V for the deformed area after 6 h of immersion [57] Simultaneously, there is a shift in the pitting current, i_p, which is the lowest for the undeformed area (after both periods of immersion). The pitting current, i_p, shifted to higher values for the deformed areas after 3 h of immersion and continued to increase over time. Afterwards, an abrupt change in the current density was observed, which indicates the position of the pitting potential (E_{pit}) [25,42,58]. The pitting potential for both undeformed samples has the same value of $E_{pit} = -0.68$ V/REF (Table 3). Slightly more positive pitting potentials were recorded for the deformed samples, with $E_{pit} = -0.64$ V/REF after 3 h of immersion. Increasing the immersion time to 6 h decreased the pitting potential of the deformed area to $E_{pit} = -0.66$ V/REF (after 6 h of immersion). The numerical values of pitting corrosion resistance (R_{pit}) may be described by the difference between pitting potential and corrosion potential, E_{pit} and E_{corr} [59,60]

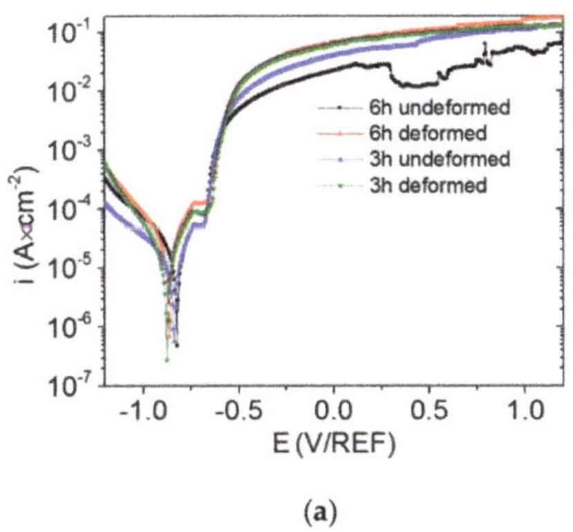

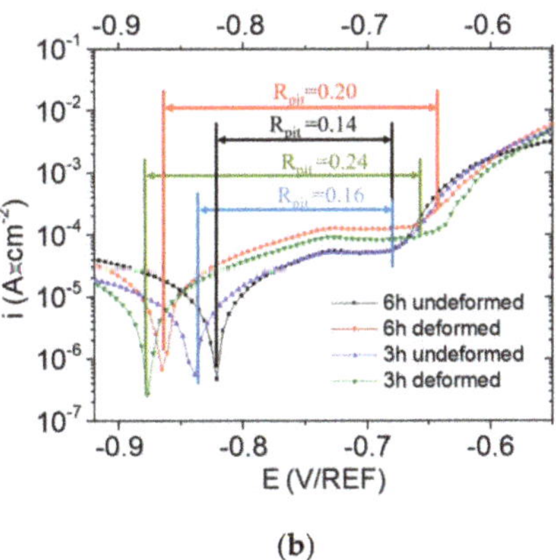

Figure 6. The electrochemical curves obtained after 3 and 6 h of immersion in 0.5 M NaCl (pH 3) for undeformed and deformed areas of the AA5754: (**a**), potentiodynamic polarization curves, and (**b**), pitting susceptibility of analyzed materials calculated using Formula (1).

Table 3. The corrosion potential (E_{corr}), pitting potential (E_{pit}), corrosion current density (i_{corr}), and corrosion rate calculated after 3 and 6 h of immersion in naturally-aerated 0.5 M NaCl (pH 3) for the undeformed and deformed areas of AA5754.

Electrochemical Parameters	Sample Area and Immersion Time			
	Undeformed	Deformed	Undeformed	Deformed
	After 3 h of Immersion		After 6 h of Immersion	
E_{corr} (V/REF)	−0.84	−0.88	−0.82	−0.86
E_{pit} (V/REF)	−0.68	−0.64	−0.68	−0.66
i_{corr} (μA·cm^{-2})	18.6	34.1	31.6	32.7
Corrosion rate (mpy)	1.8	2.6	2.6	9.5

$$R_{pit} = \left| E_{corr} - E_{pit} \right| \tag{2}$$

Considering the data presented in Figure 6b, the smallest difference between E_{corr} and E_{pit} was calculated for the undeformed area of the sample for both immersion periods (R_{pit} = 0.16 after 3 h of immersion, and R_{pit} = 0.14 after 6 h of immersion), suggesting that the undeformed area is less-resistant to localized corrosion.

Regardless of immersion time, the deformed area of the alloy showed the greatest R_{pit} = 0.22V. The higher R and i_p values of the deformed areas were attributed to the weakened passivation properties [59,60].

The extrapolated from Tafel plot data show corrosion potentials of E_{corr} = −0.84 V/REF and E_{corr} = −0.82 V/REF for the undeformed areas immersed for 3 and 6 h, respectively. Corrosion potentials registered for the deformed areas had slightly lower values than the corrosion potentials of the undeformed areas, with values of E_{corr} = −0.88 V/REF and E_{corr} = −0.86 V/REF after 3 and 6 h of immersion, respectively (Table 3). To further view the overall corrosion behavior of the analyzed materials, LPR data were recorded after 3 and 6 h of immersion. The corrosion rates after 3 h of immersion were 1.8 mpy for the undeformed region, and almost twice as high (2.6 mpy) for the deformed region of the material (Table 3). After 6 h of immersion, the corrosion rate in the undeformed area increased slightly to 2.6 mpy, and the corrosion rate of the deformed area reached a significantly higher value of 9.5 mpy.

4.2.2. SEM Observations after Immersion

A comparison of the E_{corr} measurements under open-circuit conditions and surface analysis can be used to characterize the effect of immersion time on the microstructure-dependent corrosion processes. The corrosion morphologies of the undeformed and deformed areas after immersion in 0.5 M NaCl at pH 3 are presented in Figures 7 and 8, respectively. The SEM images in Figures 7 and 8 clearly show that the corrosion of AA5754 is a complex process that involves several microstructure-dependent corrosion mechanisms. At the beginning of the experiment, the main corrosion attack on both the undeformed and deformed samples occurred around the Si-rich intermetallic particles (marked as P_2 in Figures 7a and 8b), while Fe-rich particles remained unreactive (marked as P_1 Figures 7a and 8b). The local corrosion attack around the Si-rich constituents led to cavitation around the particles, as previously observed [61]. Moreover, SEM investigations clearly indicate that after 3 h of immersion, trenching was more intense around the particles on the deformed area (compare Figures 7a and 8b). In both cases, trenching became more severe over time (Figures 7d and 8d), and when two P_2 particles were relatively close to each other, trenching spread from one particle to the other, as shown in Figure 7d.

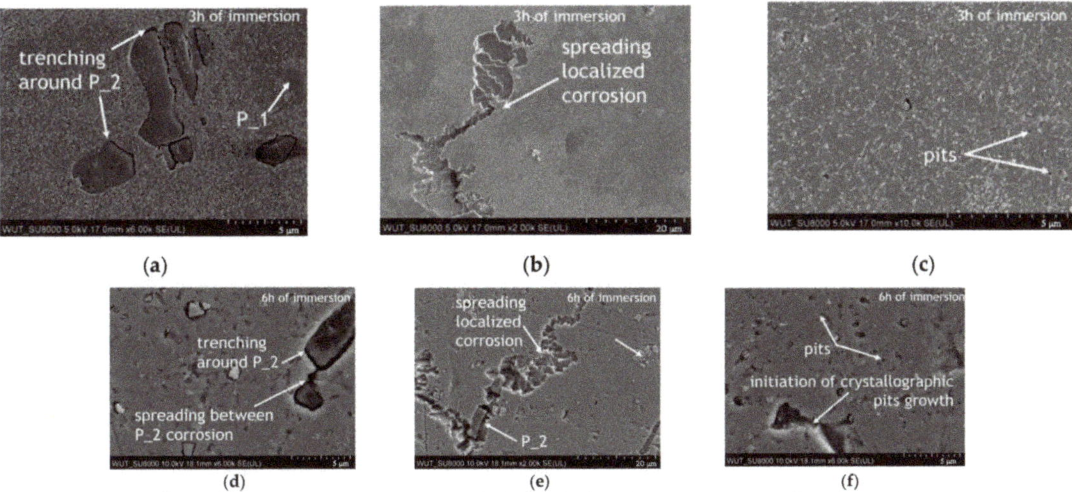

(a)   (b)   (c)

(d)   (e)   (f)

Figure 7. Post-corrosion observations of AA5754 in the undeformed area (A) after immersion under open-circuit conditions. Images taken after 3 h of immersion showing: (**a**), the propagation of crystallographically-grown pits, (**b**), trenching around Si-rich (P_2) particles, and (**c**) high-magnification image of tiny pits. Images taken after 6 h of immersion showing: (**d**), the propagation of crystallographically-grown pits, (**e**), trenching around Si-rich (P_2) particles, and (**f**) the formation of tiny pits.

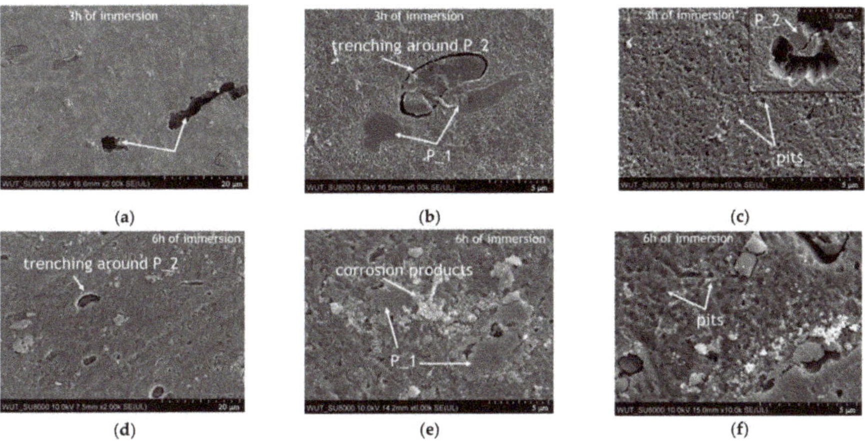

(a)   (b)   (c)

(d)   (e)   (f)

Figure 8. Post-corrosion observations of AA5754 in the deformed area (B) after immersion under open-circuit conditions. Images taken after 3 h of immersion showing: (**a**), the propagation of crystallographically-grown pits, (**b**), the trenching around Si-rich (P_2) particles, and (**c**), a high-magnification image of tiny pits. Images taken after 6 h of immersion showing: (**d**), trenching around Si-rich (P_2) particles, (**e**), corrosion products around Fe-rich (P_1) particles, and (**f**), the formation of tiny pits.

SEM observations confirmed that after immersion in 0.5 M NaCl (pH 3) trenching occurred around Si-rich particles over the entire analyzed sample surface. Besides intense local corrosion around Si-rich particles, after 3 h of immersion, highly-pronounced local pitting corrosion presented in the form of several superficial crystallographically-grown pits in the original and deformed areas of AA5754 (Figures 7b and 8a) was formed. As the experiment duration increased, more crystallographically-grown pits were observed on the original material compared with the deformed area. Crystallographically-grown pits limited by perpendicular standing walls of the crystallographic lattice have been previously observed on Al and Al alloys, and the pit growth kinetics have been previously explained [62–66]. It was also demonstrated that in acidic chloride-containing solutions, the general mode of pit propagation is corrosion tunneling [62]. Considering the deformation effect on corrosion mechanisms on AA5754 after 3 h of immersion, minor differences were observed regarding the shape of crystallographically-grown pits (compare Figures 7a and 8b), but not their amount. As the experiment duration increased, more crystallographically-grown pits were observed on the surface of the original material, and some underwent further lateral spreading, while on the deformed area after 6 h of immersion, only a few crystallographically-grown pits were observed, suggesting that no new pits were grown on the deformed area. The locally-formed crystallographic pits on the undeformed material spread laterally on the surface, whilst the pits on the deformed sample seemed to propagate into the depth of the material. Some pits were created around Si-rich particles (type P_2), as shown in Figure 7e and the inset in Figure 8c. In many areas of the observed samples, these kinds of pits were created after the trenches formed around P_2 particles.

Apart from the microgalvanic coupling between Si-rich particles and the matrix, or the large crystallographically-grown pits, very tiny metastable ellipsoidal pits formed on the surface of both areas (Figure 7c,f—undeformed areas after 3 and 6 h of immersion, respectively; Figure 8c,f—deformed areas after 3 and 6 h of immersion, respectively). These pits grew larger with the experiment duration (Figures 7f and 8f); however, when comparing Figures 7c and 8c, it is visible that more and deeper pits were created on the deformed area of the sample. Moreover, after 6 h of immersion, loosely-adhered corrosion products were formed on the deformed area of the sample, especially in the locations where two Fe-rich particles (P_1) were located relatively close to each other (Figure 8e).

5. Discussion

The results of this work show that the observed trenching around Si-rich coarse particles depends on their number and distribution, and the increased amount of these particles in the deformed area resulted in more localized corrosion initiation, which was one of the reasons for its lower corrosion resistance.

As demonstrated in this work, the corrosion that occurred around Si-rich particles in the deformed area of the material was more severe than the corrosion around the same type of particles in the undeformed area of AA5457. This phenomenon is related to the increased number of particles in the material after deformation and due to their mutual interaction with the matrix, particularly particle-dislocation interactions. In our opinion, the cumulative stress occurring during deformation was sufficient to cause particle fragmentation. The increased number of hard coarse particles provided more locations where trenching could occur, and the intensity of such trenching is related to particle-dislocation interactions. As is commonly known, the hard coarse particles are the places where deformation-induced dislocation movement is blocked, leading to the formation of a high concentration of dislocations around them [67].

The chemical composition of the coarse particles is also important, considering their corrosion behavior in chloride-containing solution. The Si-rich coarse intermetallics observed in AA5754 also contained Mg and Al. Previous reports have clearly indicated that the rapid dissolution of Mg from intermetallics occurs during exposure, which leads to the rapid dealloying of these particles [21,24,61,68,69]. The results of previous studies [68,69] explain the mechanism of Mg and Al dissolution from the coarse particles composed mainly of Mg and Si in Al-Mg-Si alloys. The results of this research show that at the beginning of immersion, the preferential and selective dissolution of Mg occurs, and thus, the Volta potential of the particle changes, which also changes the nature of Mg-containing particles from anodic to cathodic, forming a galvanic couple with the Al matrix. The results of this work show that the increased trenching around Si-rich particles in the deformed area was a result of the cracking of these particles induced by the cumulative stresses generated during deformation. During deformation, the coarse hard particles blocked the dislocation movement; therefore, a higher dislocation density was formed near such particles. As the locally-formed galvanic cells approach equilibrium, the cathodic particle forces the galvanic dissolution of the anodic matrix, leading to matrix corrosion. The increased dislocation density near the coarse particles promotes the corrosion reactions, leading to more intense trenching around the particles rich in Mg and Si. The reason for this is a higher dislocation density formed in the bent area of the sample, which promoted pit propagation in the depth due to their less-ordered structure [58,70]. This hypothesis is indisputably confirmed by the results presented in this work. It is also worth noting that Fe-rich particles remained unreactive throughout the entire experiment; however, their presence, especially when located relatively close to each other, may enhance corrosion product formation.

Another discrepancy that needs to be explained is that Aballe et al. [71] did not observe any relationship between the formation of crystallographic pitting and the existence of intermetallic particles, but Neetu et al. [42] claimed that pit initiation occurred at particles rich in Fe and Mn. We observed spontaneously-formed pits on the undeformed samples; however, in contrast to both works, our observations indicate that crystallographic pit growth is often initiated in the areas occluded by secondary particles enriched with Si. Moreover, this observation is supported by the fact that corrosion attack around Si-rich particles occurred before crystallographically-grown pits were formed. In a separate set of experiments, we observed trenching around Si-rich particles, but we did not observe any crystallographically-grown pits on the samples immersed for one hour in the same solution (data not published). Since pitting attack is considered to be an autocatalytic process in the active areas, it is reasonable to assume that the higher number of Si-rich coarse particles provided more sites were localized corrosion may be initiated. These places are prone to coalescence and form laterally-spreading local corrosion.

The second type of corrosion that occurred on the AA5754 alloy is microgalvanic corrosion between Si-rich nanoparticles and the alloy matrix. The observations showed that tiny pits were created around Si-rich nanoparticles, and the corrosion damage around them was more severe in the deformed area. Although the deformation-induced stress was not sufficient to change the size or distribution of the observed nanoparticles, our experiments clearly show that the nanoparticles in the deformed area with a high dislocation-stacking fault density around them promoted the formation of many deep tiny pits.

6. Conclusions

Based on the results of this work, the following conclusions can be drawn:

- The AA5754 alloy undergoes microstructure-dependent corrosion attack.
- The bending that was applied to obtain the desired shape of the design elements made from AA5754 affected the corrosion resistance of the alloy—the bent areas were more susceptible to corrosion than the original material.
- The lower corrosion resistance of the bent areas was related to the bending-induced microstructural changes, such as the increased density of Si-rich coarse particles and nanoparticles, the increased dislocation density around them, and their mutual interactions.

Author Contributions: Conceptualization, corrosion investigations, A.D.; surface characterization, A.D.; microstructure characterization, P.B. and A.S.; methodology, A.D. and A.S.; validation, H.G. and J.M.; investigation, A.D., A.S. and P.B.; writing—original draft preparation, A.D.; writing—review and editing, A.S.; supervision, H.G.; project administration, H.G.; funding acquisition, J.M. All authors have read and agreed to the published version of the manuscript.

Funding: This study was funded by the National Centre for Research and Development under grant STRATEGMED1/248664/7/NCBR/14.

Institutional Review Board Statement: Not applicable.

Informed Consent Statement: Not applicable.

Data Availability Statement: Data available in a publicly accessible repository.

Conflicts of Interest: The authors declare no conflict of interest.

References

1. Ding, L.; Jia, Z.; Zhang, Z.; Sanders, R.E.; Liu, Q.; Yang, G. The natural aging and precipitation hardening behaviour of Al-Mg-Si-Cu alloys with different Mg/Si ratios and Cu additions. *Mater. Sci. Eng. A* **2015**, *627*, 119–126. [CrossRef]
2. Miller, W.S.; Zhuang, L.; Bottema, J.; Wittebrood, A.J.; de Smet, P.; Haszler, A.; Vieregge, A. Recent development in aluminium alloys for the automotive industry. *Mater. Sci. Eng.* **2000**, *A280*, 37–49. Available online: www.elsevier.com/locate/msea (accessed on 20 October 2020). [CrossRef]
3. Abid, T.; Boubertakh, A.; Hamamda, S. Effect of pre-aging and maturing on the precipitation hardening of an Al-Mg-Si alloy. *J. Alloys Compd.* **2010**, *490*, 166–169. [CrossRef]
4. Lathabai, S.; Lloyd, P.G. The effect of scandium on the microstructure, mechanical properties and weldability of a cast Al-Mg alloy. *Acta Mater.* **2002**, *50*, 4275–4292. [CrossRef]
5. Hirsch, J.; Al-Samman, T. Superior light metals by texture engineering: Optimized aluminum and magnesium alloys for automotive applications. *Acta Mater.* **2013**. [CrossRef]
6. Chehreh, A.B.; Grätzel, M.; Bergmann, J.P.; Walther, F. Effect of corrosion and surface finishing on fatigue behavior of friction stir welded EN AW-5754 aluminum alloy using various tool configurations. *Materials (Basel)* **2020**, *13*, 3121. [CrossRef]
7. Conserva, M.; Leoni, M. Effect of thermal and thermo-mechanical processing on the properties of Al-Mg alloys. *Metall. Trans. A* **1975**, *6*, 189–195. [CrossRef]
8. Cavanaugh, M.K.; Birbilis, N.; Buchheit, R.G. Modeling pit initiation rate as a function of environment for Aluminum alloy 7075-T651. *Electrochim. Acta* **2012**, *59*, 336–345. [CrossRef]
9. Arrabal, R.; Mingo, B.; Pardo, A.; Mohedano, M.; Matykina, E.; Rodríguez, I. Pitting corrosion of rheocast A356 aluminium alloy in 3.5wt.% NaCl solution. *Corros. Sci.* **2013**, *73*, 342–355. [CrossRef]
10. Włodarczyk, P.P.; Włodarczyk, B. Effect of hydrogen and absence of passive layer on corrosive properties of aluminum alloys. *Materials (Basel))* **2020**, *13*, 1580. [CrossRef]
11. Yasakau, K.A.; Zheludkevich, M.L.; Ferreira, M.G.S. Role of intermetallics in corrosion of aluminum alloys. Smart corrosion protection. In *Intermetallic Matrix Composites*; Mitra, R.B.T.-I.M.C., Ed.; Woodhead Publishing: Cambridge, UK, 2018; pp. 425–462. [CrossRef]
12. Buchheit, R.G. A Compilation of Corrosion Potentials Reported for Intermetallic Phases in Aluminum Alloys. *J. Electrochem. Soc.* **1995**, *142*, 3994–3996. [CrossRef]
13. Li, S.M.; Li, Y.D.; Zhang, Y.; Liu, J.H.; Yu, M. Effect of intermetallic phases on the anodic oxidation and corrosion of 5A06 aluminum alloy. *Int. J. Miner. Metall. Mater.* **2015**, *22*, 167–174. [CrossRef]
14. Ma, Y.; Wu, H.; Zhou, X.; Li, K.; Liao, Y.; Liang, Z.; Liu, L. Corrosion behavior of anodized Al-Cu-Li alloy: The role of intermetallic particle-introduced film defects. *Corros. Sci.* **2019**, *158*. [CrossRef]
15. Yuan, D.; Chen, K.; Chen, S.; Zhou, L.; Chang, J.; Huang, L.; Yi, Y. Enhancing stress corrosion cracking resistance of low Cu-containing Al-Zn-Mg-Cu alloys by slow quench rate. *Mater. Des.* **2019**, *164*, 107558. [CrossRef]
16. Revilla, R.I.; Verkens, D.; Rubben, T.; De Graeve, I. Corrosion and corrosion protection of additively manufactured aluminium alloys—A critical review. *Materials (Basel)* **2020**, *13*, 4804. [CrossRef] [PubMed]
17. Davoodi, A.; Pan, J.; Leygraf, C.; Norgren, S. The Role of Intermetallic Particles in Localized Corrosion of an Aluminum Alloy Studied by SKPFM and Integrated AFM/SECM. *J. Electrochem. Soc.* **2008**, *155*, C211. [CrossRef]
18. Buchheit, R.G.; Boger, R.K.; Carroll, M.C.; Leard, R.M.; Paglia, C.; Searles, J.L. The electrochemistry of intermetallic particles and localized corrosion in Al alloys. *JOM* **2001**, *53*, 29–33. [CrossRef]
19. Birbilis, N.; Buchheit, R.G. Electrochemical Characteristics of Intermetallic Phases in Aluminum Alloys. *J. Electrochem. Soc.* **2005**, *152*, B140. [CrossRef]
20. Birol, F.; Birol, Y. Corrosion Behavior of Twin-Roll Cast Al-Mg and Al-Mg-Si Alloys. In Proceedings of the 9th International Conference on Aluminium Alloys, Brisbane, Australia, 2–5 August 2004; pp. 338–344.
21. Yasakau, K.A.; Zheludkevich, M.L.; Lamaka, S.V.; Ferreira, M.G.S. Role of intermetallic phases in localized corrosion of AA5083. *Electrochim. Acta* **2007**, *52*, 7651–7659. [CrossRef]
22. Serdechnova, M.; Volovitch, P.; Brisset, F.; Ogle, K. On the cathodic dissolution of Al and Al alloys. *Electrochim. Acta* **2014**, *124*, 9–16. [CrossRef]
23. Ma, Y.; Zhou, X.; Huang, W.; Thompson, G.E.; Zhang, X.; Luo, C.; Sun, Z. Localized corrosion in AA2099-T83 aluminum-lithium alloy: The role of intermetallic particles. *Mater. Chem. Phys.* **2015**, *161*, 201–210. [CrossRef]
24. Gao, W.; Wang, D.; Seifi, M.; Lewandowski, J.J. Anisotropy of corrosion and environmental cracking in AA5083-H128 Al-Mg alloy. *Mater. Sci. Eng. A* **2018**, *730*, 367–379. [CrossRef]
25. Guan, L.; Zhou, Y.; Zhang, B.; Wang, J.Q.; Han, E.-H.; Ke, W. Influence of aging treatment on the pitting behavior associated with the dissolution of active nanoscale β -phase precipitates for an Al–Mg alloy. *Corros. Sci.* **2016**, *103*, 255–267. [CrossRef]
26. Yang, W.; Shen, W.; Zhang, R.; Cao, K.; Zhang, J.; Liu, L. Enhanced age-hardening by synergistic strengthening from Mg–Si and Mg–Zn precipitates in Al-Mg-Si alloy with Zn addition. *Mater. Charact.* **2020**, *169*, 110579. [CrossRef]
27. Yi, G.; Zeng, W.; Poplawsky, J.D.; Cullen, D.A.; Wang, Z.; Free, M.L. Characterizing and modeling the precipitation of Mg-rich phases in Al 5xxx alloys aged at low temperatures. *J. Mater. Sci. Technol.* **2017**, *33*, 991–1003. [CrossRef]

28. Li, J.; Dang, J. A summary of corrosion properties of Al-Rich solid solution and secondary phase particles in al alloys. *Metals (Basel)* **2017**, *7*, 84. [CrossRef]

29. Da Ren, W.; LI, J.-F.; Zheng, Z.-Q.; Chen, W.-J. Localized corrosion mechanism associated with precipitates containing Mg in Al alloys. *Trans. Nonferrous Met. Soc. China* **2007**, *17*, 727–732. [CrossRef]

30. Lyndon, J.A.; Gupta, R.K.; Gibson, M.A.; Birbilis, N. Electrochemical behaviour of the β-phase intermetallic (Mg2Al3) as a function of pH as relevant to corrosion of aluminium-magnesium alloys. *Corros. Sci.* **2013**, *70*, 290–293. [CrossRef]

31. Yang, Y.K.; Allen, T. Direct visualization of β phase causing intergranular forms of corrosion in Al-Mg alloys. *Mater. Charact.* **2013**, *80*, 76–85. [CrossRef]

32. Vignesh, R.V.; Padmanaban, R. Intergranular corrosion susceptibility of friction stir processed aluminium alloy 5083. *Mater. Today Proc.* **2018**, *5*, 16443–16452. [CrossRef]

33. Fan, L.; Ma, J.; Zou, C.; Gao, J.; Wang, H.; Sun, J.; Guan, Q.; Wang, J.; Tang, B.; Li, J.; et al. Revealing foundations of the intergranular corrosion of 5XXX and 6XXX Al alloys. *Mater. Lett.* **2020**, *271*, 127767. [CrossRef]

34. Searles, J.L.; Gouma, P.I.; Buchheit, R.G. Stress corrosion cracking of sensitized AA5083 (Al-4.5Mg-1.0Mn). *Mater. Sci. Forum.* **2002**, *396–402*, 1437–1442. [CrossRef]

35. Brillas, E.; Cabot, P.L.; Centellas, F.; Garrido, J.A.; Pérez, E.; Rodríguez, R.M. Electrochemical oxidation of high-purity and homogeneous Al-Mg alloys with low Mg contents. *Electrochim. Acta* **1997**, *43*, 799–812. [CrossRef]

36. Sriyono, F.; Hastuti, E.P.; Sunaryo, G.R. Study on Pitting Corrosion of AlMg2 in Solution Containing Chloride. *J. Phys. Conf. Ser.* **2019**, *1198*. [CrossRef]

37. Guan, L.; Zhang, B.; Wang, J.Q.; Han, E.H.; Ke, W. The reliability of electrochemical noise and current transients characterizing metastable pitting of Al-Mg-Si microelectrodes. *Corros. Sci.* **2014**, *80*, 1–6. [CrossRef]

38. Laycock, N.J.; Newman, R.C. Localised dissolution kinetics, salt films and pitting potentials. *Corros. Sci.* **1997**, *39*, 1771–1790. [CrossRef]

39. Yi, G.; Cullen, D.A.; Derrick, A.T.; Zhu, Y.; Free, M.L. Effects of Different Temper and Aging Temperature on the Precipitation Behavior of Al 5xxx Alloy. *Light Met.* **2015**, 359–365. [CrossRef]

40. Li, Y.; Hung, Y.; Du, Z.; Xiao, Z.; Jia, G. The Effect of Homogenization on the Corrosion Behavior of Al–Mg Alloy. *Phys. Met. Metallogr.* **2018**, *119*, 339–346. [CrossRef]

41. Abo Zeid, E.F. Mechanical and electrochemical characteristics of solutionized AA 6061, AA6013 and AA 5086 aluminum alloys. *J. Mater. Res. Technol.* **2019**, *8*, 1870–1877. [CrossRef]

42. Singh, N.S.; Rao, P.N.; Jayaganathan, R.; Midathada, A.; Verma, K.; Ravella, U.K. Elevated corrosion in strain hardened Al–Mg alloy. *Vacuum* **2018**, *157*, 402–413. [CrossRef]

43. Abdi behnagh, R.; Besharati Givi, M.K.; Akbari, M. Mechanical properties, corrosion resistance, and microstructural changes during friction stir processing of 5083 aluminum rolled plates. *Mater. Manuf. Process.* **2012**, *27*, 636–640. [CrossRef]

44. Naeini, M.F.; Shariat, M.H.; Eizadjou, M. On the chloride-induced pitting of ultra fine grains 5052 aluminum alloy produced by accumulative roll bonding process. *J. Alloys Compd.* **2011**, *509*, 4696–4700. [CrossRef]

45. Kus, E.; Lee, Z.; Nutt, S.; Mansfeld, F. A comparison of the corrosion behavior of nanocrystalline and conventional Al 5083 samples. *Corrosion* **2006**, *62*, 152–161. [CrossRef]

46. Ezuber, H.; El-Houd, A.; El-Shawesh, F. A study on the corrosion behavior of aluminum alloys in seawater. *Mater. Des.* **2008**, *29*, 801–805. [CrossRef]

47. He, C.; Luo, B.; Zheng, Y.; Yin, Y.; Bai, Z.; Ren, Z. Effect of Sn on microstructure and corrosion behaviors of Al-Mg-Si alloys. *Mater. Charact.* **2019**, *156*, 109836. [CrossRef]

48. Products, M.; Plate, T.; Products, M.; Sheet, A.; Environments, S.; Alloys, A. *Standard Specification for Aluminum and Aluminum-Alloy Sheet and Plate*; ASTM International: West Conshohocken, PA, USA, 2009; Volume 1, pp. 1–29. [CrossRef]

49. Wejrzanowski, T.; Kurzydlowski, K.J. Stereology of grains in nano-crystals. *Diffus. Defect Data Pt. B Solid State Phenom.* **2003**, *94*, 221–228. [CrossRef]

50. Wejrzanowski, T.; Spychalski, W.L.; Rózniatowski, K.; Kurzydłowski, K.J. Image based analysis of complex microstructures of engineering materials. *Int. J. Appl. Math. Comput. Sci.* **2008**, *18*, 33–39. [CrossRef]

51. Norfleet, D.M.; Dimiduk, D.M.; Polasik, S.J.; Uchic, M.D.; Mills, M.J. Dislocation structures and their relationship to strength in deformed nickel microcrystals. *Acta Mater.* **2008**, *56*, 2988–3001. [CrossRef]

52. Guan, L.; Zhou, Y.; Lin, H.Q.; Zhang, B.; Wang, J.Q.; Han, E.H.; Ke, W. Detection and analysis of anodic current transients associated with nanoscale β-phase precipitates on an Al-Mg microelectrode. *Corros. Sci.* **2015**, *95*, 6–10. [CrossRef]

53. Abady, G.M.; Hilal, N.H.; El-Rabiee, M.; Badawy, W.A. Effect of Al content on the corrosion behavior of Mg-Al alloys in aqueous solutions of different pH, Electrochim. *Acta* **2010**, *55*, 6651–6658. [CrossRef]

54. Wang, Z.; Chen, P.; Li, H.; Fang, B.; Song, R.; Zheng, Z. The intergranular corrosion susceptibility of 2024 Al alloy during re–ageing after solution treating and cold–rolling. *Corros. Sci.* **2017**, *114*, 156–168. [CrossRef]

55. Meng, G.; Wei, L.; Zhang, T.; Shao, Y.; Wang, F.; Dong, C.; Li, X. Effect of microcrystallization on pitting corrosion of pure aluminium. *Corros. Sci.* **2009**, *51*, 2151–2157. [CrossRef]

56. Jilani, O.; Njah, N.; Ponthiaux, P. Transition from intergranular to pitting corrosion in fine grained aluminum processed by equal channel angular pressing. *Corros. Sci.* **2014**, *87*, 259–264. [CrossRef]

57. Mance, A.; Cerović, D.; Mihajlović, A. The effect of small additions of indium and thallium on the corrosion behaviour of aluminium in sea water. *J. Appl. Electrochem.* **1984**, *14*, 459–466. [CrossRef]
58. Brunner, J.G.; May, J.; Höppel, H.W.; Göken, M.; Virtanen, S. Localized corrosion of ultrafine-grained Al-Mg model alloys. *Electrochim. Acta* **2010**, *55*, 1966–1970. [CrossRef]
59. Benedetti, A.; Cirisano, F.; Delucchi, M.; Faimali, M.; Ferrari, M. Potentiodynamic study of Al-Mg alloy with superhydrophobic coating in photobiologically active/not active natural seawater. *Colloids Surfaces B Biointerfaces* **2016**, *137*, 167–175. [CrossRef] [PubMed]
60. Esmailzadeh, S.; Aliofkhazraei, M.; Sarlak, H. Interpretation of Cyclic Potentiodynamic Polarization Test Results for Study of Corrosion Behavior of Metals: A Review. *Prot. Met. Phys. Chem. Surfaces* **2018**, *54*, 976–989. [CrossRef]
61. Eckermann, F.; Suter, T.; Uggowitzer, P.J.; Afseth, A.; Schmutz, P. The influence of MgSi particle reactivity and dissolution processes on corrosion in Al-Mg-Si alloys. *Electrochim. Acta* **2008**, *54*, 844–855. [CrossRef]
62. Baumgärtner, M.; Kaesche, H. Aluminum pitting in chloride solutions: Morphology and pit growth kinetics. *Corros. Sci.* **1990**, *31*, 231–236. [CrossRef]
63. Xiao, R.G.; Yan, K.P.; Yan, J.X.; Wang, J.Z. Electrochemical etching model in aluminum foil for capacitor. *Corros. Sci.* **2008**, *50*, 1576–1583. [CrossRef]
64. Newman, R.C. Local chemistry considerations in the tunneling corrosion of aluminium. *Corros. Sci.* **1995**, *37*, 527–533. [CrossRef]
65. Zaid, B.; Saidi, D.; Benzaid, A.; Hadji, S. Effects of pH and chloride concentration on pitting corrosion of AA6061 aluminum alloy. *Corros. Sci.* **2008**, *50*, 1841–1847. [CrossRef]
66. Towarek, A.; Dobkowska, A.; Zdunek, J.; Jurczak, W.; Mizera, J. The influence of Mg addition and hydrostatic extrusion HE on the repassivation ability of pure Al, AlMg1 and AlMg3 model alloys in 3.5 wt% NaCl. *Corros. Eng. Sci. Technol.* **2019**, *54*, 666–672. [CrossRef]
67. Dar, S.M.; Liao, H. Creep behavior of heat resistant Al–Cu–Mn alloys strengthened by fine (θ') and coarse (Al20Cu2Mn3) second phase particles. *Mater. Sci. Eng. A* **2019**, *763*, 138062. [CrossRef]
68. Zheng, Y.Y.; Luo, B.H.; He, C.; Gao, Y.; Bai, Z.H. Corrosion evolution and behaviour of Al–2.1Mg–1.6Si alloy in chloride media. *Rare Met.* **2020**. [CrossRef]
69. Zhu, Y.; Sun, K.; Frankel, G.S. Intermetallic Phases in Aluminum Alloys and Their Roles in Localized Corrosion. *J. Electrochem. Soc.* **2018**, *165*, C807–C820. [CrossRef]
70. Pouraliakbar, H.; Jandaghi, M.R.; Khalaj, G. Constrained groove pressing and subsequent annealing of Al-Mn-Si alloy: Microstructure evolutions, crystallographic transformations, mechanical properties, electrical conductivity and corrosion resistance. *Mater. Des.* **2017**, *124*, 34–46. [CrossRef]
71. Aballe, A.; Bethencourt, M.; Botana, F.J.; Cano, M.J.; Marcos, M. Localized alkaline corrosion of alloy AA5083 in neutral 3.5% NaCl solution. *Corros. Sci.* **2001**, *43*, 1657–1674. [CrossRef]

 *materials*

Article

Microstructure and Corrosion of Cast Magnesium Alloy ZK60 in NaCl Solution

Zhen Li [1], Zeyin Peng [1], Kai Qi [1,2], Hui Li [3], Yubing Qiu [1,2,*] and Xingpeng Guo [4,5]

[1] School of Chemistry and Chemical Engineering, Huazhong University of Science and Technology, Wuhan 430074, China; fylz1989@hust.edu.cn (Z.L.); m201770328@hust.edu.cn (Z.P.); qikai@hust.edu.cn (K.Q.)

[2] Key Laboratory of Material Chemistry for Energy Conversion and Storage, Huazhong University of Science and Technology, Ministry of Education, Wuhan 430074, China

[3] Changqing Oil and Gas Technology Institute, Changqing Oil Field Company, Xi'an 710021, China; lhui2_cq@petrochina.com.cn

[4] Hubei Key Laboratory of Materials Chemistry and Service Failure, Wuhan 430074, China; guoxp@mail.hust.edu.cn

[5] School of Chemistry and Chemical Engineering, Guangzhou University, Guangzhou 510006, China

* Correspondence: qiuyubin@mail.hust.edu.cn; Tel.: +86-13-545-266-622

Received: 5 August 2020; Accepted: 27 August 2020; Published: 30 August 2020

Abstract: In this work, the effects of the microstructure and phase constitution of cast magnesium alloy ZK60 (Mg-5.8Zn-0.57Zr, element concentration in wt.%) on the corrosion behavior in aqueous NaCl (0.1 mol dm^{-3}) were investigated by weight-loss measurements, hydrogen evolution tests, and electrochemical techniques. The alloy was found to be composed of α-Mg matrix, with large second-phase particles of $MgZn_2$ deposited along grain boundaries and a Zr-rich region in the central area of the grains. The large second-phase particles and the Zr-rich regions were more stable than the Mg matrix, resulting in a strong micro-galvanic effect. A filiform corrosion was found. It originated from the second-phase particles in the grain boundary regions in the early corrosion period. The filaments gradually occupied most areas of the alloy surface, and the general corrosion rate decreased significantly. Corrosion pits were developed under filaments. The pit growth rate decreased over time; however, it was about eight times larger than the general corrosion rate. A schematic model is presented to illustrate the corrosion mechanism.

Keywords: ZK60 magnesium alloys; microstructure; filiform-like corrosion; corrosion pit

1. Introduction

Magnesium (Mg) alloys have been widely applied as lightweight engineering materials due to their unique properties [1–8]. As commercial Mg-Zn-based alloys, ZK60 (Mg–Zn–Zr) alloys [9] have attracted great interest from researchers due to their high strength [10–12]. The microstructure [13,14], mechanical properties [15–17], and biological applications [18,19] of ZK60 alloys have been studied widely in recent decades. It has been verified that microstructure evolution is essential for the mechanical properties of ZK60 alloys, grain refinement, and stable precipitates, having vital effects on improving the mechanical properties [20–23]. Nevertheless, the weak corrosion resistance of ZK60 alloys limits their further applications.

The microstructure of Mg alloys, especially their second phases, has an evident impact on their corrosion behavior [3,5]. The second phases of Mg alloys may have a dual role in their corrosion, i.e., a galvanic acceleration effect or a corrosion blocking effect, depending on their quantities and distribution [24–26]. In Mg-Al alloys, when the amount of aluminum (Al) is low (e.g., Mg-5Al), the β-phase ($Mg_{12}Al_{17}$) is relatively discontinuous in the Mg matrix and mainly acts as a cathode phase to accelerate the dissolution of the matrix. As the content of the Al element increases (e.g., Mg-10Al), the β-phase precipitates are tiny and continuously distributed along the grain boundaries, producing a barrier to prevent corrosion [27]. The similar effect of the second phase was reported in other Mg alloys [28,29]. However, the effect of the second phase in ZK60 alloys on their corrosion behaviors is barely reported. Some researchers tried to enhance the corrosion resistance of ZK60 alloys by modifying their microstructure through heat treatment [30,31], deformation processing [32], and alloying [33–36]. Even so, the relationship between the microstructure and corrosion behavior of ZK60 alloys needs more studies for it to be investigated.

Some studies reported the corrosion behavior of ZK60 alloys on different occasions. Cheng et al. [37] pointed out that the Zr element in ZK60 alloys refined the grain and purified the alloy composition, which could improve its corrosion resistance in 1 M NaCl. Zeng et al. [38] investigated the effects of the microstructure and concentration of NaCl (3.5 and 5.0 wt.%) on the corrosion behavior of an extruded ZK60 alloy. They found that an increase in the grain size of the ZK60 alloy accelerated its corrosion rate. The alloy microstructure played a crucial role in the pitting and intergranular corrosion. Xu et al. [39] reported that the corrosion rate of a cast ZK60 alloy decreased with the immersion time in solutions containing 3.5 wt.% NaCl, NaBr, and NaI, while it displayed passivation in 3.5 wt.% NaF solution. Apart from the above reports, some studies focused on the biodegradation behavior of ZK60 alloys in Hank's solution, Ringer's solution, simulated body fluid, and artificial urine for biomedical applications [40–43]. The biodegradable property of the ZK60 alloys is the interesting issue in these investigations. In general, the above research usually concentrated on the uniform corrosion of ZK60 alloys, and little attention was paid to the development of their local corrosion. The influence of microstructure on the local corrosion of ZK60 alloys is still not clearly understood, especially the effect of the second phase.

In this study, a commercial cast ZK60 alloy was selected as the test material. Its microstructure was characterized by X-ray diffraction (XRD), scanning electron microscopy (SEM), energy-dispersive X-ray spectroscopy (EDX), and scanning Kelvin probe force microscopy (SKPFM) analysis. The general and local corrosion of the alloy in 0.1 M NaCl was investigated using weight loss tests, hydrogen evolution tests, and electrochemical measurements, as well as corrosion morphology monitoring with an optical microscope and SEM. The effect of the microstructure of the cast ZK60 alloy, especially the second phase and the distribution of alloying elements, on the corrosion initiation and developmental features of the alloy were investigated, and the mechanisms involved were studied. This work will help to verify the relationship between the microstructure and the corrosion behavior of ZK60 alloys. Moreover, it may also provide a theoretical basis for improving the corrosion resistance of ZK60 alloys by adjusting the microstructure in future research.

2. Materials and Methods

2.1. Test Material and Solution

A commercial as-cast ZK60 alloy was used in this study. Table 1 presents its chemical composition, analyzed by inductively coupled plasma-atomic emission spectrometry (ICP-AES, SPECTRO, Kleve, Germany). All the solutions used in this work were prepared with analytical-grade reagents and distilled water. The test solution was 0.1 M NaCl under an open-air condition, which was controlled at 25 ± 1 °C with a water bath.

Table 1. Chemical composition of the cast ZK60 alloy (wt.%).

Zn	Al	Fe	Ni	Cu	Zr	Mg
5.80	<0.01	<0.01	<0.01	<0.01	0.57	Bal.

2.2. Microstructure Characterization

The cast ZK60 samples ($10 \times 10 \times 10$ mm^3) were sealed with epoxy resin (working area = 1.0 cm^2), ground with 2000 grit SiC paper, and polished with 3 µm diamond paste. Then, they were etched with a picric acid solution for metallographic analysis. The metallographic structure of the cast ZK60 sample was observed using a 3D optical microscope (VHX-1000, KEYENCE, Osaka, Japan), SEM (Quanta 200, FEI, Eindhoven, The Netherlands) equipped with EDX (EDAX-Genesis), and transmission electron microscopy (TEM). The TEM sample was firstly mechanically ground to a thickness of about 20 µm, and then, it was ion milled at 4 keV and 4°, cooled by liquid nitrogen. TEM observations were carried out with an FEI Talos F200X transmission electron microscope (FEI, Portland, OR, USA) operated at 200 kV. The element content and elemental distribution of the cast ZK60 sample were characterized by EDX and an electron probe micro-analyzer (EPMA-8050G, SHIMADZU, Kyoto, Japan), respectively. Phase structure analysis was performed by XRD (X'Pert PRO, PANalytical B.V., Almelo, The Netherlands) using Cu Kα radiation. The scan range of 2θ was from 20° to 90° with a scan step of 0.02°. The XRD pattern was analyzed with the X'Pert HighScore Plus software (2.0, PANalytical B.V., Almelo, The Netherlands). SKPFM (SPM-9700, SHIMADZU, Kyoto, Japan) was used to measure the relative Volta potential differences among different microstructural constituents to show their relative nobility. Meanwhile, the corresponding topography maps of the same area were also obtained.

2.3. Electrochemical Tests

All the electrochemical tests were carried out using a CS 350 Corrtest electrochemical workstation (Wuhan Corrtest, Wuhan, China). The working electrode ($10 \times 10 \times 10$ mm^3) was sealed with epoxy resin (working area = 1.0 cm^2), which was ground with 2000 grit SiC paper and rinsed in distilled water and ethanol. A saturated calomel electrode (SCE) and a platinum (Pt) electrode were used as the reference electrode and the counter electrode, respectively. Polarisation curves were generated and electrochemical impedance spectroscopy (EIS) was performed, respectively, at different corrosion times. At free corrosion potentials (E_{corr}), the EIS tests were performed with an AC voltage amplitude of 10 mV in the frequency range of 100 kHz–0.05 Hz. The EIS results were fitted with the Zview2.0 software. The polarization curves were generated at the scan rate of 0.5 mV s^{-1}, scanning towards the positive direction. The corrosion current density (i_{corr}, mA cm^{-2}) was estimated by the cathodic Tafel extrapolation method, according to [5,44]. The corresponding corrosion rate (P_i, mm y^{-1}) was converted by the equation [5,45]:

$$P_i = 22.06 i_{corr} \tag{1}$$

where the corrosion current density i_{corr} is estimated by the Tafel extrapolation of the cathodic branch of the polarization curves, and P_i is related to the average corrosion rate.

All the electrochemical tests were repeated at least three times in this study.

2.4. Weight Loss Tests

The cast ZK60 specimens ($20 \times 20 \times 4$ mm^3) were ground with 2000 grit SiC paper, rinsed in distilled water and ethanol, dried with cold air, and kept in a vacuum desiccator before the weight loss test. The specimens were corroded in 0.1 M NaCl for different times and then immersed in a CrO$_3$ solution (180 g L^{-1}, ~25 °C) for 10 min to remove corrosion products. At least three parallel tests were performed under each test condition. The average weight loss rate of the cast ZK60 alloy (ΔW, mg cm^{-2} h^{-1}) can be converted to a general corrosion rate P$_w$ (mm y^{-1}) using [5,44]:

$$P_w = 3.6524 \Delta W / \rho \tag{2}$$

where ρ is the metal density (g cm^{-3}). For the cast ZK60 alloy, $\rho = 1.8$ g cm^{-3}; thus, Equation (2) becomes:

$$P_w = 48.67 \Delta W \tag{3}$$

2.5. Hydrogen Evolution Tests

Following the work of Shi [46], plug-in specimens were employed to perform the hydrogen evolution test in 0.1 M NaCl at room temperature (~25 °C). Figure 1 presents the schematics of the test system. The cast ZK60 specimens (10 × 10 × 10 mm^3) were treated as the above weight loss samples. The evolved hydrogen was collected into a burette, and its volume (V_H, mL cm^{-2}) was recorded at different times. The hydrogen evolution rate, v_H (mL·cm^{-2} day^{-1}), can also be converted to a general corrosion rate, P_H (mm y^{-1}), using [5,44].

$$P_H = 2.2\, v_H \tag{4}$$

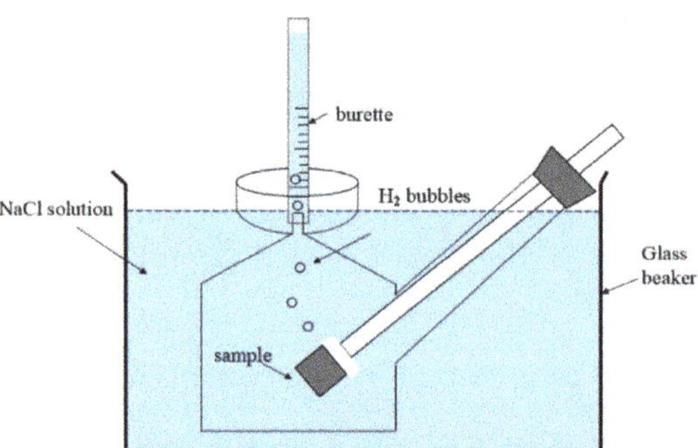

Figure 1. Schematic diagram of the test system for the hydrogen evolution measurement.

2.6. Corrosion Morphology Characterization

The etched metallographic specimens were placed in 0.1 M NaCl, where the solution thickness on the sample surface was about 1 mm, to observe the corrosion development in the initial period (0–1 h) in situ using a 3D optical microscope (KEYENCE VHX-1000). The corrosion morphologies of the cast ZK60 samples corroded in 0.1 M NaCl for different times were measured by SEM and with the 3D optical microscope. The cross-section corrosion morphologies and the elemental distributions were analyzed by EPMA (EPMA-8050G).

3. Results and Discussion

3.1. Microstructure of the Cast ZK60 Alloy

Figure 2 shows the optical and the back-scattered electron (BSE) SEM micrographs of the cast ZK60 alloy. The microstructure of the cast ZK60 alloy was composed of an α-Mg phase and large second-phase particles, which were mainly deposited along the grain boundaries (Figure 2a,b). According to [47,48], the main components of these particles are Mg and Zn and may be MgZn$_2$ and MgZn. The XRD pattern of the cast ZK60 alloy in Figure 3 only displays the presence of MgZn$_2$. No discernable diffraction peaks from MgZn were detected in this work, which suggests no MgZn phase in the studied alloy; also, there is no Zr detected, maybe due to its low content in the test sample. However, the XRD patterns cannot give accurate structural information for the second phase. Therefore, the crystal structures of the abovementioned second phases were characterized using TEM in detail as follows.

To further verify the second phase in the as-cast ZK60 alloy, Figure 4 presents the TEM micrograph of the as-cast ZK60 alloy. Block-shaped and globular second-phase particles can be observed in Figure 4. The blocky phase has a size of about 500 nm, and the globular phase is about 100 nm. Figure 4 also presents the corresponding selected area electron diffraction (SAED) patterns of the second phase. The SAED patterns proved that the two second-phase particles were MgZn$_2$ [49]. No MgZn phase was found in the as-cast ZK60 alloy.

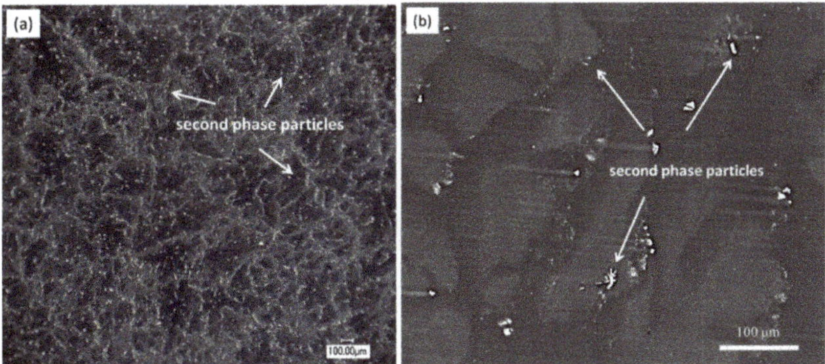

Figure 2. (**a**) Optical and (**b**) BSE-SEM micrographs of the cast ZK60 alloy.

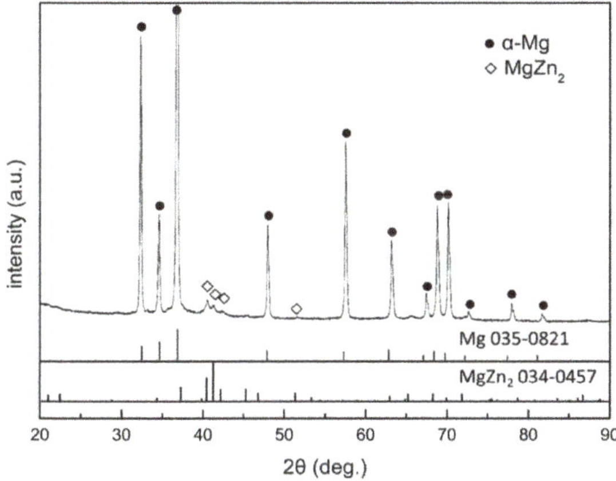

Figure 3. XRD pattern of the cast ZK60 alloy.

The BSE-SEM micrograph in Figure 2b indicates the nonuniform distribution of the alloying elements in the cast ZK60 because the brighter areas contain more elements of higher atomic weight than the darker regions [50]. Figure 5 presents the area distribution of the alloying elements in the cast ZK60, proving that the center area of the grains with light color was richer in Zr and Zn (Zr-rich region) than the neighboring darker zones (grain boundary region). Here, the "grain boundary region" of the as-cast ZK60 is denoted as the areas between the Zr-rich region, i.e., the dark areas in Figure 2b. This uneven distribution of the alloying elements in the cast ZK60 alloy may cause the inhomogeneous electrochemical activity resulting in the micro-galvanic corrosion [51].

Figure 6 presents the SKPFM maps of the cast ZK60 sample, which clearly show that the second phase had the highest potential. Furthermore, the Volta potential profiles (Figure 6c) along the line A and line B indicated that the central region of the grains (i.e., Zr-rich region) exhibited higher potential than those of the grain boundary regions, owing to the higher Zr and Zn contents in the center of grains and higher Mg content in the grain boundary regions (Figure 5). Thus, the second-phase particles and the central region of the grains should be more stable than the Mg matrix in grain boundary regions and more likely to become cathodes in the micro-galvanic cells. A second-phase particle occurred in the grain area (Figure 6a), which is consistent with Figure 2a, so the micro-galvanic corrosion may also have been initiated in the grains.

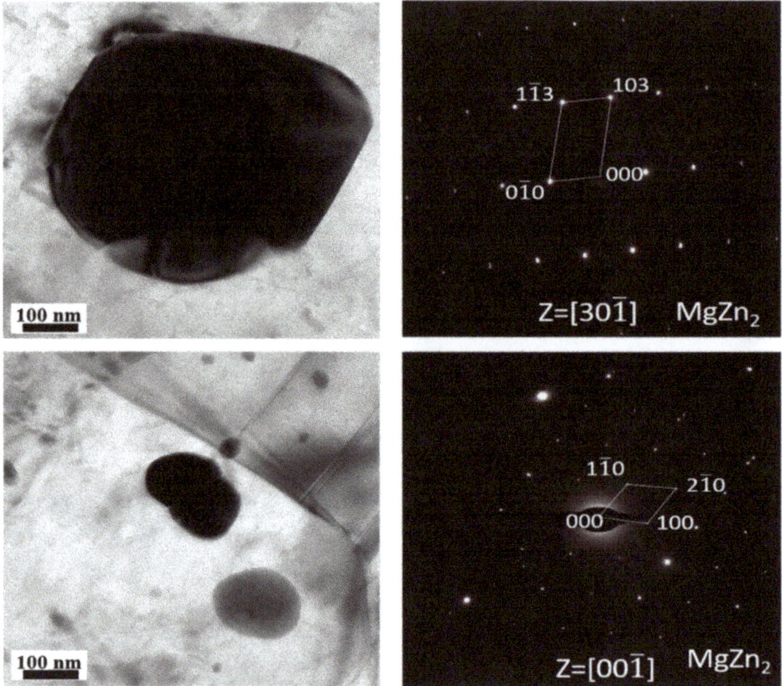

Figure 4. TEM micrograph of the cast ZK60 alloy and the corresponding selected area electron diffraction (SAED) patterns of the second-phase particles.

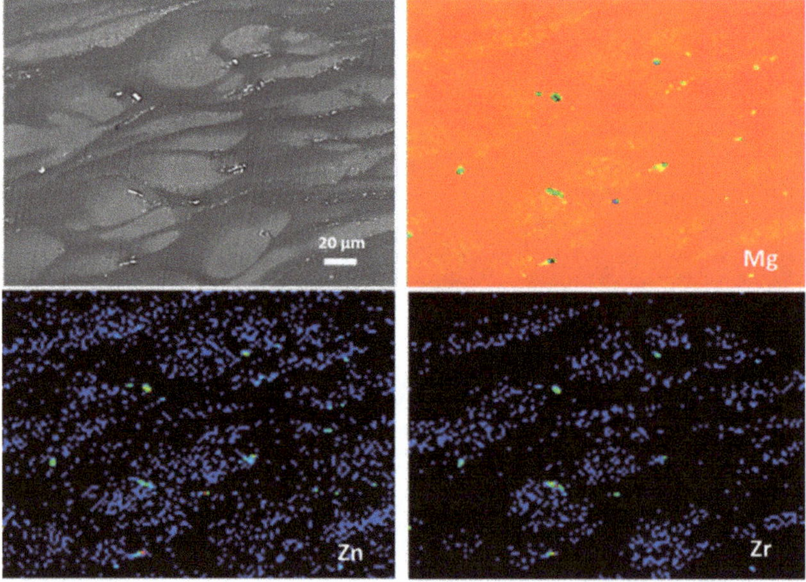

Figure 5. BSE-SEM image of the cast ZK60 alloy and the corresponding area distributions of Mg, Zn, and Zr.

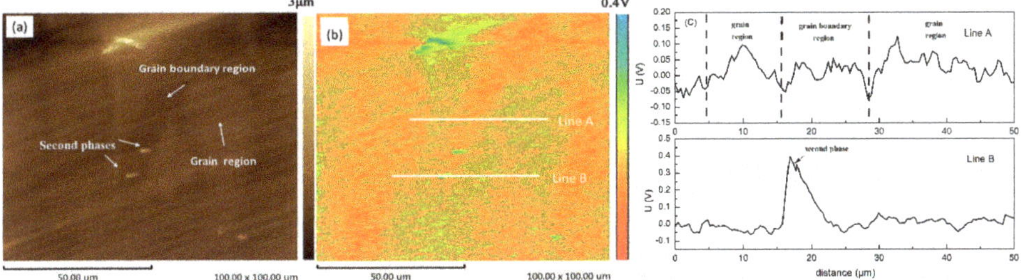

Figure 6. (**a**) Scanning Kelvin probe force microscopy (SKPFM) topography map; (**b**) Potential map of the same area; (**c**) Volta potential profiles along lines A and B in the SKPFM image.

3.2. Weight Loss Tests

Figure 7 shows the ΔW (mg cm^{-2} h^{-1}) and P_w (mm y^{-1}) values of the cast ZK60 alloy immersed in 0.1 M NaCl for different times (t). The P_w at 24 h was the largest (3.5 ± 0.1 mm y^{-1}), which is similar to that of a cast ZK60 in 0.9% NaCl (4.6 ± 0.6 mm y^{-1}) reported by Merson et al. [52]. Then, it gradually decreased by approximately half to a relatively stable value (1.4~1.7 mm y^{-1}) over 72–96 h. The decrease in P_w over 72–96 h may have been related to the increase in the corrosion product layer Mg(OH)$_2$ on its surface [53].

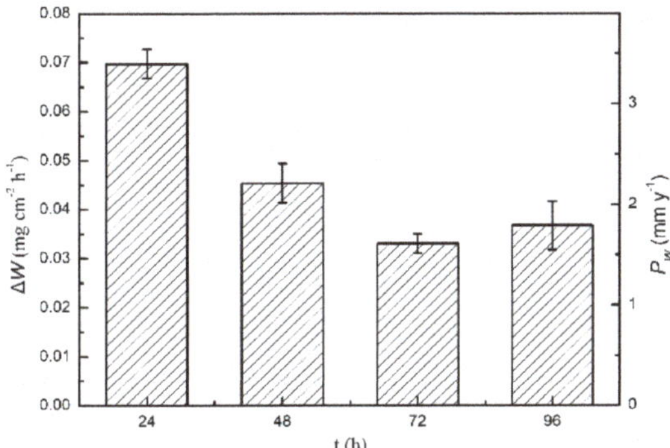

Figure 7. Corrosion rates (ΔW and P_W) of the cast ZK60 alloy after immersion in 0.1M NaCl for different times measured by weight loss tests.

3.3. Hydrogen Evolution Tests

Figure 8 presents the hydrogen evolution test results for the cast ZK60 alloy to show the change in its corrosion rate in detail, in which v_H is the differentiation of the V_H-t curve in Figure 8a and P_H is calculated by Equation (4). The V_H at 24 h (Figure 8a) was about 0.7 ± 0.03 mL cm^{-2}, which is higher than that of an extruded ZK60 (~0.5 mL cm^{-2}) [54]. The P_H value (Figure 8b) increased with time in the initial corrosion period (0–1.5 h) and then decreased over 1.5–48 h; at last, it increased again over 48–72 h. These results imply that there exist different corrosion stages in the periods of 0–2 h, 2–48 h, and 48–72 h, which may be related to the change in the alloy surface condition. According to Song [24], the total volume of hydrogen collected should equal the total amount of metal lost, and both the weight loss and hydrogen evolution tests were reliable methods. Even though the hydrogen evolution rate P_H we measured is slightly lower than the weight loss corrosion rate P_w, which may be caused by the difference in the test methods, they generally show similar change tendencies.

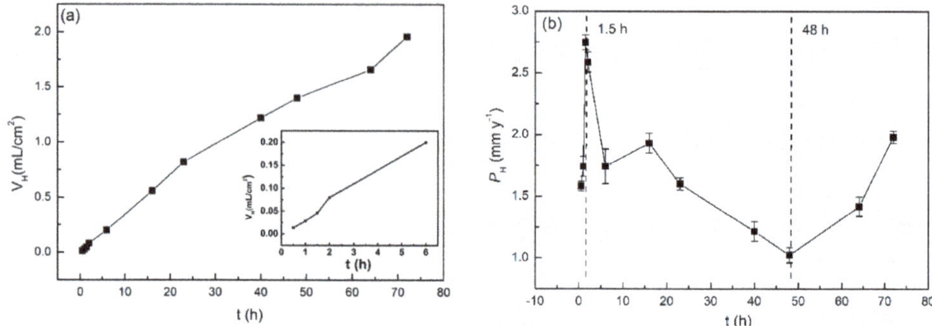

Figure 8. Changes in (**a**) V_H and (**b**) P_H of the cast ZK60 alloy as a function of time measured by the hydrogen evolution test in 0.1M NaCl.

3.4. Polarisation Curve Measurements

Figure 9 presents the polarization curves of the cast ZK60 alloy after immersion in 0.1 M NaCl for different times and the changes in its corrosion rates (i_{corr} and P_i) with time. All the polarization curves showed the typical features of activation-controlled processes [55,56]. A so-called breakdown potential (E_{break}) occurred in the anodic polarization curves (except t = 2 h), owing to the breakdown of the oxide film on the alloy [57], as shown in Figure 9b. E_{corr} (Figure 9a) and E_{break} (Figure 9) moved positively in the period of 0–24 h and then became negative again, while P_i and i_{corr} (Figure 9b) displayed a similar change with corrosion time to that of P_H as shown in Figure 8b, which is consistent with previous reports [38,58]. The i_{corr} and P_i values in Figure 9b are much smaller than the P_w (Figure 7) and P_H (Figure 8b) values. Similar results from other Mg alloys have been discussed in detail in [44]. However, they displayed a similar change tendency. All these results also suggest that there may be different corrosion stages in the corrosion periods of 0–2 h, 2–24 h, and 24–72 h.

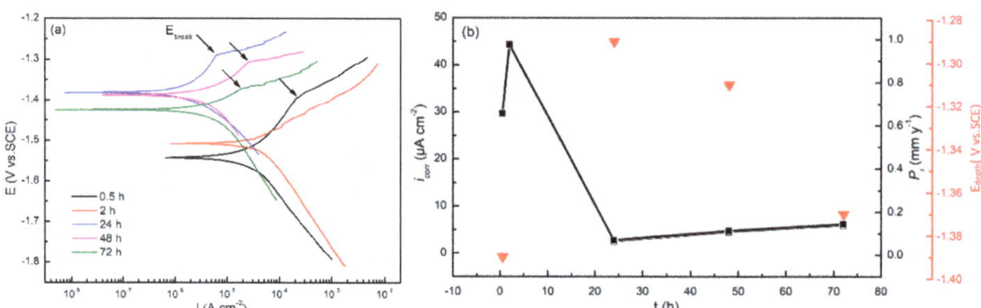

Figure 9. (**a**) Polarization curves of the cast ZK60 alloy immersed in 0.1 M NaCl for different times; (**b**) Changes in the i_{corr}, P_i, and E_{break} with time (25 °C).

3.5. EIS Measurements

Figure 10 presents the EIS of the cast ZK60 alloy immersed in 0.1 M NaCl for different times and the change in R_p (polarization resistance) as a function of time. All the Nyquist plots display two capacitive loops in the high-frequency region and an inductive loop in the low-frequency region, which is similar to the EIS features of other Mg alloys [59]. The capacitive loops are related to the processes in the surface film and the electrical double layer [60], and the inductive loop should be attributed to the initiation of the localized corrosion on the surface of the cast ZK60 alloy according to [61–63]. Based on these EIS features, Figure 11 presents an equivalent circuit to fit the EIS results in Figure 10a [58,59,64,65]. R_s is the solution resistance. CPE_f and CPE_{dl} represent the constant phase elements (CPE) for the surface film and the electrical double layer, respectively. R_f and R_{ct} represent the surface film resistance and charge-transfer resistance, respectively. R_L and L represent equivalent resistance and inductance to describe the low-frequency inductance. It should be noted that R_{ct} is the parallel of the charge-transfer resistance of the anodic process and the cathodic process ($R_{ct,a}$ and $R_{ct,c}$) at E_{corr} [63]. The fitting curves are also displayed in Figure 10a, and the fitting parameters are listed in Table 2.

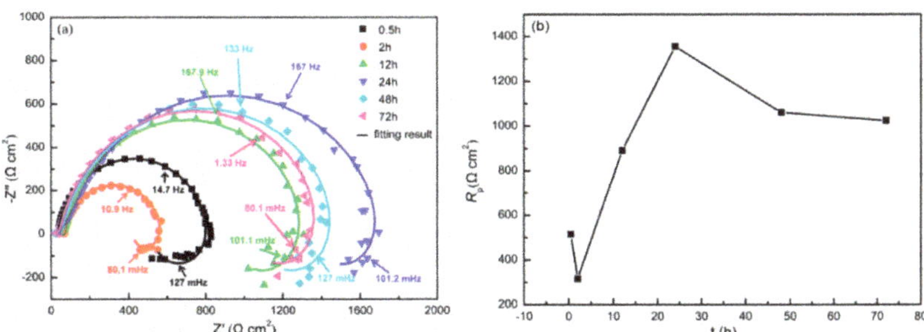

Figure 10. (**a**) Nyquist plots measured at E_{corr} for the cast ZK60 alloy immersed in 0.1 M NaCl for different times; (**b**) R_p–t curves at 25 °C.

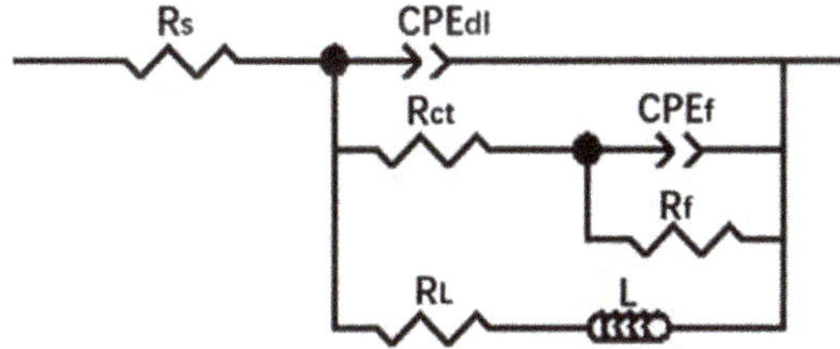

Figure 11. Equivalent circuit used to model the electrochemical impedance spectroscopy (EIS) response of the cast ZK60 alloy in 0.1 M NaCl.

Table 2. Fitting parameters of the EIS results in Figure 9a.

Time h	R_s $\Omega\cdot cm^2$	R_{ct} $\Omega\cdot cm^2$	CPE_{dl}-T $\mu Fcm^{-2}Hz^{1-n1}$	CPE_{dl}-P n1	R_f $\Omega\cdot cm^2$	CPE_f-T $\mu Fcm^{-2}Hz^{1-n2}$	CPE_f-P n2	R_L $\Omega\cdot cm^2$	L $H\cdot cm^2$	Chi-Squared Error
0.5	32	26	22	0.87	759	16	0.69	1360	2359	1.55×10^{-2}
2	65	14	67	0.87	502	21	0.96	1224	1143	1.02×10^{-2}
12	47	37	26	0.85	1254	50	0.89	2686	8005	1.33×10^{-2}
24	40	179	38	0.79	1546	22	0.88	5865	16,588	1.23×10^{-2}
48	27	58	37	0.81	1453	48	0.87	3429	9781	1.48×10^{-2}
72	28	34	38	0.83	1390	59	0.88	3557	9029	1.67×10^{-2}

According to [59,66,67] and the results in Table 2, the fitting curves can be extrapolated to the zero-frequency limit to obtain the polarization resistance (R_p) at different times. Therefore, the R_p–t curve was obtained as shown in Figure 10b. R_p decreased with time over 0–2 h, indicating corrosion acceleration, and then, R_p clearly increased between 2 and 24 h before decreasing slightly over 24–72 h. In this case, the change in R_p with t is consistent with that of P_i (Figure 9b). The R_t and R_f values in Table 2 show the same change tendencies as those for R_p. The change in the CPE_{dl}, CPE_f, R_L, and L with t should be closely related to the surface layer of the cast ZK60 alloy [59]. The specific changes in these EIS results are discussed later.

3.6. Corrosion Morphology of the Cast ZK60 Alloy

To observe the real-time corrosion development in the initial period (0–1 h), Figure 12 presents the in situ corrosion images of the etched sample after immersion in 0.1 M NaCl for different times. In the initial period (20 min, Figure 12a), there was no apparent corrosion, with a limited number of H_2 bubbles observed on the surface, implying a slight corrosion state. This slight corrosion should be related to the protective film on the alloy surface impeding the corrosion in the initial period [68]. Over 40–60 min (Figure 12b,c), some dark threads appeared on the surface, which lengthened with time and displayed filiform-like corrosion characteristics. In this period, H_2 bubbles constantly evolved at or near the leading edges of the dark threads, leaving cloudy trails behind them. The magnified picture in Figure 12d suggests that the corrosion filaments seem to traverse within the grain boundary regions and do not extend toward the center of the grains in the initial stage of expansion, which is further verified later. These corrosion characteristics are consistent with those of other Mg alloys in similar corrosion media and should be related to their microstructure features [50,65,69].

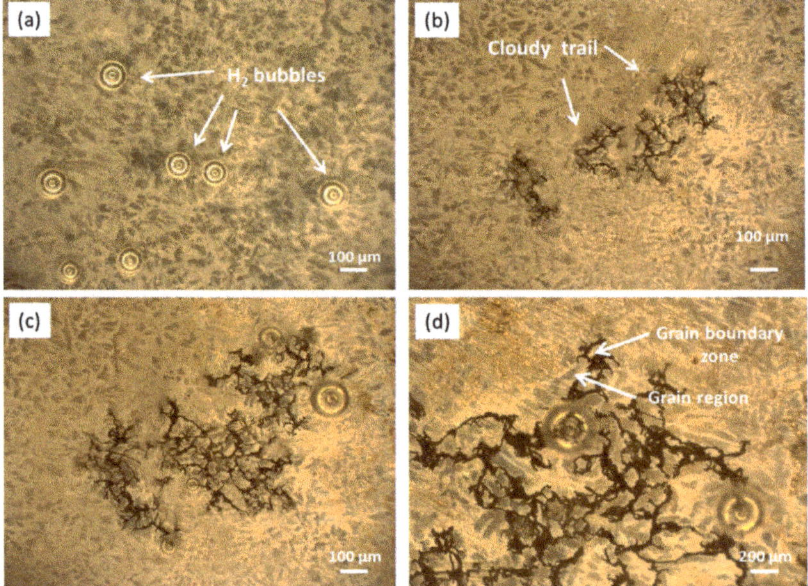

Figure 12. Optical in situ corrosion images of the cast ZK60 etched sample immersed in 0.1 M NaCl for different times: (**a**) 20 min; (**b**) 40 min; (**c,d**) 60 min.

Figure 13 presents the BSE-SEM micrographs of the etched samples after immersion in 0.1 M NaCl for 1 and 2 h. The black corrosion filaments (Figure 13a) covered by $Mg(OH)_2$ [70] mainly occurred on the "darker zones" in grain boundary regions (see Figures 2b and 5) and encompassed the second-phase particles. Some second-phase particles in the boundary regions (Figure 13b) were surrounded by a small number of black corrosion products, implying that the corrosion filaments may have started from the surrounding areas of these second-phase particles. Most of the central regions of the grains, i.e., the Zr-rich areas with a light color (Figures 2b and 5), were still uncorroded. When t = 2 h (Figure 13c), only a small number of second-phase particles occurred in the corroded areas, implying that most of them were removed due to the dissolution of their surrounding Mg matrix. Some broad corrosion areas in Figure 13 show that the corrosion could develop to the central area of grains with increasing corrosion time. Moreover, the corrosion filaments in Figure 13c,d show an apparent corrosion depth, implying that the corrosion also developed under the black corrosion product layer.

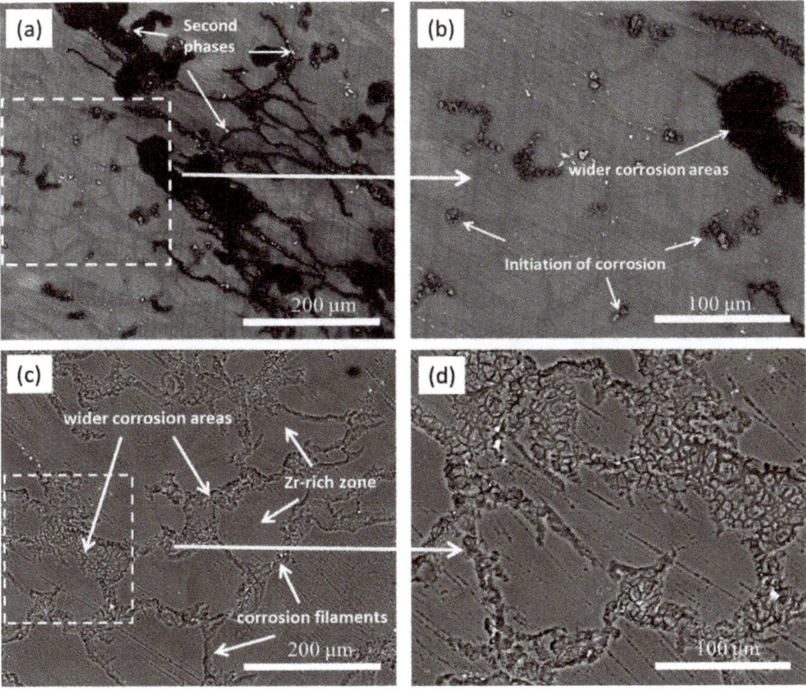

Figure 13. BSE-SEM micrographs of the cast ZK60 etched samples immersed in 0.1 M NaCl for different times: (**a**,**b**) 1 h (with corrosion products); (**c**,**d**) 2 h (without corrosion products).

As shown in Figures 12 and 13, the etched samples were used to observe the origination and developmental features of the corrosion filaments on the cast ZK60. To avoid the influence of the etching treatment on the corrosion process, Figure 14 presents the secondary electron (SE) SEM images of the raw alloy sample after immersion in 0.1 M NaCl for 0.5 and 2 h. The corrosion morphologies in Figure 14a,b also exhibit the characteristics of filiform corrosion, similar to those in Figure 12b. The polishing scratches on the surface of the test samples do not appear to influence the initiation and extension of the corrosion filaments, suggesting that they are controlled by the microstructure of the cast ZK60 alloy. When t = 2 h, the corrosion images in Figure 14c,d display features similar to those in Figure 13c and some small corrosion pits occur in the corrosion areas, which may have been the result of the loss of the second phase-particles. The raw cast ZK60 sample also displayed the filiform corrosion features in the early corrosion period (0–2 h).

Figure 15 presents the secondary electron (SE) SEM images of the cast ZK60 alloy after immersion in 0.1 M NaCl for different times to observe the corrosion development over a long period (12–72 h). The corrosion filaments gradually extended to the whole surface of the test samples after 24 h, and the number and depth of the corrosion pits increased with time. These results further prove that with increasing corrosion time, the corrosion of the cast ZK60 alloy can develop toward the central area of grains and toward the depth of the alloy matrix, which is discussed below.

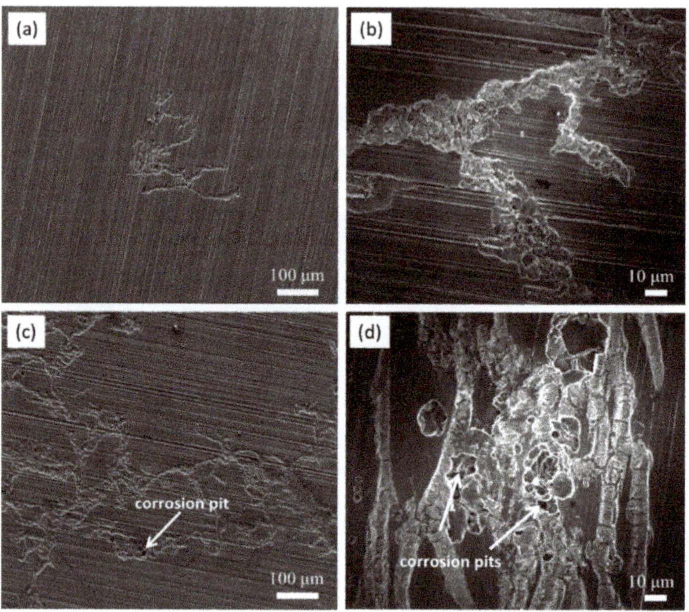

Figure 14. SE-SEM micrographs of the cast ZK60 sample immersed in 0.1 M NaCl for different times: (**a**,**b**) 30 min; (**c**,**d**) 2 h (without corrosion products).

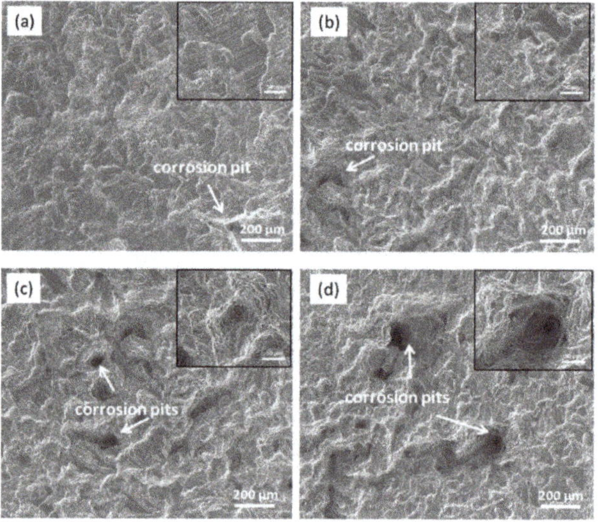

Figure 15. SE-SEM micrographs of the cast ZK60 alloy immersed in 0.1 M NaCl for different times: (**a**) 12 h; (**b**) 24 h; (**c**) 48 h; (**d**) 72 h (with the corrosion products removed).

3.7. Characterization of the Corrosion Pits on the Cast ZK60 Alloy

To characterize the development of corrosion pits, the deepest pit on the cast ZK60 sample was selected after immersion in 0.1 M NaCl for different times, and their 3D morphologies and depth profiles were measured, as shown in Figure 16. The density and depth of the corrosion pits increased over 12–72 h, and the size of the pit mouth was generally larger than the pit depth, which is similar to the corrosion morphologies in Figure 15. The corresponding pit depth at different corrosion times was employed to calculate the average penetration rate (P_{depth}, mm y^{-1}), which is also presented in Figure 17. The pit depth increased with time, but the fastest P_{depth} occurred in the period of 12–24 h (~24 mm y^{-1}) before gradually decreasing to ~16.6 mm y^{-1} in the period of 24–72 h.

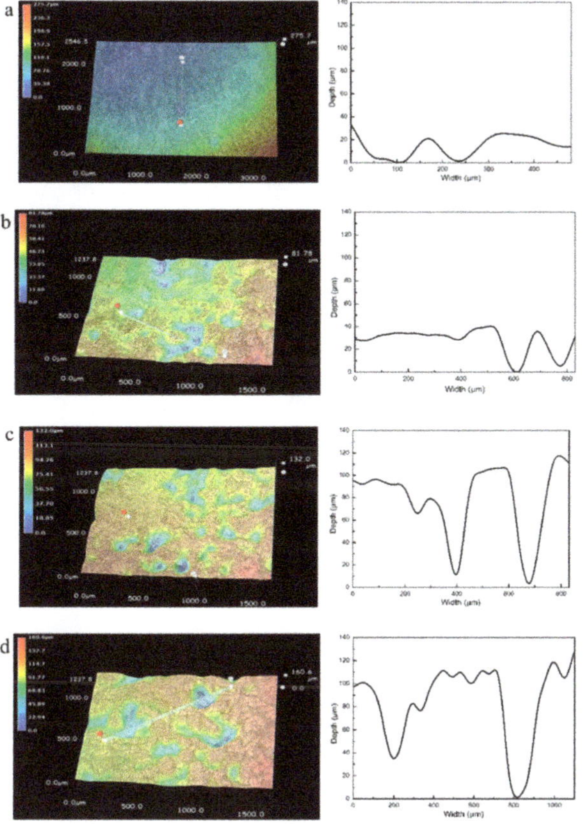

Figure 16. 3D corrosion images and depth profiles of the cast ZK60 alloy immersed in 0.1 M NaCl for different times: (**a**) 12 h; (**b**) 24 h; (**c**) 48 h; (**d**) 72 h (without corrosion products).

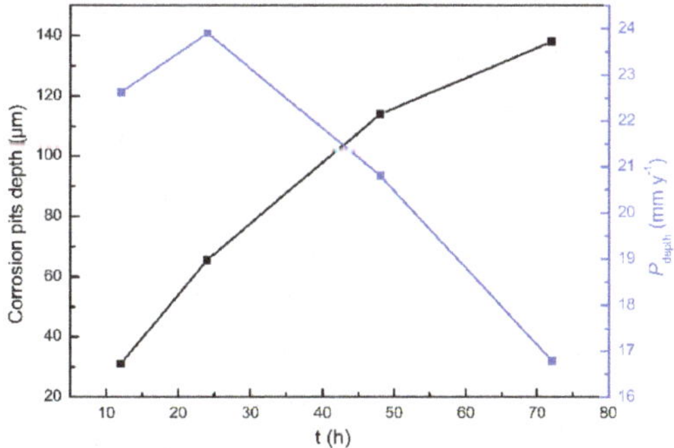

Figure 17. Depth of the deepest corrosion pit and the corresponding P_{depth} of the cast ZK60 alloy immersed in 0.1 M NaCl for different times.

Figure 18 is the cross-sectional EPMA pattern of a corrosion pit on the cast ZK60 sample immersed in 0.1 M NaCl for 72 h. The corrosion pit was covered by the corrosion product, and the second-phase particles were encompassed in it. Only Mg, O, and Cl elements were found in the corrosion product layer, which should result from $Mg(OH)_2$ and Cl^- in the test solution. Moreover, a large number of Cl^- through the whole corrosion product layer suggests that Cl^- could penetrate the product layer easily and may become enriched in the bottom of the corrosion pit to propagate the pit corrosion [71]. The second-phase particles in the corrosion pit imply that they may be related to the formation of corrosion pits, which is discussed later.

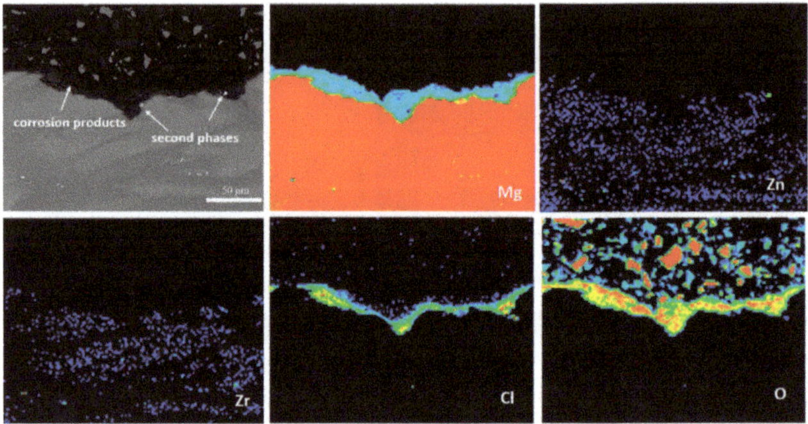

Figure 18. Electron probe micro-analyzer (EPMA) maps of the elemental distribution in the cross-section of a corrosion pit on the cast ZK60 alloy immersed in 0.1 M NaCl for 72 h.

4. Discussion

4.1. Initiation and Development of Corrosion on the Cast ZK60 Alloy

4.1.1. Corrosion Reactions on the Cast ZK60 Alloy

A thin oxide film can be formed on the surface of the cast ZK60 alloy in a moist atmosphere at room temperature according to Reactions (5) and (6), which may have a bilayer structure with an inner layer of MgO and an outer layer of $Mg(OH)_2$ [72,73].

$$2Mg + O_2 = 2MgO \tag{5}$$

$$MgO + H_2O = Mg(OH)_2 \tag{6}$$

Because MgO and $Mg(OH)_2$ are both relatively soluble in water according to Reactions (7) and (8) [3,74], where K_{sp} is the solubility product constant, MgO will be gradually dissolved and converted to $Mg(OH)_2$ when they are immersed in NaCl solution.

$$MgO + H_2O = Mg^{2+} + 2OH^- \quad K_{sp} = 10^{-6} \tag{7}$$

$$Mg(OH)_2 = Mg^{2+} + 2OH^- \quad K_{sp} = 10^{-11} \tag{8}$$

In this case, the $MgO/Mg(OH)_2$ oxide film formed on the cast ZK60 alloy was partly protective in a neutral NaCl solution. The anodic and cathodic partial reactions of the corrosion process can be written as Reactions (9) and (10), respectively, and the corrosion product is formed as Reaction (11) [75].

$$Mg = Mg^{2+} + 2e^- \tag{9}$$

$$2H_2O + 2e^- = H_2(g) + 2OH^- \tag{10}$$

$$Mg^{2+} + 2OH^- = Mg(OH)_2 \tag{11}$$

4.1.2. Initiation of Corrosion on the Cast ZK60 Alloy

Based on the microstructure of the cast ZK60 alloy, Figure 19 presents a schematic model to illustrate the initiation and development of the corrosion on the cast ZK60 alloy. As discussed above, a partly protective $MgO/Mg(OH)_2$ oxide film will be formed on the cast ZK60 alloy in neutral 0.1 M NaCl (Figure 19a). Because the large second-phase particles and the Zr-rich region in the grains were more stable than the Mg matrix in grain boundary regions, as shown in Figure 6, the oxide film in these regions should be more stable than that in their adjacent grain boundary regions (i.e., darker areas in Figure 2b), which was denoted as a stable and active oxide film in Figure 19a [76,77].

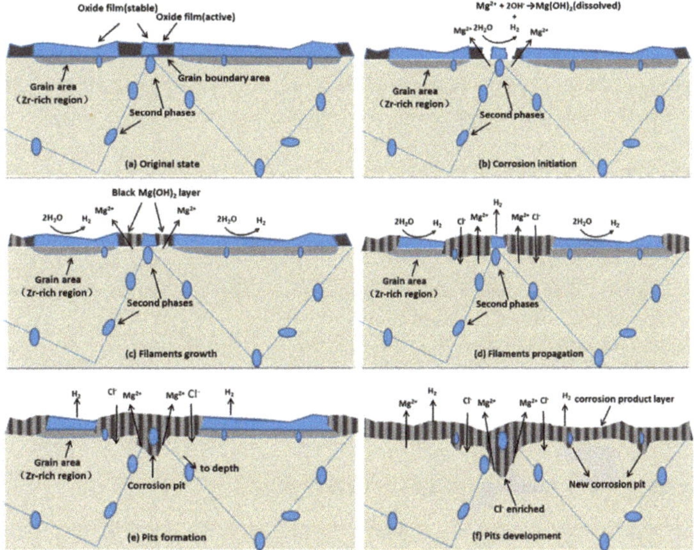

Figure 19. Schematic model for the initiation and development of the corrosion on the cast ZK60 alloy in NaCl solution. (**a**) original state; (**b**) corrosion initiation; (**c**) filaments growth; (**d**) filaments propagation; (**e**) pit formation and (**f**) pit development.

Compared to the Mg matrix in the grain boundary regions, those with stable oxide film act as micro-galvanic cathodes. Under the attack of Cl^-, the film around the second phase could be easily broken due to the higher numbers of imperfections [78]. Therefore, the micro-galvanic corrosion was initiated in the areas around the large discontinuous second-phase particles in the grain boundaries, owing to the strong galvanic effect between these second-phase particles and the Mg matrix in the grain boundaries (Figure 13b). It should be noted that there were some second-phase particles deposited in the grains (Figures 2a and 6), and therefore, the corrosion may also be initiated in the grains. However, we did not observe this phenomenon, which may be due to the stable oxide film in the grains (Zr-rich region) and the light galvanic effect between them.

After the corrosion initiation (Figure 19b), the oxide film near the second-phase particles in the grain boundaries cracked first. However, there were no black $Mg(OH)_2$ products observed in this beginning period (Figure 12a), because they were dissolved in the NaCl solution (Equation (5)). After the concentration of $Mg(OH)_2$ reached saturation in the NaCl solution, black $Mg(OH)_2$ precipitation occurred (Figure 12b).

4.1.3. Growth of Corrosion Filaments on the Cast ZK60 Alloy

Because the Zr-rich area within the grains was more stable than the Mg matrix in grain boundary regions, the corrosion preferentially propagated in the grain boundary regions in the initial corrosion period (Figure 19c), displaying the characteristics of filiform-like corrosion, as shown in Figures 12 and 13. According to [69,70], the front edges of the corrosion filaments acted as intense anodes via Equation (6), while the dark tracks behind the anodes acted as locally activated cathodes, where cathodic Equation (7) occurred. In NaCl solution, Cl^- may be enriched around the anode areas to attack and break the oxide film [28]. By contrast, the activated cathodes behind the anodes provide the driving force for the development of the corrosion filaments. Then, the precipitates of $Mg(OH)_2$ cover the surface of the active cathodes behind the anodes to make them gradually turn into inert cathodes.

The second phase in the grain boundaries and the Zr-rich region within the grains have an essential influence on the origination and propagation of the corrosion filaments on the cast ZK60. These filiform-like corrosion characteristics of the cast ZK60 were quite different from those of Mg-3Zn and Mg-8Li alloys reported previously [65,71], in which the corrosion filaments seem not to be affected by the second phases and can originate from the center of the grains and spread to various directions. These differences should be ascribed to their different microstructures, especially the quantity and size of the second phase and the distribution of alloying elements in these alloys.

As the corrosion time was increased, the active "dark areas" along the grain boundaries were gradually covered with a black $Mg(OH)_2$ layer (Figure 19c) and potentially became cathodic areas [79]. This change would impede the anodic corrosion process in these areas and make their potential positive, reducing and even eliminating the potential difference between these areas and the Zr-rich regions. Therefore, the corrosion filaments could gradually develop to the central area of grains (Figure 19d) to form some broad corrosion areas, as shown in Figures 13 and 14. Additionlly, the black $Mg(OH)_2$ layer was porous and imperfect and could not prevent the entry of the NaCl solution (Figure 18). Thus, the anodic process still existed beneath the $Mg(OH)_2$ layer, developing an apparent corrosion depth under the corrosion filaments, as shown in Figures 13 and 14. When t > 24 h, most of the alloy surface was gradually covered with the $Mg(OH)_2$ layer (Figure 15), and finally, it would be covered completely (Figure 19e,f).

4.1.4. Development of Corrosion Pits on the Cast ZK60 Alloy

With the development of the corrosion filaments, the corrosion pits will be formed when the Mg matrix surrounding the large second-phase particles is corroded completely (Figure 19e). Some small corrosion pits occurred in the corrosion filaments when t = 2 h (Figure 14c,d), which may be due to the loss of the second-phase particles when the corrosion product was removed. The visible corrosion pits occurred after 12 h of immersion (Figure 15), and their number and depth increased with time over 24–72 h, in which P_{depth} was about one magnitude order higher than those corrosion rates (P_w, P_H, and P_i), as shown in Figures 15–17. The high P_{depth} of the cast ZK60 alloy indicates a severe localized corrosion state and warrants further investigation.

The development of the corrosion pits was related to the occluded environment under the corrosion filaments, covered by the second-phase particles and $Mg(OH)_2$ precipitate. The solution in the occluded corrosion pit would be alkalized to increase the pH to around 10.5 through Reaction (5). At the same time, Cl^- would migrate in due to the charge neutralization and become enriched at the bottom of the pit (Figure 18). These effects would accelerate the anodic process in the corrosion pits and make them develop deeply [80].

According to the above discussion, the microstructure of the cast ZK60 alloy had an essential influence on the corrosion initiation and propagation in NaCl solution. Thus, it is necessary to improve its corrosion resistance by microstructure modification, which will be studied later.

4.2. Change in the Corrosion Rate in Different Corrosion Periods

The P_H, P_i, and R_p values of the cast ZK60 alloy generally showed different change tendencies in different immersion periods (Figures 8, 9b and 10b). In the early corrosion period (0–2 h), the corrosion rates (P_H and P_i) increased with time (Figures 8b and 9b), while R_t, R_f, and R_p decreased (Table 2 and Figure 10b). In this period, the filiform corrosion was initiated and developed quickly on the cast ZK60 alloy without visible corrosion pits (Figures 12–14), corresponding to the states in Figure 19a–d. In this case, with the development of the corrosion filaments, the broken oxide film increased steadily to accelerate the anodic and cathodic processes, as shown in Figure 9, and made the corrosion rates increase. The increase in the broken oxide film may be the reason why no E_{break} occurred in the anodic polarization curve when t = 2 h (Figure 9). Furthermore, the increase in the broken oxide film enlarged the reactive areas on the cast ZK60 surface to increase CPE_{dl} and CPE_f, meanwhile decreasing R_L and L in this period (0–2 h, Table 2).

The second corrosion period was determined to be within 2–24 h, according to the above discussion about the corrosion development on the cast ZK60 alloy. In this period, the corrosion filaments gradually occupied most of the active oxide film, and a small number of corrosion pits occurred and developed quickly (Figures 15 and 16), which corresponds to the state in Figure 19e. Because most of the active oxide film was corroded and covered with a black $Mg(OH)_2$ precipitate, the anodic process, Reaction (9), was inhibited significantly to make E_{corr} more positive, and E_{break} occurs again with a more positive value (Figure 9). In this case, the corrosion rate (P_H and P_i) decreased significantly (Figures 8b and 9b), while R_t, R_f, and R_p increased (Table 2 and Figure 10b). The difference is that P_i and $1/R_p$ reached the lowest value at t = 24 h, but P_H decreased persistently until t = 48 h, while P_w as shown in Figure 7 displayed the highest value at t = 24 h, which may be ascribed to the differences among these test methods. In this period, the decrease in the broken oxide film may also result in a clear increase in R_L and L and decrease in CPE_{dl} (Table 2). Meanwhile, CPE_f firstly increased and then decreased, which may have been due to the increase in the area and thickness of the corrosion product layer. Even though very few corrosion pits occurred in this period (Figures 15 and 16), their P_{depth} increased significantly and reached a large value of about 24 mm y^{-1} (Figure 17), which was much larger than the general corrosion rates of P_w, P_H, and P_i in this period (Figures 7, 8 and 9b). Therefore, the pitting corrosion became more critical than the filiform corrosion in this period.

Over 24–72 h, most of the surface of the cast ZK60 was gradually covered with $Mg(OH)_2$ layer, but the number and depth of the corrosion pits increased (Figures 15–17), corresponding to the state in Figure 19f. Because of the increase in corrosion pits, the anodic process was accelerated again. In this case, the E_{break} and E_{corr} of the cast ZK60 alloy became negative, and P_i increased again (Figure 9), while R_t, R_f, and R_p decreased (Table 2 and Figure 10b). However, P_w and P_{depth} decreased with time in this period (Figures 7 and 17), indicating that the increase in the corrosion product layer still impeded the growth rate of the pitting depth and the general weight loss, possibly due to the alkalizing effect of the cathodic reaction and the low solubility of $Mg(OH)_2$. In this period, the increase in CPE_f may have been due to the dissolution of the corrosion product layer reducing its thickness; meanwhile, the decrease in R_L and L may have been related to the acceleration of the pitting corrosion process, similar to in the processes on passive metals [61]. In general, pitting corrosion became more severe in this period and dominated the corrosion behavior of the cast ZK60 alloy. In this case, P_{depth} should be adequate to describe the corrosion rate of the cast ZK60 alloy in this period.

5. Conclusions

1. The microstructure of the cast ZK60 alloy was composed of an α-Mg phase and large second-phase particles ($MgZn_2$), which mainly deposited along the grain boundaries, and a Zr-rich region existed in the central area of the grains. The grain boundaries and their adjacent regions (noted as grain boundary regions) had relatively higher Mg contents.

2. The second-phase particles and the central area of the grains (Zr-rich region) were more electrochemically stable than the grain boundary regions, resulting in a strong micro-galvanic effect between them.

3. The corrosion of the cast ZK60 alloy in 0.1 M NaCl solution originated from the areas around the second-phase particles in grain boundaries and firstly developed in the grain boundary regions, showing filiform-like corrosion characteristics owing to the strong micro-galvanic effect in its microstructure.

4. The general corrosion rate increased in the early corrosion period (about 0–2 h). Then, the black corrosion filaments covered with $Mg(OH)_2$ gradually occupied most of the alloy surface to inhibit the corrosion process and decrease its general corrosion rate. However, corrosion pits occurred under the corrosion filaments and had a high growth rate (P_{depth}) in this period. After 24 h, the number of corrosion pits increased, and P_{depth} decreased with time (17–24 mm y^{-1}) but was still eight times larger than the general corrosion rates (P_w, P_H, and P_i), which should be paid more attention. The microstructure of the cast ZK60 alloy has an essential influence on the initiation and development of its corrosion.

Author Contributions: Conceptualization, Z.L. and Y.Q.; methodology, Z.L.; software, Z.L., Z.P. and K.Q.; validation, Y.Q., K.Q. and X.G.; formal analysis, Z.L.; investigation, Z.L.; resources, Z.L., Z.P. and H.L.; data curation, Z.L., Z.P., and H.L.; writing—original draft preparation, Z.L.; writing—review and editing, K.Q., Y.Q., and X.G.; visualization, Z.L.; supervision, Y.Q.; funding acquisition, Y.Q. All authors have read and agreed to the published version of the manuscript.

Funding: This research received no external funding.

Acknowledgments: The authors are thankful for the analysis support of the Key Laboratory of Material Chemistry for Energy Conversion and Storage and the Analytical and Testing Center, Huazhong University of Science and Technology.

Conflicts of Interest: The authors declare no conflict of interest.

References

1. Xu, T.C.; Yang, Y.; Peng, X.D.; Song, J.F.; Pan, F.S. Overview of advancement and development trend on magnesium alloy. *J. Magnesium. Alloy* **2019**, *7*, 536–544. [CrossRef]

2. Zhang, L.L.; Zhang, Y.T.; Zhang, J.S.; Zhao, R.; Zhang, J.X.; Xu, C.X. Effect of alloyed mo on mechanical properties, biocorrosion and cytocompatibility of as-cast Mg-Zn-Y-Mn alloys. *Acta Metall. Sin.* **2020**, *33*, 500–513. [CrossRef]

3. Esmaily, M.; Svensson, J.E.; Fajardo, S. Fundamentals and advances in magnesium alloy corrosion. *Prog. Mater. Sci.* **2017**, *89*, 92–193. [CrossRef]

4. You, S.H.; Huang, Y.D.; Kainer, K.U.; Hort, N. Recent research and developments on wrought magnesium alloys. *J. Magnesium. Alloy* **2017**, *5*, 239–253. [CrossRef]

5. Atrens, A.; Song, G.L. Review of recent developments in the field of magnesium corrosion. *Adv. Eng. Mater.* **2015**, *17*, 400–453. [CrossRef]

6. Song, G.L.; Atrens, A. Recent insights into the mechanism of magnesium corrosion and research suggestions. *Adv. Eng. Mater.* **2007**, *9*, 177–183. [CrossRef]

7. Mordike, B.L.; Ebert, T. Magnesium properties-applications-potential. *Mater. Sci. Eng. A* **2001**, *302*, 37–45. [CrossRef]

8. Sezer, N.; Evis, Z.; Kayhan, S.M.; Tahmasebifar, A.; Koç, M. Review of magnesium-based biomaterials and their applications. *J. Magnesium. Alloy* **2018**, *6*, 23–43. [CrossRef]

9. Bettles, C.; Gibson, M.A. Current wrought magnesium alloys: Strengths and weaknesses. *JOM* **2005**, *57*, 46–49. [CrossRef]

10. Dharmendra, C.; Rao, K.P.; Jain, M.K.; Prasad, Y.V.R.K. Role of loading direction on compressive deformation behavior of extruded ZK60 alloy plate in a wide range of temperature. *J. Alloys Compd.* **2018**, *744*, 289–300. [CrossRef]

11. Torbati-Sarraf, S.A.; Sabbaghianrad, S.; Figueiredo, R.B.; Langdon, T.G. Orientation imaging microscopy and microhardness in a ZK60 magnesium alloy processed by high-pressure torsion. *J. Alloys Compd.* **2017**, *712*, 185–193. [CrossRef]

12. He, Y.B.; Pan, Q.L.; Qin, Y.J. Microstructure and mechanical properties of ZK60 alloy processed by two-step equal channel angular pressing. *J. Alloys Compd.* **2010**, *492*, 605–610. [CrossRef]

13. Yang, Y.; Wang, Z.; Jiang, L.H. Evolution of precipitates in ZK60 magnesium alloy during high strain rate deformation. *J. Alloys Compd.* **2017**, *705*, 566–571. [CrossRef]

14. Wang, C.; Luo, T.J.; Zhou, J.X.; Yang, Y.S. Effects of solution and quenching treatment on the residual stress in extruded ZK60 magnesium alloy. *Mater. Sci. Eng. A* **2018**, *722*, 14–19. [CrossRef]

15. Wang, W.; Zhang, W.C.; Chen, W.Z.; Cui, G.R.; Wang, E.D. Effect of initial texture on the bending behavior, microstructure and texture evolution of ZK60 magnesium alloy during the bending process. *J. Alloys Compd.* **2018**, *737*, 505–514. [CrossRef]

16. Liu, W.C.; Dong, J.; Zhang, P.; Yao, Z.Y.; Zhai, C.Q.; Ding, W.J. High cycle fatigue behavior of as-extruded ZK60 magnesium alloy. *J. Mater. Sci.* **2009**, *44*, 2916–2924. [CrossRef]

17. Lin, J.B.; Wang, Q.D.; Peng, L.M.; Roven, H.J. Microstructure and high tensile ductility of ZK60 magnesium alloy processed by cyclic extrusion and compression. *J. Alloys Compd.* **2009**, *476*, 441–445. [CrossRef]

18. Xia, K.D.; Pan, H.; Wang, T.L. Effect of Ca/P ratio on the structural and corrosion properties of biomimetic Ca–P coatings on ZK60 magnesium alloy. *Mater. Sci. Eng. C* **2017**, *72*, 676–681. [CrossRef]

19. Chen, J.X.; Tan, L.L.; Yang, K. Effect of heat treatment on mechanical and biodegradable properties of an extruded ZK60 alloy. *Biol. Mater.* **2017**, *2*, 19–26. [CrossRef]

20. Cho, J.H.; Han, S.H.; Jeong, H.T.; Choi, S.H. The effect of aging on mechanical properties and texture evolution of ZK60 alloys during warm compression. *J. Alloys Compd.* **2018**, *743*, 553–563. [CrossRef]

21. Torbati–Sarraf, S.A.; Langdon, T.G. Properties of a ZK60 magnesium alloy processed by high–pressure torsion. *J. Alloys Compd.* **2014**, *613*, 357–363. [CrossRef]

22. He, Y.B.; Pan, Q.L.; Qin, Y.J.; Liu, X.Y.; Li, W.B. Microstructure and mechanical properties of ultrafine grain ZK60 alloy processed by equal channel angular pressing. *J. Mater. Sci.* **2010**, *45*, 1655–1662. [CrossRef]

23. Liu, Z.; Xin, R.L.; Wu, X.; Liu, D.J.; Liu, Q. Improvement in the strength of friction-stir-welded ZK60 alloys via post-weld compression and aging treatment. *Mater. Sci. Eng. A* **2018**, *712*, 493–501. [CrossRef]

24. Song, G.L.; Atrens, A. Understanding magnesium corrosion–A framework for improved alloy performance. *Adv. Eng. Mater.* **2010**, *5*, 837–858. [CrossRef]

25. Song, G.L.; Atrens, A.; Dargusch, M. Influence of microstructure on the corrosion of diecast AZ91D. *Corros. Sci.* **1998**, *41*, 249–273. [CrossRef]

26. Song, G.L.; Atrens, A.; St John, D.; Li, Z. *Magnesium Alloys and Their Applications*; Wiley–VCH: Weinheinm, Germany, 2000; pp. 426–431.

27. Lunder, O.; Lein, J.E.; Aune, T.K.; Nisancioglu, K. The role of $Mg_{17}Al_{12}$ phase in the corrosion of Mg alloy AZ91. *Corrosion* **1989**, *45*, 741–748. [CrossRef]

28. Lafront, A.M.; Zhang, W.; Jin, S.; Tremblay, R.; Dube, D.; Ghali, E. Pitting corrosion of AZ91D and AJ62x magnesium alloys in alkaline chloride medium using electrochemical techniques. *Electrochim. Acta* **2015**, *51*, 489–501. [CrossRef]

29. Hara, N.; Kobayashi, Y.; Kagaya, D.; Akao, N. Formation and breakdown of surface films on magnesium and its alloys in aqueous solutions. *Corros. Sci.* **2007**, *49*, 166–175. [CrossRef]

30. Choi, H.Y.; Kim, W.J. Effect of thermal treatment on the bio-corrosion and mechanical properties of ultrafine-grained ZK60 magnesium alloy. *J. Mech. Behav. Biomed. Mater.* **2014**, *37*, 307–322. [CrossRef]

31. Li, X.; Jiang, J.H.; Zhao, Y.H.; Ma, A.B.; Wen, D.J.; Zhu, Y.T. Effect of equal-channel angular pressing and aging on corrosion behavior of ZK60 Mg alloy. *Trans. Nonferrous. Met. Soc. China* **2015**, *25*, 3909–3920. [CrossRef]

32. Orlov, D.; Ralston, K.D.; Birbilis, N.; Estrin, Y. Enhanced corrosion resistance of Mg alloy ZK60 after processing by integrated extrusion and equal channel angular pressing. *Acta Mater.* **2011**, *59*, 6176–6186. [CrossRef]

33. Zengin, H.; Turen, Y.; Ahlatci, H.; Sun, Y. mechanical properties and corrosion behavior of as-cast Mg-Zn-Zr-(La) magnesium alloys. *J. Mater. Eng. Perform.* **2018**, *27*, 389–397. [CrossRef]

34. Baek, S.M.; Kim, B.; Park, S.S. Influence of intermetallic particles on the corrosion properties of extruded ZK60 Mg alloy containing Cu. *Metals* **2018**, *8*, 323. [CrossRef]

35. Biancardi, O.V.; Rosa, V.L.; Abreu, L.B.; Tavares, A.P.R.; Corrêa, R.G.; Cavalcanti, P.H.; Napoleão, B.I.; Pereira, S.E. Corrosion behavior of as-cast ZK60 alloy modified with rare earth addition in sodium sulfate medium. *Corros. Sci.* **2019**, *158*. [CrossRef]

36. Zhang, T.T.; Cui, H.W.; Cui, X.L.; Zhao, E.T.; Feng, R.; Pan, Y.K. Investigations on microstructures, mechanical and corrosion properties of Mg-5.5Zn-0.8Zr alloys with Sm addition. *Mater. Res. Express* **2019**, *6*. [CrossRef]

37. Cheng, Y.L.; Qin, T.W.; Wang, H.M.; Zhang, Z. Comparison of corrosion behaviors of AZ31, AZ91, AM60 and ZK60 magnesium alloys. *Trans. Nonferrous. Met. Soc. China* **2009**, *19*, 517–524. [CrossRef]

38. Zeng, R.C.; Kainer, K.U.; Blawer, C.; Dietzel, W. Corrosion of an extruded magnesium alloy ZK60 component—The role of microstructural features. *J. Alloys Compd.* **2011**, *509*, 4462–4469. [CrossRef]

39. Xu, H.Y.; Diwu, J.T.; Liu, X.; Yang, Y.Q. Corrosion behavior of ZK60 magnesium alloy in sodium halide solutions. *J. Chin. Soc. Corros. Prot.* **2015**, *35*, 245–251.

40. Gu, X.N.; Li, N.; Zheng, Y.F.; Ruan, L.Q. In vitro degradation performance and biological response of a Mg–Zn–Zr alloy. *Mater. Sci. Eng. B* **2011**, *176*, 1778–1784. [CrossRef]

41. Huan, Z.G.; Leeflang, M.A.; Zhou, J.; Fratila-Apachitei, L.E.; Duszczyk, J. In vitro degradation behavior and cytocompatibility of Mg–Zn–Zr alloys. *J. Mater. Sci. Mater. Med.* **2010**, *21*, 2623–2635. [CrossRef]

42. Zhang, S.Y.; Bi, Y.Z.; Li, J.Y.; Wang, Z.G.; Yan, J.M.; Song, J.W.; Sheng, H.B.; Guo, H.Q.; Li, Y. Biodegradation behavior of magnesium and ZK60 alloy in artificial urine and rat models. *Biol. Mater.* **2017**, *2*, 53–62. [CrossRef] [PubMed]

43. Jamesh, M.I.; Wu, G.S.; Zhao, Y.; McKenzie, D.R.; Bilek, M.M.M.; Chu, P.K. Electrochemical corrosion behavior of biodegradable Mg–Y–RE and Mg–Zn–Zr alloys in ringer's solution and simulated body fluid. *Corros. Sci.* **2015**, *91*, 160–184. [CrossRef]

44. Shi, Z.M.; Liu, M.; Atrens, A. Measurement of the corrosion rate of magnesium alloys using tafel extrapolation. *Corros. Sci.* **2010**, *52*, 579–588. [CrossRef]

45. Zhao, M.C.; Schmutz, P.; Brunner, S.; Liu, M.; Song, G.L.; Atrens, A. An exploratory study of the corrosion of Mg alloys during interrupted salt spray testing. *Corros. Sci.* **2009**, *51*, 1277–1292. [CrossRef]

46. Shi, Z.M.; Atrens, A. An innovative specimen configuration for the study of Mg corrosion. *Corros. Sci.* **2011**, *53*, 226–246. [CrossRef]

47. Wei, L.Y.; Dunlop, G.L.; Westengen, H. The intergranular microstructure of cast Mg–Zn and Mg–Zn–Rare earth alloys. *Metall. Mater. Trans.* **1995**, *26*, 1947–1955. [CrossRef]

48. Pan, F.S.; Mao, J.J.; Chen, X.H.; Peng, J.; Wang, J.F. Influence of impurities on microstructure and mechanical properties of ZK60 magnesium alloy. *Trans. Nonferrous. Met. Soc. China* **2010**, *20*, 1299–1304. [CrossRef]

49. He, Y.B.; Pan, Q.L.; Qin, Y.J. Effect of heat treatment on microstructure and mechanical properties of wrought ZK60 magnesium alloy. *Heat Treat. Met.* **2011**, *36*, 52–57.

50. Song, Y.W.; Shan, D.Y.; Chen, R.S.; Han, E.H. Effect of second phases on the corrosion behaviour of wrought Mg–Zn–Y–Zr alloy. *Corros. Sci.* **2010**, *52*, 1830–1837. [CrossRef]

51. Song, G.L. *Corrosion Protection of Magnesium Alloys*; Chemical Industry Press: Beijing, China, 2006.

52. Merson, D.; Vasiliev, E.; Markushev, M.; Vinogradov, A. On the corrosion of ZK60 magnesium alloy after severe plastic deformation. *Lett. Mater.* **2017**, *7*, 421–427. [CrossRef]

53. Cao, F.Y.; Shi, Z.M.; Hofstetter, J.; Uggowitzer, P.J.; Song, G.L.; Liu, M.; Atrens, A. Corrosion of ultra-high-purity Mg in 3.5% NaCl solution saturated with Mg(OH)$_2$. *Corros. Sci.* **2013**, *75*, 78–99. [CrossRef]

54. Sarraf, H.T.; Sarraf, S.A.T.; Poursaee, A.; Langdon, T.G. Electrochemical behavior of a magnesium ZK60 alloy processed by high-pressure torsion. *Corros. Sci.* **2019**, *154*, 90–100. [CrossRef]

55. Song, G.L. Recent progress in corrosion and protection of magnesium alloys. *Adv. Eng. Mater.* **2005**, *7*, 563–587. [CrossRef]

56. Song, G.L.; Unocic, K.A. The anodic surface film and hydrogen evolution on Mg. *Corros. Sci.* **2015**, *9*, 758–765. [CrossRef]

57. Song, Y.W.; Shan, D.Y.; Chen, R.S.; Han, E.H. Investigation of surface oxide film on magnesium lithium alloy. *J. Alloys Compd.* **2009**, *29*, 1039–1045. [CrossRef]

58. Gandel, D.S.; Easton, M.A.; Gibson, M.A.; Abbott, T.; Birbilis, N. The influence of zirconium additions on the corrosion of magnesium. *Corros. Sci.* **2014**, *81*, 27–35. [CrossRef]

59. King, A.D.; Birbilisa, N.; Scully, J.R. Accurate electrochemical measurement of magnesium corrosion rates; A combined impedance, mass-loss and hydrogen collection study. *Electrochim. Acta* **2014**, *121*, 394–406. [CrossRef]

60. Liu, W.J.; Cao, F.H.; Chen, A.N.; Chang, L.R.; Zhang, J.Q.; Cao, C.N. Corrosion behaviour of AM60 magnesium alloys containing Ce or La under thin electrolyte layers. Part 1: Microstructural characterization and electrochemical behaviour. *Corros. Sci.* **2010**, *52*, 627–638. [CrossRef]

61. Cao, C.N.; Zhang, J.Q. *An Introduction of Electrochemical Impedance Spectroscopy*; Science Press: Beijing, China, 2002.

62. Song, G.L.; Atrens, A.; John, D.S.; Wu, X.; Nairn, J. The anodic dissolution of magnesium in chloride and sulphate solutions. *Corros. Sci.* **1997**, *39*, 1981–2004. [CrossRef]

63. Song, Y.W.; Han, E.H. The effect of Zn concentration on the corrosion behavior of Mg–xZn alloys. *Corros. Sci.* **2012**, *65*, 322–330. [CrossRef]

64. Baril, G.; Galicia, G.; Deslouis, C.; Pebere, N.; Tribollet, B.; Vivier, V. An impedance investigation of the mechanism of pure magnesium corrosion in sodium sulfate solutions. *J. Electrochem. Soc.* **2007**, *154*, 108–113. [CrossRef]

65. Song, Y.W.; Han, E.H.; Shan, D.Y. Corrosion characterization of Mg–8Li alloy in NaCl solution. *Corros. Sci.* **2009**, *51*, 1087–1094. [CrossRef]

66. Shi, Z.M.; Cao, F.Y.; Song, G.L.; Atrens, A. Low apparent valence of Mg during corrosion. *Corros. Sci.* **2014**, *88*, 434–443. [CrossRef]

67. Gomes, M.P.; Costa, I.; Pebere, N.; Rossi, J.L.; Tribollet, B.; Vivier, V. On the corrosion mechanism of Mg investigated by electrochemical impedance spectroscopy. *Electrochim. Acta* **2019**, *306*, 61–70. [CrossRef]

68. Jönsson, M.; Persson, D.; Thierry, D. Corrosion product formation during NaCl induced atmospheric corrosion of magnesium alloy AZ91D. *Corros. Sci.* **2007**, *49*, 1540–1558. [CrossRef]

69. Williams, G.; Dafydd, H.L.; Grace, R. The localised corrosion of Mg alloy AZ31 in chloride containing electrolyte studied by a scanning vibrating electrode technique. *Electrochim. Acta* **2013**, *109*, 489–501. [CrossRef]

70. Williams, G.; Grace, R. Chloride–induced filiform corrosion of organic–coated magnesium. *Electrochim. Acta* **2011**, *56*, 1894–1903. [CrossRef]

71. Wang, H.X.; Song, Y.W.; Yu, J.; Shan, D.Y.; Han, E.H. Characterization of filiform corrosion of Mg–3Zn Mg Alloy. *J. Electrochem. Soc.* **2017**, *1649*, 574–580. [CrossRef]

72. Atrens, A.; Song, G.L.; Shi, Z.M.; Soltan, A.; Johnston, S.; Dargusch, M.S. Understanding the corrosion of Mg and Mg alloys. *Encycl. Interfacial Chem.* **2018**, 515–534.

73. Song, Y.W.; Han, E.H.; Shan, D.Y.; Yim, C.D.; You, B.S. Microstructure and protection characteristics of the naturally formed oxide films on Mg–x Zn alloys. *Corros. Sci.* **2013**, *72*, 133–143. [CrossRef]

74. Nordlien, J.H.; Ono, S.; Masuko, N.; Nisancioglu, K. Morphology and structure of oxide films formed on magnesium by exposure to air and water. *J. Electrochem. Soc.* **1995**, *142*, 3320–3322. [CrossRef]

75. Dinodi, N.; Shetty, A.N. Electrochemical investigations on the corrosion behaviour of magnesium alloy ZE41 in a combined medium of chloride and sulphate. *J. Magnesium. Alloys* **2013**, *1*, 201–209. [CrossRef]

76. Kaya, A.; Ben–Hamu, G.; Eliezer, D.; Shin, K.S.; Cohen, S. Corrosion and oxidation of alloys of the Mg–Y–Zr–REM system. *Met. Sci. Heat. Treat.* **2006**, *48*, 518–523. [CrossRef]

77. Ben–Hamu, G.; Eliezer, D.; Shin, K.S.; Cohen, S. The relation between microstructure and corrosion behavior of Mg–Y–RE–Zr alloys. *J. Alloys Compd.* **2007**, *431*, 269–276. [CrossRef]

78. Lindström, R.; Johansson, L.G.; Thompson, G.E.; Skeldon, P.; Svensson, J.E. Corrosion of magnesium in humid air. *Corros. Sci.* **2004**, *46*, 1141–1158. [CrossRef]

79. Song, G.L.; Atrens, A. Corrosion mechanisms of magnesium alloys. *Adv. Eng. Mater.* **1999**, *1*, 11–33. [CrossRef]

80. Zeng, R.C.; Zhang, J.; Huang, W.J.; Kainer, K.U.; Blawer, C.; Dietzel, W.; Ke, W. Review of studies on corrosion of magnesium alloys. *Trans. Nonferrous. Met. Soc. China* **2006**, *16*, 763–771. [CrossRef]

© 2020 by the authors. Licensee MDPI, Basel, Switzerland. This article is an open access article distributed under the terms and conditions of the Creative Commons Attribution (CC BY) license (http://creativecommons.org/licenses/by/4.0/).

Article

Microstructure, Micro-Mechanical and Tribocorrosion Behavior of Oxygen Hardened Ti–13Nb–13Zr Alloy

Alicja Łukaszczyk [1,*], Sławomir Zimowski [2,*], Wojciech Pawlak [3], Beata Dubiel [4] and Tomasz Moskalewicz [4]

[1] Faculty of Foundry Engineering, AGH University of Science and Technology, Reymonta 23, 30-059 Kraków, Poland

[2] Faculty of Mechanical Engineering and Robotics, AGH University of Science and Technology, Mickiewicza Av. 30, 30-059 Kraków, Poland

[3] Institute of Materials Science and Technology, Lodz University of Technology, Stefanowskiego 1/15, 90-924 Łódź, Poland; Wojciech.pawlak@p.lodz.pl

[4] Faculty of Metals Engineering and Industrial Computer Science, AGH University of Science and Technology, Czarnowiejska 66, 30-054 Kraków, Poland; bdubiel@agh.edu.pl (B.D.); tmoskale@agh.edu.pl (T.M.)

*   Correspondence: alicjal@agh.edu.pl (A.Ł.); zimowski@agh.edu.pl (S.Z.); Tel.: +48-(12)-617-2763 (A.Ł.); +48-(12)-617-3060 (S.Z.)

Abstract: In the present work, an oxygen hardening of near-β phase Ti–13Nb–13Zr alloy in plasma glow discharge at 700–1000 °C was studied. The influence of the surface treatment on the alloy microstructure, tribological and micromechanical properties, and corrosion resistance is presented. A strong influence of the treatment on the hardened zone thickness, refinement of the α' laths and grain size of the bulk alloy were found. The outer hardened zone contained mainly an oxygen-rich Ti α' (O) solid solution. The microhardness and elastic modulus of the hardened zone decreased with increasing hardening temperature. The hardened zone thickness, size of the α' laths, and grain size of the bulk alloy increased with increasing treatment temperature. The wear resistance of the alloy oxygen-hardened at 1000 °C was about two hundred times, and at 700 °C, even five hundred times greater than that of the base alloy. Oxygen hardening also slightly improved the corrosion resistance. Tribocorrosion tests revealed that the alloy hardened at 700 °C was wear-resistant in a corrosive environment, and when the friction process was completed, the passive film was quickly restored. The results show that glow discharge plasma oxidation is a simple and effective method to enhance the micromechanical and tribological performance of the Ti–13Nb–13Zr alloy.

Keywords: oxygen hardening; Ti–13Nb–13Zr alloy; microstructure; tribocorrosion behavior; micromechanical properties

Citation: Łukaszczyk, A.; Zimowski, S.; Pawlak, W.; Dubiel, B.; Moskalewicz, T. Microstructure, Micro-Mechanical and Tribocorrosion Behavior of Oxygen Hardened Ti–13Nb–13Zr Alloy. *Materials* **2021**, *14*, 2088. https://doi.org/10.3390/ma14082088

Academic Editor: Marián Palcut

Received: 11 March 2021
Accepted: 16 April 2021
Published: 20 April 2021

Publisher's Note: MDPI stays neutral with regard to jurisdictional claims in published maps and institutional affiliations.

Copyright: © 2021 by the authors. Licensee MDPI, Basel, Switzerland. This article is an open access article distributed under the terms and conditions of the Creative Commons Attribution (CC BY) license (https:// creativecommons.org/licenses/by/ 4.0/).

1. Introduction

β-phase titanium alloys are characterized by high mechanical strength, good corrosion resistance, high biocompatibility and relatively low Young's modulus [1–6]. In addition, references [7,8] show that Nb and Zr are valuable alloying elements. Nb and Zr have an advantage over possibly toxic elements in body fluid, such as aluminum and vanadium, as they do not generate allergic reactions or genetic changes [9]. Consequently, β or near-β titanium alloys, such as Ti–13Nb–13Zr, have been developed for biological applications [1,10,11].

β-phase titanium alloys are referred to as second-generation titanium biomaterials. One of the rationales behind their development is to address the stress shielding associated with the high Young's modulus of the implant [2–6]. A lower elastic modulus is best suited for a biomaterial implant, as discussed by different authors [12–14]. Mohan and coworkers [15] indicated that the percentage of β phase greatly affects the mechanical properties of the Ti–Mo and Ti–Mo–Zr alloy systems. They showed that Mo improved the β phase stability when the percentage of Mo was increased from 12% to 15%. It was further noted that the elastic modulus of the alloys was lower compared to Ti–6Al–4V and commercially pure Ti. The Ti–15Mo–6Zr alloy had the lowest elastic modulus, thereby confirming that the higher percentage of β phase is the main reason for the lower elastic modulus observed. However, the main drawback of all titanium alloys is their poor tribological performance, as demonstrated by the high coefficient of friction (COF) combined with a tendency for scuffing and adhesive wear [2,3]. Moreover, the friction of titanium alloys in a corrosive environment increases their wear rate [16,17], which is disadvantageous in their application as orthopedic bioimplants. Surface treatment is usually applied to enhance the properties of titanium alloys as bioimplants. Several techniques, including oxidizing [18], nitriding [19], ion implantation [20] and physical vapour deposition (PVD) coating [21], selective laser melting [22], Ar arc melting [12], spark plasma sintering [23], anodization [14], mechanical blending process [15] have been investigated. It is known that oxygen hardening can increase the hardness and tribocorrosion resistance [24–30].

The characterization of the oxygen diffusion zone and the impact of this zone on macroscopic properties are still subjects of intensive investigation. Dong and Li [28] reported that the boost diffusion oxidation (BDO) process was an effective way to harden a titanium surface. The BDO process involved two steps. In the first step, titanium samples were thermally oxidized in the air. During the second step, the air was removed from the reaction chamber, and pre-oxidized samples were further diffusion treated in a vacuum. The method was successfully employed by Zabler and coworkers [30] for the formation of oxygen diffusion hardening (ODH) zones in the commercially pure Ti (grade 2) and Ti–6Al–4V alloy. A well-controlled hardness and hardened zone depth were achieved. A disadvantage of the process is the formation of an oxide scale on the alloy surface. Thicker titanium oxide layers tend to deteriorate the alloy properties. Jamesh et al. [31] reported that the thermal oxidation process of commercially pure Ti above 850 °C led to oxide scale spallation. This behavior is dangerous in frictional contact, where the detached hard titanium oxide particles accelerate the abrasion.

Januszewicz and Siniarski [32] showed that the assistance of plasma glow discharge during oxidation is of great benefit as it prevents the formation of brittle rutile phase on the treated alloy surface by mechanically damaging the oxide scale by high-energy ions. In our previous work [33], we showed that diffusion hardening of the near-surface zone by interstitial oxygen atoms with using glow discharge plasma in Ar + O_2 atmosphere is a very effective method to improve the mechanical and tribological properties of two-phase (α + β) Ti–6Al–4V titanium alloy. It was found that the treatment increased the surface hardness about three times, from 3.4 GPa to 10.6 GPa. The wear resistance in dry sliding contact with the Al_2O_3 ball was also significantly improved.

The oxygen diffusion hardening of the near-β Ti–13Nb–13Zr was performed by thermal oxidation in an oxygen-containing atmosphere [34]. The results suggest that zirconium plays a key role in the effective oxygen diffusion hardening at 500 °C for alloys of the Ti–Nb-Zr system. It was found that a sufficient Zr concentration (>6 wt.%) permits oxygen diffusion into the alloys. The near-surface hardness appears to increase further as Nb content increases. The high hardness of oxygen hardened Ti–13Nb–13Zr alloy is a result of an outer oxide layer composed of mixed metal (i.e., Ti, Nb, Zr) oxides on top of interstitial oxygen hardened alloy. This dense oxide layer is not only highly passive from a chemical/corrosion point of view but also resistant to abrasive wear, which is attributed to the mechanical stability of the oxide/substrate interface.

To our best knowledge, there are no data on the influence of oxygen plasma glow-discharge on the hardening of the Ti–13Nb–13Zr alloy and the influence of the surface treatment on microstructure and tribological properties in corrosive environments. Therefore, detailed studies on the effect of oxygen plasma glow discharge on the alloy usage properties are necessary. The friction wear corrosion results in metal degradation due to the simultaneous action of mechanical wear and (electro)chemical oxidation. The two mechanisms do not proceed separately and depend on each other in a complex way. In many cases, corrosion is accelerated by wear, and similarly, wear may be affected by corrosion phenomena. The nature of the passive film and the electrochemical conditions play a significant role in friction, wear and corrosion mechanism [35]. When the implants are to be placed in bone tissue, bone ingrowth into porous implant surfaces starts and osteoblast cell adhesion on the substrate increases, which may improve osseointegration. Improved osseointegration provides mechanical stability by interlocking the surrounding bone tissue with the implant. Tribocorrosion is a material degradation process and involves chemical and/or electrochemical interactions between material and its environment during a tribological process. Friction wear corrosion is a degradation process resulting from the combined action of small movements between contacting parts and the environment's corrosivity [36].

The main purpose of this work is to investigate the effect of oxygen plasma glow discharge temperature on the microstructure development and micromechanical properties of the near-β Ti–13Nb–13Zr alloy. An influence of the applied surface treatment on tribocorrosion behavior of the alloy in Ringer's solution is also studied.

2. Materials and Methods

A near-β phase titanium alloy with a chemical composition of 13.6 Nb, 13.6 Zr, 0.06 Fe, 0.04 C, 0.01–0.02 N, 0.001 H, 0.11 O (in wt.%), provided by Xi'an Saite Metal Materials Development Co., Ltd., Xi'an, China was used for oxygen hardening. The β transformation temperature of this alloy is 735 °C [37]. The samples were cut from the bar (27 mm diameter) into discs with a thickness of 1–2 mm. The samples were ground with sandpaper up to 3000-grit until a flat surface was obtained. Next, they were pre-polished with Al_2O_3 suspension (grain size 1 μm) and finally polished with silica suspension (0.04 μm) until a mirror finish. The prepared alloy had a fine acicular martensitic morphology composed of α' (hexagonal close-packed; hcp) and some α'' (orthorhombic, Cmcm space group) laths in β (body-centered cubic; bcc) grains (Figure 1). A detailed description of the microstructure of the as-received alloy has been presented in our previous works [24,25]. The grain size, estimated by image analysis, was in the range of 20–80 μm.

Figure 1. Microstructure of the as-received alloy.

The samples were placed in a quartz tube furnace. After air evacuation, argon of 99.999% purity was purged into the tube. The pressure was adjusted to 10 Pa by flowing argon (flow rate of 30 standard cm^3/min). When the samples reached the desired temperature of oxidation, i.e., 700 °C or 850 °C or 1000 °C, a glow discharge was started. The glow discharge was generated at 1250 V and an electric current of 45 mA. During discharge, high purity oxygen (99.999%) was mixed with argon and pumped to the furnace in 20 cycles of 1 min duration (flow rate of 6 standard cm^3/min). During oxygen pumping, the pressure was increased to 12 Pa. After oxidation, cooling to room temperature with working Ar glow discharge was provided.

The microstructure of the oxygen hardened alloy was characterized by light microscopy (LM, ZEISS Axio Imager M1m microscope, Oberkochen, Germany), scanning electron microscopy (SEM, FEI Nova NanoSEM 450 microscope, (Eindhoven, The Netherlands) and transmission electron microscopy (TEM, JEOL JEM-2010 ARP microscope, (Tokyo, Japan) techniques. The phase constitution was studied by selected area electron diffraction (SAED). The Java Electron Microscopy Software (JEMS, version 4.4131U2016, Pierre Stadelmann, Switzerland) was used to interpret the SAED patterns. The lamella from the cross-section of the coated alloy was prepared for the TEM investigation. The samples were thinned using FEI Quanta 3D 200i scanning electron microscope equipped with a Ga+ ion gun, Pt precursor gas injection systems (GIS) and OmniProbe micromanipulator for in situ lift-out [38]. Ion beam accelerating voltage of 30 kV and ion currents in the range of 15 nA–30 pA were applied. The lamella was transferred via a micromanipulator to a TEM half ring, where a focused ion beam (FIB) milling to electron transparency was performed (ion currents of 500–30 pA). FIB deposition process from Pt precursor was used to attach the manipulator probe to the lamella and attach it to the grid. The phase identification was supplemented by energy-dispersive X-ray spectroscopy (SEM-EDS, TEM-EDS) microanalysis and STEM-EDS line analysis.

The open-circuit potential (OCP), linear sweep voltamperometry and electrochemical impedance spectroscopy (EIS) were carried out using a potentiostat Autolab PGSTAT302N (Metrohm Autolab B.V., Utrecht, The Nederlands). The reference electrode was a saturated calomel electrode (SCE), and a platinum plate was used as the counter electrode. Ringer's solution was used as the electrolyte for the corrosion study. The chemical composition of the Ringer's solution was: 8.6 g/L NaCl, 0.3 g/L KCl, 0.25 g/L CaCl$_2$. Measurements were performed at pH 7.4 in deaerated solutions at 37 °C. The polarization test was performed at a scan rate of 1 mV/s from -1.3 V to $+2.2$ V vs. SCE. For the EIS measurements, the amplitude was 10 mV; the frequency was 10^5 Hz to 10^{-3} Hz. The EIS measurements were performed at the OCP potential. The EIS data were fitted using ZView software.

The hardness and elastic modulus of the as-received and oxygen hardened Ti–13Nb–13Zr alloys were determined by micro-combitester (CSM Instruments). The indentation tests were performed using a Vickers diamond indenter. The indenter was loaded with a force of 200 mN and kept for 5 s before un-loading. An analysis of the mechanical properties of the alloy was carried out using the loading/unloading curve by Oliver and Pharr method [39]. It allowed for the determination of the elastic modulus (E_{IT}) and hardness (H_{IT}). For each sample, ten consecutive experiments at randomly selected places were performed, and the average of 10 measurements was calculated.

The abrasive wear resistance and the friction coefficient of the as-received and treated alloy were determined by friction tests based on the ISO 20808 standard [40]. The tests were carried out in dry sliding contact with an Al$_2$O$_3$ ball (6 mm diameter) using a ball-on-disc tribotester (ITeE Radom, Radom, Poland). In a typical ball-on-disc arrangement, a counter-element in the form of a ball was pressed against a rotating disc (sample) made of titanium alloy. The tribological tests were repeated three times, with the same parameters: ball load $F_n = 5$ N, sliding speed v = 0.07 m/s, sliding distance s = 1000 m, room temperature 23 °C and relative humidity 55%. The wear rate $W_v = V/F_n \times s$ was determined as the ratio of the volume of material removed during friction (V) to the load (F_n) and the sliding distance (s). The volume was calculated based on the size of the cross-sectional area of the wear groove. The groove profile was measured with a stylus profilometer in six places around the wear track.

Friction-wear tests with simultaneous measurement of corrosion potential during friction in Ringer's solution were also performed. To record the changes in the corrosion potential, a 2-electrode system with the working electrode (titanium alloy) and reference electrode (calomel electrode in 3 M KCl) was designed. The schematic of the tribocorrosion device is shown in Figure 2. A specially designed system ensuring an efficient and stable electric contact with the alloy sample was used. In this system, the ball with the holder was moving, whereas the titanium alloy sample was stationary. To fix the ball and sample positions, special polymer holders were used. The same measurement parameters were applied as in the case of dry friction tests. The friction process began after a stable corrosion potential for the titanium alloy was reached. When the friction was activated, a change in the potential was recorded as a result of the wear of the alloys' surface layer.

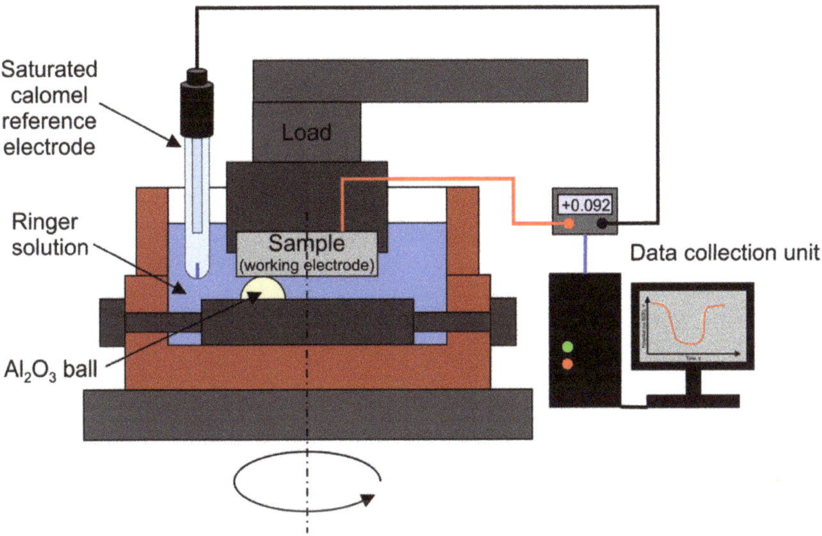

Figure 2. Schematic of the tribological device with parallel corrosion potential measurement during wear tests of the titanium alloy in sliding contact with a ball.

Before each test, the alumina ball and titanium alloy were ultrasonically cleaned in ethanol and left to dry. Additionally, the ball used in the tribocorrosion test was also washed in Ringer's solution. The holders were subjected to an identical cleaning procedure. The whole measurement took place in a plastic vessel filled with Ringer's solution.

3. Results

3.1. Microstructure of the Oxygen Hardened Alloy

LM and SEM images of the alloy treated at 700 °C, 850 °C and 1000 °C are shown in Figure 3. The alloy microstructure was dependent on the hardening temperature. The differences in the thickness of the oxygen-rich hardened zone and the grain size of the bulk of the material were observed. The highest thickness of the hardened zone, up to 470 µm, was found in the sample treated at 1000 °C (Figure 3e,f). In the case of alloy treated at 850 °C, the hardened zone thickness was about 160 µm (Figure 3c,d), while in the alloy treated at 700 °C it had the lowest thickness, up to 120 µm (Figure 3a,b). A coarsening of the α' laths in the oxygen-hardened zone with increasing treatment temperature was also observed. The voids in the hardened zone of the sample treated at 1000 °C were formed at a distance of ~20–30 µm from the surface. It should be noted that the presence of pores usually contributes to the formation of microcracks, which were also sporadically observed in this sample. A typical microcrack is marked with an arrow in Figure 3f. The observed microstructural defects exclude the sample hardened at 1000 °C from the intended biological applications.

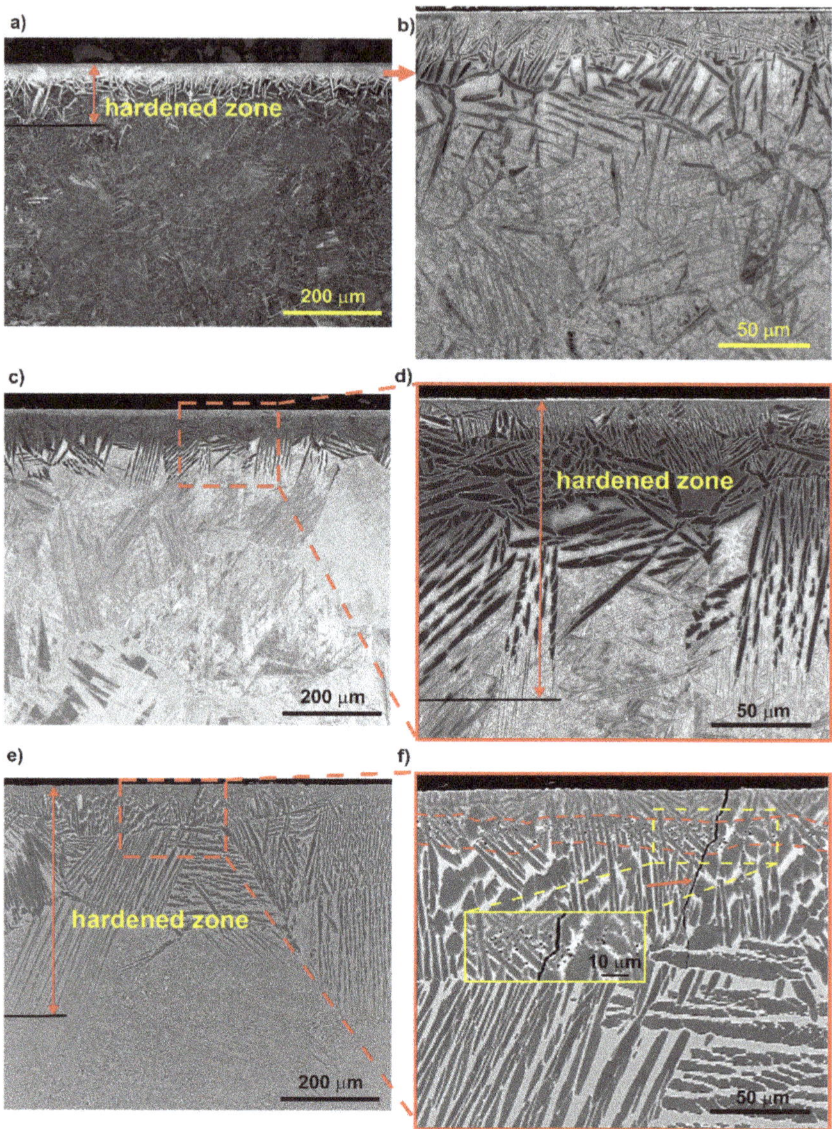

Figure 3. Microstructure of the Ti–13Nb–13Zr alloy after oxygen hardening at 700 °C (**a**,**b**), 850 °C (**c**,**d**) and 1000 °C (**e**,**f**). LM (**a**) and SEM (**b**–**f**), cross-section samples. The zone with voids is marked with a dashed line in Figure 3f. An arrow in Figure 3f indicates a microcrack developed in the near-surface zone.

Investigation of the cross-section showed a significant grain growth resulting from the hardening process. The grain size estimated from LM and SEM images was in the range of 40–100 μm, 100–250 μm and up to 700 μm for the alloys treated at 700 °C, 850 °C and 1000 °C, respectively.

The sample treated at 700 °C was selected for a detailed microstructure characterization by TEM and STEM. The TEM and STEM images are shown in Figures 4 and 5. In the near-surface region (depth up to 10 μm from the surface), a high fraction of the Ti α′

(O) solid solution and low amount of fine laths of the Ti α″ (O) solid solution in β phase were found (Figure 4). In the SAED pattern no. 2, the diffraction spots from crystal planes belonging to particular three [100] Tiα″ zone axes corresponding to three sets of the Ti α″ laths inclined by the angle of 60° are marked with black, green and red color, respectively.

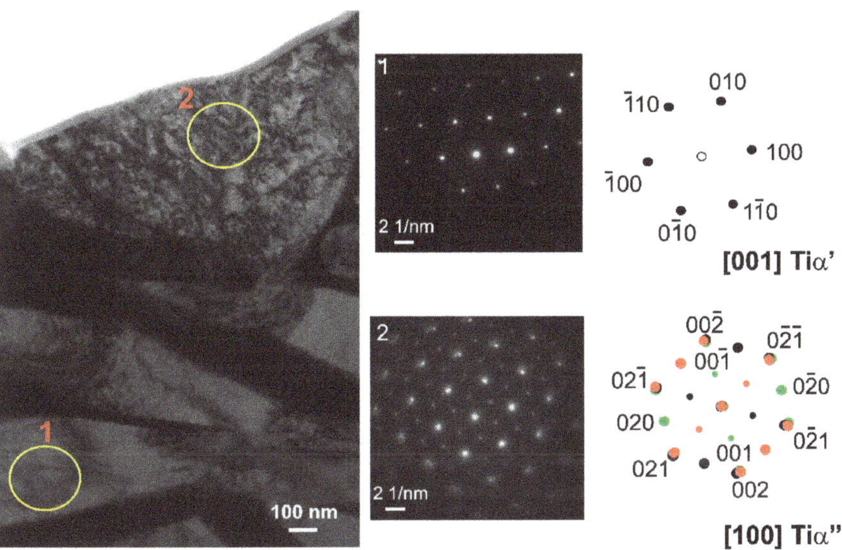

Figure 4. TEM image of the near-surface region in the Ti–13Nb–13Zr alloy after oxygen hardening at 700 °C. SAED patterns of α′ (hcp) and α″ (orthorhombic, Cmcm) were taken from areas marked with 1 and 2, respectively. In the SAED pattern no. 2, the spots belonging to the three [100] α″ zone axes are marked with black, green and red color. Indices of green spots are given.

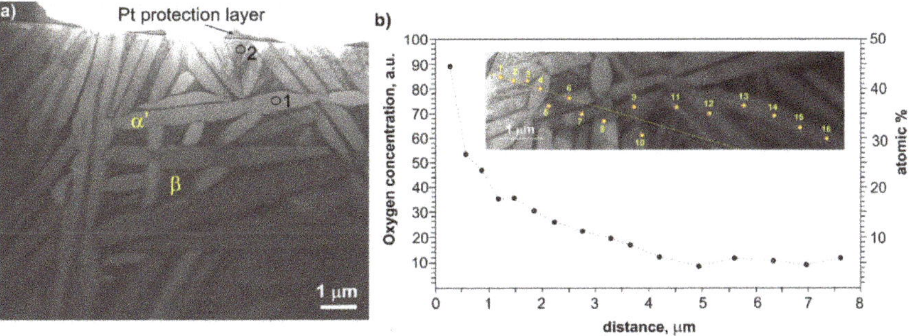

Figure 5. STEM image of the microstructure of the near-surface region in the Ti–13Nb–13Zr alloy cross-section after oxygen hardening at a temperature of 700 °C (**a**) and concentration profile of oxygen obtained by TEM-EDS microanalysis performed in points 1–16 (**b**). The exemplary grains of the α′ and β phases, as well as the areas 1 and 2 given in Figure 4, are marked in Figure 5a.

To examine the concentration profile of oxygen in the near-surface region, a TEM-EDS microanalysis was performed at 16 points located in the α′ phase at a distance from 0.3 μm to 8 μm. The oxygen concentration profile is presented in Figure 5b. The results confirmed the increased content of oxygen in the zone close to the surface. At a point located 0.3 μm from the outer surface edge of the lamellae, the oxygen concentration was about 47 at.%, and with the increasing distance to 6 μm, it dropped gradually to about

6 at.%. At greater depths in the sample, the oxygen concentration in the α' phase remained constant at ~6 at.%. The obtained results of the oxygen distribution should be treated as approximate values since EDS microanalysis gives only a rough indication of light elements concentration. Nevertheless, the result is an indication that the Ti α' (O) solid solution is enriched in oxygen in the near-surface zone at a depth of up to 6 μm.

In our previous study [33], we showed that the near-surface region of the oxygen diffusion hardened two-phase ($\alpha + \beta$) Ti–6Al–4V alloy consisted of Ti α (O) solid solution enriched with oxygen mainly. Oxygen is a strong interstitial solid solution strengthening element of titanium [36]. It is an α phase stabilizer and has a high solubility in the hcp α phase, up to 31.9 at.%. The solubility of oxygen in the β phase is much lower, maximum 8 at.% [41]. Therefore, the presence of the Ti α' (O) phase in the near-surface zone is preferred due to the diffusion of interstitial oxygen atoms.

According to [39], the plasma glow discharge strengthens the oxygen diffusion into the metallic substrate. It is likely due to an increase in the number of point defects formed during the first stage of the process. In addition, the plasma glow discharge inhibits the formation of the rutile layer on the titanium alloy surface. The oxides formed on the alloy surface act as limited reservoirs of oxygen atoms, which are then forced to diffuse into the alloy matrix and form a solid solution [32]. Therefore, the surface of the alloy investigated in this work was not covered by titanium oxide.

3.2. Micromechanical and Tribological Properties

Table 1 shows the micromechanical and tribological properties of the as-received and oxygen hardened Ti–13Nb–13Zr alloy. The surface treatment of the alloy resulted in a significant increase in hardness and elastic modulus than the as-received alloy. Two effects can explain this strengthening: (i) the crystallographic strains generated by the strong lattice deformation that expand the c/a ratio, and (ii) the long-range ordering of the interstitial atoms in the hcp-structure of the host [36]. However, based on the present investigation results, it was found that both the hardness and elastic modulus decreased significantly in the surface layer of hardened alloy with increasing hardening temperature (Table 1). The tendency to lower the hardness with increasing the titanium alloy's treatment temperature was also noticed elsewhere [42]. The Ti–13Nb–13Zr alloy hardened at 700 °C had the highest microhardness and modulus of elasticity, 12.8 GPa and 180 GPa, respectively. The microhardness of the alloy treated at 1000 °C reached only 6.9 GPa. This behavior is related to the microstructure of the hardened alloy surface layer, particularly to the refinement of the α' laths at 700 °C. The largest grain size was found in the oxygen-enriched layer of the titanium alloy hardened at 1000 °C. The large grain size does not favor the material strengthening. Additionally, for the alloy treated at 1000 °C, the formation of voids in the hardened zone, located ~20–30 μm from the surface, was observed (Figure 3f). Such microstructure features facilitate a plastic deformation during indentation and result in lower hardness.

Table 1. Microhardness (H_{IT}), elastic modulus (E_{IT}), the penetration depth of the indenter (h_{max}), and wear rate (W_v) of the Ti–13Nb–13Zr alloy.

Sample of Ti–13Nb–13Zr Alloy	h_{max} (nm)	H_{IT} (GPa)	E_{IT} (GPa)	$W_v \cdot 10^{-6}$ (mm^3/Nm)
Hardened at 700 °C	880 ± 37	12.8 ± 0.8	180 ± 16	2.3 ± 0.3
Hardened at 850 °C	925 ± 52	9.8 ± 0.9	175 ± 20	3.1 ± 0.3
Hardened at 1000 °C	1407 ± 131	6.9 ± 0.7	152 ± 17	5.8 ± 0.7
As-received	1662 ± 82	3.9 ± 0.2	79 ± 9	1250 ± 46

The hardening improved the mechanical properties of the Ti–13Nb–13Zr alloy, and both the hardness and elastic modulus increased about three times compared to the baseline alloy. Interstitial oxygen diffusion hardening of the alloy carried out at 700 °C allowed to achieve a hardness comparable to the hardness of the alloy treated by plasma electrolytic oxidation (PEO) in an electrolyte with and without the addition of zirconia nanoparticles [43].

The wear resistance of the titanium alloy was tested in dry sliding contact. Figure 6 shows the average COF of the as-received and hardened alloy. The COF of the hardened alloy samples in dry sliding contact with the Al_2O_3 ball was in the range of 0.63–0.78. A significantly lower COF = 0.50 occurred during the friction of the baseline titanium alloy, and the cooperation with the Al_2O_3 counterpart was more stable.

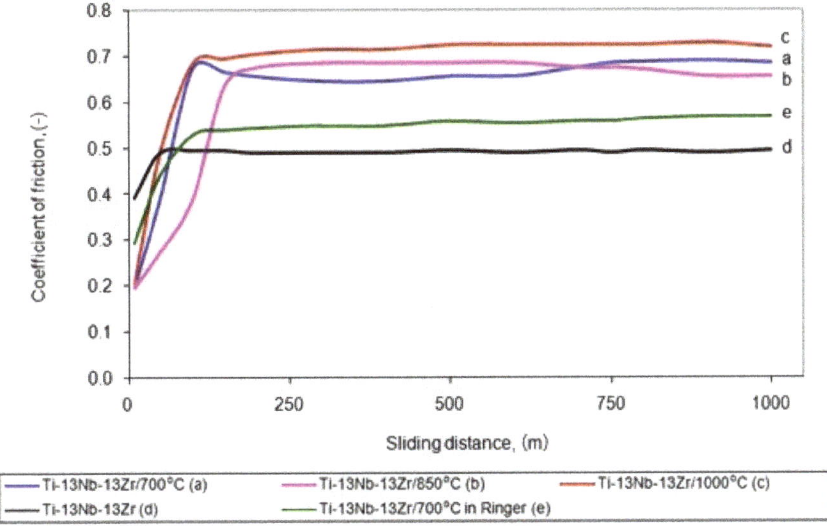

Figure 6. COF of the alloy oxygen hardened at 700 °C (**a**), 850 °C (**b**) and 1000 °C (**c**) compared to as-received alloy (**d**) in dry friction condition as well as the alloy hardened at 700 °C in Ringer's solution (**e**) against alumina ball.

The unstrengthened titanium alloy showed a more significant deformation in the sliding point contact than the hardened alloy since their moduli of elasticity differed significantly. Less deformation should reduce the mechanical component of the friction force and thus the friction coefficient for the hardened alloy. However, such an effect has not been observed as the surface interactions in the friction microcontact had the primary influence on the resistance to motion. The hardened surface layer of the titanium alloy constitutes difficult cooperation conditions in a non-lubricated contact with the hard ceramic Al_2O_3 ball. A microstructural investigation has shown that the surface of the alloy examined in this work was not covered by titanium oxide. TiO_2 could reduce the resistance to motion. The improved friction and wear properties can be attributed to the low-friction TiO_2 rutile layer [44]. Moreover, the mean contact pressure (p_m) in the initial period of friction test of the 700 °C hardened alloy was 0.82 GPa, i.e., much higher than the untreated alloy (0.54 GPa). As a result of the sliding interaction of such hard oxide materials, so-called severe friction developed [45], which caused a high resistance to motion. The obtained results are characteristic for this type of material during dry friction [46].

The wear resistance of the oxygen-hardened alloy was strongly dependent on the treatment temperature and was at least two hundred times greater than that of the base alloy (Figure 7). The wear rate of the as-received Ti–13Nb–13Zr alloy reached the value of 1250×10^{-6} mm^3/Nm. In comparison, the wear rate for the alloy hardened at 700 °C, 850 °C, and 1000 °C was 2.3×10^{-6} mm^3/Nm, 3.1×10^{-6} mm^3/Nm and 5.8×10^{-6} mm^3/Nm, respectively. Based on the microscopic analysis of the wear track surface, an abrasive wear nature of the hardened titanium alloy and abrasive-adhesive wear of the as-received alloy were found. The wear process of the Ti–13Nb–13Zr alloy is typical of dry friction in contact with a hard counterpart and has already been analyzed in detail elsewhere [47].

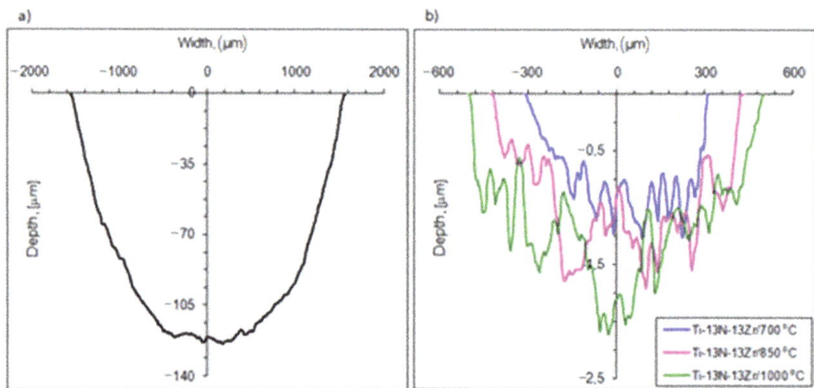

Figure 7. Cross-section profile of wear track of as-received Ti–13Nb–13Zr alloy (**a**) and hardened Ti–13Nb–13Zr alloy (**b**) after dry friction.

The best mechanical and tribological properties were found for the Ti–13Nb–13Zr alloy hardened at 700 °C. Therefore, this alloy was selected for further corrosion resistance tests. Figure 8a shows the evolution of the OCP for the as-received and the alloy hardened at 700 °C. The results of E_{ocp} show that the as-received alloy has a less noble potential than the alloy heat-treated at 700 °C, indicating that the as-received alloy is more susceptible to corrosion. The OCP slightly increased for the hardened alloy and reached a stable value after about 2000s.

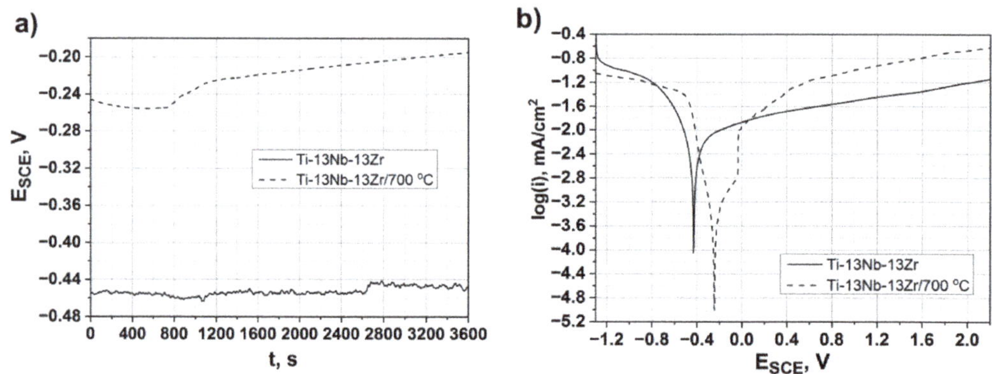

Figure 8. Electrochemical measurements of as-received and oxygen hardened alloy (Ti–13Nb–13Zr/700 °C) in Ringer's solution at 37 °C, (**a**) evolution of the corrosion potential vs. time and (**b**) polarization curves at 1 mV/s.

The potentiodynamic polarization testing was conducted to understand the corrosion properties of the untreated and treated alloy (Ti–13Nb–13Zr/700 °C) (Figure 8b). Because of an absence of linear regions, the Tafel extrapolation was not applicable to interpret the electrochemical response. In such cases, the corrosion rate could be defined by the limiting current density, which passes through the passivating film, thus becoming a measure of the film protective performance [48]. The passive current density (i_p) was reduced from 68 $\mu A/cm^2$ for the treated alloy to 20 $\mu A/cm^2$ for the as-received alloy, while the cathodic–

anodic transition increased from about -0.43 V up to -0.25 V. These results indicate that the as-received alloy has a slightly smaller corrosion rate than the treated one.

Figure 9 shows the EIS graphs presented as a Bode plot (Figure 9a) and a Nyquist plot (Figure 9b) of the as-received and hardened alloy in the Ringer's solution. From Figure 9a, the Z modulus at a lower frequency in the Bode impedance plot indicated a comparable corrosion resistance of the investigated samples. The as-received and treated alloys showed a highly capacitive behavior from medium to low frequencies. The equivalent circuit, as shown in Figure 10, was used to fit the EIS data. According to the double-layer model for the oxygen hardened alloy, the equivalent circuit consisted of the electrolyte resistance (R1), the treated resistance (R2) and the constant phase elements (CPE). A good fitting between the experimental and simulated results was achieved, and the parameters are listed in Table 2.

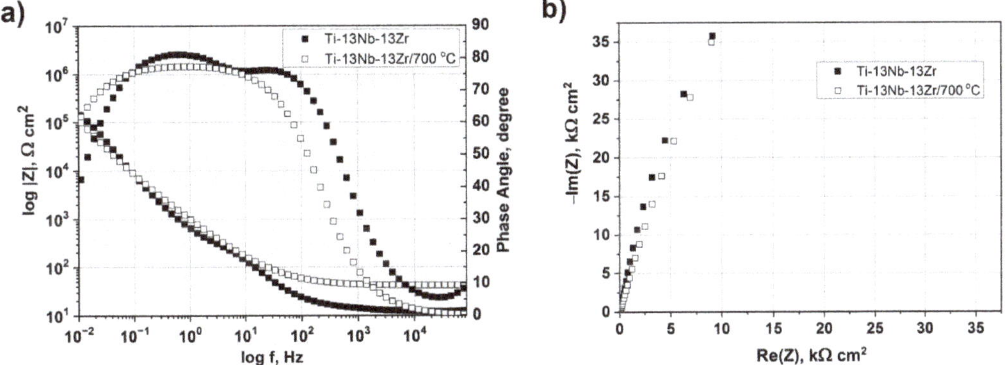

Figure 9. Electrochemical impedance curves of the as-received and oxygen hardened alloy in Ringer's solution. (**a**) Bode impedance and phase angle plot, (**b**) Nyquist impedance plot.

Figure 10. Equivalent circuit used for fitting EIS data.

Table 2. Chi-squared (χ^2) values obtained by fitting equivalent electrical circuit with Z View software and electrochemical parameters for the as-received and treated titanium alloy.

Sample	χ^2	R_1 ($\Omega * cm^2$)	CPE-T ($Fs^{n-1}\ cm^{-2}$)	CPE-P	R_2 ($\Omega * cm^2$)
As-received alloy	0.005	12.30 ± 0.12	$4.31 \times 10^{-5} \pm 0.03 \times 10^{-5}$	$8.96 \times 10^{-1} \pm 0.02 \times 10^{-1}$	$3.22 \times 10^5 \pm 0.06 \times 10^5$
Alloy hardened at 700 °C	0.004	41.99 ± 0.25	$4.35 \times 10^{-5} \pm 0.02 \times 10^{-5}$	$8.70 \times 10^{-1} \pm 0.01 \times 10^{-1}$	$7.26 \times 10^5 \pm 0.11 \times 10^5$

To investigate the passivation kinetics of the hardened alloy in a condition where the oxygen-rich hardened layer of the alloy is abraded, tribological tests were performed in the presence of Ringer's solution. The tests were coupled with simultaneous measurement of the OCP (Figure 2). Figure 11 shows the change in the corrosion potential over time. In the diagram, we can distinguish 3 characteristic stages:

i) Stage 1—increase and stabilization of the stationary potential;
ii) Stage 2—a drop of the corrosion potential, resulting from the wear of the alloy's surface, with the assumed sliding distance equaling 1000 m;
iii) Stage 3—an increase of the stationary potential after the interruption of the friction process of the alloy's surface.

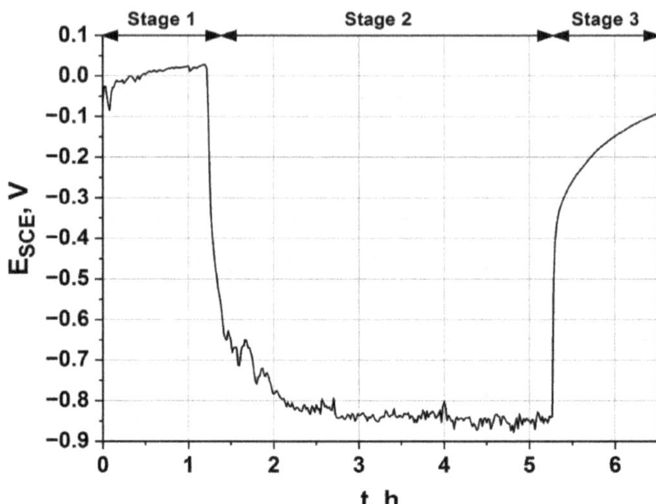

Figure 11. Change in the corrosion potential during the tribological test in Ringer's solution at 25 °C of the alloy hardened at 700 °C.

The friction process was activated after 1 h 20 min of stabilization in Ringer's solution. A significant drop in the potential corrosion value was observed, resulting from the change in the measurement conditions. As a result of friction, a groove was formed, with significant surface roughness and less passivating, thereby showing a considerable drop of the corrosion potential. The potential reached about -846 ± 50 mV vs. SCE and was maintained at this value during the whole friction process. When the friction process was terminated (Stage 3), an increase in the corrosion potential was observed. The increase in the potential took place abruptly. The strengthened outer layer of the alloy treated at 700 °C provided good protection against tribological wear in a corrosive environment, and the passive film was quickly restored.

4. Conclusions

In this work, the oxygen hardening of Ti–13Nb–13Zr alloy by plasma glow discharge at 700–1000 °C was studied.

1. The hardening temperature had a significant influence on the alloy microstructure and thickness of the hardened zone. A refinement of the α' laths of the near β-phase Ti alloy was observed. The thickness of the hardened zone and grain size of the bulk alloy both increased with increasing temperature. In addition, voids and microcracks were observed in the near-surface zone of the alloy treated at 1000 °C;
2. The outer hardened zone consisted mainly of the Ti α' (O) solid solution with small amounts of fine laths of the Ti α'' (O) solid solution in the β phase. Oxygen enrichment in a depth of up to 6 μm was found;
3. The oxidation of the Ti–13Nb–13Zr alloy under glow discharge conditions resulted in a significant increase of hardness and elastic modulus compared with the base alloy. The best results were found for the alloy hardened at 700 °C. With an increasing temperature, a decrease in both hardness and modulus of elasticity in the hardened zone were observed;
4. The hardened titanium alloy zone significantly reduces abrasive wear, and the wear resistance is proportional to the hardness of the alloy;
5. Oxygen hardened alloy does not adversely affect the corrosion resistance;
6. The friction reduces the corrosion resistance of the oxygen-hardened Ti–13Nb–13Zr alloy. However, when the friction process was stopped, the corrosion potential was

quickly restored. The strengthened outer layer of the alloy treated at 700 °C provides good protection against tribological wear in a corrosive environment.

Author Contributions: Conceptualization, A.Ł., S.Z. and T.M.; formal analysis, A.Ł., S.Z., W.P., B.D. and T.M.; funding acquisition, T.M. and S.Z.; investigation, A.Ł., S.Z., W.P., B.D. and T.M.; methodology, A.Ł., S.Z., W.P., B.D. and T.M., Project administration, T.M. and S.Z.; resources, T.M.; supervision, T.M. and S.Z.; validation, A.Ł., S.Z., B.D. and T.M.; visualization, A.Ł., S.Z., B.D. and T.M; writing—original draft, A.Ł., S.Z., W.P., B.D. and T.M.; writing—review and editing, A.Ł., S.Z. and T.M. All authors have read and agreed to the published version of the manuscript.

Funding: The study was supported by the National Science Centre, Poland (decision no. DEC-2016/21/B/ST8/00238). Part of this work concerning the mechanical and tribocorrosion investigation was supported by AGH University of Science and Technology, Faculty of Mechanical Engineering and Robotics, project no. 16.16.130.942/2021.

Institutional Review Board Statement: Not applicable.

Informed Consent Statement: Not applicable.

Data Availability Statement: The data presented in this study are available on request from the corresponding author.

Acknowledgments: The authors appreciate the valuable contributions of M. Gajewska (ACMiN AGH) for FIB lamella preparation and Ł. Cieniek, for SEM investigation.

Conflicts of Interest: The authors declare no conflict of interest. The funders had no role in the design of the study; in the collection, analyses, or interpretation of data; in the writing of the manuscript, or in the decision to publish the results.

References

1. Slokar, L.; Matković, T.; Matković, P. Alloy design and property evaluation of new Ti–Cr–Nb alloys. *Mater. Des.* **2012**, *33*, 26–30. [CrossRef]
2. Cojocaru, V.D.; Raducanu, D.; Cinca, I.; Dan, I.; Ivanescu, S.; Jalba, E. The mechanical properties evaluation for an as-cast Ti-Ta-Zr alloy. *Metal. Int.* **2011**, *16*, 35–38. [CrossRef]
3. Raducanu, D.; Vasilescu, E.; Cojocaru, V.D.; Cinca, I.; Drob, P.; Vasilescu, C.; Drob, S.I. Mechanical and corrosion resistance of a new nanostructured Ti-Zr-Ta-Nb alloy. *J. Mech. Behav. Biomed. Mater.* **2011**, *4*, 1421–1430. [CrossRef]
4. Tong, Y.X.; Guo, B.; Zheng, Y.F.; Chung, C.Y.; Ma, L.W. Effects of Sn and Zr on the Microstructure and Mechanical Properties of Ti-Ta-Based Shape Memory Alloys. *J. Mater. Eng. Perform.* **2011**, *20*, 762–766. [CrossRef]
5. Niinomi, M. Mechanical properties of biomedical titanium alloys. *Mater. Sci. Eng. A* **1998**, *243*, 231–236. [CrossRef]
6. Niinomi, M. Mechanical biocompatibilities of titanium alloys for biomedical applications. *J. Mech. Behav. Biomed. Mater.* **2008**, *1*, 30–42. [CrossRef]
7. Cremasco, A.; Messias, A.D.; Esposito, A.R.; de RezendeDuek, E.A.; Caram, R. Effects of alloying elements on the cytotoxic response of titanium alloys. *Mater. Sci. Eng. C* **2011**, *31*, 833–839. [CrossRef]
8. Robin, A.; Carvalho, O.A.S.; Schneider, S.G.; Schneider, S. Corrosion behavior of Ti-xNb-13Zr alloys in Ringer's solution. *Mater. Corros.* **2008**, *59*, 929–933. [CrossRef]
9. Khan, M.A.; Williams, R.L.; Williams, D.F. The corrosion behaviour of Ti-6Al-4V, Ti-6Al-7Nb and Ti-13Nb-13Zr in protein solutions. *Biomaterials* **1999**, *20*, 631–637. [CrossRef]
10. Zhao, C.; Zhang, X.; Cao, P. Mechanical and electrochemical characterization of Ti–12Mo–5Zr alloy for biomedical application. *J. Alloys Compd.* **2011**, *509*, 8235–8238. [CrossRef]
11. Geetha, M.; Singh, A.K.; Asokamani, R.; Gogia, A.K. Ti based biomaterials, the ultimate choice for orthopaedic implants A review. *Prog. Mater. Sci.* **2009**, *54*, 397–425. [CrossRef]
12. Ke, Z.; Yi, C.; Zhang, L.; Yuan, Z.; He, Z.; Tan, J.; Jiang, Y. Characterization of a new Ti-13Nb-13Zr-10Cu alloy with enhanced antibacterial activity for biomedical applications. *Mater. Lett.* **2019**, *253*, 335–338. [CrossRef]
13. Mohammed, M.; Khan, Z.; Siddiquee, A. Beta Titanium Alloys: The Lowest Elastic Modulus for Biomedical Applications: A review. *Int. J. Chem. Mol. Nucl. Mater. Metall. Eng.* **2014**, *8*, 822–827. [CrossRef]
14. Barjaktarević, D.; Medjo, B.; Gubeljak, N.; Cvijović-Alagić, I.; Štefane, P.; Djokic, V.; Rakin, M. Experimental and numerical analysis of tensile properties of Ti-13Nb-13Zr alloy and determination of influence of anodization process. *Procedia Struct. Integr.* **2020**, *28*, 2187–2194. [CrossRef]
15. Mohan, P.; Rajak, D.K.; Pruncu, C.I.; Behera, A.; Amigó-Borrás, V.; Abou Bakr Elshalakany, A. Influence of β-phase stability in elemental blended Ti-Mo and Ti-Mo-Zr alloys. *Micron* **2021**, *142*, 102992. [CrossRef]

16. Dong, H. Tribological properties of titanium-based alloys. In *Woodhead Publishing Series in Metals and Surface Engineering. Surface Engineering of Light Alloys. Aluminium, Magnesium and Titanium Alloys*; Dong, H., Ed.; Woodhead Publishing: Cambridge, UK, 2010; pp. 58–80. [CrossRef]

17. Garbacz, H.; Wieciński, P.; Ossowski, M.; Ortore, G.; Wierzchoń, T.; Kurzydłowski, K.J. Surface engineering techniques used for improving the mechanical and tribological properties of the Ti6A14V alloy. *Surf. Coat. Technol.* **2008**, *202*, 2453–2457. [CrossRef]

18. Dong, H.; Bell, T. Enhanced wear resistance of titanium surfaces by a new thermal oxidation treatment. *Wear* **2000**, *238*, 131–137. [CrossRef]

19. Muraleedharan, T.M.; Meletis, E.I. Surface modification of pure titanium and Ti-6A1-4V by intensified plasma ion nitriding. *Thin Solid Films* **1992**, *221*, 104–113. [CrossRef]

20. Hutchings, R.; Oliver, W.C. A study of the improved wear performance of nitrogen-implanted Ti-6Al-4V. *Wear* **1983**, *92*, 143–153. [CrossRef]

21. Wilson, A.D.; Leyland, A.; Matthews, A. A comparative study of the influence of plasma treatments, PVD coatings and ion implantation on the tribological performance of Ti–6Al–4V. *Surf. Coat. Technol.* **1999**, *114*, 70–80. [CrossRef]

22. Zhou, L.; Tiechui Yuan, T.; Ruidi Li, R.; Lanbo Li, L. Two ways of evaluating the wear property of Ti-13Nb-13Zr fabricated by selective laser melting. *Mater. Lett.* **2019**, *242*, 9–12. [CrossRef]

23. Kong, Q.; Lai, X.; An, X.; Feng, W.; Lu, C.; Wu, J.; Wu, C.; Wu, L.; Wang, Q. Characterization and corrosion behaviour of Ti-13Nb-13Zr alloy prepared by mechanical alloying and spark plasma sintering. *Mater. Today Commun.* **2020**, *23*, 101130. [CrossRef]

24. Jugowiec, D.; Kot, M.; Moskalewicz, T. Electrophoretic deposition and characterization of chitosan coating on near-beta titanium alloy. *Arch. Metall. Mater.* **2016**, *61*, 657–664. [CrossRef]

25. Sak, A.; Moskalewicz, T.; Zimowski, S.; Cieniek, Ł.; Dubiel, B.; Radziszewska, A.; Kot, M.; Łukaszczyk, A. Influence of polyetheretherketone coatings on the Ti-13Nb-13Zr titanium alloy's bio-tribological properties and corrosion resistance. *Mater. Sci. Eng. C* **2016**, *63*, 52–61. [CrossRef] [PubMed]

26. Hornberger, H.; Randow, C.; Fleck, C. Fatigue and surface structure of titanium after oxygen diffusion hardening. *Mater. Sci. Eng. A* **2015**, *630*, 51–57. [CrossRef]

27. Grabarczyk, J.; Batory, D.; Kaczorowski, W.; Pązik, B.; Januszewicz, B.; Burnat, B.; Czerniak-Reczulska, M.; Makówka, M.; Niedzielski, P. Comparison of Different Thermo-Chemical Treatments Methods of Ti-6Al-4V Alloy in Terms of Tribological and Corrosion Properties. *Materials* **2020**, *13*, 5192. [CrossRef]

28. Dong, H.; Li, X.Y. Oxygen boost diffusion for the deep-case hardening of titanium alloys. *Mater. Sci. Eng. A* **2000**, *280*, 303–310. [CrossRef]

29. Zhang, Z.X.; Dong, H.; Bell, T.; Xu, B.S. The effect of deep-case oxygen hardening on the tribological behaviour of a-C:H DLC coatings on Ti6Al4V alloy. *J. Alloys Compd.* **2008**, *464*, 519–525. [CrossRef]

30. Zabler, S. Interstitial Oxygen diffusion hardening—A practical route for the surface protection of titanium. *Mater. Charact.* **2011**, *62*, 1205–1213. [CrossRef]

31. Jamesha, M.; Sankara Narayanan, T.S.N.; Chu, P.K. Thermal oxidation of titanium: Evaluation of corrosion resistance as a function of cooling rate. *Mater. Chem. Phys.* **2013**, *138*, 565–572. [CrossRef]

32. Januszewicz, B.; Siniarski, D. The glow discharge plasma influence on the oxide layer and diffusion zone formation during process of thermal oxidation of titanium and its alloys. *Vacuum* **2006**, *81*, 215–220. [CrossRef]

33. Moskalewicz, T.; Wendler, B.; Zimowski, S.; Dubiel, B.; Czyrska-Filemonowicz, A. Microstructure, micro-mechanical and tribological properties of the nc-WC/a-C nanocomposite coatings magnetron sputtered on non-hardened and oxygen hardened Ti–6Al–4V alloy. *Surf. Coat. Technol.* **2010**, *205*, 2668–2677. [CrossRef]

34. Poggie, R.A.; Kovacs, P.; Davidson, J.A. Oxygen Diffusion Hardening of Ti-Nb-Zr Alloys. *Mater. Manuf. Process.* **1996**, *11*, 185–197. [CrossRef]

35. Barril, S.; Mischler, S.; Landolt, D. Electrochemical effects on the fretting corrosion behaviour of Ti6Al4V in 0.9% sodium chloride solution. *Wear* **2005**, *259*, 282–291. [CrossRef]

36. Albayrak, Ç.; Hacısalihoğlu, İ.; Yenalvangölü, S.; Alsaran, A. Tribocorrosion behavior of duplex treated pure titanium in Simulated Body Fluid. *Wear* **2013**, *302*, 1642–1648. [CrossRef]

37. Davidson, J.A.; Mishra, A.K.; Kovacs, P.; Poggie, R.A. New surface-hardened, low-modulus, corrosion-resistant Ti-13Nb-13Zr alloy for total hip arthroplasty. *Biomed. Mater. Eng.* **1994**, *4*, 231–243. [CrossRef]

38. Langford, R.M.; Clinton, C. In situ lift-out using a FIB-SEM system. *Micron* **2004**, *35*, 607–611. [CrossRef] [PubMed]

39. Oliver, W.C.; Pharr, G.M. Measurement of hardness and elastic modulus by instrumented indentation: Advances in understanding and refinements to methodology. *J. Mater. Res.* **2004**, *19*, 3–20. [CrossRef]

40. ISO 20808:2016 Fine Ceramics (Advanced Ceramics, Advanced Technical Ceramics)—Determination of Friction and Wear Characteristics of Monolithic Ceramics by Ball-on-Disc Method. Available online: https://www.iso.org/standard/65415.html (accessed on 10 April 2021).

41. Massalski, T.B.; Okamoto, H.; Subramanian, P.R.; Kacprzak, L. *Binary Alloy Phase Diagrams*; ASM International: Materials Park, OH, USA, 1990; Volume 3. [CrossRef]

42. Lee, T. Variation in Mechanical Properties of Ti-13Nb-13Zr Depending on Annealing Temperature. *Appl. Sci.* **2020**, *10*, 7896. [CrossRef]

43. Lederer, S.; Lutz, P.; Fürbeth, W. Surface modification of Ti 13Nb 13Zr by plasma electrolytic oxidation. *Surf. Coat. Technol.* **2018**, *335*, 62–71. [CrossRef]

44. Krishna, D.S.R.; Brama, Y.L.; Sun, Y. Thick rutile layer on titanium for tribological applications. *Tribol. Int.* **2007**, *40*, 329–334. [CrossRef]

45. Kolubaev, A.V.; Tarasov, S.Y. Studies on formation and destruction of surface layers under severe friction. *Facta Univ.* **1997**, *1*, 429–432.

46. Majumdar, P.; Singh, S.B.; Chakraborty, M. Wear response of heat-treated Ti–13Zr–13Nb alloy in dry condition and simulated body fluid. *Wear* **2008**, *264*, 1015–1025. [CrossRef]

47. Mohan, L.; Anandan, C. Wear and corrosion behavior of oxygen implanted biomedical titanium alloy Ti–13Nb–13Zr. *Appl. Surf. Sci.* **2013**, *282*, 281–290. [CrossRef]

48. Mc Cafferty, E.; Hubler, G.K. Electrochemical behavior of palladium implanted titanium. *J. Electrochem. Soc.* **1978**, *125*, 1892–1893. [CrossRef]

 *materials*   **MDPI**

Article

The Effect of Sn Addition on Zn-Al-Mg Alloy; Part I: Microstructure and Phase Composition

Peter Gogola *, Zuzana Gabalcová, Martin Kusý and Henrich Suchánek

Institute of Materials Science, Faculty of Materials Science and Technology in Trnava, Slovak University of Technology in Bratislava, Ulica Jána Bottu 25, 917 24 Trnava, Slovakia; zuzana.gabalcova@stuba.sk (Z.G.); martin.kusy@stuba.sk (M.K.); henrich.suchanek@stuba.sk (H.S.)
* Correspondence: peter.gogola@stuba.sk

Abstract: In this study, the addition of Sn on the microstructure of Zn 1.6 wt.% Al 1.6 wt.% Mg alloy was studied. Currently, the addition of Sn into Zn-Al-Mg based systems has not been investigated in detail. Both as-cast and annealed states were investigated. Phase transformation temperatures and phase composition was investigated via DSC, SEM and XRD techniques. The main phases identified in the studied alloys were η(Zn) and α(Al) solid solutions as well as Mg_2Zn_{11}, $MgZn_2$ and Mg_2Sn intermetallic phases. Addition of Sn enabled the formation of Mg_2Sn phase at the expense of Mg_xZn_y phases, while the overall volume content of intermetallic phases is decreasing. Annealing did not change the phase composition in a significant way, but higher Sn content allowed more effective spheroidization and agglomeration of individual phase particles.

Keywords: Zn-based alloy; phase composition; XRD; DSC; microstructure formation; Sn-addition; intermetallic phases

Citation: Gogola, P.; Gabalcová, Z.; Kusý, M.; Suchánek, H. The Effect of Sn Addition on Zn-Al-Mg Alloy; Part I: Microstructure and Phase Composition. *Materials* **2021**, *14*, 5404. https://doi.org/10.3390/ma14185404

Academic Editor: Jana Bidulská

Received: 27 July 2021
Accepted: 14 September 2021
Published: 18 September 2021

Publisher's Note: MDPI stays neutral with regard to jurisdictional claims in published maps and institutional affiliations.

Copyright: © 2021 by the authors. Licensee MDPI, Basel, Switzerland. This article is an open access article distributed under the terms and conditions of the Creative Commons Attribution (CC BY) license (https://creativecommons.org/licenses/by/4.0/).

1. Introduction

Zn-Al alloys are used as corrosion protection coatings for a series of applications including steel strands used to reinforce overhead power lines. Such strands ensure the overall mechanical rigidity of aluminum conductors and thus reduce the number of supporting towers needed for a specific distance of overhead power lines [1]. These power lines are designed to operate at about 180 °C, while due to an increased current load, they may heat up to 300 °C. [1,2] Pure Zn coatings are not suitable in such conditions. Above 200 °C, the pure Zn coating starts to react with the steel [1,3,4] substrate and continues to form ZnFe intermetallics. Such a reaction can reduce the actual steels cross section area and thus reduce the cables' mechanical properties. This limitation can be overcome by alloying [5,6].

Nowadays, various Zn-Al-Mg alloy system coatings are available, such as the commercially well-known Magizinc (MZ) with a chemical composition of Zn 1.6 wt.% Al 1.6 wt.% Mg [7–14].

Mg is added to Zn-based coatings to further increase their corrosion protection capabilities. Mg is useful especially for increasing the galvanic protection offered by a coating at cut-edges and mechanically damaged spots. Additional alloying may further improve the properties of Zn-Al-Mg alloys as reported by several literature sources [15–25]. Unfortunately, there is still lack of information about Sn addition into these systems [26].

In recent years, the effects of Sn addition on the microstructure of Mg- and Al-based alloys has been studied. This includes microstructural stability upon thermal exposure. Sn is known to have a high affinity to Mg, creating mainly the Mg_2Sn intermetallic phase. All in all, a positive influence of Sn on Mg and Al-based alloys has been reported depending on the amount of Sn added to these alloys. Alloying enabled the formation of Mg_2Sn in all these alloys [27–33]. The aim of this research is to confirm if Mg_2Sn phase is also preferred compared to Mg_xZn_y phases in the current alloy system as suggested by studied literature sources [32,34]. Phase composition and overall microstructure character will be investigated on as-cast samples.

As one of the potential applications involves a long-term thermal exposure, microstructure and phase composition will be investigated after a representative annealing treatment of each alloy as well. Experimental annealing temperature will be set to 310 °C to be clearly above potential exposure temperatures, but below the melting temperatures (~340 °C) of the investigated alloys.

Mg_2Zn_{11} and $MgZn_2$ are the most common intermetallic phases in the currently investigated Zn-based systems. $MgZn_2$ is reported to be less ductile, while both are reported to be slightly less ductile compared to Mg_2Sn [18–20,29,35,36]. The possibility to replace Mg_xZn_y phases by Mg_2Sn and thus reduce the overall volume content of intermetallic phases in a Zn 1.6 wt.% Al 1.6 wt.% Mg based alloy would indicate an interesting research path for further extensive investigation of corrosion and mechanical properties of such alloys.

2. Materials and Methods

Five different alloys with the designed nominal composition of Zn-1.6Al-1.6Mg-xSn (wt.%), where x = 0.0, 0.5, 1.0, 2.0 and 3.0 wt.%, respectively, were prepared by melting pure Zn at 470 °C and mixing in the appropriate amount of a 50 wt.% Al and 50 wt.% Mg master alloy. These raw materials were preheated to 400 °C to facilitate their rapid melting. Due to the low melting point of Sn, it was added in the last step. Table 1 indicates that the measured bulk chemical compositions of the alloys by glow discharge optical emission spectroscopy (GDOES, Spectruma GDA 750, Spectruma Analytik GmbH, Hof, Germany) are in good agreement with the nominal ones.

Table 1. Chemical composition of the studied alloys (wt.%).

Alloy	Al	Mg	Sn	Zn
MZ + 0.0Sn	1.56 ± 0.07	1.40 ± 0.01	0.07 ± 0.02	bal.
MZ + 0.5Sn	1.64 ± 0.02	1.41 ± 0.01	0.52 ± 0.01	bal.
MZ + 1.0Sn	1.62 ± 0.03	1.45 ± 0.02	1.06 ± 0.02	bal.
MZ + 2.0Sn	1.57 ± 0.01	1.44 ± 0.01	1.95 ± 0.01	bal.
MZ + 3.0Sn	1.57 ± 0.12	1.43 ± 0.05	2.69 ± 0.06	bal.

Casting was done from 470 °C of melt temperature into a water-cooled copper mold with a diameter of 30 mm and depth of 20 mm. During casting, the sample temperature was continuously measured at a sampling frequency of 25 Hz with K-type thermocouples attached to the mold surface. The cooling rate of 60 °C/s was established.

Two types of cylindrical samples were prepared for each alloy: (i) as-cast samples and (ii) cast and subsequently solution annealed at 310 °C for 1 h. The annealing step was finished by ice-water quenching. A cooling rate of 75 °C/s was recorded. The selected solution annealing temperature corresponds to the $\gamma + \eta$ region of the Zn-Al system [37], while it is clearly below the melting point of all chosen alloys. Cooling in cold water ensured a very good control of the annealing time.

Vickers hardness tests were carried out in line with ISO 6507-1 [38] on polished surfaces of as-cast and annealed samples via a BUEHLER Indentamet 1105 (Buehler Ltd., Lake Bluff, IL, USA) hardness tester at an applied load of 9.8 N, holding time at the point of load application was 10 s.

DSC measurements were carried out by the Perkin Elmer Diamond DSC (Perkin Elmer Inc., Billerica, MA, USA) device. The DSC samples were cut from as-cast samples to a target weight of 5 mg. The samples were heated to the temperature of 500 °C at a heating rate of 10 °C/min and then cooled to ambient temperature at a cooling rate of 10 °C/min under the protective argon atmosphere.

The XRD analysis was carried out on metallic filings of the as-cast and annealed samples by the PANalytical Empyrean X-ray diffractometer (XRD) (Malvern Panalytical

Ltd., Malvern, UK). The procedure to measure on metallic filings instead of bulk castings was chosen to limit the influence of casting texture on the recorded XRD pattern. The casting texture added additional complexity to the XRD measurements by influencing the theoretical relative intensities for the individual crystallographic planes. This made the quantitative analysis very unreliable due to the complex texture corrections needed. The measurements were performed in Bragg–Brentano geometry. Theta-2Theta angle range between 10° and 148° 2Theta was chosen. The XRD source was set to 40 kV and 40 mA. The incident beam was modified by 0.04 rad soller slit, 1/4° divergence slit and 1/2° anti-scatter slit. The diffracted beam path was equipped with a 1/2° anti-scatter slit, 0.04 rad soller slit, Ni beta filter and PIXcel3D position sensitive detector operated in 1D scanning mode. The phase quality was analyzed using PANalytical Xpert High Score program (HighScore Plus version 3.0.5) with the ICSD FIZ Karlsruhe database. Quantitative results were determined from XRD patterns using the Rietveld refinement-based program MAUD version 2.84 [39]. The program uses an asymmetric pseudo-Voight function to describe experimental peaks. Instrument broadening was determined by measuring the NIST660c LaB_6 (The National Institute of Standards and Technology, Gaithersburg, MD, USA) line position and line broadening standard and introduced to the Rietveld refinement program via the Caglioti equation. Anisotropy size-strain model was applied to Zn solid solution while other phases were treated by isotropic models. A minor discrepancy between the nominal and measured peak intensities was corrected using the spherical harmonic functions with fibre symmetry. The quality of the fit was in all analyzed samples below 10% R_{wp}.

The metallographic preparation of DSC, as-cast and annealed samples consisted of standard grinding using abrasive papers and polishing on diamond pastes with various grain sizes of down to 0.25 μm.

The microstructure evaluation was performed by the JEOL JSM 7600F scanning electron microscopy (SEM, Jeol Ltd., Tokyo, Japan) with a Schottky field emission electron source operating at 20 kV and 90 μA. The samples were placed at a working distance of 15 mm and documented using a backscattered electron detector. The chemical element analysis was performed via the Oxford Instruments X-Max silicon drift detector, energy dispersive X-ray spectrometer (EDS, Oxford Instruments plc, Abingdon, UK).

Image analysis was performed on at least 15 sites for each sample by ImageJ FIJI 1.53c software [40]. Area ratio of η(Zn) based areas and other microstructure components were established.

All results are listed as the average values of multiple measurements with ±standard deviation error bars.

3. Results

The DSC curves of MZ + xSn alloys system in Figure 1a–e show charts of DSC heating runs. Figure 1 shows the most relevant section of the recorded data, while measurements were done from 20 to 500 °C at a heating rate of 10 °C/min. The first heating runs recorded are presented to observe the reactions in the as-cast samples during heating and subsequent melting, including eutectoid reactions (not visible in DSC cooling runs).

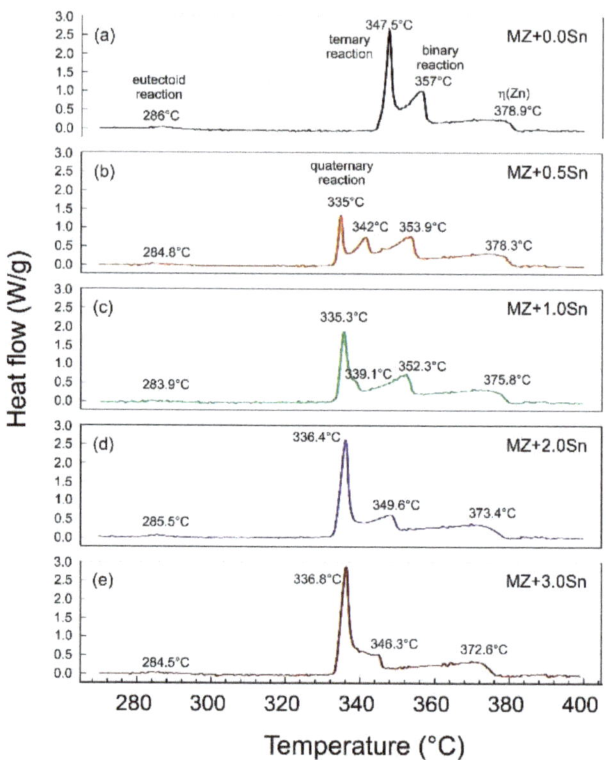

Figure 1. Comparison of the DSC curves for all investigated alloys.

Hence, the individual microstructure features in the as-cast samples cannot be clearly distinguished; the DSC samples after the cooling run were investigated via SEM. These observations enabled us to identify the individual reactions for each recorded peak.

Peaks were recorded describe the reactions of the following phases: η(Zn)–hcp Zn-based solid solution; γ(Al)–fcc Al-based solid solution present above the eutectoid reaction temperature reported at 275 °C in the Zn-Al system [37] up to melting temperature; α(Al)–fcc Al-based solid solution present below the Zn-Al system eutectoid reaction; Mg_2Zn_{11}, $MgZn_2$ and Mg_2Sn intermetallic phases of respective systems.

The curve in Figure 1a corresponds to the MZ + 0.0Sn alloy. The first peak observed at about 285.0 °C corresponds to the eutectoid transformation α(Al) + η(Zn) → γ(Al). This peak is repeated for all alloys observed. Melting of pure MZ + 0.0Sn starts at 344.0 °C (peak maximum at 347.5 °C) with melting of the ternary eutectic consisting of η(Zn), γ(Al) and Mg_2Zn_{11} phases. An example of such areas is provided in Figure 2a. This is followed by the melting of the Zn/Mg_xZn_y binary eutectic (peak maximum at 357.0 °C). A clear example can be observed in Figure 2b. Mg_xZn_y corresponds to a mixture of Mg_2Zn_{11} and $MgZn_2$ phases as documented in Figure 2c. A closer detail of this area is given in Figure 2d) showing also the α(Al) + η(Zn) eutectoid particles in detail. Chemical composition of the individual phases is documented in Table 2. The last peak corresponds with the melting of the Zn rich dendrites (peak maximum at 378.9 °C) [41].

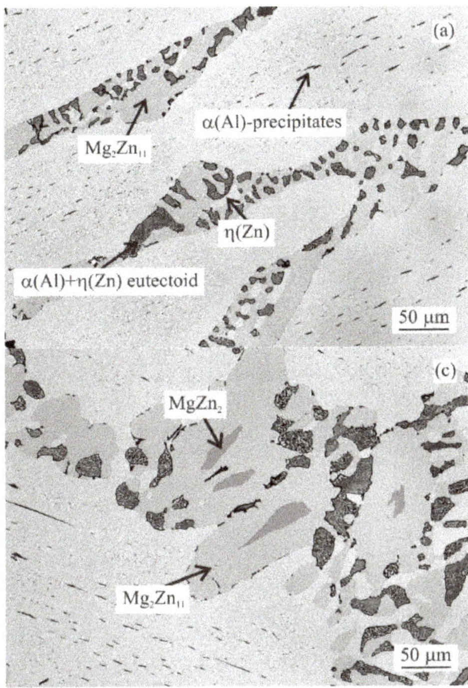
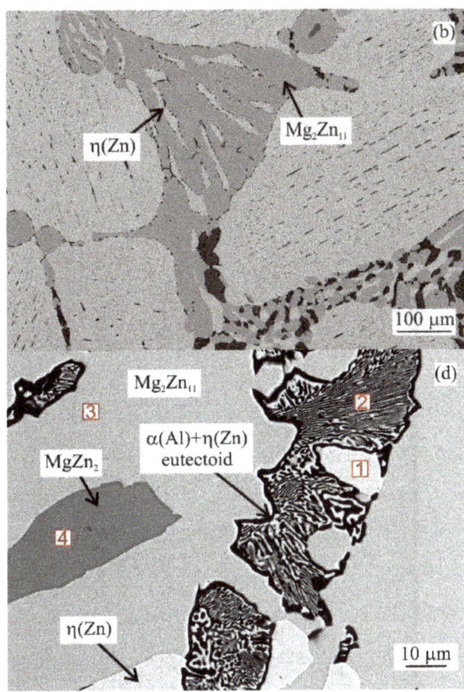

Figure 2. Selected features of MZ + 0.0Sn alloy microstructure after DSC measurement: (**a**) example of ternary eutectic; (**b**) example of binary eutectic; (**c**) example of Mg_2Zn_{11} and $MgZn_2$ mixture; (**d**) area with EDS measurement points 1–4 listed in Table 2.

Table 2. EDS chemical composition of selected sites (at.%).

Chemical Element (at.%)	Site No.				
	1	2	3	4	5
Zn	99.20	54.50	83.70	66.30	66.10
Al	0.80	45.50	-	-	-
Mg	-	-	16.30	33.70	-
Sn	-	-	-	-	33.90
Phase/Region	η(Zn)	α(Al) + η(Zn) eutectoid	Mg_2Zn_{11}	$MgZn_2$	Mg_2Sn

As indicated in Figure 2a, η(Zn) dendrites are always decorated by a needle like α(Al) particles. These are formed as a result of the decreasing solubility of Al in Zn in the temperature range below ~285 °C (see Figure 1). These particles are observed for all alloys investigated.

As observed in Figure 1b, the addition of 0.5 wt.% of Sn enables the formation of a peak at 335 °C. This effect corresponds to newly emerging quaternary eutectic areas consisting of η(Zn), γ(Al), Mg_2Zn_{11} and Mg_2Sn phases. A typical such area is shown in Figure 3a) with a selected detail in Figure 3b. The chemical composition of the Mg_2Sn phase was measured and listed in Table 2. These areas start to melt at about 334 °C (peak maximum at 335.0 °C). The next peak corresponds to the melting of the ternary eutectic (342.0 °C) composed of η(Zn), γ(Al) and Mg_2Zn_{11}. The adjacent peak indicates the melting of the η(Zn) + Mg_2Zn_{11} binary eutectic (353.9 °C). The peak at 378.3 °C indicates the melting of the remaining η(Zn) dendrites.

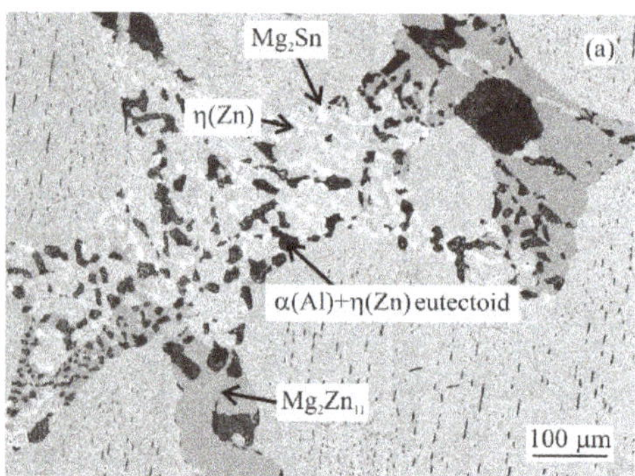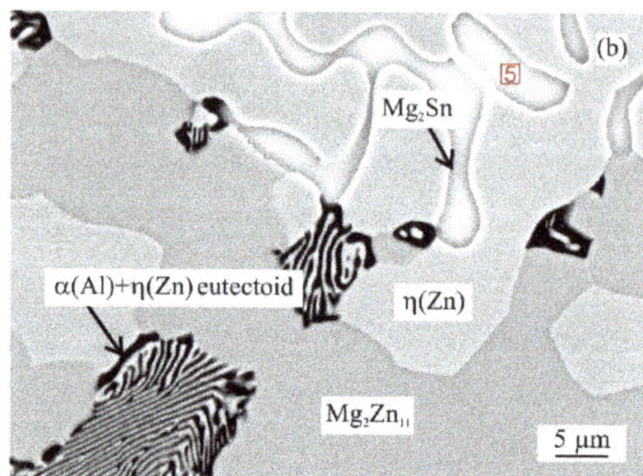

Figure 3. Quaternary eutectic area in MZ + 0.5Sn alloy microstructure after DSC measurement: (**a**) example of quaternary eutectic; (**b**) closer detail of such area including EDS measurement point 5 listed in Table 2.

The alloy with 1 wt.% of Sn (Figure 1c) has a minor peak left corresponding to the ternary eutectic (339.1 °C), while the peak corresponding to the quaternary eutectic (335.3 °C) increased further in peak area. Both other peaks correspond to the same reactions as described above.

With 2 wt.% of Sn (Figure 1d), the melting starts at 333.5 °C. This reaction melts the complex eutectic area shown in Figure 4. The peak corresponding to the ternary eutectic reaction is not resolved separately anymore. The peak at 349.6 °C in this alloy represents the melting of two binary eutectics: $\eta(Zn) + Mg_2Zn_{11}$ as well as $\eta(Zn) + Mg_2Sn$. The last peak in this DSC curve at 373.4 °C corresponds again to $\eta(Zn)$ dendrites.

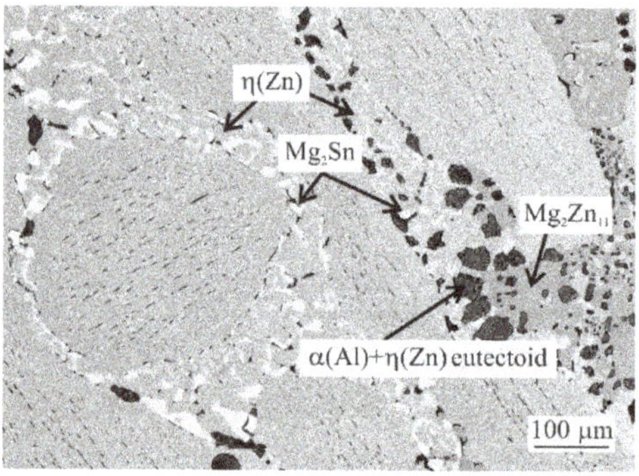

Figure 4. Binary and quaternary eutectic areas in MZ + 2.0Sn alloy microstructure after DSC measurement.

Adding 3 wt.% of Sn (Figure 1e) changes the peak of the quaternary eutectic reaction only slightly (peak maximum at 336.8 °C). The peak observed at 346.3 °C corresponds according to microstructure observations solely to the melting of the binary $\eta(Zn) + Mg_2Sn$ eutectic. Melting of the remaining $(\eta)Zn$ is indicated by the peak at 372.6 °C.

XRD measurements were performed on the metallic filings prepared from the bulk samples. Figures 5 and 6 shows the XRD patterns for selected alloys in the as-cast and annealed state, respectively. A quantitative analysis using the Rietveld method was performed considering the phases listed in Table 3 characterized in the ICSD FIZ Karlsruhe database. These phases enabled the identification of all significant peaks in the measured XRD patterns.

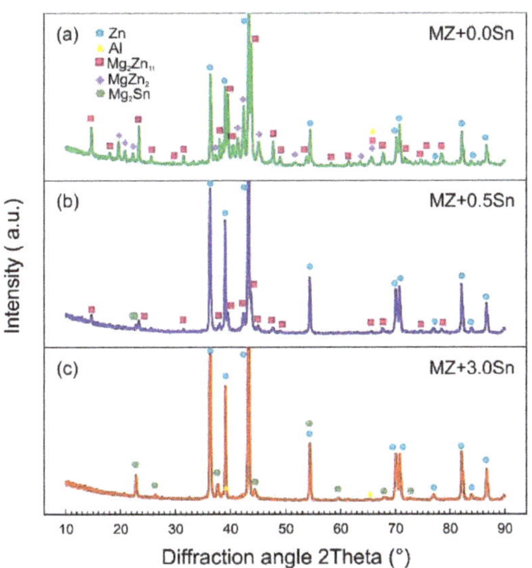

Figure 5. XRD diffraction patterns recorded on powder samples of selected alloys in as-cast state (**a**) MZ + 0.0Sn, (**b**) MZ + 0.5Sn, (**c**) MZ + 3.0Sn.

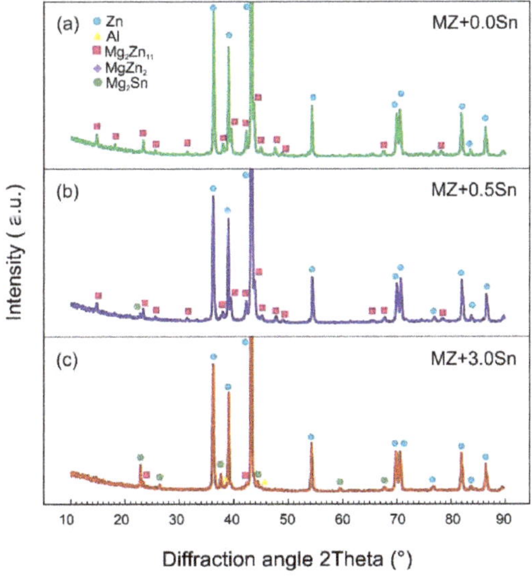

Figure 6. XRD diffraction patterns recorded on powder samples of selected alloys in annealed state (**a**) MZ + 0.0Sn, (**b**) MZ + 0.5Sn, (**c**) MZ + 3.0Sn.

Table 3. Phases identified during XRD analysis.

Phase Chemical Formula	Reference Code–ICSD Database FIZ Karlsruhe	Crystal System	Space Group	Space Group Number
$\eta(Zn) = Zn + 2$ at.% Al	98-024-7160 modified according [42]	Hexagonal	$P6_3/mmc$	194
$\alpha(Al) = Al + 14$ at.% Zn	98-060-6001 modified according [42]	Cubic	$Fm\overline{3}m$	225
$MgZn_2$	98-010-4897	Hexagonal	$P6_3/mmc$	194
Mg_2Zn_{11}	98-010-4898	Cubic	$Pm\overline{3}$	200
Mg_2Sn	98-064-2855	Cubic	$Fm\overline{3}$	225

As the cooling speed in all experiments was rather high at 60–75 °C/s for both as-cast and annealed samples, it was assumed that the solubility changes below the eutectoid transformation [$\gamma(Al) \rightarrow \alpha Al + \eta(Zn)$ at 275 °C] will be significantly limited. Such a phenomenon was reported for $\alpha(Al)$ as well as $\eta(Zn)$ phases by Gogola et al. [42] based on XRD measurements of Zn-Al based samples. To enable the correct quantitative analysis of $\alpha(Al)$ as well as $\eta(Zn)$ phases, their chemical composition had to be changed by adding 14 at.% of Zn and 2 at.% of Al, respectively, as suggested in this publication. The chemical composition of the phases was changed in MAUD software before the quantitative analysis of each XRD pattern. The soundness of this approach was double checked comparing the GDOES chemical composition data with the chemical composition calculated from XRD quantitative analysis for each sample.

The volume content of individual phases in the as-cast samples evolved as shown in Figure 7.

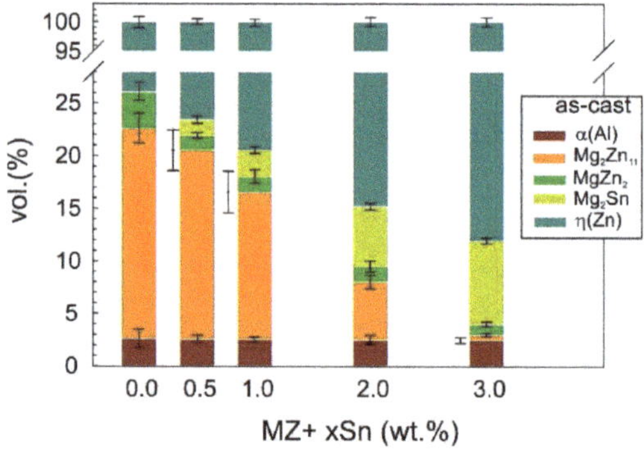

Figure 7. Phase composition in vol.% in metallic bulk samples as measured on metallic filings from as-cast samples.

In all alloys, both Mg_2Zn_{11} as well as the non-equilibrium $MgZn_2$, the intermetallic phases can be detected. The overall content of Mg_xZn_y intermetallic phases was reduced from 23.5 vol.% to about 1.5 vol.% by adding 3 wt.% of Sn into the basic Zn 1.6 wt.% Al 1.6 wt.% Mg (MZ) alloy. At the same time, the Mg_2Sn phase occupied about 8 vol.% of the as-cast MZ + 3.0Sn alloy.

The content of $MgZn_2$ was reduced from 3.5 vol.% to ~1.5 vol.% by adding 0.5 wt.% of Sn. The further addition of Sn did not change the content of this phase significantly. Its content was gradually further reduced to ~1 vol.% by adding up to 3 wt.% of Sn into the alloy. However, at ~1 vol.% of $MgZn_2$, the detectability limit of $MgZn_2$ was likely reached in the current alloy with the applied measurement setup.

Mg_2Zn_{11} phase was detected in all alloys. Its content was gradually reduced from ~20 vol.% down to below 1 vol.% by adding up to 3 wt.% of Sn.

Peaks corresponding to Mg_2Sn can be already clearly identified in the as-cast MZ + 0.5Sn sample representing as low as 1.5 vol.% of this phase. Its volume content clearly gradually increased up to ~8 vol.% when 3 wt.% of Sn was added.

The content of $\alpha(Al)$ is calculated to be 2.5 vol.% on average across all alloys investigated. Addition of Sn did not change the content of $\alpha(Al)$ in a significant way.

Annealing the investigated alloys clearly influenced their phase composition (Figure 8) as calculated from XRD measurements (Figure 6).

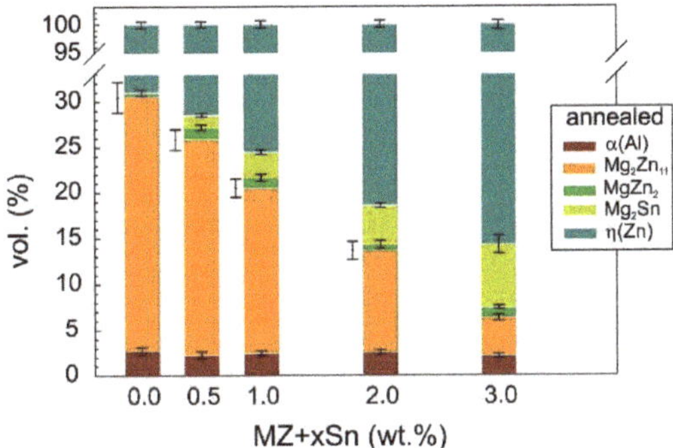

Figure 8. Phase fractions in vol.% in metallic bulk samples as measured on metallic filings of annealed samples.

In the MZ + 0.0Sn alloy, the content of non-equilibrium $MgZn_2$ is significantly reduced after annealing. Its content is reduced from ~3.5 vol.% to below 1 vol.%. For all the other alloys, the content of $MgZn_2$ is rather similar in both as-cast and annealed states.

The content of the Mg_2Zn_{11} phase increases to ~28 vol.% after annealing the MZ + 0.0Sn alloy, hence indicating that this is the equilibrium phase for this alloy [34,37]. Additionally, for all the other compositions, the volume content of Mg_2Zn_{11} increases after annealing. Annealing changes the content of Mg_2Sn only slightly. Most noticeably, its content reduces at 2 and 3 wt.% of Sn, probably in favor of Mg_2Zn_{11}. The content of $\alpha(Al)$ remains basically unchanged by the annealing process.

Comparison of microstructure images for the most important edge cases is given in Figure 9a–f. Figure 9a,c,e correspond to as-cast states, while Figure 9b,d,f correspond to the annealed state.

The microstructure of MZ + 0.0Sn samples is formed by $\eta(Zn)$ phase dendrites, where the interdendritic areas are formed by a mixture of binary $\eta(Zn)/Mg_xZn_y$ eutectic and ternary $\eta(Zn)/\alpha(Al)/Mg_xZn_y$ eutectic [43,44]. Adding 0.5 wt.% of Sn changes the microstructure appearance in an insignificant way (Figure 9a vs. Figure 9c). On the other hand, in Figure 9e, we can clearly observe the presence of $\eta(Zn)/Mg_2Sn$ binary eutectic regions. Gradual addition of Sn reduces the amount of $(\eta)Zn/Mg_xZn_y$ eutectic regions and gives rise to $\eta(Zn)/Mg_2Sn$ eutectic regions. Additionally, areas formed by $\eta(Zn)/\alpha(Al)/Mg_xZn_y$ ternary eutectic are reduced in favor of probably $\eta(Zn)/\alpha(Al)/Mg_xZn_y/Mg_2Sn$ quaternary eutectics.

Figure 9b,d,f shows the microstructure of selected alloys after annealing. $\eta(Zn)$ loses its dendritic character as well as all interdendritic areas being spheroidized, while areas with common chemical compositions are connected. None of the previously described eutectic regions can be recognized. Based on XRD, the vast majority of Mg_xZn_y particles are corresponding to Mg_2Zn_{11}. As annealing was done above the eutectoid temperature of the Zn-Al system (285 °C as reported by DSC measurements, Figure 1), Al rich particles were spheroidized as $\gamma(Al)$ particles. Hence, the outer shape of Al rich particles remained frozen while decomposition to $\alpha(Al) + \eta(Zn)$ eutectoid particles took place upon cooling from annealing temperature.

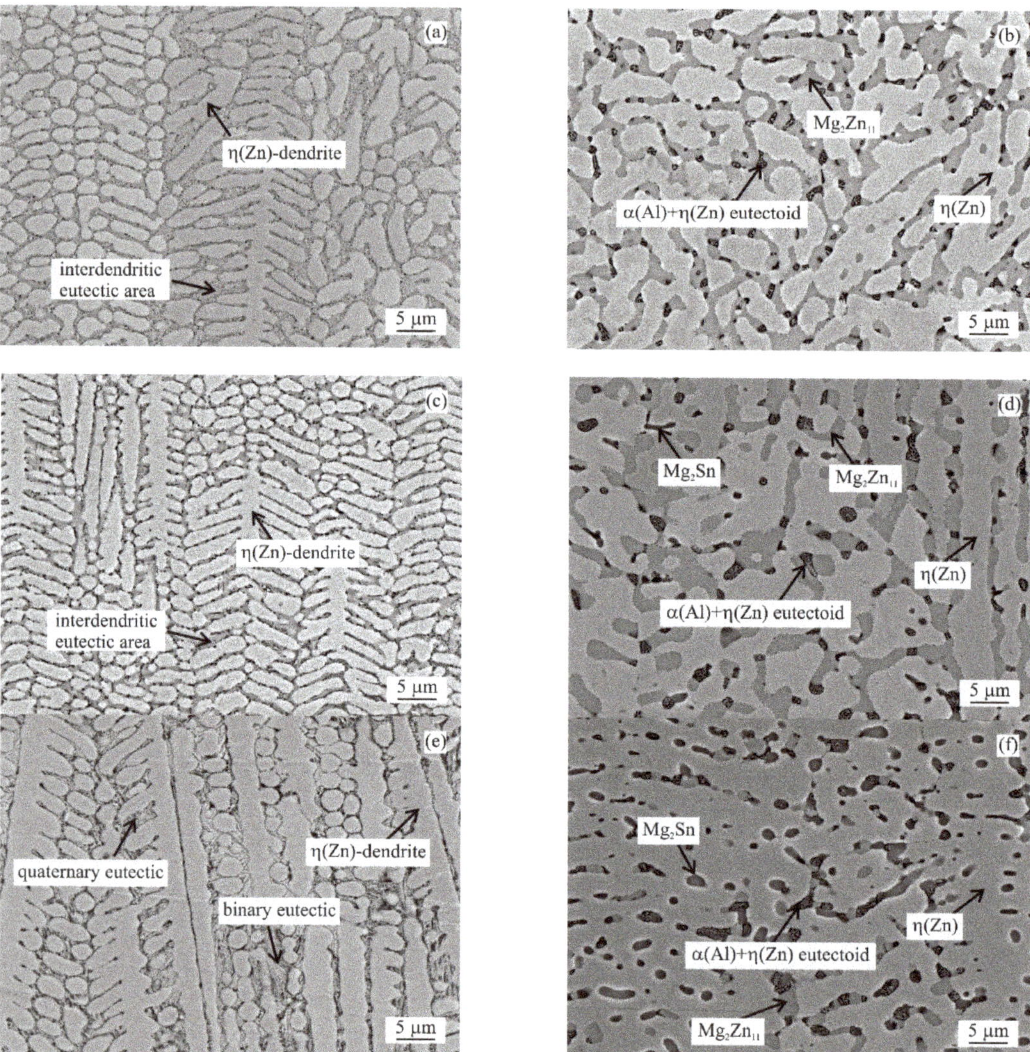

Figure 9. Microstructure images for selected samples, longitudinal section, near sample surface: (**a**) MZ + 0.0Sn as-cast, (**b**) MZ + 0.0Sn annealed, (**c**) MZ + 0.5Sn as-cast, (**d**) MZ + 0.5Sn annealed, (**e**) MZ + 3.0Sn as-cast, (**f**) MZ + 3.0Sn annealed.

Figure 10 summarizes the vol.% of all other microstructure components apart from η(Zn).

For as-cast samples, this represents the interdendritic spaces which are formed mainly by various eutectics including a certain portion of η(Zn) phase solidified within them as well as α(Al) + η(Zn) eutectoid particles.

For the annealed samples, it was possible to clearly distinguish between η(Zn) matrix and all intermetallic phase particles along with α(Al) + η(Zn) eutectoid particles.

The difference between data for as-cast and annealed samples is mainly caused by the fact that η(Zn) solidified in the interdendritic spaces of the as-cast samples cannot be separately identified, while during annealing, these small η(Zn) particles are allowed to connect to the larger primary η(Zn) areas forming a uniform matrix.

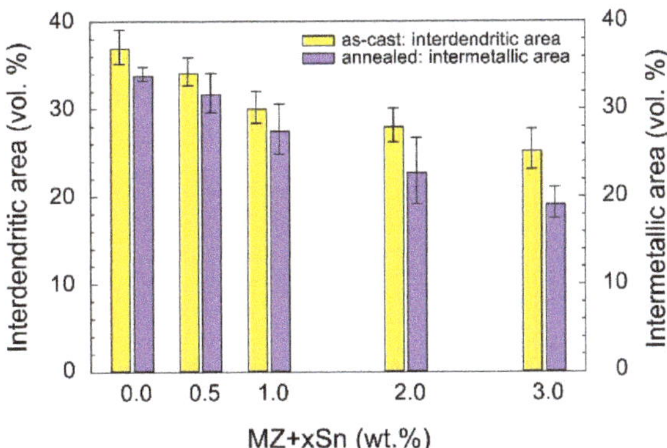

Figure 10. Vol.% of microstructural components as determined from SEM image analysis.

In general, the vol.% of all other microstructure components outside of η(Zn) is decreasing with the addition of Sn into the investigated alloys.

Microhardness was measured on the as-cast and the corresponding annealed samples as well. All samples showed an over twice higher hardness compared to pure Zn. MZ + 0.0Sn showed the hardness of ~113 HV, while the gradual addition of Sn was almost linearly decreasing the alloy's hardness down to ~85 HV measured on the MZ + 3.0Sn as-cast sample. Annealing of these samples further decreased their hardness, however, by only 3 to 7% compared to respective as-cast states of each alloy.

4. Discussion

Sn was chosen to supplement the composition of ZnAlMg based alloys. Similar Zn-based alloy compositions have not been reported in the literature so far, probably due to the concerns related to the corrosion properties of such alloys. These properties will be investigated in adjacent research.

Different aspects of the alloy's microstructure were investigated. The main phases identified were in the general agreement with the published data as follows: η(Zn), α(Al), Mg_2Zn_{11}, $MgZn_2$ [7–18,20–24] and Mg_2Sn [26,27,32,33].

DSC curves can be clearly described only by investigating the microstructure of DSC samples after cooling in the DSC equipment (Figures 2–4). Copper mold as-cast microstructure is an order of magnitude finer and hence less likely to be clearly identified. A clear comparison can be for example given by comparing the size Mg_2Sn/η(Zn) eutectic particles in Figure 4 (DSC sample of MZ + 3.0Sn) and Figure 9e (as-cast sample of MZ + 3.0Sn). Furthermore, the kinetic of solidification may also affect the order in which the phase or phase mixtures are formed. DSC curves show that the ternary η(Zn)/α(Al)/Mg_xZn_y eutectics of MZ + 0.0Sn alloy are replaced by quaternary η(Zn)/α(Al)/Mg_xZn_y/Mg_2Sn eutectics by adding 1 wt.% of Sn. By gradually adding Sn from 0.0 to 1.0 wt.%, the peak of the quaternary eutectic areas is formed at ~335 °C, while the peak of the ternary eutectic areas, found at temperatures in the range from 347.5 to 339.1 °C, is being gradually suppressed. Further addition of Sn (2.0 and 3.0 wt.%) causes the ternary reaction peak to shift towards even lower temperatures and being completely overlapped by the quaternary reaction peak. This is in line with available assessment of liquidus projection for the Zn-Mg-Sn ternary system [32,45]. These systems also predict a decrease in liquidus temperature for less complex eutectics when Sn concentration is approaching a more complex eutectic point near the Zn-rich corner of this system.

Peaks at 357–346.3 °C represent the binary eutectics. While DSC curves suggest only a gradual peak shift of binary eutectic reaction, the microstructure investigation showed that Mg_2Zn_{11}/η(Zn) binary eutectics is being replaced by Mg_2Sn/η(Zn) in case of the MZ + 3.0Sn alloy. Based on available ternary Zn-Mg-Sn assessments [32,45], it is hypothesized, that Sn supports the preferential formation of $Mg_2Sn/\eta(Zn)$ binary eutectic instead of $Mg_2Zn_{11}/\eta(Zn)$ eutectic mixture. This is observed in the currently investigated system as well, despite the presence of Al as an additional alloying element. It is also worth mentioning that the temperature difference between the binary and ternary eutectic points calculated [32,45] is only 1 °C. This may cause difficulties to reveal the real order of solidification reactions since even a slight local chemical difference or temperature heterogeneity may cause local fluctuation and a competitive formation of $Mg_2Sn/\eta(Zn)$ and $Mg_2Zn_{11}/\eta(Zn)$ binary eutectic areas as indicated for the MZ + 2.0Sn alloy.

Heating curves shown in Figure 1 depicted also the peaks corresponding to the $\alpha(Al) + \eta(Zn) \rightarrow \gamma(Al)$ eutectoid transformation at ~285 °C. The corresponding reaction cannot be observed during a cooling DSC run since this eutectoid transformation is rather sluggish; therefore, it is without a detectable heat release. The microstructure investigation shows that the $\gamma(Al) \rightarrow \alpha(Al) + \eta(Zn)$ reaction clearly occurs; however, probably over a much broader temperature range compared to the heating curves. This reaction might be finished even at an ambient temperature [46].

Adding Sn reduced the volume content of Mg_xZn_y intermetallic phases, and these were replaced by Mg_2Sn particles. This behavior is in line with the literature findings on similar systems [27,32]. The addition of Sn mainly affects the volume content of Mg_2Zn_{11} (Figure 7).

For similar Zn-Al-Mg alloys, the sources report the same two Mg_xZn_y phases to be present [21,22,25,47]. Vlot et al. [47] identified only $MgZn_2$ in similar alloys, while other literature sources confirmed the presence of both phases mentioned [21,22,25]. In our samples, Mg_2Zn_{11} is the primary phase; however, $MgZn_2$ was also clearly identified by both XRD and even SEM/EDX. At 3.5 vol.%, its content was highest in the as-cast MZ + 0.0Sn sample. As $MgZn_2$ is a non-equilibrium phase in the current system, its content is significantly reduced by annealing the basic MZ + 0.0Sn alloy at 310 °C for 1 h.

Mg_2Zn_{11} and $MgZn_2$ are competing phases and their final ratio is complex to predict and control even in a simple Mg-Zn alloy as reported by several sources [23,24]. All in all, their presence will depend on several factors like exact alloy composition or cooling rate [48].

The amount and distribution of intermetallic particles appears to have a direct influence on the microhardness of the studied alloys. Overall volume content of intermetallic phases is decreasing with the increasing wt.% of Sn as measured by XRD (Figures 7 and 8) as well as the SEM image analysis (Figure 10). This is reflected in the decreasing alloy hardness summarized in Figure 11.

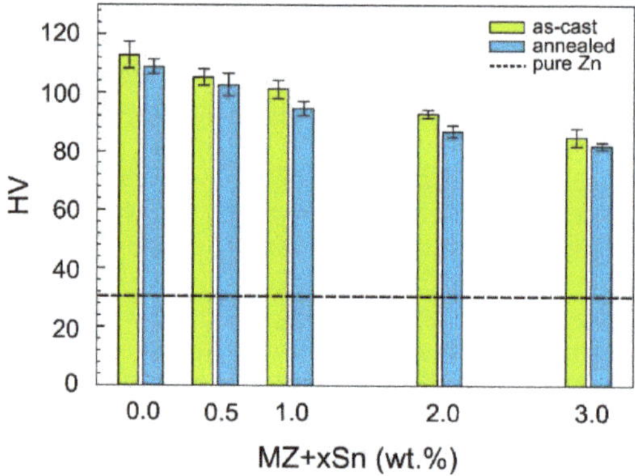

Figure 11. HV1 microhardness of the investigated samples.

Formation of Mg_2Sn particles and the gradual increase of their vol.% is reported to cause an increase in hardness for specific Mg-based alloys with similar Sn content [29–31].

The same mechanism does not apply to our Zn-based alloys, as in our samples, the overall vol.% of intermetallics is decreasing.

The hardness of as-cast samples is marginally higher compared to annealed samples (Figure 11). This is most probably caused mainly by the change in shape and distribution of the intermetallic particles. This can be observed when comparing the images of as-cast vs annealed conditions in Figure 9. The annealing allows $\eta(Zn)$ to diffuse from eutectics in the interdendritic areas towards the primary $\eta(Zn)$ dendrites, hence changing the eutectic nature of the interdendritic areas. For the MZ + 0.0Sn and MZ + 0.5Sn alloys, the original dendritic character of the microstructure can still be recognized even after annealing Figure 9a vs. Figure 9b,c vs. Figure 9d. For higher Sn content, this is not possible. The addition of 1 to 3 wt.% of Sn into this alloy system, enabled a more effective spheroidization and agglomeration of individual phase particles. Hence, the annealing had a more apparent influence on the microstructure of these alloys (MZ + 1.0Sn, MZ + 2.0Sn, MZ + 3.0Sn).

For as-cast and annealed samples a different ratio of microstructural components was established for the same alloys by SEM. Nevertheless, both as-cast and annealed samples show the same trend compared to the XRD quantitative analysis. Additionally, for the annealed samples, the SEM image analysis and XRD analysis are in even better agreement.

5. Conclusions

Melting of MZ + 0.0Sn starts at 344 °C, while with the addition of 0.5–3.0 wt.% of Sn, melting starts already at 334 °C. Melting is finished at 382 °C for the MZ + 0.0Sn and this temperature is being continuously decreased to 376 °C by the addition of up to 3 wt.% of Sn.

Main phases identified in the MZ + 0.0Sn alloy were $\eta(Zn)$ and $\alpha(Al)$ solid solutions as well as Mg_2Zn_{11} and $MgZn_2$ intermetallic phases. Addition of Sn enabled the formation of Mg_2Sn intermetallic phase at the expense of Mg_xZn_y phases, while mainly affecting the vol.% of Mg_2Zn_{11}.

The microstructure is dendritic for all as-cast alloys. The interdendritic areas are formed by the binary, ternary and quaternary eutectics specific for each alloy. Alloying with Sn causes the following changes of microstructural components: ternary eutectics consisting of $\eta(Zn)$, $\alpha(Al)$ and Mg_xZn_y phases are gradually replaced by quaternary $\eta(Zn)$, $\alpha(Al)$, Mg_xZn_y and Mg_2Sn eutectics. Binary $\eta(Zn)$ + Mg_xZn_y eutectics are gradually replaced by binary $\eta(Zn)$ + Mg_2Sn eutectics.

For the MZ + 0.0Sn and MZ + 0.5Sn alloys, the original dendritic character of the microstructure can still be recognized even after annealing. At the same time, the individual phases from the eutectics are connected to discrete particles, and thus the original eutectics are not recognizable anymore. Introducing 1 to 3 wt.% of Sn into this alloy system enabled a more effective spheroidization and agglomeration of individual phase particles significantly changing even the shape of the primary $\eta(Zn)$ dendrites.

Annealing causes slight changes in the phase composition. For MZ + 0.0Sn mainly $MgZn_2$ is transformed to Mg_2Zn_{11}. For the alloys with Sn, the volume content of Mg_2Zn_{11} is partially increased mainly at the expense of Mg_2Sn.

The microhardness is decreasing with the increasing of Sn content. The annealing changes the microhardness only slightly.

Based on microstructure observation, these alloys are overall suitable for coatings exposed to extended high temperature exposure. As coatings of steel substrates, their corrosion properties will be at least maintained as reported in part two of this research: The effect of Sn addition on Zn-Al-Mg alloy-Part II.

Author Contributions: Conceptualization, P.G. and M.K.; methodology, P.G., Z.G. and M.K.; validation, P.G., Z.G., M.K. and H.S.; formal analysis, P.G., Z.G.; investigation, P.G., Z.G., M.K. and H.S.; resources, P.G., Z.G. and H.S.; data curation, P.G., Z.G. and M.K.; writing—original draft preparation, P.G. and Z.G.; writing—review and editing, P.G. and M.K.; visualization, P.G. and Z.G.; supervision, P.G. and Z.G.; project administration, P.G.; funding acquisition, M.K. All authors have read and agreed to the published version of the manuscript.

Funding: This research was supported by the Grant Agency VEGA of the Slovak Ministry of Education, Research, Science and Sport, Project No. 1/0490/18: "The effect of microstructure and phase composition on corrosion resistance of hot dip alloys" and by the Slovak Research and Development Agency under the Contract no. APVV-20-0124.

Institutional Review Board Statement: Not applicable.

Informed Consent Statement: Not applicable.

Data Availability Statement: Data sharing is not applicable.

Conflicts of Interest: The authors declare no conflict of interest.

References

1. Coffin, C.; Depamelaere, H.; King, D.; Van Raemdonck, W. Evaluation of High Temperature Behavior of Zn and ZnAl Coatings on Core Wires and Strands for ACSR, ACSS and Alike Overhead Power Conductors, WJI 2010, Monterrey ITC Preview. pp. 68–75. Available online: https://issuu.com/wirejournal/docs/de-aug10-reduced (accessed on 24 February 2021).
2. Kiessling, F.; Nefzger, P.; Nolasco, J.F.; Kaintzyk, U. *Overhead Power Lines—Planning, Design, Construction*, 1st ed.; Springer: Berlin/Heidelberg, Germany, 2003; p. 250. [CrossRef]
3. Mingyuan, G.U.; Notis, M.R.; Marder, A.R. The Effect of Continuous Heating on the Phase Transformations in Zinc-Iron Electrodeposited Coatings. *Metall. Trans. A* **1991**, *22*, 1737–1743. [CrossRef]
4. Onishi, M.; Wakamatsu, Y.; Miura, H. Formation and Growth Kinetics of Intermediate Phases in Fe-Zn Diffusion Couples. *Trans. JIM* **1974**, *15*, 331–337. [CrossRef]
5. Rico, Y.; Carrasquero, E.J. Microstructural Evaluation of Double-Dip Galvanized Coatings on Carbon Steel. *MRS Adv.* **2017**, *2*, 3917–3923. [CrossRef]
6. Wright, R.N. Wire coatings. In *Wire Technology—Process Engineering and Metallurgy*, 2nd ed.; Elsevier—Butterworth Heinemann Books: Oxforfd, UK, 2011; pp. 245–256, ISBN 9780128026786.
7. Tanaka, S.; Honda, K.; Takahashi, A.; Morimoto, Y.; Kurosaki, M.; Shindo, H.; Nishimura, K.; Sugiyama, M. The performance of a Zn-Al-Mg-Si hot-dip galvanized steel sheet. In Proceedings of the 5th International Conference on Zinc and Zinc Alloy Coated Steel Sheet (GALVATECH 2001), Brussels, Belgium, 26–28 June 2001; Lamberights, M., Ed.; Verl. Stahleisen: Düsseldorf, Germany, 2001; p. 153, ISBN 3-514-00673-3.
8. Morimoto, Y.; Honda, K.; Nishimura, K.; Tanaka, S.; Takahashi, A.; Shindo, H.; Kurosaki, M. Excellent Corrosion-resistant Zn-Al-Mg-Si Alloy Hot-dip Galvanized Steel Sheet "SUPER DYMA". *Nippon Steel Tech. Rep.* **2003**, *87*, 24–26.
9. Nishimura, K.; Kato, K.; Shindo, H. Highly Corrosion-resistant Zn-Mg Alloy Galvanized Steel Sheet for Building Construction Materials. *Nippon. Steel Tech. Rep.* **2000**, *81*, 85–88. [CrossRef]
10. Shindo, H.; Nishimura, K.; Okado, T.; Nishimura, N.; Asai, K. Developments and Properties of Zn-Mg Galvanized Steel Sheet "DYMAZING" Having Excellent Corrosion Resistance. *Nippon. Steel Tech. Rep.* **1999**, *79*, 63–67.
11. Nishimura, K.; Shindo, H.; Kato, K.; Morimoto, Y.; Funaki, S.O. Microstructure and corrosion behaviour of Zn–Mg–Al hot-dip galvanized steel sheet. In Proceedings of the 4th International Conference on Zinc and Zinc Alloy Coated Steel Sheet (GALVATECH '98), Chiba, Japan, 20–23 September 1998; Masuko, N., Ed.; ISIJ: Tokyo, Japan, 1998; pp. 437–442.
12. Shindo, H.; Nishimura, K.; Kato, K. Anti-Corrosion in Atmospheric Exposure of Zn–Mg–Al Hot-Dip Galvanized Steel Sheet. In Proceedings of the 4th International Conference on Zinc and Zinc Alloy Coated Steel Sheet (GALVATECH '98), Chiba, Japan, 20–23 September 1998; Masuko, N., Ed.; ISIJ: Tokyo, Japan, 1998; pp. 433–4363.
13. Tsujimura, T.; Komatsu, A.; Andoh, A. Influence of Mg content in coating layer and coating structure on corrosion resistance of hot-dip Zn–Al–Mg–Si alloy coated steel shee. In Proceedings of the 5th International Conference on Zinc and Zinc Alloy Coated Steel Sheet (GALVATECH 2001), Brussels, Belgium, 26–28 June 2001; Lamberights, M., Ed.; Verl. Stahleisen: Düsseldorf, Germany, 2001; pp. 145–152, ISBN 3-514-00673-3.
14. Kittaka, T.; Andoh, A.; Komatsu, A.; Tsujimura, T.; Yamaki, N.; Watanabe, K. Hot-Dip Zn-Al-Mg Coated Steel Sheet Excellent in Corrosion Resistance and Surface Appearance and Process for the Production Thereof. U.S. Patent 6,235,410, 22 May 2001.
15. Tokuda, S.; Muto, I.; Sugawara, Y.; Takahashi, M.; Matsumoto, M.; Hara, N. Micro-electrochemical investigation on the role of Mg in sacrificial corrosion protection of 55mass%Al-Zn-Mg coated steel. *Corros. Sci.* **2017**, *129*, 126–135. [CrossRef]
16. Vida, T.A.; Brito, C.; Lima, T.S.; Spinelli, J.E.; Cheung, N. Near-eutectic Zn-Mg alloys: Interrelations of solidification thermal parameters, microstructure length scale and tensile/corrosion properties. *Curr. Appl. Phys.* **2009**, *51*, 2355–2363. [CrossRef]
17. Krystýnová, M.; Doležal, P.; Fintová, S.; Zapletal, J.; Marada, T.; Wasserbauer, J. Characterization of Brittle Phase in Magnesium Based Materials Prepared by Powder Metallurgy. *Key Eng. Mater.* **2018**, *784*, 61–66. [CrossRef]
18. Vida, T.A.; Soares, T.; Septimio, R.S.; Brito, C.C.; Cheung, N.; Garcia, A. Effects of Macrosegregation and Microstructure on the Corrosion Resistance and Hardness of a Directionally Solidified Zn-5.0wt.%Mg Alloy. *Mater. Res.* **2019**, *22*, 1–13. [CrossRef]
19. Pinc, J.; Čapek, J.; Kubásek, J.; Veřtát, P.; Hosová, K. Microstructure and mechanical properties of the potentially biodegradable ternary system Zn-Mg0.8-Ca0.2. *Procedia Struct. Integr.* **2019**, *23*, 21–26. [CrossRef]
20. De Bruycker, E.; Zermout, Z.; De Cooman, B.C. Zn-Al-Mg Coatings—Thermodynamic Analysis and Microstructure Related Properties. *Mater. Sci. Forum* **2007**, *539–543*, 1276–1281. [CrossRef]

21. De Bruycker, E.; De Cooman, B.C.; De Meyer, M. Experimental study and microstructure simulation of Zn-Al-Mg coatings. *Rev. Metall—CIT* **2005**, *102*, 543–550. [CrossRef]

22. Akdeniz, V.M.; Wood, J.V. Microstructures and phase selection in rapidly solidified Zn-Mg alloys. *J. Mater. Sci.* **1996**, *31*, 545–550. [CrossRef]

23. Liu, H.Y.; Jones, H. Solidification Microstructure Selection and Characteristics in the Zinc-Based Zn-Mg System. *Acta Metall. Mater.* **1992**, *40*, 229–239. [CrossRef]

24. Prosek, T.; Nazarov, A.; Goodwin, F.; Šerák, J.; Thierry, D. Improving corrosion stability of Zn-Al-Mg by alloying for protection of car bodies. *Surf. Coat. Technol.* **2016**, *306*, 439–447. [CrossRef]

25. Farahany, S.; Tat, L.H.; Hamzah, E.; Bakhsheshi-Rad, H.R.; Cho, M.H. Microstructure development, phase reaction characteristics and properties of quaternary Zn-0.5Al-0.5Mg-xBi hot dipped coating alloy under slow and fast cooling rates. *Surf. Coat. Technol.* **2017**, *315*, 112–122. [CrossRef]

26. Gondek, J.; Babinec, M.; Kusý, M. The corrosion performance of Zn-Al-Mg based alloys with tin addition in neutral salt spray environment. *J. Achiev. Mater. Manuf. Eng.* **2015**, *70*, 70–77.

27. Chen, J.; Chen, Z.; Yan, H.; Zhang, F.; Liao, K. Effects of Sn addition on microstructure and mechanical properties of Mg–Zn–Al alloys. *J. Alloys Compd.* **2008**, *461*, 209–215. [CrossRef]

28. Chen, L.; Yan, A.; Liu, H.; Li, X. Strength and fatigue fracture behaviour of Al-Zn-Mg-Cu-Zr(-Sn) alloys. *Trans. Nonferrous Met. Soc. China* **2013**, *223*, 2817–2825. [CrossRef]

29. Kim, B.; Do, J.S.; Lee, H.; Park, I. In situ fracture observation and fracture toughness analysis of squeeze cast AZ51–xSn magnesium alloys. *Mater. Sci. Eng. A* **2010**, *527*, 6745–6757. [CrossRef]

30. Turen, Y. Effect of Sn addition on microstructure, mechanical and casting properties of AZ91 alloy. *Mater. Des.* **2013**, *49*, 1009–1015. [CrossRef]

31. Wang, X.-Y.; Wang, Y.-F.; Wang, C.; Xu, S.; Rong, J.; Yang, Z.-Z.; Wang, J.-G.; Wang, H.-Y. A simultaneous improvement of both strength and ductility by Sn addition in as-extruded Mg-6Al-4Zn alloy. *J. Mater. Sci. Technol.* **2020**, *49*, 117–125. [CrossRef]

32. Ghosh, P.; Mezbahul-Islam, M.; Medraj, M. Critical assessment and thermodynamic modeling of Mg-Zn, Mg-Sn, Sn-Zn and Mg-Sn-Zn systems. *Calphad* **2012**, *36*, 28–43. [CrossRef]

33. Guangyin, Y.; Yangshan, S.; Wenjiang, D. Effects of Sn addition on the microstructure and mechanical properties of AZ91 magnesium alloy. *Scripta Mater.* **2001**, *308*, 34–38. [CrossRef]

34. Mezbahul-Islam, M.; Mostafa, A.O.; Medraj, M. Essential Magnesium Alloys Binary Phase Diagrams and Their Thermochemical Data. *J. Mater. Hindawi* **2014**, *2014*, 704283. [CrossRef]

35. Agarwal, G. (RWTH Aachen, Germany). Personal communication, 2014.

36. Kevorkijan, V.; Škapin, S.D. Preparation and Study of Mg2Sn-based Composites with Different Compositions. *Mater. Tehnol.* **2010**, *44*, 251–259.

37. Durmus, Y.E.; Montiel Guerrero, S.S.; Tempel, H.; Hausen, F.; Kungl, H.; Eichel, R.-A. Influence of Al alloying on the electrochemical behavior of Zn electrodes for Zn–Air batteries with neutral sodium chloride electrolyte. *Front. Chem.* **2019**, *7*, 800. [CrossRef]

38. ISO. *Metallic Materials—Vickers Hardness Test*; 6507-1:2018; International Organization for Standardization: Geneva, Switzerland, 2018.

39. Lutterotti, L.; Matthies, S.; Wenk, H.R. MAUD (Material Analysis Using Diffraction): A user friendly Java program for rietveld texture analysis and more. In Proceedings of the 12th International Conference on Textures of Materials (ICOTOM-12), Montreal, QC, Canada, 9–13 August 1999; Volume 1, p. 1599. Available online: http://hdl.handle.net/11572/57067 (accessed on 25 March 2021).

40. Schindelin, J.; Arganda-Carreras, I.; Frise, E.; Kaynig, V.; Longair, M.; Pietzsch, T.; Preibisch, S.; Rueden, C.; Saalfeld, S.; Schmid, B.; et al. Fiji: An open-source platform for biological-image analysis. *Nat. Methods* **2012**, *9*, 676–682. [CrossRef]

41. De Bruycker, E. Zn-Al-Mg coatings: Thermodynamic Analysis and Microstructure-Related Properties. Ph.D. Thesis, Ghent University, Ghent, Belgium, 2006.

42. Gogola, P.; Gabalcová, Z.; Suchánek, H.; Babinec, M.; Bonek, M.; Kusý, M. Quantitative x-ray diffraction analysis of Zn-Al based alloys. *Arch. Metall. Mater.* **2020**, *65*, 959–966. [CrossRef]

43. Prosek, T.; Persson, D.; Stoulil, J.; Thierry, D. Composition of corrosion products formed on Zn–Mg, Zn–Al and Zn–Al–Mg coatings in model atmospheric conditions. *Corros. Sci.* **2014**, *86*, 231–238. [CrossRef]

44. Raghavan, V. Al-Mg-Zn (Aluminum-Magnesium-Zinc). *JPED* **2007**, *28*, 203–208. [CrossRef]

45. Meng, F.G.; Wang, J.; Liu, L.B.; Jin, Z.P. Thermodynamic modeling of the Mg–Sn–Zn ternary system. *J. Alloys Compd.* **2010**, *508*, 570–581. [CrossRef]

46. Larsson, L.E. Pre-precipitation and precipitation phenomena in the Al-Zn system. *Acta Metall.* **1967**, *15*, 35–45. [CrossRef]

47. Vlot, M.; Zuijderwijk, M.; Toose, M.; Elliot, L.; Bleeker, R.; Maalman, T. Hot dip ZnAlMg coatings: Microstructure and forming properties. In Proceedings of the 7th International Conference on Zinc and Zinc Alloy Coated Steel Sheet (Galvatech '07), Osaka, Japan, 19–21 November 2007; Tsuru, T., Ed.; ISIJ: Tokyo, Japan, 2007; pp. 574–579.

48. Kim, J.N.; Lee, C.S.; Jin, Y.S. Structure and Stoichiometry of MgxZny in Hot-Dipped Zn–Mg–Al Coating Layer on Interstitial-Free Steel. *Met. Mater. Int.* **2018**, *24*, 1090–1098. [CrossRef]

Article

The Effect of Sn Addition on Zn-Al-Mg Alloy; Part II: Corrosion Behaviour

Zuzana Gabalcová *, Peter Gogola, Martin Kusý and Henrich Suchánek

Faculty of Materials Science and Technology in Trnava, Institute of Materials Science, Slovak University of Technology in Bratislava, Ulica Jána Bottu 25, 917 24 Trnava, Slovakia; peter.gogola@stuba.sk (P.G.); martin.kusy@stuba.sk (M.K.); henrich.suchanek@stuba.sk (H.S.)

* Correspondence: zuzana.gabalcova@stuba.sk

Abstract: Corrosion behaviour of Sn (0.0, 0.5, 1.0, 2.0 and 3.0 wt.%)-doped Zn 1.6 wt.% Al 1.6 wt.% Mg alloys exposed to salt spray testing was investigated. Intergranular corrosion was observed for all alloys in both as-cast and annealed states. However, due to microstructure spheroidisation in the annealed samples, potential intergranular corrosion paths are significantly reduced. Samples with 0.5 wt.% of Sn showed the best corrosion properties. The main corrosion products identified by XRD analysis for all samples were simonkolleite and hydrozincite. Occasionally, ZnO and AlO were identified in limited amounts.

Keywords: Zn-based alloy; Sn-addition; corrosion products; salt spray test; intergranular corrosion; corrosion penetration depth; weight loss

Citation: Gabalcová, Z.; Gogola, P.; Kusý, M.; Suchánek, H. The Effect of Sn Addition on Zn-Al-Mg Alloy; Part II: Corrosion Behaviour. *Materials* **2021**, *14*, 5290. https://doi.org/10.3390/ma14185290

Academic Editor: Daoguang He

Received: 2 August 2021
Accepted: 11 September 2021
Published: 14 September 2021

Publisher's Note: MDPI stays neutral with regard to jurisdictional claims in published maps and institutional affiliations.

Copyright: © 2021 by the authors. Licensee MDPI, Basel, Switzerland. This article is an open access article distributed under the terms and conditions of the Creative Commons Attribution (CC BY) license (https://creativecommons.org/licenses/by/4.0/).

1. Introduction

A wide range of commercial Zn-based hot-dip coatings are used for corrosion protection. These also include Zn-Al-Mg-based coatings such as Magizinc (MZ) with Zn 1.6 wt.% Al and 1.6 wt.% Mg. It is widely used in the coating industry including steel sheet production for building, energetics, and the automotive industry [1–12].

Neutral salt spray testing (NSST) is used as an industry standard for corrosion resistance testing. Zn-Al-Mg coatings perform notably better compared to conventional hot-dip Zn coatings. The presence of Mg in the Zn-Al-Mg coatings enables the stabilisation of protective corrosion products like simonkolleite and hydrozincite [13–16]. Regarding the microstructure, Mg addition to binary Zn-Al alloys results in the formation of intermetallic phases such as Zn_2Mg and $Zn_{11}Mg_2$. These phases are more corrosion active even compared to the η(Zn) phase, hence enabling the more effective cathodic protection of steel substrates [17]. They are formed within eutectics in the interdendritic areas of primary η(Zn) dendrites. Unfortunately, these phases are also enabling the cathodic protection of this Zn-based matrix, hence overall corrosion attack starts as the intergranular (IG) corrosion. Sources have reported this phenomenon, however only on the coatings with a limited thickness of up to 50 μm. In all these corrosion test results, substantial parts of the coatings were affected by IG corrosion locally, even across the entire coating [18–20].

The potentials of additional alloying of Zn-Al-Mg systems by Cr, Zr, Ti Mo, Mn, Si, etc. have been already studied in the literature [10]. Sn is also an interesting candidate due to its high affinity to Mg [21]. The preliminary research [12] into the development of microstructure and corrosion resistance of the Zn-Al-Mg + Sn alloy system has shown that Sn can affect the phase composition, and consequently the corrosion properties of MZ. In the follow-up to these results, this system is being investigated with an extended experimental scope in Parts I and II of the current articles. The main aim of these additional experiments is to observe if long time exposure to rather high temperatures (1 h at 310 °C) have a significant influence on the corrosion properties of these alloys. Based on these inputs, bulk samples were chosen for our research. This enabled to investigate the IG

179

corrosion phenomena for these alloys in both as-cast and annealed states without the limit of a coating's thickness.

2. Materials and Methods

As already described in Part I of this article [22], five different alloys with the designed nominal composition of Zn-1.6Al-1.6Mg-xSn (wt.%), where x = 0.0, 0.5, 1.0, 2.0 and 3.0 wt.%, respectively, were prepared by melting pure Zn at 470 °C and mixing in the appropriate amount of a 50 wt.% Al + 50 wt.% Mg master alloy. These raw materials were preheated to 400 °C to facilitate their rapid melting. Due to the low melting point of Sn, it was added in the last step. Table 1 indicates that the measured chemical compositions of the alloys by glow discharge optical emission spectroscopy (GDOES, Spectruma GDA 750, Spectruma Analytik GmbH, Hof, Germany) are in a good agreement with the nominal ones.

Table 1. Chemical composition of the studied alloys (wt.%).

Alloy	Al	Mg	Sn	Zn
MZ + 0.0Sn	1.56 ± 0.07	1.40 ± 0.01	0.07 ± 0.02	bal.
MZ + 0.5Sn	1.64 ± 0.02	1.41 ± 0.01	0.52 ± 0.01	bal.
MZ + 1.0Sn	1.62 ± 0.03	1.45 ± 0.02	1.06 ± 0.02	bal.
MZ + 2.0Sn	1.57 ± 0.01	1.44 ± 0.01	1.95 ± 0.01	bal.
MZ + 3.0Sn	1.57 ± 0.12	1.43 ± 0.05	2.69 ± 0.06	bal.

As a reference material for the corrosion test, 4N5 purity Zn-samples were cast. Two types of cylindrical samples were prepared for each alloy: (i) as-cast samples and (ii) cast and subsequently solution annealed at 310 °C for 1 h.

Casting was done from 470 °C of melt temperature into a water-cooled copper mould with a diameter of 30 mm and depth of 20 mm. During casting, the sample temperature was continuously measured and an average cooling rate of 60 °C/s was established. The annealing step was finished by quenching it in a water bath below 10 °C at an average cooling rate of 75 °C/s.

The investigated surface of the as-cast and annealed samples was subjected to grinding using up to 4000 grit abrasive papers. The surface topography was determined using a ZEISS LSM700 laser scanning confocal microscope (LSCM, Carl Zeiss AG, Oberkochen, Germany). The 405 nm light source was used, which in combination with a Epiplan-Apochromat 50×/0.95 objective enabled to reach step sizes of 250 nm on the X and Y axes as well as 200 nm on the Z axis. These surfaces were subjected to the corrosion in the salt chamber.

The investigated samples were coated with Lacomit Varnish to prevent the corrosion of the entire sample and limit the exposed area. The exposed surface was digitally scanned to double check the exposed area. These data, together with the surface topography data, made it possible to calculate the real surface area exposed to the corrosion on each sample.

The neutral salt spray corrosion test (NSST) was performed in a Co.Fo.Me.Gra 400E (CO.FO.ME.GRA. Srl, Milano, Italy) corrosion chamber according to the ISO 9227:2017 Standard [23]. The NSST samples were immediately exposed in the cabinet to a 5 wt.% NaCl solution. The air pressure of the atomized saline solution was maintained in the range of 95–105 kPa, and the temperature inside the cabinet was 35 ± 2 °C, pH level was 6.6–7.1, and the salt solution deposition rate 125–200 mL/h/m^2. Custom holders were used to keep the prescribed sample orientation of 15° from the vertical position.

Exposure times for all types of samples were 250, 500, 750 and 1000 h. Three samples were prepared for all as-cast and annealed conditions for all exposure times. All in all, 144 individual samples were exposed at the same time. After the salt spray testing, the samples were dried at room temperature for 24 h at minimum before being further processed. After drying, loose corrosion products were removed and collected separately. It was of upmost importance to prevent any kind of a mechanical damage to the metallic sample surface. The bulk samples were cleaned by acetone and dried on air. The initial weight of the specimen was measured (w_0) by using the Mettler Toledo XPR205 weighing balance (Mettler-Toledo International Inc., Columbus, OH, USA). According to the ASTM G31 Standard [24], the specimens were immersed in the chromate acid (CrO_3) to ensure that the corrosion products were removed. Samples were cleaned in 60 s intervals. After each cleaning interval, the samples were repeatedly weighed. This process was considered finished when less than 5 mg of weight was lost after a cleaning cycle for all three repeats of a condition [23–25]. The final weight for each sample was recorded (w_n). The recorded weight difference was normalized by the exposed area of each sample (A_n) corrected by the sample topography coefficient (k). The topography coefficient is retrieved from LSCM software as the ratio between real surface area, incorporating surface topography, and the ideal surface. This value was 1.09 on average. These data enabled the calculation the average weight change (w') for each condition in mg/mm^2 according to equation:

$$w' = \frac{w_0 - w_n}{A_n\, k} \tag{1}$$

The metallographic preparation on the longitudinal section (along the cylinder axis) of corroded as-cast and annealed samples consisted of standard grinding using abrasive papers, polishing on diamond pastes with various grain sizes down to 0.25 μm.

The microstructure evaluation was performed by the JEOL JSM 7600F scanning electron microscopy (SEM, Jeol Ltd., Tokyo, Japan) with a Schottky field emission electron source operating at 20kV and 90 μA. The samples were placed at a working distance of 15 mm and documented using a backscattered electron detector.

The quantitative analysis of IG corrosion depth was performed by ImageJ 1.53c software [26] along the longitudinal section for each condition. At least 150 individual values were recorded for each data point.

The weight measurements are displayed with +/− standard deviation error bars and the depth of IG corrosion measurements are given with +/− standard error.

The X-ray diffraction (XRD) analysis was carried out by the PANalytical Empyrean X-ray diffractometer (Malvern Panalytical Ltd., Malvern, UK) with configurations as detailed in Table 2. The measurements were performed on the samples after 1000 h of NSST with Ni filtered Cu radiation. X-ray diffraction data were further analysed qualitatively using the PANalytical Xpert High Score program (HighScore Plus 3.0.5 version) with ICSD FIZ Karlsruhe database. These findings were confirmed and enhanced using the Rietveld refinement-based program, MAUD version 2.84 [27]. The program uses an asymmetric pseudo-Voight function to describe the experimental peaks. The instrument broadening was determined by measuring the NIST660c LaB$_6$ (The National Institute of Standards and Technology, Gaithersburg, MD, USA) line position and line broadening standard and introduced to the Rietveld refinement program (MAUD version 2.84) via the Caglioti equation. An anisotropic size-strain model was applied to the majority of corrosion products, while the other phases were treated by isotropic models. A minor discrepancy between nominal and measured peak intensities was corrected using the spherical harmonic functions with fibre symmetry. The quality of the fit was in all analysed patterns achieved below 10% R$_{wp}$.

Table 2. List of XRD measurements settings.

Sample	XRD Device Geometry	Angle Range	Incident Beam	Diffracted Beam	Detector
Powder of loose corrosion products scraped from the surface of the bulk samples	Theta-2Theta, Bragg-Brentano geometry	5°–90° 2Theta	Divergence slit: 1/4° Soller slit: 0.04 rad Anti-scatter slit: 1/2°	Anti-scatter slit: 1/2° Soller slit: 0.04 rad	PIXcel3D detector in 1D scanning mode
Corroded surface of bulk samples after loosely attached corrosion products were removed	Grazing incident (GI) with 0.5° incident angle	5°–80° 2Theta	Parallel beam optics with: Divergence slit: 1/16° Soller slit: 0.04 rad	Parallel plate collimator: 0.27° Soller slit: 0.04 rad	Scintillation detector

3. Results

As mentioned before, the weight changes for each sample were calculated according to Equation (1) and the obtained data are plotted in Figures 1 and 2. Reference Zn samples showed a gradual weight loss for both as-cast and annealed conditions as expected. It can be observed that for several as-cast samples, a weight gain rather than a weight loss was recorded. The annealed samples showed the weight loss for all conditions as expected. MZ + 3.0Sn showed the best results at even 40% lower values compared to MZ + 0.0Sn.

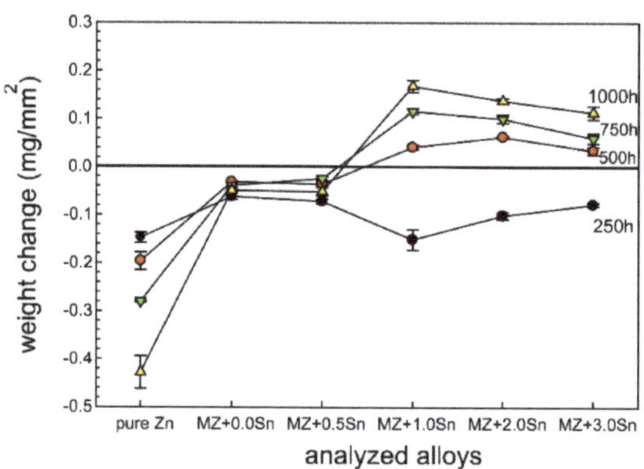

Figure 1. Weight change after corrosion measured on as-cast samples.

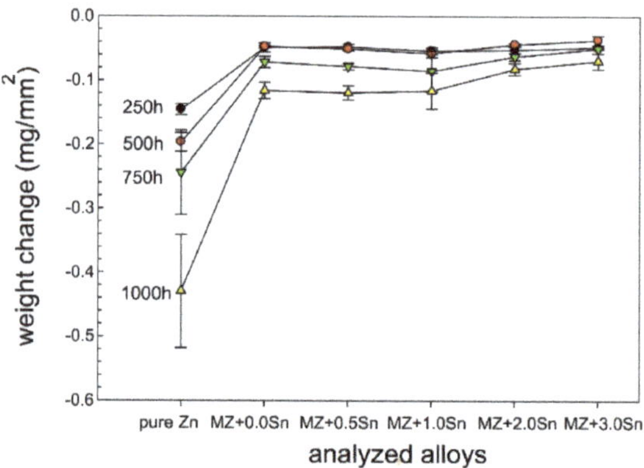

Figure 2. Weight change after corrosion measured on annealed samples.

Since the weight gain instead of the weight loss was recorded for several as-cast conditions, it was decided to prepare longitudinal cuts of the samples and investigate potential reasons of this phenomena. The intergranular corrosion was present in most samples to a significant extent. Most phases present in the interdendritic spaces were corroded. Such corrosion products could not be cleaned by CrO_3 acid solution [24]. These corrosion products, anchored among the still mainly intact η(Zn) dendrites, were increasing the total weight of the samples even after the cleaning process (Figure 3a). Their presence is visualised by chemical element distribution maps in Figure 3b.

Backscattered-electron scanning electron microscopy (BSEM) images of the longitudinal sections for representative as-cast samples with 0.0, 0.5 and 3.0 wt.% of Sn after 1000 h of NSST are given in Figure 4. Corresponding quantitative analysis results of the intergranular corrosion penetration depth are summarized in Figure 5. The same is available for the annealed samples in Figures 6 and 7.

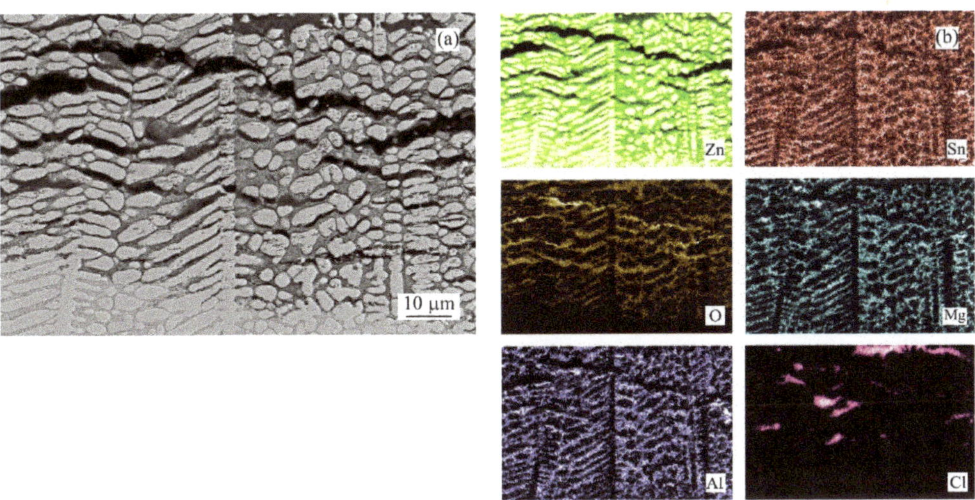

Figure 3. Anchoring effect of η(Zn) dendrites with corroded interdendritic spaces (MZ + 2.0Sn, as-cast): (**a**) overview BSEM image; (**b**) chemical element distribution maps of Zn, Sn, O, Mg, Al and Cl.

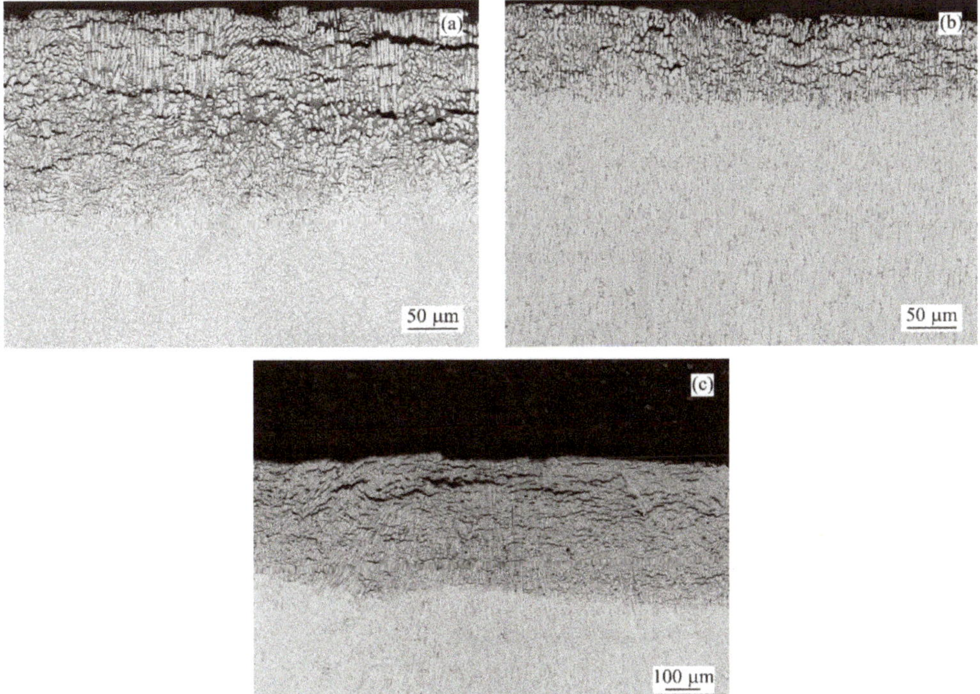

Figure 4. BSEM images indicating the extend of IG corrosion observed for the as-cast samples after 1000 h of NSST: (**a**) MZ + 0.0Sn (**b**) MZ + 0.5Sn (**c**) MZ + 3.0Sn.

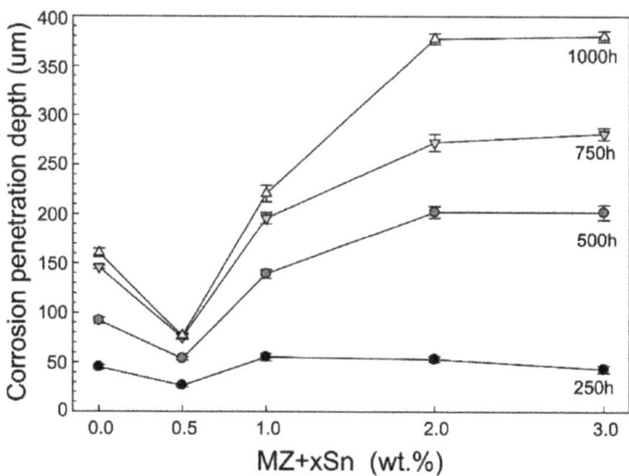

Figure 5. Depth of IG corrosion—as-cast samples.

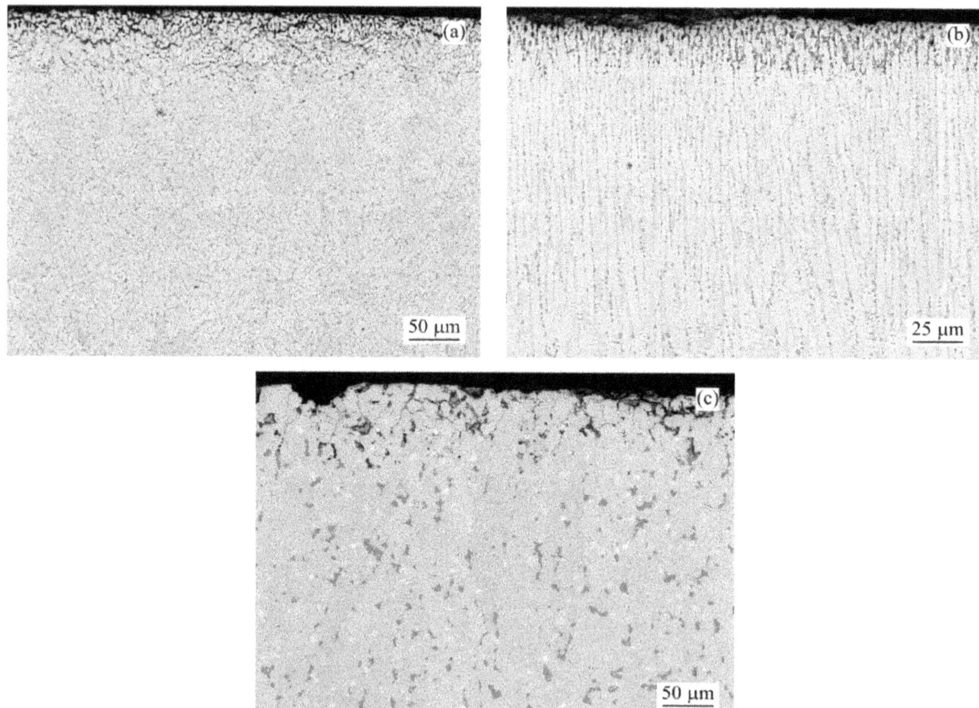

Figure 6. BSEM images indicating the extent of IG corrosion observed for the annealed samples: (a) MZ + 0.0Sn (b) MZ + 0.5Sn (c) MZ + 3.0Sn.

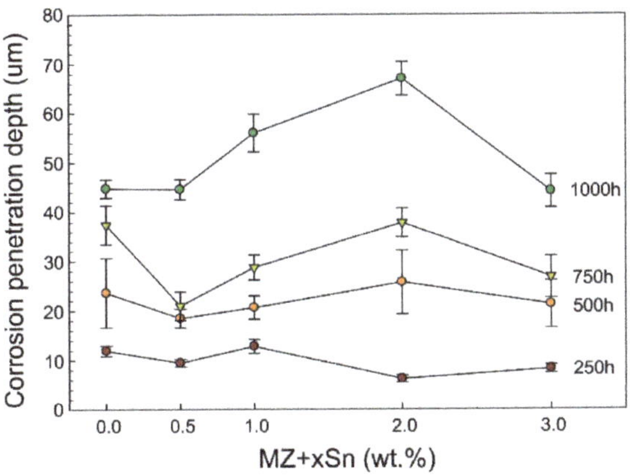

Figure 7. Depth of IG corrosion—annealed samples.

SEM investigation of as-cast MZ + 0.0Sn samples after 250 h of NSST revealed a significant portion of the microstructure being affected by the intergranular corrosion reaching as deep as ~50 μm (Figure 5). This effect is even more pronounced on the as-cast samples with 1–3 wt.% of Sn. IG corrosion can be formed as deep as ~150 μm for the as-cast MZ + 3.0Sn samples (Figure 5). This effect is further emphasised during longer exposures in the salt spray chamber. The IG corrosion can reach depths of over 370 μm on average for the as-cast MZ + 2.0Sn and MZ + 3.0Sn samples exposed for 1000 h (Figures 4c and 5). Complex ZnAlMg interdendritic areas were affected preferentially by the IG corrosion (Figure 8).

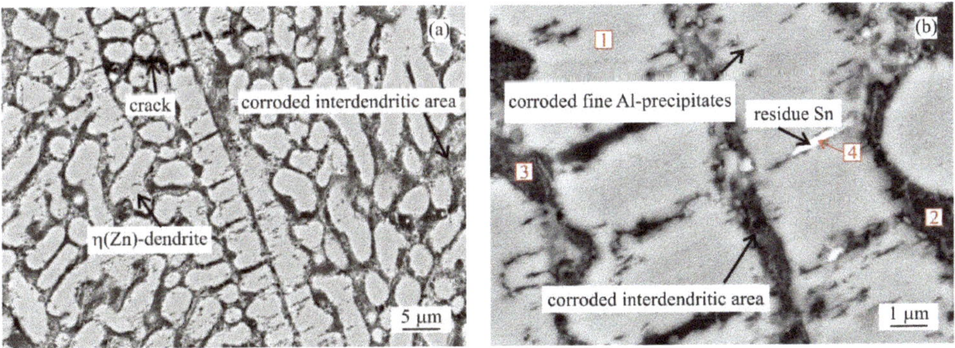

Figure 8. MZ + 2.0Sn as-cast microstructure after 1000 h in NSST affected by IG corrosion: (**a**) overview, (**b**) detail.

The depth of IG corrosion is significantly lower for the annealed samples with maximums reaching only about 70 μm even after 1000 h of NSST. For 250 and 500 h, all alloys behaved rather similar with IG corrosion depths of 10 and 22 μm, respectively. MZ + 1.0Sn and MZ + 2.0Sn seem to be more susceptible to the IG corrosion when comparing the samples after the full 1000 h test. On the contrary, annealed MZ + 3.0Sn samples showed values comparable even to MZ + 0.0Sn, or MZ + 0.5Sn reaching a maximum depth of about 45 μm.

The examples of areas affected by the intergranular corrosion are given in Figures 8 and 9 for the as-cast and annealed samples, respectively. The EDS chemical analysis of the microstructure in Table 3 confirms the intergranular corrosion attack. Figure 8a shows the η(Zn) dendritic microstructure affected by the corrosion along the interdendritic areas.

$\eta(Zn)$ primary dendrites also showed the signs of corrosion in the form of fine cracks. These can be attributed to the presence of fine, sub-micron Al-rich particles observed within the $\eta(Zn)$ primary dendrites. In a more detailed image (Figure 8b) it can be seen that $\alpha(Al)$ and Mg_xZn_y particles were corroded. Mg_2Sn particles were subject to the process of dealloying [28–31], leaving thus pure metallic Sn particles behind.

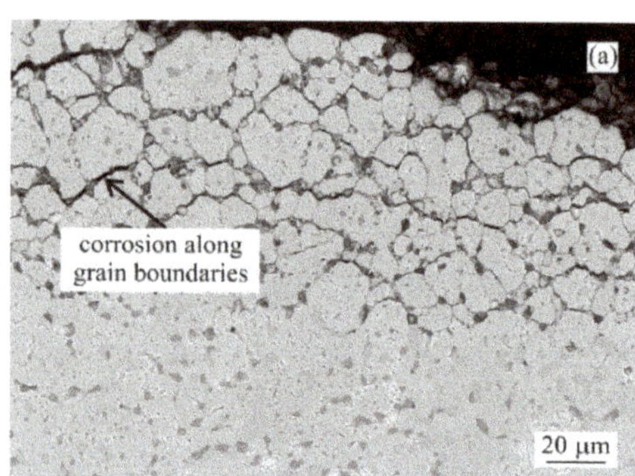

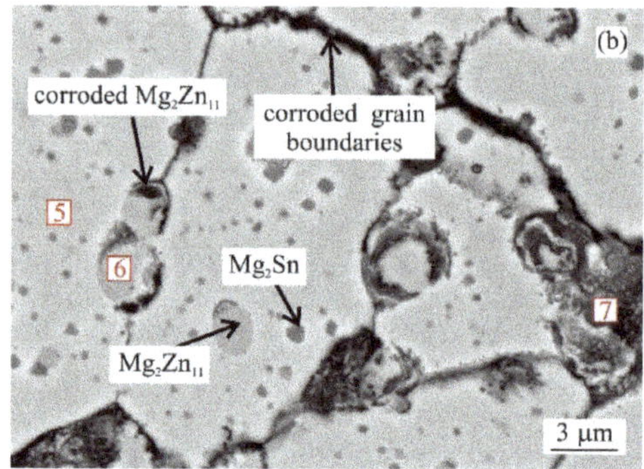

Figure 9. MZ + 2.0Sn annealed microstructure after 1000 h in NSST affected by IG corrosion: (**a**) overview, (**b**) detail.

Table 3. EDS chemical composition of selected sites (at.%).

	Site No. (as Labeled in Figures 8 and 9)						
Chemical Element (at.%)	**1**	**2**	**3**	**4**	**5**	**6**	**7**
Zn	97.31	55.32	29.66	46.12	96.08	77.88	21.29
Al	2.69	2.69	5.31	2.83	2.53	2.54	30.42
Mg	0.00	5.11	13.23	2.88	0.00	17.38	1.00
Sn	0.00	0.38	0.25	14.05	0.00	0.00	0.98
O	0.00	33.15	59.95	33.97	1.39	2.20	46.13
Cl	0.00	3.34	0.60	0.14	0.00	0.00	0.19
Phase/Region	$\eta(Zn)$	Corrosion product	Corrosion product	Residue Sn from Mg_2Sn particle	$\eta(Zn)$	Mg_2Zn_{11}	Corrosion product

As reported in the first part of this research [22], the basic dendritic character of the microstructure was still rather well maintained for the MZ + 0.0Sn and MZ + 0.5Sn alloys even after annealing. Hence, the IG corrosion is observed to propagate preferably along the interdendritic areas. For the annealed samples with 1 and more wt.% of Sn the microstructure is more spheroidized. The individual intermetallic phases were coalesced into coarse, discrete particles, while $\eta(Zn)$ dendrites were reshaped and new grains are formed within the microstructure. The boundaries of these grains contained a significant portion of intermetallic phase particles. As such, they were more susceptible to the IG corrosion. The propagation of the IG corrosion is documented in Figure 9a and the grain boundaries decorated by intermetallic particles are shown in closer detail in Figure 9b.

The XRD analysis was performed on all samples after NSST. As described, loose corrosion products were gathered and investigated. The XRD was used to determine the phase composition of the corrosion products formed on the samples during NSST. The XRD patterns for all corrosion products retrieved from the as-cast and annealed samples are summarized in Figures 10 and 11, respectively. The presence of the identified phases was also confirmed using the Rietveld method (Table 4). Despite the differences between the microstructure of the as-cast and annealed samples, their corrosion products showed an identical qualitative phase composition. The semi-quantitative results from these calculations indicate that the majority of the corrosion products were formed by a hydrozincite for all samples. About 20 vol.% of simonkolleite was measured for all pure

Zn samples (Figures 10a and 11a). The corrosion products of MZ-based samples contained only about 10 vol.% of simonkolleite on average (Figures 10b–d and 11b–d).

ZnO was identified solely in the corrosion products of the pure as-cast Zn sample (Figure 10a), representing only about 2 vol.%.

NaCl was identified in randomly varying amounts in the corrosion products as a remainder of the corrosion environment.

Next to hydrozincite and simonkolleite, the sources indicated that other phases might also be formed [17–20,32–39]. Therefore, the corroded surfaces of the bulk metallic samples were investigated after the loose corrosion products were removed. The measurement in grazing incident diffraction mode with 0.5° incident angle was chosen to limit the signal from the substrate (mainly Zn) as much as possible. Additionally, to previously identified phases, zincite (ZnO) and aluminium (II) oxide (AlO) were identified as present directly attached to the sample surface. An example of such a pattern is given in Figure 12 for the MZ + 3.0Sn annealed sample surface after 1000 h of NSST. However, only about 2 and 5 vol.% of ZnO and AlO, respectively, were identified using the Rietveld method.

Table 4. Phases identified during XRD analysis.

Phase Name	Phase Chemical Formula	Reference Code—ICSD FIZ Karlsruhe Database	Crystallography Open Database COD ID	Crystal System	Space Group	Space Group Number
Hydrozincite	$Zn_5(OH)_6(CO_3)_2$	01-072-1100	9007481	Monoclinic	$C2/m$	12
Simonkolleite	$Zn_5(OH)_8Cl_2 \cdot H_2O$	98-003-4904	9004683	Hexagonal	$R\bar{3}m$	166
Zincite	ZnO	98-015-4487	9004178	Hexagonal	$P6_3/mc$	186
Aluminium (II) Oxide	AlO	98-002-8920	-	Cubic	$Fm\bar{3}m$	225
Sodium Chloride	NaCl	01-075-0306	1000041	Cubic	$Fm\bar{3}m$	225

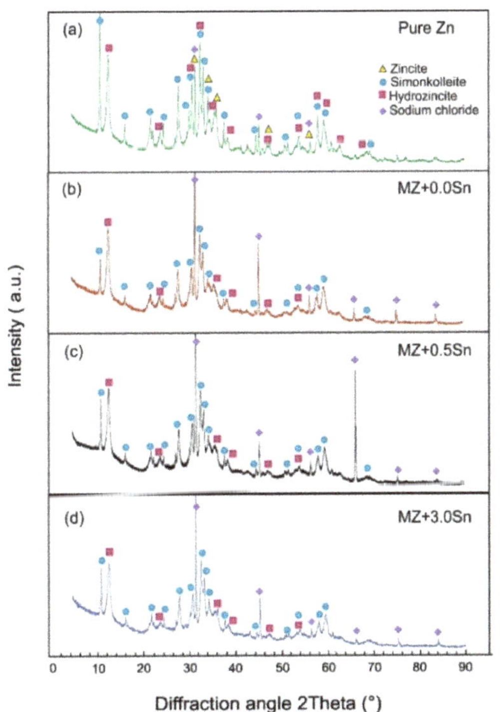

Figure 10. XRD patterns of corrosion product powders retrieved from the as-cast samples after 1000 h of NSST: (**a**) pure Zn (**b**) MZ + 0.0Sn (**c**) MZ + 0.5Sn (**d**) MZ + 3.0Sn.

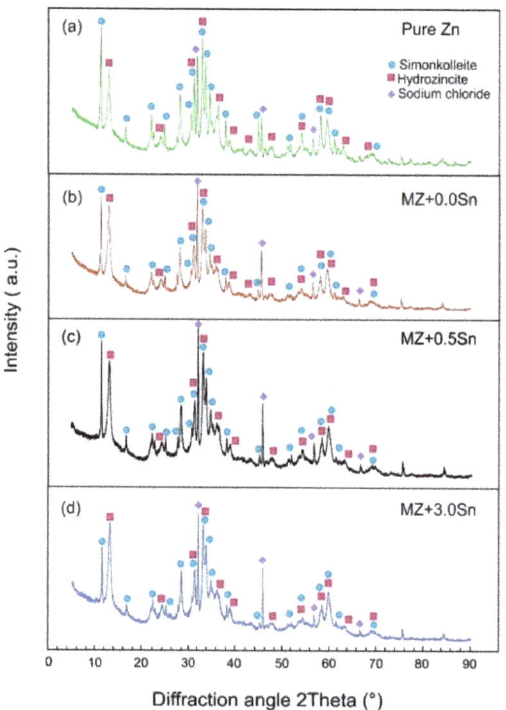

Figure 11. XRD patterns of corrosion product powders retrieved from the annealed samples after 1000 h of NSST. (**a**) pure Zn (**b**) MZ + 0.0Sn (**c**) MZ + 0.5Sn (**d**) MZ + 3.0Sn.

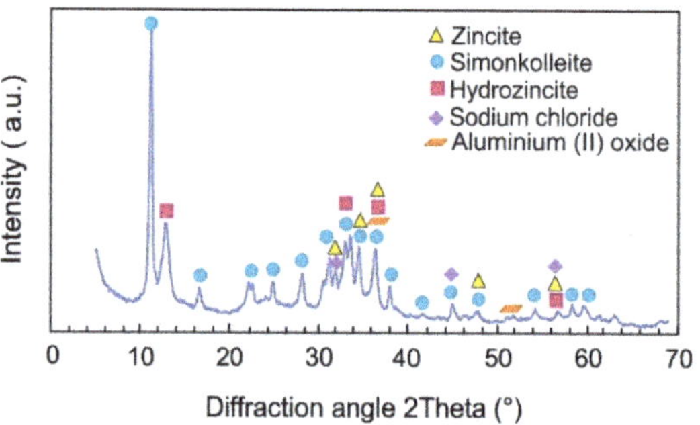

Figure 12. XRD pattern on MZ + 3.0Sn annealed sample surface after 1000 h NSST, measured in grazing incident geometry.

4. Discussion

4.1. SEM vs. Mass Balance after NSST

Both weight loss and weight gain were observed for a significant portion of the as-cast samples due to several related properties of the ZnAlMg alloy system. During directional cooling of these alloys, the η(Zn) phase forms the primary dendrites, while the interdendritic spaces are formed by a fine mixture of various phases including α(Al) solid solution, $MgZn_2$, Mg_2Zn_{11}, and Mg_2Sn intermetallic phases. There is an inherent difference in the open circuit potential (OCP) of these phases mainly compared to the η(Zn) phase (Figure 13). Consequently, the interdedritic phases seem to offer the galvanic protection to the η(Zn) phase dendrites. Due to this phenomenon, the interdendritic spaces corrode prior to the η(Zn) phase. The still intact η(Zn) phase dendrites act as anchors holding these corrosion products in place. These corrosion products cannot be removed by the environment during the NSST, nor by chemical cleaning done in the preparation for the sample weighing after the test. Naturally, the total weight of such corrosion products is greater as the weight of the original metallic phases. This phenomenon will cause weight gain for several samples even after the corrosion products were removed as much as possible before weighing. This increase in weight is also followed by an increase in volume. Following the BSEM images, it is clear that the η(Zn) phase dendrites are cracking as seen in Figure 8. This could be attributed to volume expansion-induced cracking (Figures 3a and 8a).

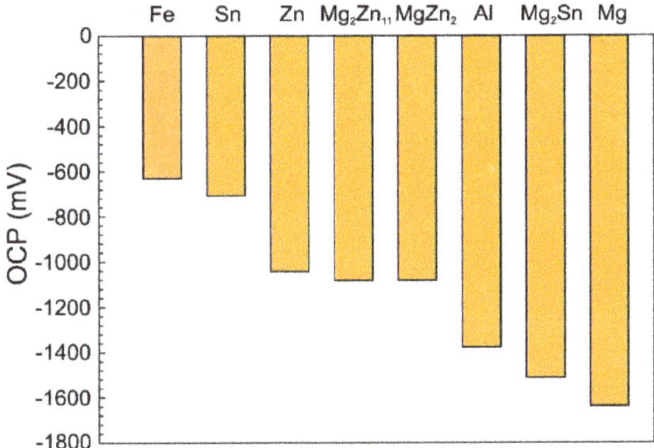

Figure 13. Overview of OCP values for phases present in the investigated system [34,40–44].

4.2. Corrosion of Individual Phases

Based on the SEM investigation, it can be concluded that individual phases are corroding in the following order: α(Al) → Mg$_x$Zn$_y$ → Mg$_2$Sn (if present) → η(Zn).

For α(Al), Mg$_x$Zn$_y$ and η(Zn) this order corresponds with their respective corrosion potentials reported in the literature as seen in Figure 13. On the contrary, Mg$_2$Sn behaves as a more noble phase compared to α(Al) and Mg$_x$Zn$_y$ phases despite having a lower OCP compared to these phases. A clear example is given in Figure 14a, where already in an early stage of the corrosion attack, an Mg$_2$Zn$_{11}$ particle is affected by the corrosion when in contact with an Mg$_2$Sn particle. Such a phenomenon can be caused by several factors, such as for example a local change in pH, or local change in chemical composition of these Mg$_2$Sn particles. The second phenomenon was regularly observed in all Sn-containing alloys. During the corrosion, the local dealloying of Mg$_2$Sn particles occurs. Mg$_2$Sn particles are separated into Mg and Sn atoms. Mg is most probably immediately forming new corrosion products, while Sn resides in the form of metallic particles. These can be observed on most BSEM images of the areas affected by the IG corrosion. As shown in Figure 14b, even the formation of an Sn-rich shell can be observed on larger particles found in the annealed samples. The Mg content of former Mg$_2$Sn particles is being gradually reduced, hence, the remaining metallic particle will have locally a higher potential compared to neighbouring microstructure components. As a final stage, pure Sn particles are formed in the place of Mg$_2$Sn particles. This process is even described by several authors [28–31] as a potential energy storage system for batteries. The final stage of this process is documented in Figure 15a. The corresponding EDS maps in Figure 15b confirm the presence of Sn-based metallic particles. Mg and O maps are overlapping, indicating that Mg is forming corrosion products.

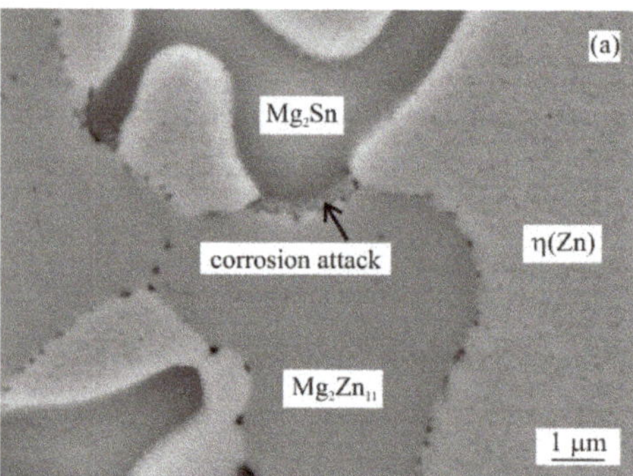
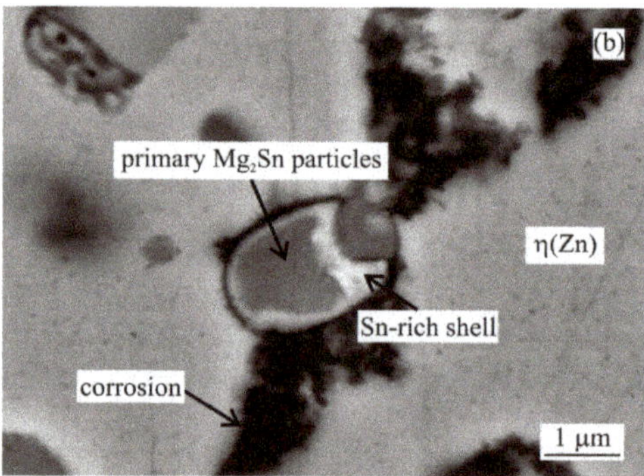

Figure 14. Behaviour of Mg₂Sn intermetallic phase during corrosion (MZ + 2.0Sn, annealed): (**a**) Influence of Mg₂Sn on initial stage of corrosion of Mg₂Zn₁₁ intermetallic particle; (**b**) dealloying of Mg₂Sn particle.

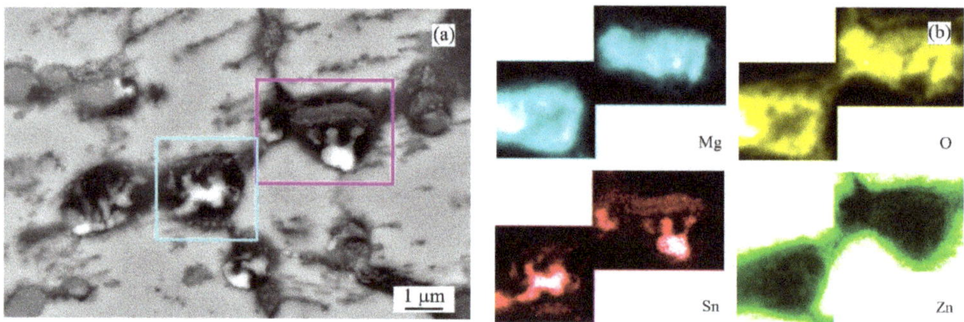

Figure 15. Details of Mg₂Sn particles affected by dealloying (MZ + 3.0Sn, annealed): (**a**) overview BSEM image; (**b**) chemical element distribution maps of Mg, O, Sn and Zn.

4.3. Phase Composition of Corrosion Products

Hydrozincite and simonkolleite are the most common corrosion products reported by several authors for similar alloy systems [18–20,32,33,36–38]. This is in good agreement with the current results.

The semi-quantitative analysis showed that a slightly higher portion of simonkolleite was found in the corrosion products of both as-cast and annealed pure Zn samples compared to MZ + xSn alloys. This is in line with the observation of Prosek et al. [32], where simonkolleite was more likely to be identified for pure Zn coatings. As the XRD analysis could not give the data on the chemical composition of these phases, the presence of Mg and Al in the corrosion products of MZ-based alloys was measured by the SEM EDS analysis with up to 2 wt.% of Mg and up to 1 wt.% of Al. From the two main corrosion products, the hydrozincite can accommodate Mg as a metallic ion in its structure [45,46]. This would support our observation, where the increased amount of the hydrozincite was identified on the MZ + xSn samples compared to pure Zn.

Additionally, layered double hydroxide (LDH) phases were identified, where LDH can represent a group of similar phases [18–20,32,36,39,47]. Azevedo et al. [20] identified LDH within the corrosion products formed on a Zn3.7Al3.0Mg alloy coating after 100 h of NSST (5% NaCl). Applying the Rietveld method refinement to their XRD pattern revealed that about 2–3 vol.% of the corrosion products were formed by LDH (ICSD FIZ Karlsruhe database 98-015-5051). Similarly, a low amount of LDH was indicated by the semi-quantitative results of Prosek et al. [32] within the corrosion products of

Zn1.5Al1.5Mg alloy coatings exposed to model atmospheric conditions. However, in the studied system, LDH was not confirmed in any of our measurements, not even during GI XRD measurements performed directly on the corroded surface with the loose corrosion products removed. When comparing our experiments to the literature, there are two probable causes: we had rather low Al and Mg content compared to NSST done by Azevedo et al. [20]. Prosek et al. [32] used the same coating, however, in a very different corrosion environment. We might have a specific combination of parameters, which are not favourable for the creation of LDH.

5. Conclusions

Based on the experimental results discussed in part I and part II of this research, the following conclusions can be drawn:

- Weight change cannot be correlated with alloy composition nor NSST exposure time due to presence of IG corrosion.
- Increasing the exposure time in NSST from 250 h to 1000 h increases the intergranular corrosion penetration depth, regardless of the chemical composition and heat treatment.
- As-cast samples were more susceptible to the IG corrosion as interdendritic areas are forming a connected network of less noble phases. These include α(Al), Mg_xZn_y and Mg_2Sn, while dendrites are formed mainly by η(Zn).
- Adding 0.5 wt.% Sn has almost no effect on the weight change of the as-cast samples after NSST compared to MZ + 0.0Sn, while being significantly less susceptible to the IG corrosion. As a result, the as-cast MZ + 0.5Sn samples show the most favourable corrosion behaviour.
- Adding 1 to 3 wt.% of Sn yields in the weight gain instead of the weight loss as well as a significant increase in the IG corrosion depth.
- Annealed alloys are less susceptible to the IG corrosion as intermetallic phases are coalesced, spheroidised, and more uniformly distributed within the η(Zn) matrix or at newly formed grain boundaries of η(Zn).
- Changing the alloy composition of the annealed samples has only a slight effect on the weight change. Nevertheless, the samples with 3 wt.% of Sn showed the most favourable results. Meanwhile, the IG corrosion depth is comparable to MZ + 0.0Sn, resulting in overall best performance of the annealed MZ + 3.0Sn samples.
- Hydrozincite and simonkolleite were identified as the main corrosion products on all samples. A small portion of ZnO was identified only on pure Zn samples. GI XRD measurements indicated a small amount of AlO formed on most MZ-based samples.
- Current results show that even the high temperature exposure of up to 310 °C does not negatively affect the corrosion performance of these alloys. It could be noted that such exposure even provides a beneficial effect and enhances the corrosion resistance of the coating by suppressing the IG corrosion.

Author Contributions: Conceptualization, P.G. and Z.G.; methodology, P.G., Z.G. and M.K.; validation, P.G., Z.G., M.K. and H.S.; formal analysis, P.G. and Z.G.; investigation, P.G., Z.G., M.K. and H.S.; resources, Z.G. and P.G.; data curation, P.G., Z.G. and M.K.; writing—original draft preparation, P.G. and Z.G.; writing—review and editing, P.G., Z.G. and M.K.; visualization, Z.G. and P.G.; supervision, P.G. and Z.G.; project administration, M.K.; funding acquisition, M.K. All authors have read and agreed to the published version of the manuscript.

Funding: This research was supported by the Grant Agency VEGA of the Slovak Ministry of Education, Research, Science and Sport Project No. 1/0490/18: "The effect of microstructure and phase com-position on corrosion resistance of hot dip alloys" and by the Slovak Research and Development Agency under the Contract No. APVV-20-0124.

Institutional Review Board Statement: Not applicable.

Informed Consent Statement: Not applicable.

Data Availability Statement: Data sharing is not applicable.

Conflicts of Interest: The authors declare no conflict of interest.

References

1. Vida, T.A.; Brito, C.; Lima, T.S.; Spinelli, J.E.; Cheung, N. Near-eutectic Zn-Mg alloys: Interrelations of solidification thermal parameters, microstructure length scale and tensile/corrosion properties. *Curr. Appl Phys.* **2009**, *51*, 2355–2363. [CrossRef]
2. Tokuda, S.; Muto, I.; Sugawara, Y.; Takahashi, M.; Matsumoto, M.; Hara, N. Micro-electrochemical investigation on the role of Mg in sacrificial corrosion protection of 55mass%Al-Zn-Mg coated steel. *Corros. Sci.* **2017**, *129*, 126–135. [CrossRef]
3. Krystýnová, M.; Doležal, P.; Fintová, S.; Zapletal, J.; Marada, T.; Wasserbauer, J. Characterization of Brittle Phase in Magnesium Based Materials Prepared by Powder Metallurgy. *Key Eng. Mater* **2018**, *784*, 61–66. [CrossRef]
4. Vida, T.A.; Soares, T.; Septimio, R.S.; Brito, C.C.; Cheung, N.; Garcia, A. Effects of Macrosegregation and Microstructure on the Corrosion Resistance and Hardness of a Directionally Solidified Zn-5.0wt.%Mg Alloy. *Mater. Res.* **2019**, *22*, 1–13. [CrossRef]
5. Pinc, J.; Čapek, J.; Kubásek, J.; Veřtát, P.; Hosová, K. Microstructure and mechanical properties of the potentially biodegradable ternary system Zn-Mg0.8-Ca0.2. *Procedia Struct. Integr.* **2019**, *23*, 21–26. [CrossRef]
6. De Bruycker, E.; Zermout, Z.; De Cooman, B.C. Zn-Al-Mg Coatings—Thermodynamic Analysis and Microstructure Related Properties. *Mater. Sci. Forum* **2007**, *539–543*, 1276–1281. [CrossRef]
7. De Bruycker, E.; De Cooman, B.C.; De Meyer, M. Experimental study and microstructure simulation of Zn-Al-Mg coatings. *Rev. Metall-CIT* **2005**, *102*, 543–550. [CrossRef]
8. Akdeniz, V.M.; Wood, J.V. Microstructures and phase selection in rapidly solidified Zn-Mg alloys. *J. Mater. Sci.* **1996**, *31*, 545–550. [CrossRef]
9. Liu, H.Y.; Jones, H. Solidification Microstructure Selection and Characteristics in the Zinc-Based Zn-Mg System. *Acta Metall. Mater.* **1992**, *40*, 229–239. [CrossRef]
10. Prosek, T.; Nazarov, A.; Goodwin, F.; Šerák, J.; Thierry, D. Improving corrosion stability of Zn-Al-Mg by alloying for protection of car bodies. *Surf. Coat. Technol.* **2016**, *306 Pt B*, 439–447. [CrossRef]
11. Farahany, S.; Tat, L.H.; Hamzah, E.; Bakhsheshi-Rad, H.R.; Cho, M.H. Microstructure development, phase reaction characteristics and properties of quaternary Zn-0.5Al-0.5Mg-xBi hot dipped coating alloy under slow and fast cooling rates. *Surf. Coat. Technol.* **2017**, *315*, 112–122. [CrossRef]
12. Gondek, J.; Babinec, M.; Kusý, M. The corrosion performance of Zn-Al-Mg based alloys with tin addition in neutral salt spray environment. *J. Achiev. Mater. Manuf. Eng.* **2015**, *70*, 70–77.
13. Vargel, C. *Corrosion of Aluminium*, 1st ed.; Elsevier Science: Amsterdam, The Netherlands, 2004; pp. 167–182. [CrossRef]
14. Volovitch, P.; Allely, C.; Ogle, K. Understanding corrosion via corrosion product characterization: I Case study of the role of Mg alloying in Zn–Mg coating on steel. *Corros. Sci.* **2009**, *51*, 1251–1262. [CrossRef]
15. Odnevall, W.; Leygraf, C. A Critical Review on Corrosion and Runoff from Zinc and Zinc-Based Alloys in Atmospheric Environments. *Corros. J. Sci. Eng.* **2017**, *73*, 1060–1077. [CrossRef]
16. De la Fuente, D.; Castaño, J.G.; Morcillo, M. Long-term atmospheric corrosion of zinc. *Corros. Sci.* **2007**, *51*, 1420–1436. [CrossRef]
17. Volovitch, P.; Vu, T.N.; Allély, C.; Abdel Aal, A.; Ogle, K. Understanding corrosion via corrosion product characterization: II Role of alloying elements in improving the corrosion resistance of Zn–Al–Mg coatings on steel. *Corros. Sci.* **2011**, *53*, 2437–2445. [CrossRef]
18. Thierry, D.; Persson, D.; Luckeneder, G.; Stellnberger, K.-H. Atmospheric corrosion of ZnAlMg coated steel during long term atmospheric weathering at different worldwide exposure sites. *Corros. Sci.* **2019**, *148*, 338–354. [CrossRef]
19. LeBozec, N.; Thierry, D.; Persson, D.; Riener, C.K.; Luckeneder, G. Influence of microstructure of zinc-aluminium-magnesium alloy coated steel on the corrosion behavior in outdoor marine atmosphere. *Surf. Coat. Technol.* **2019**, *374*, 897–909. [CrossRef]
20. Azevedo, M.S.; Allély, C.; Ogle, K.; Volovitch, P. Corrosion mechanisms of Zn(Mg, Al) coated steel in accelerated tests and natural exposure 1. The role of electrolyte composition in the nature of corrosion products and relative corrosion rate. *Corros. Sci.* **2015**, *90*, 472–481. [CrossRef]
21. Ghosh, P.; Mezbahul-Islam, M.; Medraj, M. Critical assessment and thermodynamic modeling of Mg-Zn, Mg-Sn, Sn-Zn and Mg-Sn-Zn systems. *Calphad* **2012**, *36*, 28–43. [CrossRef]
22. Gogola, P.; Gabalcová, Z.; Kusý, M.; Suchánek, H. The effect of Sn addition on Zn-Al-Mg alloy; Part I: Microstructure and phase composition. *Materials* **2021**, under review.
23. *Corrosion Tests in Artificial Atmospheres-Salt Spray Tests*; ISO 9227:2017; International Organization for Standardization: Geneva, Switzerland, 2017.
24. *Standard Guide for Laboratory Immersion Corrosion Testing of Metals*; ASTM G31-21; ASTM International: West Conshohocken, PA, USA, 2021.
25. Jokar, M.; Aliofkhazraei, M. *Comprehensive Materials Finishing*, 1st ed.; Elsevier Science: Amsterdam, The Netherlands, 2017; pp. 306–335. [CrossRef]
26. Schneider, C.A.; Rasband, W.S.; Eliceiri, K.W. NIH Image to ImageJ: 25 Years of image analysis. *Nat. Methods* **2012**, *9*, 671–675. [CrossRef]
27. Lutterotti, L.; Matthies, S.; Wenk, H.R. MAUD (Material Analysis Using Diffraction): A user friendly Java program for rietveld texture analysis and more. In Proceedings of the 12th International Conference on Textures of Materials (ICOTOM-12), McGill University Montreal, Montréal, QC, Canada, 9–13 August 1999; pp. 1599–1604. Available online: http://hdl.handle.net/11572/57067 (accessed on 28 July 2021).

28. Nguyen, G.T.H.; Nguyen, D.-T.; Song, S.-W. Unveiling the Roles of Formation Process in Improving Cycling Performance of Magnesium Stannide Composite Anode for Magnesium-Ion Batteries. *Adv. Mater. Interfaces* **2018**, *5*, 1801039. [CrossRef]

29. Nguyen, D.-T.; Song, S.-W. Magnesium stannide as a high-capacity anode for magnesium-ion batteries. *J. Power Sources* **2017**, *368*, 11–17. [CrossRef]

30. Singh, N.; Arthur, T.S.; Ling, C.; Matsui, M.; Mizuno, F. A high energy-density tin anode for rechargeable magnesium-ion batteries. *Chem. Commun.* **2013**, *49*, 149–151. [CrossRef]

31. Yaghoobnejad Asl, H.; Fu, J.; Kumar, H.; Welborn, S.S.; Shenoy, V.B. In Situ Dealloying of Bulk Mg_2Sn in Mg-Ion Half Cell as an Effective Route to Nanostructured Sn for High Performance Mg-Ion Battery Anodes. *Chem. Mater.* **2018**, *30*, 1815–1824. [CrossRef]

32. Prosek, T.; Persson, D.; Stoulil, J.; Thierry, D. Composition of corrosion products formed on Zn–Mg, Zn–Al and Zn–Al–Mg coatings in model atmospheric conditions. *Corros. Sci.* **2014**, *86*, 231–238. [CrossRef]

33. Zhang, X.G. *Corrosion and Electrochemistry of Zinc*, 1st ed.; Springer: Boston, MA, USA, 1996; pp. 157–181. [CrossRef]

34. Buyn, J.M.; Yu, J.M.; Kim, D.K.; Kim, T.-Y.; Jung, W.-S.; Kim, Y.D. Corrosion Behavior of Mg_2Zn_{11} and $MgZn_2$ Single Phases. *Korean J. Met. Mater.* **2012**, *51*, 413–419. [CrossRef]

35. McMahon, M.E.; Burns, T.J.; Scully, J.R. Development of new criteria for evaluating the effectiveness of Zn-rich primers in protecting Al-Mg alloys. *Prog. Org. Coat.* **2019**, *135*, 392–409. [CrossRef]

36. Persson, D.; Thierry, D.; LeBozec, N.; Prosek, T. In situ infrared reflection spectroscopy studies of the initial atmospheric corrosion of Zn–Al–Mg coated steel. *Corros. Sci.* **2013**, *72*, 54–63. [CrossRef]

37. Li, B.; Dong, A.; Zhu, G.; Chu, S.; Qian, H.; Hu, C.; Sun, B.; Wang, J. Investigation of the corrosion behaviors of continuously hot-dip galvanizing Zn–Mg coating. *Surf. Coat. Technol.* **2012**, *206*, 3989–3999. [CrossRef]

38. Zhu, Z.; Li, A.; Xu, B. Study on corrosion mechanism of arc sprayed Zn-Al-Mg coatings by XRD and EIS. *Adv. Mater. Res.* **2011**, *230–232*, 85–88. [CrossRef]

39. Ennadi, A.; Legrouri, A.; De Roy, A.; Besse, J.P. X-ray Diffraction Pattern Simulation for Thermally Treated [Zn-Al-Cl] Layered Double Hydroxide. *J. Solid State Chem.* **2000**, *152*, 568–572. [CrossRef]

40. Singh, I.B.; Singh, M.; Das, S. A comparative corrosion behavior of Mg, AZ31 and AZ91 alloys in 3.5 NaCl solution. *J. Magnesium Alloys* **2015**, *3*, 142–148. [CrossRef]

41. Špoták, M.; Drienovský, M.; Rízeková-Trnková, L.; Palcut, M. Corrosion of Candidate Lead-Free Solder Alloys in Saline Solution. In Proceedings of the 24th International Conference on Metallurgy and Materials METAL 2015, Brno, Czech Republic, 3–5 June 2015; pp. 1650–1656.

42. Hu, C.-C.; Wang, C.-K. Effects of composition and reflowing on the corrosion behavior of Sn–Zn deposits in brine media. *Electrochim. Acta* **2006**, *51*, 4125–4134. [CrossRef]

43. Calabrese, L.; Bonaccorsi, L.; Capri, A.; Proverbio, E. Electrochemical behavior of hydrophobic silane-zeolite coatings for corrosion protection of aluminum substrate. *J. Coat. Technol. Res.* **2014**, *11*, 883–898. [CrossRef]

44. Gogola, P.; Gabalcová, Z.; Palcut, P. Experimental determination of the corrosion potential for the intermetallic Mg_2Sn phase. (manuscript in preparation).

45. Mindat.org Hydrozincite. Available online: https://www.mindat.org/min-1993.html (accessed on 28 July 2021).

46. Mindat.org Simonkolleite. Available online: https://www.mindat.org/min-3668.html (accessed on 28 July 2021).

47. Zhitova, E.S.; Krivovichev, S.V.; Pekov, I.; Greenwell, H.C. Crystal chemistry of natural layered double hydroxides. 5. Single-crystal structure refinement of hydrotalcite, $[Mg_6Al_2(OH)_{16}](CO_3)(H_2O)_4$. *Mineral. Mag.* **2019**, *83*, 269–280. [CrossRef]

 *materials*

Article

Influence of Degradation Product Thickness on the Elastic Stiffness of Porous Absorbable Scaffolds Made from an Bioabsorbable Zn–Mg Alloy

Jannik Bühring [1,*], Maximilian Voshage [2], Johannes Henrich Schleifenbaum [2] and Holger Jahr [3] and Kai-Uwe Schröder [1]

1   Institute of Structural Mechanics and Lightweight Design, RWTH Aachen University, 52062 Aachen, Germany; kai-uwe.schroeder@sla.rwth-aachen.de
2   Digital Additive Production, RWTH Aachen University, 52074 Aachen, Germany; maximilian.voshage@dap.rwth-aachen.de (M.V.); johannes.henrich.schleifenbaum@dap.rwth-aachen.de (J.H.S.)
3   Institute of Anatomy and Cell Biology, University Hospital, RWTH Aachen University, 52074 Aachen, Germany; hjahr@ukaachen.de
*   Correspondence: jannik.buehring@sla.rwth-aachen.de; Tel.: +49-241-80-96842

Citation: Bühring, J.; Voshage, M.; Schleifenbaum, J.H.; Jahr, H.; Schröder, K.-U. Influence of Degradation Product Thickness on the Elastic Stiffness of Porous Absorbable Scaffolds Made from an Bioabsorbable Zn–Mg Alloy. *Materials* **2021**, *14*, 6027. https://doi.org/10.3390/ma14206027

Academic Editor: Marián Palcut

Received: 19 August 2021
Accepted: 8 October 2021
Published: 13 October 2021

Publisher's Note: MDPI stays neutral with regard to jurisdictional claims in published maps and institutional affiliations.

Copyright: © 2021 by the authors. Licensee MDPI, Basel, Switzerland. This article is an open access article distributed under the terms and conditions of the Creative Commons Attribution (CC BY) license (https://creativecommons.org/licenses/by/4.0/).

Abstract: For orthopaedic applications, additive manufactured (AM) porous scaffolds made of absorbable metals such as magnesium, zinc or iron are of particular interest. They do not only offer the potential to design and fabricate bio-mimetic or rather bone-equivalent mechanical properties, they also do not need to be removed in further surgery. Located in a physiological environment, scaffolds made of absorbable metals show a decreasing Young's modulus over time, due to product dissolution. For magnesium-based scaffolds during the first days an increase of the smeared Young's modulus can be observed, which is mainly attributed to a forming substrate layer of degradation products on the strut surfaces. In this study, the influence of degradation products on the stiffness properties of metallic scaffolds is investigated. For this, analytical calculations and finite-element simulations are performed to study the influence of the substrate layer thickness and Young's modulus for single struts and for a new scaffold geometry with adapted polar cubic face-centered unit cells with vertical struts (f2cc,z). The finite-element model is further validated by compression tests on AM scaffolds made from Zn1Mg (1 wt% Mg). The results show that even low thicknesses and Young's moduli of the substrate layer significantly increases the smeared Young's modulus under axial compression.

Keywords: additive manufacturing; scaffolds; bioabsorbable metals; biodegradation; lattice structures; stiffness properties

1. Introduction

The increasingly elderly population and the accompanying rising number of bone fractures have led to a significant rise in physical disabilities. The healing of larger bone defects is still a challenging task in orthopaedics. Using degradable implants eliminates the need for revision surgery, which may be required for some permanent medical devices. Thus, using such implants would not only benefit the patient, but also reduce healthcare costs [1]. Ideally, the implants should present a fully interconnected porous structure and should show equivalent mechanical properties, especially regarding the stiffness [2]. Such a biodegradable bone implant would allow fully natural bone regeneration, while the material gradually disappears in the body through absorption. These requirements can be fulfilled i.e., by additive manufactured (AM) lattice structures. Due to the large number of available materials and design parameters, almost any mechanical and material requirement profile can be set. However, biocompatibility and an interconnected porous structure can be fulfilled by a wide range of materials, reaching equivalent mechanical

195

properties at the same time is still challenging. Biocompatible materials can be found in a wide variety of material classes [3]. One example are polymer-based materials, which offer great advantages in terms of customized biodegradation and design [4]. Further to mention are ceramic materials, which also exhibit the aforementioned biodegradation and offer particularly good healing properties for bone defects [5]. However, for fully load-bearing applications only metals fulfill the needed properties, especially regarding strength and stiffness [6]. The Laser Powder Bed Fusion (LPBF) process enables the individualized production of high-resolution lattice structures with very fine struts (<250 μm) [7,8] at reasonable costs, and is thus ideal for the production of personalized implants [9]. In particular, the use of zinc (Zn), magnesium (Mg), iron (Fe) and their alloys, are increasingly coming into focus for orthopaedic applications [10,11]. Although Fe-based implants would biomechanically, and with respect to their corrosion speed [12,13], gain most from increased porosity [14], their limited cytocompatibility is a concern [15]. Nevertheless, in comparison to pure zinc and magnesium, iron has the highest values regarding yield strength and Young's modulus ($\sigma_{y,Fe} \approx$ 200–352 MPa, $E_{Fe} \approx$ 188–215 GPa [16–18]; $\sigma_{y,Zn} \approx$ 12–32 MPa, $E_{Zn} \approx$ 43–150 GPa [12,18,19]; $\sigma_{y,Mg} \approx$ 51 MPa, $E_{Mg} \approx$ 27–35 GPa [20–22]) and offers a large margin for introducing a controlled porosity, which directly influences the strength and stiffness properties of the material. Alloying can further improve the mechanical properties. Adding Zn to Mg-based alloys increases the yield strength and Young's modulus of the material [3,13,23]. Same goes for Zn alloyed with Mg [3,19,24], whereas adding aluminum to Zn-based alloys leads to a decrease in stiffness and strength [19].

Examples for Mg- and Zn-based studies on porous scaffolds can be found, e.g., in Witte et al. [25], who show the feasibility of producing AM open-porous, biodegradable and biocompatible Mg scaffolds. Li et al. [2] produced AM WE43 (Mg alloy with 4 wt% yttrium and 3 wt% rare earth elements) scaffolds based on diamond unit cells, to show the in vitro biodegradation behavior, mechanical properties and biocompatibility. Furthermore, Kopp et al. [26] showed that the pore size of Mg scaffolds influences the long-term stability, while heat treatment especially effects the degradation and mechanical stability. Cockerill et al. [27] used a casting approach to produce porous structures made of pure Zn and studied the topology, mechanical properties, biodegradation and biocompatibility. Another example is shown by Li et al. [28], who produced scaffolds from Zn with a diamond lattice structure via LPBF and studied the static and dynamic biodegradation behavior.

In a physiological environment biodegradable metals usually show a decreasing Young's modulus during the degradation process, due to the progressive absorption of the metallic surface, which consequently leads to a reduction of the strut cross section [29–32]. Since the strut thickness is directly related to the stiffness, the latter will also decrease. Interestingly, during the first days of in vitro corrosion of Mg-based (WE43) scaffolds, an increase of around 40% in the Young's modulus was recently reported [2]. This increase in stiffness is mainly attributed to the formation of a composite cross section, consisting of the base strut and an adherend layer of degradation products. A brief review of the literature shows [3,10,29,31] that the compound of degradation products, which adheres to the surface of the struts, consists for the most parts of hydroxides, phosphates and carbonates, for which only insufficient mechanical properties can be found. The phosphates and carbonates form a compound of usually unspecified chemical composition that further changes over time. Furthermore, a hydroxide layer is forming on the metallic surface. The basic biochemical processes, responsible for this, can be summarized as followed [29,31,32]:

Anodic reaction	$\text{Metal} \rightarrow \text{Metal}^{n+} + n(e^-)$
Cathodic reaction	$2H_2O + 2e^- \rightarrow 4OH^- + H_2$
	$2H_2O + O_2 + 4e^- \rightarrow 4OH^-$
Product formation	$\text{Metal}^{n+} + n(OH^-) \rightarrow \text{Metal}(OH)_n$
Product dissolution	$\text{Metal}(OH)_n + 2Cl^- \rightarrow \text{Metal}(Cl)_2 + 2OH^-$

Figure 1 shows a simplified schematic of the degradation process. The human body fluid releases an anodic reaction, and the free electrons undergo a cathodic reaction under the release of hydrogen and hydroxide ions, which form together with the metal a hydroxide layer on the surfaces of the struts. From equivalent reactions, phosphates and carbonates form on the strut surfaces [29]. These processes are responsible for an increase in stiffness during the early phases of the corrosion process [2]. Later, chloride ions start the dissolution of the biodegradable metal to cause a decrease of the cross-sectional strut diameter of the scaffold.

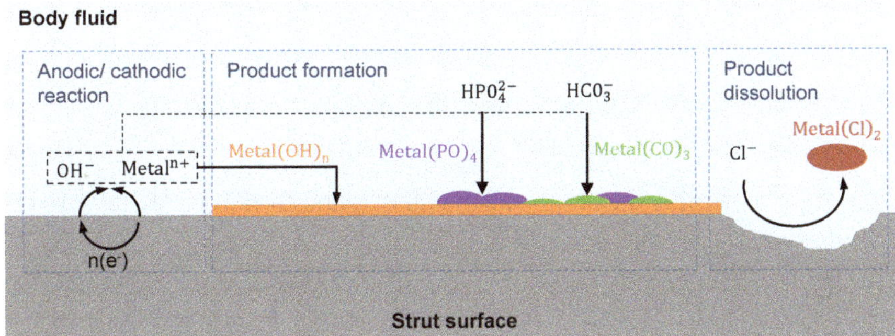

Figure 1. Schematic process sketch of the degradation process of absorbable metals according to Han et al. and Li [29,31].

We now used Zn1Mg (1 wt% Mg) as an example to investigate the influence of degradation products on the elastic stiffness properties of metallic scaffolds using analytical calculations and finite-element (FE) simulations. For this, first, we focused on the direct influence of the forming substrate layer of degradation products on the axial and bending stiffness of single struts. The corroded strut is modeled as a composite beam with a solid Zn1Mg base strut and a thin-walled layer of corrosion products of unspecified chemical composition. Instead of using concretely quantified values for the Young's modulus for the compound of degradation products, hypothetical multiples of the Zn1Mg Young's modulus are used. Afterwards, a new scaffold geometry, based on a polar modeling of a f2cc,z unit cell is produced and tested, to validate the FE model. Using the validated model, a FE parametric study is done to investigate the influence of the substrate layer thickness and Young's modulus of the compound on the smeared Young's modulus of the scaffold.

2. Materials and Methods

2.1. Scaffold Manufacturing

The LPBF (Laser Powder Bed Fusion) experiments were performed on an AconityMINI system designed by Aconity3D (Herzogenrath, Germany), which is specifically developed for laboratory use. This system is characterized by an adapted gas flow management to remove the resulting process fume for materials with low melting and evaporating temperature (i.e., zinc: 692 K, 1180 K). These materials tend to produce a large amount of process fume during manufacture. The beam source is a single-mode fiber laser (wave length of 1064 nm) with up to 400 W of power output. Samples were manufactured on a zinc baseplate using a bidirectional scanning strategy with 90° rotations between consecutive layers. The energy input during exposure was controlled by the selected process parameters (laser power (P_L), layer thickness (D_s), scanning speed (v_s), and hatch distance (Δy_s)). The volume energy density (E_V) was calculated as followed [33]:

$$E_V = \frac{P_L}{D_S v_s \Delta y_s} \tag{1}$$

Within the scope of this work, all AM scaffolds were manufactured with a constant layer thickness of 30 μm and E_V was set for all scaffolds to 133 J/mm^3. The scaffolds were afterwards sandblasted with 2.5 bar, to remove adhering powder particles.

2.2. Scaffold Geometry

Figure 2 shows the scaffold geometry, which was used for the FE study and validation tests. A modified polar f2cc,z unit cell was used. A total number of four cells in radial direction ($n_{1,2} = 4$), a total number of 17 cells in circumferential direction ($m = 17$) and a total number of 12 cells in height direction ($n_3 = 12$) was chosen. The scaffold has a total height of $h = 12$ mm and a diameter of $d = 10$ mm. The nominal strut radius is $r_s = 0.1$ mm. The cell width b results from $b = (d - 2R_m)/(2n_{1,2}) = 0.9$ mm, where $R_m = 1.4$ mm is the radius of central cavity, or rather the inner radius of the first cell ring, measured at the cells edges. Since the cells only approximate a circle, the radial position of the midpoint of the cells side faces lies at $r_{m,i} = r_i \cos(\varphi/2)$ for the inner face and $r_{m,a} = r_a \cos(\varphi/2)$ for the outer face, where r_i is the inner radius of the cell edges and r_a is the outer radius of the cell edges and $\varphi = 2\pi/m$ is the proportion that a cell has in the total circumference. The strut inclination ω of the circumferential diagonal struts can be calculated for the inner diagonals of each cell ring (ω_i) and for the outer diagonals of each cell ring (ω_o) as followed:

$$\omega_i = \arctan(h/b_i); \omega_o = \arctan(h/b_o) \tag{2}$$

The radial orientated diagonals strut inclination is equal for all cell rings and results from $\omega_r = \tan(h/b)$. Table 1 sums the resulting geometric parameters of the scaffold. It should be noticed that for the outer rings, the strut inclinations of the diagonals become lower 45°, which usually leads to unfavorable conditions in the AM process. By an optimization of the manufacturing parameters, see Section 2.1 for reference, and the good processability of the material, it was nevertheless possible to produce flat angles, as shown in Figure 3. The strut diameters of the manufactured scaffolds were measured at random positions, resulting in $r_s \approx 0.092$–0.106 mm, which lies in an acceptable tolerance range of the nominal strut radius.

Table 1. Resulting geometric parameters of the scaffold used for this study ($R_m = 1.4$ mm); i defines the actual ring, starting from the middle with $i = 1$ according to Figure 2.

i	r_i [mm]	$r_{m,i}$ [mm]	b_i [mm]	ω_i [°]	r_o [mm]	$r_{m,o}$ [mm]	b_o [mm]	ω_o [°]
1	1.4	1.376	0.515	62.8	2.3	2.261	0.845	49.8
2	2.3	2.261	0.845	49.8	3.2	3.146	1.176	40.4
3	3.2	3.146	1.176	40.4	4.1	4.030	1.507	33.6
4	4.1	4.030	1.507	33.6	5.0	4.917	1.838	28.6

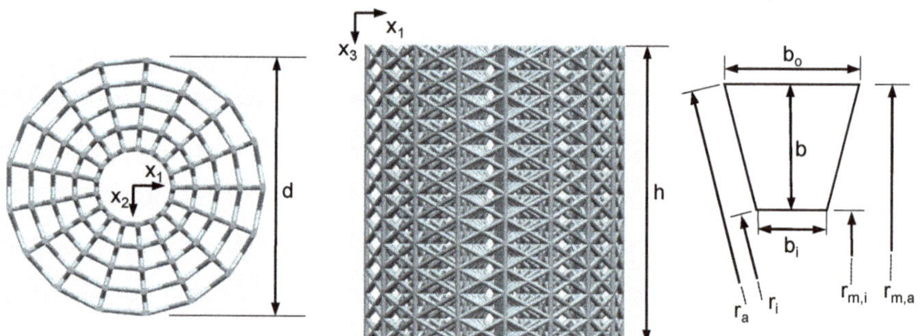

Figure 2. Scaffold geometry used for the study.

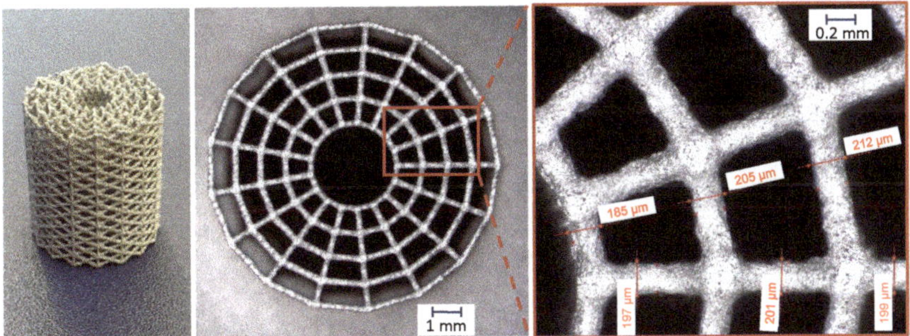

Figure 3. Resulting LPBF produced scaffold used for the physical evaluation.

2.3. Materials and Mechanical Properties

This study focuses a non-commercial Zinc-Magnesium alloy (Zn1Mg—1 wt% Mg), atomized by Nanoval GmbH (Berlin, Germany). The elastic material properties used for the numerical and analytical studies are based on literature data [19,34–36]. Validation tests are done on additively manufactured Zn1Mg scaffolds. Furthermore, this study is based on a previous study using Mg-based (WE43) scaffolds [2]. Young's modulus and yield strength of Zn1Mg were reported by Yang et al. [19]. Young's modulus of Zn1Mg is documented to be $E_{Zn1Mg} \approx 19$ GPa and yield strength $\sigma_{y,Zn1Mg} \approx 74$ MPa. The mechanical properties for Zn1Mg have been extracted via tensile tests. Both the zinc content as well as the magnesium content will take part in the biochemical reaction process. Material properties for $Zn(OH)_2$, $Mg(OH)_2$, $ZnCO_3$ and $MgCO_3$ from degradation processes are not sufficiently documented in the literature, but can be approximated by extrapolating data i.e., from Ulutan et al. [34], who reported values for the Young's modulus of $Mg(OH)_2$ of $E_{Mg(OH)_2} = 64$ GPa, Ulian et al. [35] reporting throughout anisotropic behavior an $E_{Mg(OH)_2} \approx 64$–180 GPa, or Yao et al. [36] reporting the Young's modulus of $MgCO_3$ to be $E_{MgCO_3} \approx 150$–260 GPa. For $Mg(PO)_4$ and the degradation products of Zn, insufficient data were found. Due to the poor data concerning material properties and proportions of the composite material, hypothetical Young's moduli were defined by multiples of the base materials Young's modulus, which is adequate for the analytical and numerical investigations concerning the general influence.

2.4. Analytical Model

The metallic strut and the enclosing compound of degradation products can be modeled as a composite beam. Here, the metallic core is surrounded by a thin-walled mineral cross section, which is idealized to be perfectly round in the following, and is demonstrated in Figure 4. Afterwards, the axial and bending stiffness of a composite strut can be calculated by a summation of the individual layer stiffnesses. The resulting equivalent composite axial stiffness $\overline{EA}$ can be calculated as followed:

$$\overline{EA} = \sum E_j A_j = E_s r_s^2 \pi + E_{\text{sub}} \left(2 r_s t_{\text{sub}} + t_{\text{sub}}^2\right) \pi, \tag{3}$$

where E_s is the base materials Young's modulus, E_{sub} is the Young's modulus of the compound of degradation products in the substrate layer, r_s is the inner radius of the substrate layer, or rather the base strut radius, and t_{sub} is the thickness of the substrate layer. For the equivalent composite bending stiffness $\overline{EJ}$ results:

$$\overline{EJ} = \sum E_j J_j = E_s \frac{\pi}{4} r_s^4 + E_{\text{sub}} \frac{\pi}{4} \left((r_s + t_{\text{sub}})^4 - r_s^4\right). \tag{4}$$

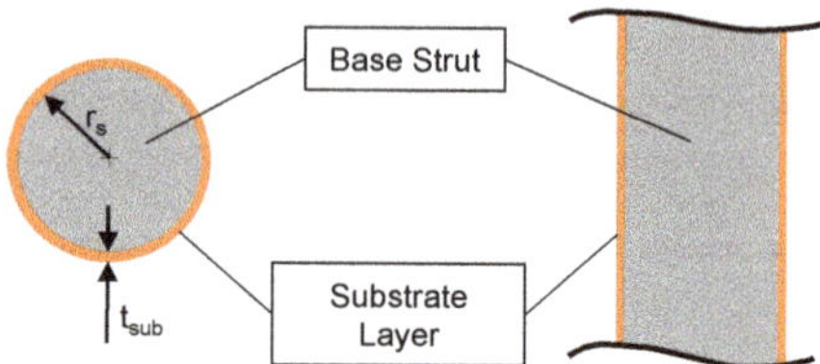

Figure 4. Cross section of the idealized corroded strut; in grey: base strut, in orange: compound of degradation/reaction products.

2.5. Finite-Element Model

For the FE calculations Abaqus/Standard with python scripting for model creation was used. The scaffolds were meshed using 3-node quadratic beam elements (B32). A convergence study showed that using five elements per strut gives sufficient results. Linear elastic material behavior and a static, displacement-controlled step was used. A displacement of $u = 1$ mm in axial direction (x_3−direction) of the scaffold was applied. The summation of the nodal reaction forces in axial direction F_3 was measured. The resulting stiffness can be calculated from $(EA) = F_3 h / u_3$. Since for beam elements no composite cross section can be defined in Abaqus/Standard, a generalized beam profile was used. Stiffnesses were defined according to Equations (3) and (4). To validate the beam formulation, a single strut under compression and bending was modeled using (a) a 3D-volume mesh with a hybrid meshing strategy using 10-node quadratic tetrahedron (C3D10) and 20-node quadratic hexagonal (C3D20) elements and (b) the aforementioned beam modeling strategy. For good mesh quality 48 elements in circumferential direction and five elements in radial direction plus one additional element for the substrate layer were used. According to the scaffold mesh, for the single strut beam model a total number of ten 3-node quadratic beam elements (B32) was used. Both models are show in Figure 5. The base strut radius was set to $r_s = 0.1$ mm and the substrate layer thickness to $t_{sub} = 0.01$ mm (see Figure 4 for reference). The strut length is $l = 5$ mm. The beam model cross section was defined with a generalized beam section according to the scaffold model. Struts under both axial compression and bending were examined. For the axial loaded strut, a simply supported beam and for the bending model a cantilever beam model was used.

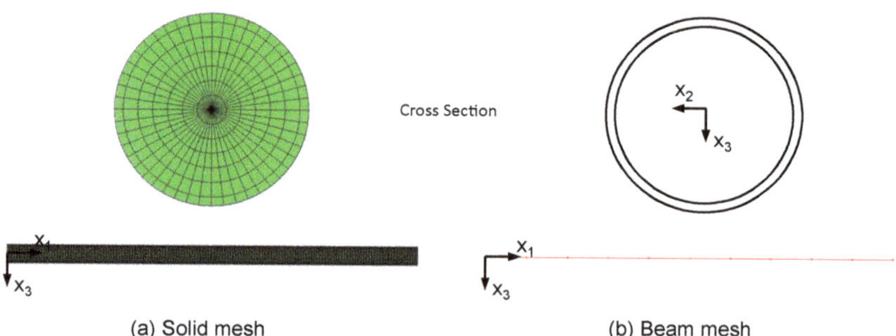

(a) Solid mesh (b) Beam mesh

Figure 5. Finite-Element Mesh; (**a**) solid model, (**b**) beam model with schematic cross section.

2.6. Compression Testing

To validate the FE model, compression tests on equivalent LPBF (Laser Powder Bed Fusion) produced Zn1Mg polar scaffolds were done. A total number of two specimens was tested. The tests were done on an Instron 5567 electric tensile/compression testing machine with 30 kN load cell. The tests were performed displacement controlled with a crosshead speed of $\dot{u} = 0.2$ mm/min. The crosshead displacement and load were documented. Since small shifts in the test setup lead to differences between the real and the crosshead displacement, the tests were monitored via DIC-technique (Direct Image Correlation) using an Aramis 4M system by GOM. By this, the real displacement of the specimen can be measured. Figure 6 shows the used setup for the compression tests.

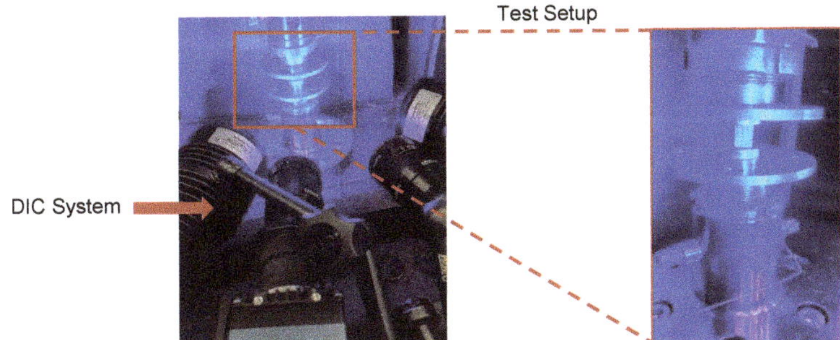

Figure 6. Experimental setup for compression tests on AM Zn1Mg scaffolds.

3. Results

3.1. Analytical Results

Figure 7 shows the results of the analytical calculations for the Zn1Mg single struts under axial compression. Shown is resulting composite Young's modulus E as a function of the substrate layer thickness t_{sub} for different strut radii r_s (50 µm–250 µm). Furthermore, Figure 7 (a) shows the resulting absolute composite Young's modulus (left axis) and relative increase E/E_{Zn1Mg} (right axis) for a Young's modulus twice as high, (b) three times as high, (c) four times as high and (d) five times as high as the base materials Young's modulus. It can be noticed that the thinner the struts and the thicker the substrate layer, the higher the resulting composite axial stiffness of the strut. Especially for smaller strut radii, i.e., $r_s = 50$ µm, as well as for small substrate thicknesses, the effect of an increase in axial stiffness is clearly visible. Same applies for high Young's moduli of the substrate layer.

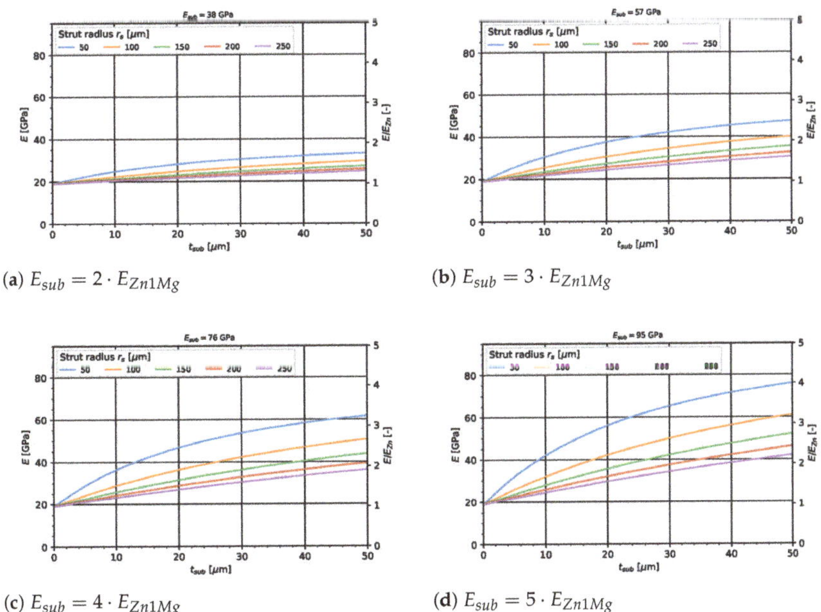

(a) $E_{sub} = 2 \cdot E_{Zn1Mg}$

(b) $E_{sub} = 3 \cdot E_{Zn1Mg}$

(c) $E_{sub} = 4 \cdot E_{Zn1Mg}$

(d) $E_{sub} = 5 \cdot E_{Zn1Mg}$

Figure 7. Analytical calculation of the axial stiffness of a single composite strut for varying parameters of the substrate layer.

Figure 8 shows the results for the analytical observations of the Zn1Mg single struts under bending. Shown is the resulting composite bending stiffness EJ as a function of the substrate layer thickness t_{sub} for different substrate Young's moduli E_{sub}, which is set to 2–5 times the base materials Young's modulus E_{Zn1Mg}. Furthermore, Figure 8 shows the resulting absolute composite bending stiffness EJ (left axis) and relative increase $EJ/(EJ)_{Zn1Mg}$ (right axis) for (a) a base strut radius $r_s = 50$ µm, (b) $r_s = 100$ µm, (c) $r_s = 150$ µm and (d) $r_s = 200$ µm. With increasing substrate layer thickness and higher substrate Young's modulus, a higher increase in bending stiffness can be observed. Especially for small strut radii, such as $r_s = 50$ µm, very high increases in bending stiffness can be achieved. This is not only the case for high moduli of the substrate layer, but also in the case when the composite of degradation products has the same Young's modulus.

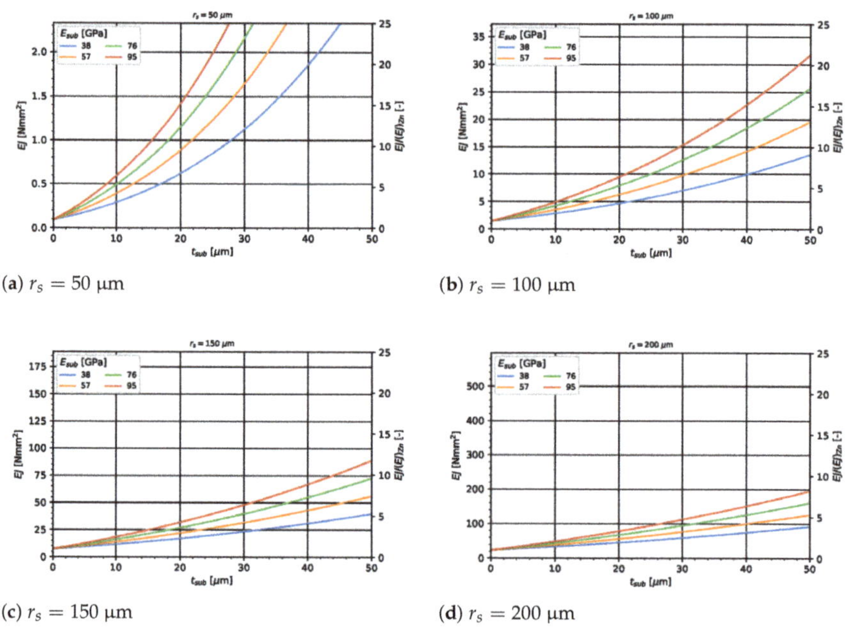

Figure 8. Analytical calculation of the bending stiffness of a single composite strut for varying parameters of the substrate layer.

3.2. Finite-Element Results

3.2.1. Single Strut Simulations

Figure 9 shows the results of the FE simulations of single struts under axial compression. For both the base strut and the corroded strut, the axial reaction force $RF1$ shows nearly equal values and the difference lies under 0.02%. In Figure 10 the comparison for the bending loaded struts is presented. In both cases the results of the beam model and the solid model are in good agreement. For the base struts, the difference regarding reaction force RF and reaction moment RM lies at around 1%. For the corroded struts, the difference is lower than 0.03%. Furthermore, especially in the case of the corroded strut, the calculation time can be massively decreased using a beam modeling approach.

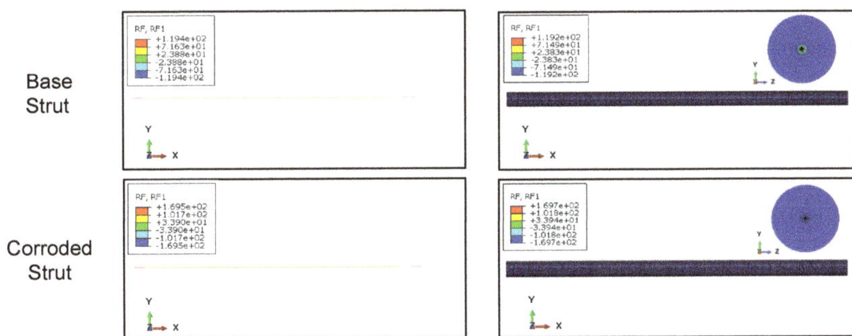

Base Strut / Corroded Strut

Figure 9. Reaction force (RF) comparison of modeling approaches for base and corroded strut under axial compression; (**left**) beam elements, (**right**) solid elements.

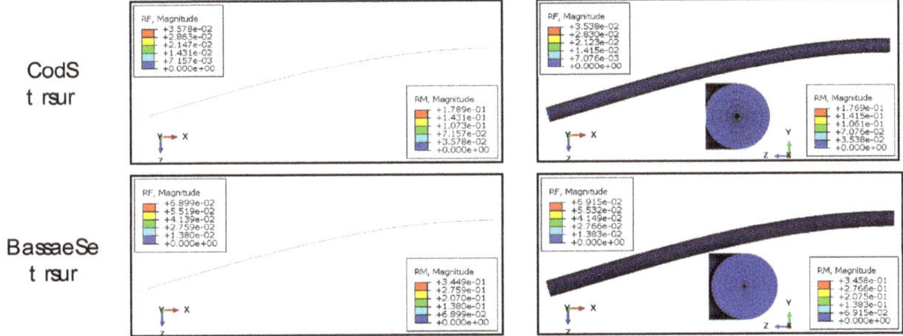

CodSt rsur / BassaeSet rsur

Figure 10. Reaction force (RF) and reaction moment (RM) comparison of modeling approaches for base and corroded strut under bending; (**left**) beam elements, (**right**) solid elements.

3.2.2. Whole Scaffold Modeling

Figure 11 shows the results of the FE Scaffold parametric study. Shown is the resulting smeared Young's modulus E of the scaffold, which results from dividing the axial reaction forces by the projected cross section of the whole scaffold A, as a function of the base strut radius r_s for different thicknesses of the substrate layer t_{sub} and (a) a compound Young's modulus of the substrate layer of $E_{sub} = 19$ GPa (equal to E_{Zn1Mg}), (b) $E_{sub} = 38$ GPa, (c) $E_{sub} = 57$ GPa and (d) $E_{sub} = 76$ GPa. The stiffness grows exponentially as a function of the strut diameter and is clearly more pronounced the higher the Young's modulus of the compound of the substrate. A significant increase in the axial stiffness of the scaffolds can be observed from all hypothetical Young's moduli of the substrate. Table 2 sums the quantitative results for the respective Young's moduli. It can be seen that already for a base materials equivalent Young's modulus of the substrate, small substrate thicknesses of a few microns and small strut radii lead to an increase in stiffness of 22–85%. The effect increases significantly when considering higher layer thicknesses and higher stiffnesses of the substrate layer.

Table 2. Percentage increase of the smeared Young's modulus E for varying substrate Young's moduli E_{sub} and layer thicknesses t_{sub}.

E_{sub} [GPa]	t_{sub} [µm]	r_s [mm]				
		0.05	0.1	0.15	0.2	0.25
19	5	22%	11%	8%	6%	4%
	10	46%	23%	15%	11%	9%
	20	102%	49%	35%	23%	18%
38	5	43%	22%	14%	11%	8%
	10	91%	44%	29%	22%	17%
	20	201%	95%	61%	45%	35%
57	5	64%	32%	21%	16%	13%
	10	136%	66%	43%	32%	25%
	20	300%	140%	90%	66%	52%
76	5	85%	42%	28%	21%	17%
	10	180%	87%	57%	42%	34%
	20	353%	164%	106%	78%	61%

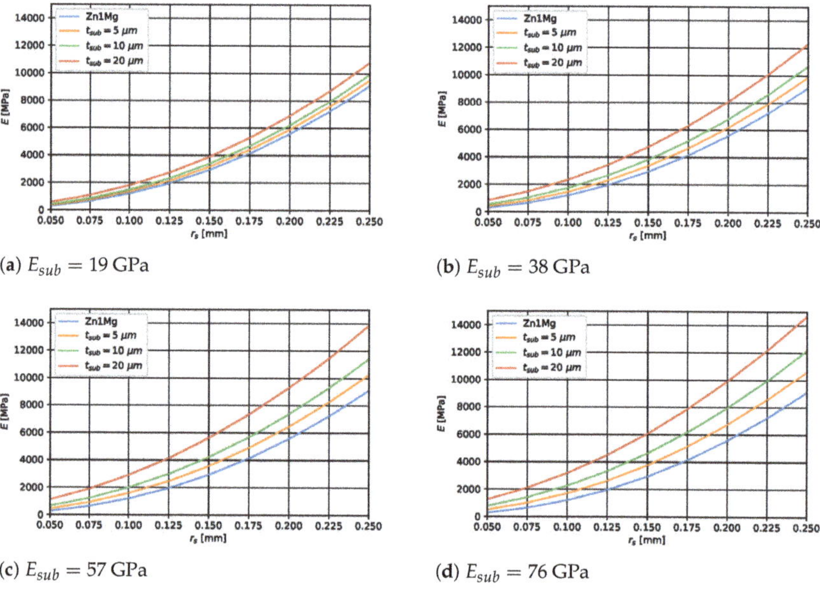

(**a**) $E_{sub} = 19\,\text{GPa}$

(**b**) $E_{sub} = 38\,\text{GPa}$

(**c**) $E_{sub} = 57\,\text{GPa}$

(**d**) $E_{sub} = 76\,\text{GPa}$

Figure 11. Results of the FE simulations of corroded scaffolds for varying parameters.

3.3. Confirmation by Physical Evaluation

Figure 12 shows the results of the two tested scaffolds under axial compression in comparison to the FE result. The tests show reproducible behavior regarding the stiffness. The smeared Young's modulus of the scaffolds can be calculated in the linear region of the load-displacement curves by $E = Fh/(Au)$, where F is the measured force in the machines load cell, h is the total height of the scaffold, A is the projected smeared cross section of the scaffold and u is the displacement associated with the measured force. From the tests, a Youngs's modulus of approximately $E_{test} \approx 1125\,\text{MPa}$ can be determined, measured in the area between 600–800 N. From the FE model a Young's modulus of $E_{FE} = 1258\,\text{MPa}$ can be extracted. Furthermore, the FE model shows that for loads smaller 800 N, nowhere the strut-stresses have exceeded the yield point. The slight differences could be attributed to local deviations in the strut diameter of the AM scaffolds, as shown in Section 2.2 respectively

Figure 3. Furthermore, the modeling using beam elements neglects the accumulation of material in the nodes of the real scaffold. Furthermore, the used Young's modulus is based on literature data and it is well known that Young's moduli of AM materials tend to show slight differences (see also Section 1). Nevertheless, the tests show that the FE model based on beam elements provides sufficiently accurate results in terms of the resulting smeared axial stiffness and can be used for the parametric study.

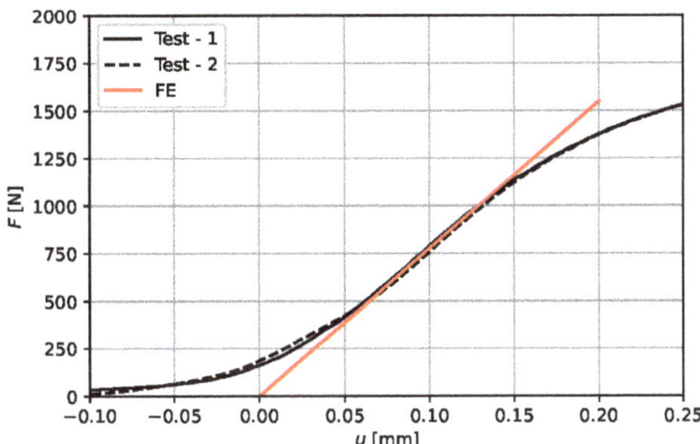

Figure 12. Validation of FE model: Resulting load-displacement curve of two tested LPBF produced scaffolds and equivalent FE model.

4. Discussion

We investigated the influence of degradation products on the elastic stiffness properties of biodegradable metallic scaffolds. For this, a hypothetical compound of degradation products was modeled as a thin-walled layer with a homogeneous cross section. The compound of degradation products consists for the most parts of hydroxides, phosphates and carbonates [29–32]. Since there is no sufficient database, yet, for the mechanical properties of the degradation products, hypothetical Young's moduli were defined using multiples of the Young's modulus of the base material, which was obtained from literature data [3,12,13,16–24]. By this, the influence of the degradation products on the elastic stiffness properties as a function of the layer thickness and Young's modulus could be investigated. This was done using analytical models and finite-element simulations for single struts, to show the direct influence of the layer of degradation products on the axial and bending stiffness, as well as for whole scaffold geometries, to show the superposed influence on the axial smeared Young's modulus of a specific scaffold geometry. Two modeling approaches were contrasted for the FE simulations, first a meshing strategy using a 3D volume mesh and second using beam elements. Both approaches show concurring results. For this reason, the beam model was used for a parametric study on whole lattice scaffold geometries, due to the enormous difference regarding the calculation time. To validate the FE model, scaffolds were produced via LPBF and compression tests on two scaffolds were done.

From the single strut investigations can be concluded that depending on the substrates Young's modulus and the ratio of strut radius to thickness of the substrate layer, significant increases of the composite axial and bending stiffness is expected. The effect intensifies, the smaller the base strut radius in the initial state is. This applies as well as for relatively low Young's moduli of the substrate layer as for very high Young's moduli. In comparable studies [2,14,15,25,26], mentioned in the introduction part, strut diameters of 300–400 μm were used for orthopaedic scaffolds. Even for low layer thicknesses (i.e., 10 μm) and low Young's moduli, for single struts with diameters in this range, depending on the thickness of the substrate layer and the composite module, an increase of more than 10% for the Young's modulus under axial compression and more than 40% in bending stiffness can be expected, which is not to be confused with the Young's modulus in the bending load case. To validate the base FE model, physical test results were compared to an equivalent FE simulation, using beam elements for meshing. As presented in the results section, the beam modeling shows similar results, compared to a much more numerically expensive meshing strategy with solid elements. The compression tests on LPBF produced scaffolds show reproducible results and furthermore equivalent smeared Young's moduli in the FE model and physical tests. For this reason, a FE parametric study on the tested geometry was done by varying the substrate layer thickness and the Young's moduli of the compound of the degradation products in the substrate layer, to study the influence of the substrate layer on the smeared Young's modulus of complex scaffold geometries. Our results show that an enormous increase in stiffness can be expected even for complex geometries, which was also observed by Li et al. [2] for diamond lattice structures made from WE43. For the previously mentioned example of strut diameters of 300–400 μm, the investigations on the scaffolds show that a much stronger effect can be observed due to the superposition of the axial and bending stiffness increase. As presented in the results section, the increase of the smeared axial Young's modulus under compression can be quantified to approximately 10–40% for a layer thickness of 10 μm and varying Young's moduli. The effect intensifies to values of approximately 20–80%, if i.e., a layer thickness of 20 μm is assumed. From this can be followed that compared to the separated reflection of the influence of the substrate layer on axial and bending stiffness of single struts, the effect of a stiffness increase is clearly more pronounced in the case of scaffold geometries. This is mainly attributable to the combined loading in compression and bending of the struts, which both ultimately have a direct effect on the smeared Young's modulus of the scaffold. Nevertheless, the investigations on single struts give clear indications about the formation of the effect. Furthermore, the analytical expressions show the direct influence of the thickness and Young's modulus of the degradation products.

5. Conclusions

In conclusion, our analytical and numerical modeling approach basically confirmed earlier assumptions by Li et al. [2] that the increase in stiffness of corrosion product layer-coated AM WE43 is indeed due to formation of a composite beam of base strut and substrate layer. As shown in this discussion, even for low thicknesses and Young's moduli of the degradation product layer, axial stiffness increases of more than 40% can be achieved. Even though the geometry of the scaffold is different at the investigations of Li et al., this study clearly shows the influence on the stiffness. Nevertheless, our results must be validated by further investigations on corroded single struts or equal, to validate the formation of an almost homogeneous layer of degradation products and to obtain more knowledge about the real composite Young's modulus, or rather the Young's modulus of the compound of degradation products.

Author Contributions: J.B. performed most of the analytical analyses, J.B. modeling, physical testing, and drafted the paper. J.B., M.V. and H.J. contributed to the design of the scaffold, while M.V. produced them H.J. initiated and supervised the study, including the analyses. J.H.S. and K.-U.S. contributed their extensive experience, gave advice regarding the content and the manuscript. All authors have read and agreed to the published version of the manuscript.

Funding: Parts of this work were supported by the Federal Ministry of Education and Research (BMBF) and the Ministry of Culture and Science of the State of North Rhine-Westphalia (MKW) under the Excellence Strategy of the Federal Government and the Länder (OPSF597).

Institutional Review Board Statement: Not applicable.

Informed Consent Statement: Not applicable.

Data Availability Statement: The data presented in this study are available on request from the corresponding author.

Acknowledgments: The authors thank the BMBF for funding of the work.

Conflicts of Interest: The authors declare no conflict of interest.

References

1. Böstman, O.; Pihlajamäki, H. Clinical biocompatibility of biodegradable orthopaedic implants for internal fixation: A review. *Biomaterials* **2000**, *21*, 2615–2621. [CrossRef]
2. Li, Y.; Zhou, J.; Pavanram, P.; Leeflang, M.A.; Fockaert, L.I.; Pouran, B.; Tümer, N.; Schröder, K.U.; Mol, J.M.C.; Weinans, H.; et al. Additively manufactured biodegradable porous magnesium. *Acta Biomater.* **2018**, *67*, 378–392. [CrossRef] [PubMed]
3. Jahr, H.; Li, Y.; Zhou, J.; Zadpoor, A.A.; Schröder, K.U. Additively Manufactured Absorbable Porous Metal Implants—Processing, Alloying and Corrosion Behavior. *Front. Mater.* **2021**, *8*, 292. [CrossRef]
4. Liu, X.; Ma, P.X. Polymeric scaffolds for bone tissue engineering. *Ann. Biomed. Eng.* **2004**, *32*, 477–486. [CrossRef] [PubMed]
5. Seitz, H.; Rieder, W.; Irsen, S.; Leukers, B.; Tille, C. Three-dimensional printing of porous ceramic scaffolds for bone tissue engineering. *J. Biomed. Mater. Res. Part B—Appl. Biomater.* **2005**, *74B*, 782–788. [CrossRef]
6. Chen, Q.; Thouas, G.A. Metallic implant biomaterials. *Mater. Sci. Eng. R Rep.* **2015**, *87*, 1–57. [CrossRef]
7. Cutolo, A.; Engelen, B.; Desmet, W.; van Hooreweder, B. Mechanical properties of diamond lattice Ti–6Al–4V structures produced by laser powder bed fusion: On the effect of the load direction. *J. Mech. Behav. Biomed. Mater.* **2020**, *104*, 103656. [CrossRef]
8. Gümrük, R.; Mines, R.; Karadeniz, S. Static mechanical behaviours of stainless steel micro-lattice structures under different loading conditions. *Mater. Sci. Eng. A* **2013**, *586*, 392–406. [CrossRef]
9. Yan, Q.; Dong, H.; Su, J.; Han, J.; Song, B.; Wei, Q.; Shi, Y. A Review of 3D Printing Technology for Medical Applications. *Engineering* **2018**, *4*, 729–742. [CrossRef]
10. Li, Y.; Jahr, H.; Zhou, J.; Zadpoor, A.A. Additively manufactured biodegradable porous metals. *Acta Biomater.* **2020**, *115*, 29–50. [CrossRef] [PubMed]
11. Wang, J.L.; Xu, J.K.; Hopkins, C.; Chow, D.H.K.; Qin, L. Biodegradable Magnesium-Based Implants in Orthopedics-A General Review and Perspectives. *Adv. Sci.* **2020**, *7*, 1902443. [CrossRef]
12. Wen, P.; Voshage, M.; Jauer, L.; Chen, Y.; Qin, Y.; Poprawe, R.; Schleifenbaum, J.H. Laser additive manufacturing of Zn metal parts for biodegradable applications: Processing, formation quality and mechanical properties. *Mater. Des.* **2018**, *155*, 36–45. [CrossRef]
13. Qin, Y.; Wen, P.; Guo, H.; Xia, D.; Zheng, Y.; Jauer, L.; Poprawe, R.; Voshage, M.; Schleifenbaum, J.H. Additive manufacturing of biodegradable metals: Current research status and future perspectives. *Acta Biomater.* **2019**, *98*, 3–22. [CrossRef] [PubMed]
14. Li, Y.; Jahr, H.; Lietaert, K.; Pavanram, P.; Yilmaz, A.; Fockaert, L.I.; Leeflang, M.A.; Pouran, B.; Gonzalez-Garcia, Y.; Weinans, H.; et al. Additively manufactured biodegradable porous iron. *Acta Biomater.* **2018**, *77*, 380–393. [CrossRef] [PubMed]
15. Li, Y.; Jahr, H.; Pavanram, P.; Bobbert, F.S.L.; Paggi, U.; Zhang, X.Y.; Pouran, B.; Leeflang, M.A.; Weinans, H.; Zhou, J.; et al. Additively manufactured functionally graded biodegradable porous iron. *Acta Biomater.* **2019**, *96*, 646–661. [CrossRef] [PubMed]
16. Song, B.; Dong, S.; Deng, S.; Liao, H.; Coddet, C. Microstructure and tensile properties of iron parts fabricated by selective laser melting. *Opt. Laser Technol.* **2014**, *56*, 451–460. [CrossRef]
17. Song, B.; Dong, S.; Liu, Q.; Liao, H.; Coddet, C. Vacuum heat treatment of iron parts produced by selective laser melting: Microstructure, residual stress and tensile behavior. *Mater. Des.* **2014**, *54*, 727–733. [CrossRef]
18. Montani, M.; Demir, A.; Mostaed, E.; Vedani, M.; Previtali, B. Processability of pure Zn and pure Fe by SLM for biodegradable metallic implant manufacturing. *Rapid Prototyp. J.* **2017**, *23*, 514–523. [CrossRef]
19. Yang, Y.; Yuan, F.; Gao, C.; Feng, P.; Xue, L.; He, S.; Shuai, C. A combined strategy to enhance the properties of Zn by laser rapid solidification and laser alloying. *J. Mech. Behav. Biomed. Mater.* **2018**, *82*, 51–60. [CrossRef]
20. Zhou, Y.; Wu, P.; Yang, Y.; Gao, D.; Feng, P.; Gao, C.; Wu, H.; Liu, Y.; Bian, H.; Shuai, C. The microstructure, mechanical properties and degradation behavior of laser-melted Mg Sn alloys. *J. Alloys Compd.* **2016**, *687*, 109–114. [CrossRef]
21. Ng, C.C.; Savalani, M.M.; Lau, M.L.; Man, H.C. Fabrication of magnesium using selective laser melting technique. *Rapid Prototyp. J.* **2011**, *17*, 479–490. [CrossRef]
22. Ng, C.C.; Savalani, M.M.; Lau, M.L.; Man, H.C. Microstructure and mechanical properties of selective laser melted magnesium. *Appl. Surf. Sci.* **2011**, *257*, 7447–7454. [CrossRef]
23. Wei, K.; Zeng, X.; Wang, Z.; Deng, J.; Liu, M.; Huang, G.; Yuan, X. Selective laser melting of Mg-Zn binary alloys: Effects of Zn content on densification behavior, microstructure, and mechanical property. *Mater. Sci. Eng. A* **2019**, *756*, 226–236. [CrossRef]
24. Kubásek, J.; Dvorský, D.; Čapek, J.; Pinc, J.; Vojtěch, D. Zn-Mg Biodegradable Composite: Novel Material with Tailored Mechanical and Corrosion Properties. *Materials* **2019**, *12*, 3930. [CrossRef] [PubMed]
25. Frank, W.; Lucas, J.; Wolfgang, M.; Zienab, K.; Kristin, S.; Tanja, S. Open-porous biodegradable magnesium scaffolds produced by selective laser melting for individualized bone replacement. *Front. Bioeng. Biotechnol.* **2016**, *4*. [CrossRef]
26. Kopp, A.; Derra, T.; Müther, M.; Jauer, L.; Schleifenbaum, J.H.; Voshage, M.; Jung, O.; Smeets, R.; Kröger, N. Influence of design and postprocessing parameters on the degradation behavior and mechanical properties of additively manufactured magnesium scaffolds. *Acta Biomater.* **2019**, *98*, 23–35. [CrossRef] [PubMed]
27. Cockerill, I.; Su, Y.; Sinha, S.; Qin, Y.X.; Zheng, Y.; Young, M.L.; Zhu, D. Porous zinc scaffolds for bone tissue engineering applications: A novel additive manufacturing and casting approach. *Mater. Sci. Eng. C Mater. Biol. Appl.* **2020**, *110*, 110738. [CrossRef] [PubMed]
28. Li, Y.; Pavanram, P.; Zhou, J.; Lietaert, K.; Taheri, P.; Li, W.; San, H.; Leeflang, M.A.; Mol, J.M.C.; Jahr, H.; et al. Additively manufactured biodegradable porous zinc. *Acta Biomater.* **2020**, *101*, 609–623. [CrossRef]

29. Han, H.S.; Loffredo, S.; Jun, I.; Edwards, J.; Kim, Y.C.; Seok, H.K.; Witte, F.; Mantovani, D.; Glyn-Jones, S. Current status and outlook on the clinical translation of biodegradable metals. *Mater. Today* **2019**, *23*, 57–71. [CrossRef]
30. Zheng, Y.F.; Gu, X.N.; Witte, F. Biodegradable metals. *Mater. Sci. Eng. R Rep.* **2014**, *77*, 1–34. [CrossRef]
31. Li, P. Absorbable Zinc-based alloy for craniomaxillofacial osteosynthesis implants. In *Faculty of Medicine*; Eberhard Karls University Tübingen: Tübingen, Germany, 2020.
32. Kannan, M.B.; Moore, C.; Saptarshi, S.; Somasundaram, S.; Rahuma, M.; Lopata, A.L. Biocompatibility and biodegradation studies of a commercial zinc alloy for temporary mini-implant applications. *Sci. Rep.* **2017**, *7*, 15605. [CrossRef] [PubMed]
33. Meiners, W. Direktes Selektives Laser-Sintern Einkomponentiger Metallischer Werkstoffe. Ph.D. Thesis, RWTH Aachen University, Aachen, Germany, 1999.
34. Ulutan, S.; Gilbert, M. Mechanical properties of HDPE/magnesium hydroxide composites. *J. Mater. Sci.* **2000**, *35*, 2115–2120. [CrossRef]
35. Ulian, G.; Valdrè, G. Anisotropy and directional elastic behavior data obtained from the second-order elastic constants of portlandite $Ca(OH)_2$ and brucite $Mg(OH)_2$. *Data Brief* **2018**, *21*, 1375–1380. [CrossRef] [PubMed]
36. Yao, C.; Wu, Z.; Zou, F.; Sun, W. Thermodynamic and Elastic Properties of Magnesite at Mantle Conditions: First-Principles Calculations. *Geochem. Geophys. Geosyst.* **2018**, *19*, 2719–2731. [CrossRef]

 *materials* 

Article

Corrosion Susceptibility and Allergy Potential of Austenitic Stainless Steels

Lucien Reclaru [1] and Lavinia Cosmina Ardelean [2,*]

[1] Scientific Independent Consultant Biomaterials and Medical Devices, 103 Paul-Vouga,
     2074 Marin-Neuchâtel, Switzerland; lreclaru@gmail.com
[2] Department of Technology of Materials and Devices in Dental Medicine, "Victor Babes" University of
     Medicine and Pharmacy Timisoara, 2 Eftimie Murgu sq, 300041 Timisoara, Romania
*   Correspondence: lavinia_ardelean@umft.ro

Received: 27 July 2020; Accepted: 14 September 2020; Published: 21 September 2020

Abstract: Although called stainless steels, austenitic steels are sensitive to localized corrosion, namely pitting, crevice, and intergranular form. Seventeen grades of steel were tested for localized corrosion. Steels were also tested in general corrosion and in galvanic couplings (steels–precious alloys) used in watchmaking applications. The evaluations have been carried out in accordance with the ASTM standards which specifically concern the forms of corrosion namely, general (B117-97, salt fog test), pitting (G48-11, FeCl$_3$), crevice (F746-87) and intergranular (A262-15, Strauss chemical test and G108-94, Electrochemical potentiodynamic reactivation test). All tests revealed sensitivity to corrosion. We have noticed that the transverse face is clearly more sensitive than the longitudinal face, in the direction of rolling process. The same conclusion has been drawn from the tests of nickel release. It should be pointed out that, despite the fact that the grade of steel is in conformity with the classification standards, the behavior is very different from one manufacturer to another, due to parameters dependent on the production process, such as casting volume, alloying additions, and deoxidizing agents. The quantities of nickel released are related to the operations involved in the manufacturing process. Heat treatments reduce the quantities of nickel released. The surface state has little influence on the release. The hardening procedures increase the quantities of nickel released. The quantities of released nickel are influenced by the inclusionary state and the existence of the secondary phases in the steel structure. Another aspect is related to the strong dispersion of results concerning nickel release and corrosion behavior of raw materials.

Keywords: austenitic steels; general (uniform) corrosion; pitting corrosion; crevice corrosion; intergranular corrosion; galvanic couplings; nickel release; contact with skin; medical devices; watchmaking

---

1. Introduction

Corrosion represents an important factor in the design and selection of metals and alloys for different purposes, as various corrosion mechanisms can lead to failure [1,2]. Corrosion resistance is an important criterion for selecting materials used, because the cost of their degradation due to corrosion and the associated environmental impact are quite substantial [3]. Like all metals, stainless steels can undergo chemical corrosion over time [4–6].

Corrosion manifests in different forms and depends on a multitude of physico-chemical factors (chemical composition and microstructure of the alloy, temperature, pH, chemical composition of the environment) and mechanical factors (stresses, friction) [7]. The relationship between corrosion rate and grain size has been revealed in numerous studies [8–11]. Its importance lies in the fact that this parameter can be tailored by the producers [8–11].

The austenitic steels belong to the stainless steels family and are being characterized by high Ni$_{eq}$ and Cr$_{eq}$ [12,13]. Over time, the chemical composition, mechanical properties, resistance to corrosion, machinability and polish ability of the austenitic steels have evolved considerably, and new production processes have been developed by the steel manufacturers [14]. Each chemical element in their composition plays an important role in their properties [15], including corrosion resistance, and can be substantially modified by adding certain elements as Cu, Ti, Nb, Al, Si and Ca. Generally, the composition of austenitic stainless steels is adjusted to meet service requirements in various corrosive environments [8]. The corrosion sensitivity of austenitic steels mainly takes the form of pitting, crevice and intergranular type [16].

An important aspect which concerns the austenitic steels is the release of nickel in contact with the skin. The role of nickel in the biological response to alloys is significant with regard to toxicology and biological performance. The current trend is to eliminate nickel from alloys for medical applications. However, this needs a careful evaluation since no compromise is acceptable concerning the mechanical properties, corrosion resistance or any other possible undesirable consequences due to the substitution of nickel [7,17].

Nickel allergy is the most widespread of all contact allergies. In the European population, the prevalence of nickel allergy is of 10%–15% of adult females and 1%–3% of adult males [18–22]. Of nickel-sensitive people in the general population, 30% develop hand eczema. Teenagers and young adults tend to have a higher prevalence due to frequent body piercing.

In Europe, for objects containing nickel, intended for permanent contact with skin, Directive 94/27/EC imposed a ban if the rate of nickel release exceeds 0.5 μg/cm^2·week. The subsequent Directive 2004/96/EC: "Piercing in the Human Body" specifies that the limit rate of nickel release, for these cases, is 0.2 μg/cm^2·week [23–25].

The aim of this study is to evaluate the sensitivity, under the same conditions, of 17 austenitic steels of the 304, 316 and 904 series, for uniform, pitting, crevice and galvanic corrosion. Our interest was also to assess the behavior differences of the transverse surface of the samples compared to the longitudinal one.

2. Materials and Methods

Table 1 shows the composition of the austenitic stainless steels (exception #16 and #17) which were used to prepare the samples for the corrosion evaluation tests.

Table 1. Chemical composition (wt.%) of the grades of austenitic steels used in corrosion tests.

Code	DIN	AISI	C	Si	Mn	P	S	Cr	Mo	Ni	Other
#1	1.4306	304L	<0.03	<1.5	<1.5	<0.035	<0.02	17.0–20.0	-	8.0–12.0	N 0.1–0.2
#2	1.4427So	-	<0.03	<1.0	<2.0	<0.045	0.10–0.13	16.5–18.5	2.0–2.7	-	-
#3	1.4435	316L	<0.03	<1.0	<2.0	<0.045	<0.025	17.0–18.5	2.5–3.0	12.5–15.0	-
#4	1.4435	316LUgim	<0.03	<1.0	<2.0	<0.045	~0.018	17.0–18.5	2.5–3.0	12.5–15.0	N < 0.11
#5	1.4435	316LVal	<0.03	<1.0	<2.0	<0.045	~0.018	17.0–18.5	2.5–3.0	12.5–15.0	N < 0.11
#6	1.4435	316LPM	<0.03	<1.0	<2.0	<0.045	~0.018	17.0–18.5	2.5–3.0	12.5–15.0	N < 0.11
#7	1.4435	316LSW	<0.03	<1.0	<2.0	<0.045	~0.018	17.0–18.5	2.5–3.0	12.5–15.0	N < 0.11
#8	1.4441	316LMed	<0.03	<1.0	<2.0	<0.025	<0.01	17.0–0.19	2.5–3.2	13.0–15.5	N < 0.10 Cu < 0.12
#9	1.4571	316Ti	<0.08	<1.0	<2.0	<0.045	<0.03	16.5–18.5	2.0–2.5	-	-
#10	1.4539	904L	<0.02	<0.7	<2.0	<0.03	<0.015	19.0–21.0	4.0–5.0	24.0–26.0	Cu 1.0–2.0; N 0.04–0.15
#11	1.4057	431	0.14–0.23	<1.0	<1.0	<0.045	<0.03	15.5–17.5	-	1.5–2.5	-
#12	1.4460	329	<0.05	<1.0	<2.0	<0.045	<0.03	25.0–28.0	1.3–2.0	4.5–6.5	N 0.05–0.2
#13	1.4462	2205	<0.03	<1.0	<2.0	-	<0.02	21.0–23.0	2.5–3.5	4.5–6.5	N 0.08–0.2
#14	1.4542	630	<0.07	<1.0	<1.0	<0.045	<0.03	15.0–17.0	-	3.0–5.0	Nb 0.15–0.45
#15	1.4841	310/314	<0.20	1.5–2.5	<2.0	<0.045	<0.03	24.0–26.0	-	19.0–22.0	-
#16	1.4876	B 163	<0.12	<1.0	<2.0	<0.03	<0.02	19.0–23.0	-	30.0–34.0	-
#17	2.4816	Inconel600	<0.15	<0.5	<1.0	<0.02	<0.015	14.0–17.0	-	>72.0	Ti < 0.3; Al < 0.3; B <0.006; Cu <0.5; Fe 6.00–10.0

According to the classification of Fontana [26], the evaluation of their behavior in uniform, pitting, crevice, intergranular corrosion was presented, as well as in galvanic couplings.

2.1. Salt Fog Test

To illustrate uniform corrosion, four austenitic steels—#2-1.4427, #3-1.4435/316L, #4-316L/1.4435Ugim and #5-1.4435/316LVal—were tested by using the salt fog test, according to ASTM B117-97 [27].

The samples were of cylindrical shape—10 mm diameter, 5 cm long. To create a reference state, as-received samples were first annealed and recrystallized at 1050 °C. The thermal treatment was carried out in a continuous industrial oven under hydrogen protection with gas cooling.

Half of the annealed cylinders were cold-worked to diameter 7.7 mm. The purpose of this cold-working operation was to increase the samples' sensitivity to corrosion [28,29]. The entire surface of the samples was "mirror" polished. The test was carried out over a period of 12 days, 6 days in 5% NaCl medium and 6 days in artificial sweat medium diluted 40 times (Table 2). Artificial sweat medium ISO 3160-2 has the following composition: NaCl 0.5 g/L; NH_4Cl 0.4375 g/L; Acetic Acid 0.063 g/L; Urea 0.125 g/L; Lactic Acid 0.375 g/L; NaOH solid, necessary quantity to induce a pH of 4.7.

The test in question was an adaptation of the ASTM B117-97 standard [27] which is in current use for metallic objects in contact with the skin (ISO 3160-2). Three samples were used for each state, making a total of 48 samples.

Table 2. The salt fog test conditions.

Test	Conditions	
Electrolyte	✓	NaCl
	✓	Artificial sweat
Temperature	35 °C	
Total duration	12 days	
NaCl 5%	6 days	
Artificial sweat Dilution x40	6 days	
Operating cycle salt spraying exposing	✓	15 min
	✓	45 min

2.2. Pitting Corrosion

Steels #1–#17 (Table 1) were tested for pitting corrosion. The test samples, in the form of wires, 10 mm diameter, 5 cm long, were fixed in a resin and "mirror" polished.

The tests were carried out in the most commonly used electrolytes: $FeCl_3$ and NaCl, and artificial sweat, according to ISO 3160-2:2015 [30]. The different electrolytes were used to verify the observations. The chemical compositions of the media used and the experimental conditions are presented in Table 3.

Table 3. Types of electrolytes used and the experimental conditions for testing pitting corrosion [31–33].

Test Medium	Concentration	Temperature	Test Duration
$FeCl_3$	0.5 M	50 °C	2 h
$FeCl_3$	0.1 M	37 °C	15 days
NaCl	0.5 M	37 °C	26 days
Artificial sweat ISO 3160-2 Diluted 20x	NaCl 0.5 g/L; NH4Cl 0.4375 g/L; Acetic Acid 0.063 g/L; Urea 0.125 g/L; Lactic Acid 0.375 g/L; NaOH solid, necessary quantity to induce a pH of 4.7	37 °C	30 days

The immersion times were therefore different: 15 days for FeCl$_3$ 0.1 M, 26 days for NaCl 0.5 M and 30 days for artificial sweat (Table 3), because of the difference in aggressiveness of the media (chemical composition, concentration) towards the evaluated materials [34].

2.3. Crevice Corrosion

The crevice corrosion test was carried out according to ASTM F746-87 [35]. The samples, of cylindrical shape—6.35 mm diameter, 5 cm long—obtained by machining of 10-mm-diameter steel profiles, "mirror" polished, were embedded in a polytetrafluoroethylene (PTFE) collar at one of the extremities (Figure 1). Both transverse and longitudinal surfaces were subjected to the corrosion test.

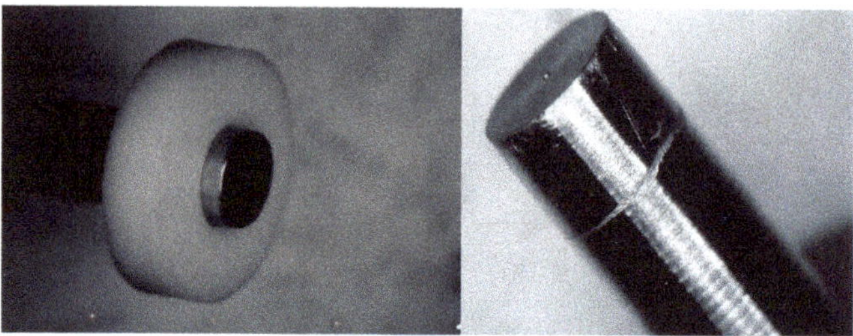

Figure 1. Specific assembly for the crevice corrosion test. Crevice corrosion aspect of the test sample.

The sample was mounted in the rotating electrode of the measuring cell (Figure 2).
The test consisted of two stages:

- In the first stage, anodic excitation of the sample to be evaluated was carried out at 800 mV vs. SCE (saturated calomel electrode) for 10 s;
- In the second stage, the potentiost at imposed the abandon potential value for 15 min. As a result, the current variation was plotted as a function of time for an imposed potential.

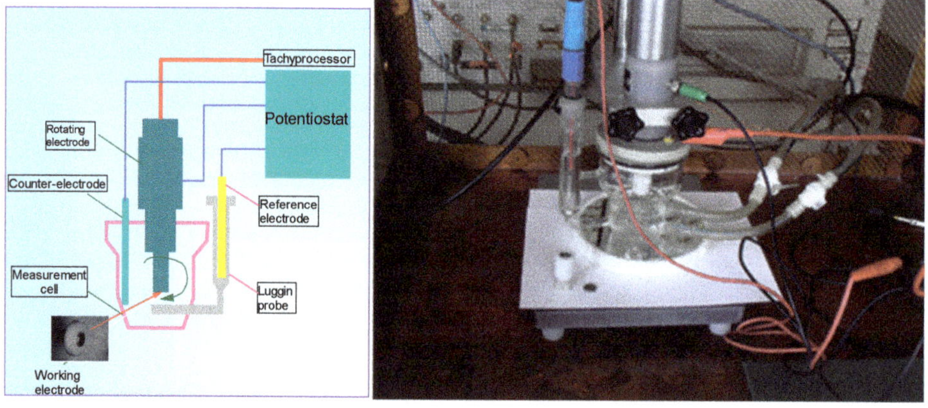

Figure 2. The electrochemical assembly and the electrochemical measuring cell.

If the current recorded stayed within the cathodic domain (negative values), a fresh measurement cycle was started: excitation for 10 s at 800 mV and current measurement for a potential set at $E_{abandon}$ +25 mV (ASTM recommends +50 mV). The cycles were repeated, each time at a higher potential, until the current measured moved into the anodic domain (positive values). As a result, the crevice potential was determined, which corresponded to the last-but-one measurement for which the current was positive. Once the Teflon ring was removed, a groove of crevice corrosion was noted on the surface of the metal sample.

2.4. Intergranular Corrosion

This type of corrosion occurs preferentially at grain boundaries and may be due to the presence of precipitates.

For assessing the intergranular corrosion morphology, chemical and electrochemical tests may be used. Two examples of evaluating intergranular corrosion susceptibility of austenitic steel tubes are being presented:

- Chemical evaluation of tubes #3, 1.4435/316L used in medical devices and endoscopic applications;
- Electrochemical evaluation of tubes #1, 1.44306/304L used in medical devices and endoscopes applications.

2.4.1. Chemical Tests

The standardized tests [36–38] to assess the sensitivity of stainless steel to intergranular corrosion, according to ASTM A262-15 [36], have been summarized in Table 4.

Table 4. ASTM A262-15 [36] Standard practice for detecting susceptibility to intergranular attack in austenitic stainless steels.

Designation	Test	Temperature	Testing Time	Applicability	Evaluation Method
Practice A	Oxalic Acid Etch Screening Test	Ambient	1.5 min	Chromium Carbide sensitization only	Microscopic Examination
B	Ferric Sulfate and 50% Sulfuric Acid	Boiling	120 h	Chromium Carbide	Weight loss/ Corrosion Rate
C	65% Nitric Acid	Boiling	4 h	Chromium Carbide	Weight loss/ Corrosion Rate
D	10% Nitric-3% Hydro Fluoric Acid (This test has been removed from A 262-15)	70 °C	4 h	Chromium carbide in 316, 316L, 317, 317L	Corrosion Rates of "unknown" over that of solution annealed specimen
E	6% Copper Sulfate 16% Sulfuric Acid with metallic copper	Boiling	24 h	Chromium Carbide	Examination for fissures after bending
F	Copper Sulfate 50% Sulfuric Acid with metallic copper	Boiling	120 h	Chromium Carbide in 316 and 316L	Weight loss/ Corrosion rate

Evaluation of intergranular corrosion according to standard ASTM A262-15-the Strauss test [36].

The test medium was a solution of copper sulphate-50% sulfuric acid in the presence of metallic copper, brought to the boiling point (125 °C) for 120 h. This test investigated the intergranular corrosion behavior of steel in the potential range between 110–350 mV. The test setup was in accordance with ASTM A262-15 [36]. The samples to be tested, tubes of #3, 1.4435/316L steel, were placed in specific glass cradle.

2.4.2. Electrochemical Tests

Electrochemical Potentiodynamic Reactivation (EPR) measurements, single or double loop, are methods of examining and assessing the corrosion sensitivity of austenitic steels [37–47].

The ASTM G108-94 (2015) [48] was practiced with the single loop method. Using the single loop test, a sample polished to a 1 μm finish, was polarized for two minutes at 200 mV vs. SCE in a solution of 0.5 M H_2SO_4 + 0.01 M KSCN. Subsequently, the potential was decreased, at a rate of 6 V/h, to the corrosion potential, E_{corr}. This decrease resulted in reactivation of the specimen, involving breakdown of the passive film covering chromium depleted areas of material. The area under the large loop generated in the curve of potential vs. current (Figure 3) was proportional to the electric charge Q, that depends on surface area and grain size. In non-sensitized material, the passive film was intact and the loop size was small.

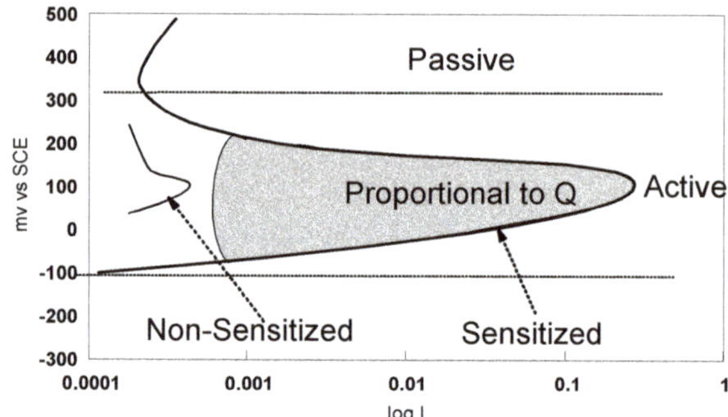

Figure 3. Procedures of single loop EPR test method according to ASTM G108–94 (2015) [48].

The samples tested were tubes of 304L AISI, used for manufacturing medical endoscopes (Table 5). The tubes #Test 1 and #Test 2 were suspected to be sensitized in intergranular corrosion. For the evaluation test, reference samples #Brut 1 and #Brut 2 and three samples from the supplier stock, subjected to heat treatments at 500, 620 and 750 °C, respectively, were used (Table 5).

Table 5. Test samples #1, 14306/304L tubes.

Code	Description
#Test 1	Tube returned from user
#Test 2	Tube returned from user
#Brut 1	Tube from supplier stock (reference)
#Brut 2	Tube from supplier stock (second reference)
#500	Tube from supplier stock+ 500 °C heat treatment, 1 h
#620	Tube from supplier stock+ 621 °C heat treatment, 1 h
#750	Tube from supplier stock+ 750 °C heat treatment, 1 h

The sample (Table 5), transversely cut, was incorporated into a resin and "mirror" polished. The resin was machined to be adapted to the working electrode. The mounting of the electrodes, the electrochemical cell and the EPR measurement conditions were those of ASTM G108-94 (2015) [48]. The ASTM G108-94 method allows a quantitative evaluation of the intergranular corrosion sensitization of steels AISI 304 and AISI 304L. The purpose of the test was to evaluate the intergranular corrosion of the transverse surface by an electrochemical scanning method from +200 mV to −400 mV vs. SCE. In the mathematical calculation for evaluating the sensitivity to intergranular corrosion, another parameter, the corresponding value of the grain index, has to be considered, according to ASTM E112-13 [49].

2.5. Galvanic Corrosion

In the first stage, to establish a galvanic series in artificial sweat the open circuit potentials of eighteen alloys were measured:

Precious metal alloys used in jewelry: Pt 950CoNi (950‰Pt, 18‰Co, 32‰Ni), AuPdCu150 (750‰Au, 100‰Pd, 150‰Cu), AuAgNi109 (750‰Au, 141‰Ag, 109‰Ni), AuCuNi130 (750‰Au, 120‰Cu, 130‰Ni), AuNiCu142 (750‰Au, 108‰ Ni, 142‰Cu), AuNiCu112 (750‰Au, 138‰Ni, 112‰Cu) and AuCuZn374 (585‰Au 41‰Cu, 374‰Zn).

Steels: 1.4441, 1.4435, 316 L F, 316 L F Cu, 1.4301, 1.4305, Sandvik 1802, 1.4104. 1.4105, 1.4539 and 12/12.

The samples, in form of 10-mm-diameter discs, were "mirror" polished, washed with a mixture of acetone and ethanol, and rinsed with deionized water 18 MΩ·cm. After drying with hot air, the samples were introduced into the PTFE sample holder, specially designed for the rotating electrode test. The electrochemical measurements were made with a potentiostatic assembly of three electrodes: a working electrode (rotating electrode), a platinum counter-electrode and a reference SCE electrode. Given that diffusion phenomena play a major role with regard to the changes produced at the metal/solution interface and consequently to the state and composition of the metal surfaces layers, readings were made in a laminar system (criterion of Re = 3200) with a limit current iL = 56 mA, and a rotational velocity of 300 rpm, to control the mass transfer phenomena. The open circuit potentials (E_{oc}) were measured after 24 h of immersion.

In the second stage, our interest was focused on galvanic couplings in the assembly of steel watch strap links with precious metal alloys (18K gold). The evaluation was indirectly made, by measuring the quantities of nickel released after 7 days of immersion in artificial sweat, according to standard EN 1811-2011+A1:2015 [50]. The tests were carried out on 27 gold-steel links (Figure 4), 5N18 (18K gold alloy)-1.4441 (316L) and 5N18 (18K alloy)-1.4539 (904L).

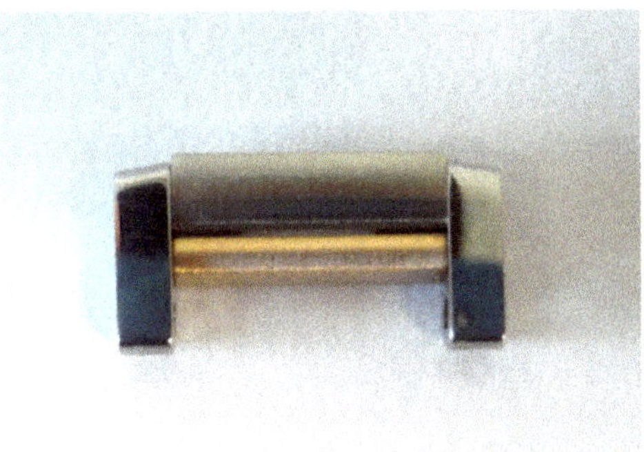

Figure 4. Link, steel-gold assembly with pins.

The microscopy investigations (scanning electron microscopy/energy-dispersive X-ray spectroscopy SEM/EDX) were carried out using a JEOL JSM-6300 SEM (JEOL, Peabody, MA, USA) equipped with an Oxford INCA EDS system (Oxford Instruments, Abingdon, UK) for local phase analysis.

3. Results and Discussion

3.1. Uniform Corrosion

After 12 days, the salt fog test revealed that here was a difference in the corrosion susceptibility of transverse and longitudinal surfaces (Figure 5). In general, the transversal surfaces were corroded, with the exception of steel #3, the intensity depending on the steel grade and the manufacturing process. Samples #3, #4 and #5 were made of grade 316L, from three different steelmakers (Germany, France, Italy). The longitudinal surfaces showed no signs of corrosion.

Figure 5. Uniform corrosion salt fog test, according to ASTM B 117-97. Samples #2-1.4427 So, #3-1.4435/316L, #4-316LUgim and #5-316L Val.

According to Zanotto et al., the test has limitations, it becomes non-discriminating in case of steels with excellent corrosion resistance [51].

In case of steels #8, #10, #6 and #17, the test, carried out under the same experimental conditions, showed no signs of corrosion of the transversal or longitudinal surfaces, similar to sample #3 (Figure 6).

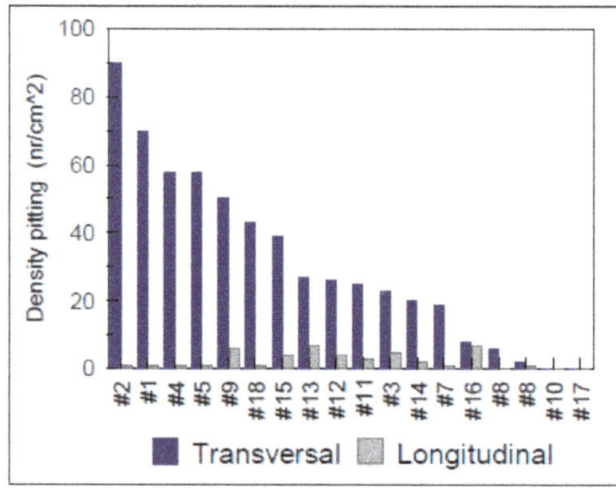

Figure 6. Pitting test results of the transverse and longitudinal surfaces for various grades of steel alloys (0.5 M $FeCl_3$ test medium at 50° for 2 h).

The results obtained are presented in Table 6. In general, the transverse surfaces were corroded with the exception of steel #3. Examination of steel #3 did not reveal any traces of corrosion either in the cold-worked state.

Table 6. Observations after the salt fog test (samples tested in annealed, cold-worked state).

Code	State	T *-Face	L **-Face
#2	Annealed	Corrosion	No corrosion
	Cold worked	Corrosion	No corrosion
#3	Annealed	No corrosion	No corrosion
	Cold worked	No corrosion	No corrosion
#4	Annealed	Corrosion	No corrosion
	Cold worked	Corrosion	No corrosion
#5	Annealed	Corrosion	No corrosion
	Cold worked	Corrosion	No corrosion

* T: transverse surface, ** L: longitudinal surface, with respect to the rolling direction.

3.2. Pitting Corrosion

For the evaluation of steels sensitivity in pitting corrosion, available standards allow characterization of the formation, the shapes and the density of pits per unit area [52]. For the establishment of Figures 6–9, it was necessary to determine the density of pitting corrosion (counting the number of pits per unit area).

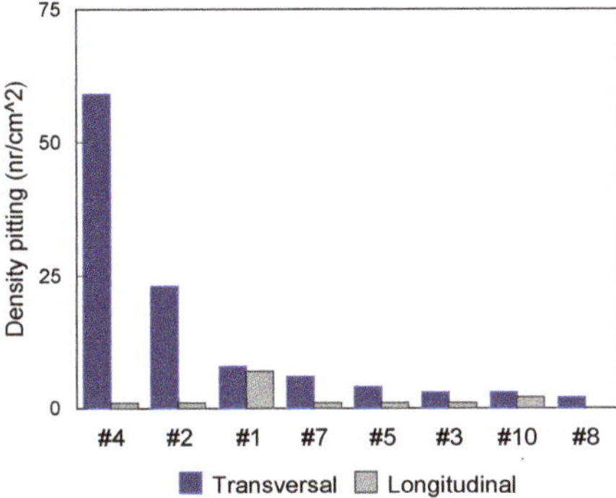

Figure 7. Pitting test results of the transverse and longitudinal surfaces for various grades of steels (0.1 M $FeCl_3$ test medium at 37 °C for 15 days).

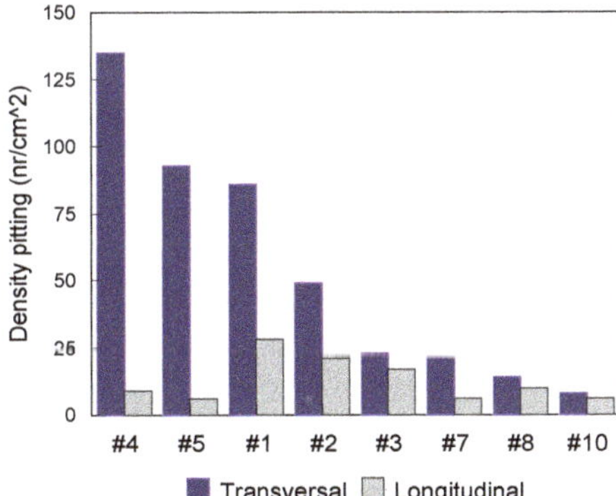

Figure 8. Pitting test results of the transverse and longitudinal surfaces for various grades of steels (0.5 M NaCl test medium at 37 °C for 26 days).

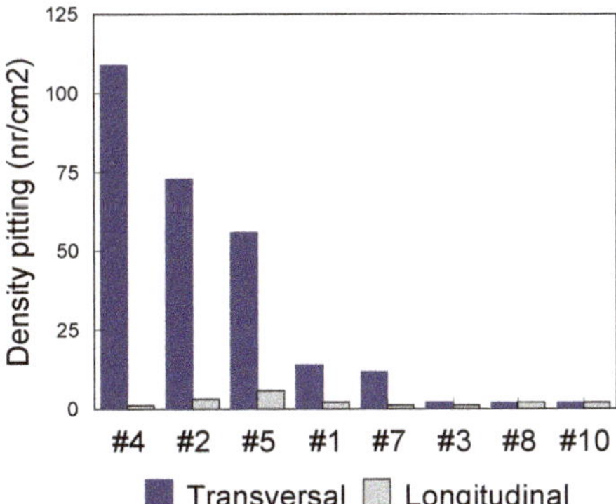

Figure 9. Pitting test results of the transverse and longitudinal surfaces for various grades of steels (artificial sweat test medium at 37 °C for 30 days).

In all test environments, the transverse surfaces showed a higher pitting density compared to the longitudinal ones.

Using a Kontron KS 300 Version 1.2 image analysis program, the cross-sectional area of alloys #1, #2, #3, #4, #5, #7, #8 and #10, tested in 0.5 M $FeCl_3$ at 50 °C for 2 h was analyzed statistically in relation to the area size of the pits. The following classes were established accordingly: <20, 20–50, 50–150, 150–500, 500–1000 and >1000 μm^2. The densities (number of pits/cm^2) according to the above mentioned classes are presented in Figure 10.

The image analysis revealed a density of pits which can be significant (more than 10,000 for sample #2, Figure 10). Under these conditions, it was necessary to define criteria which enable easier identification of pitting corrosion. Examination of the surfaces revealed numerous cavities which were not necessarily pitting.

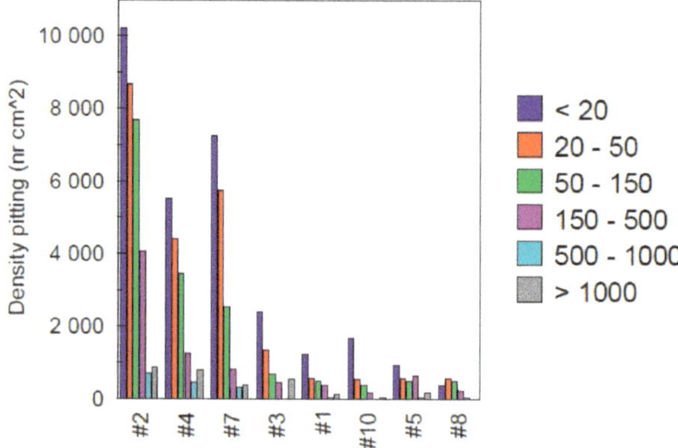

Figure 10. Number of pits counted by class, according to the size of the pit surface area.

At this point, it is essential to clarify the definition of pitting corrosion. According to the ASTM, a pitting corrosion is electrochemically active if an anodic dissolution of the alloy occurs within the cavity. Thus, by definition, a pitting corrosion releases cations in the electrolyte. It is therefore sufficient to carry out the corrosion test in an electrolyte containing traces of an analytical reagent which forms, with one of the cations released by the active corrosion pits, an insoluble colored compound which will deposit near the pit. The common element for all the alloys in question is iron, the iron dissolution mechanism being based on ferrous ions (Fe^{2+}). Tests carried out on several reagents showed that potassium ferricyanide ($K_3[Fe(CN_6)]$) is suitable to form an insoluble colored complex. Corrosion pits, revealed as blue discs encircling the pits (Figure 11), were used to determine the number per unit area.

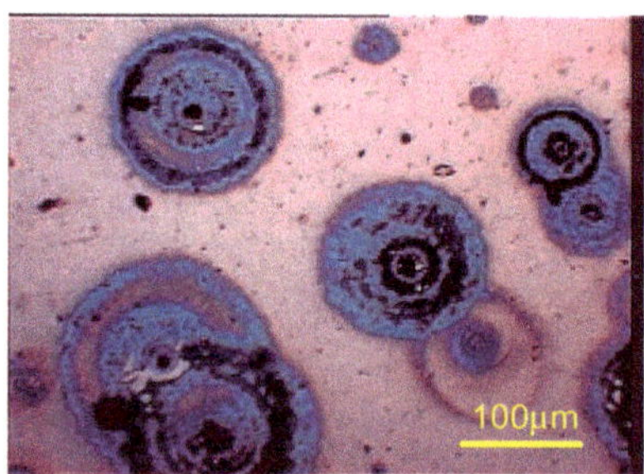

Figure 11. Corrosion pits revealed as blue discs (cross section sample #2).

In conclusion, the $FeCl_3$ solution, according to the ASTM G48-11 [31] standard, was aggressive with respect to the pitting corrosion behavior of the alloys studied; within two hours, most steel grades show readily identifiable macroscopic corrosion pits. On the other hand, in case of NaCl 0.5 M or artificial sweat ISO 3160-2 [30], the evaluation of the pits density was problematic given the difficulty of identifying the effectively electrochemically active pits. The use of potassium ferricyanide, which precipitates in the form of Turnbull blue in the presence of Fe^{2+} ions, greatly facilitated the evaluation of the density of pits after immersion in this type of electrolyte.

According to Blackwood [2], pitting corrosion was a common problem with the early 304 stainless steels. In the case of 316L stainless steels, the addition of 2–3 wt% Mo has greatly reduced the number of failures due to pitting corrosion [2].

After initiation, pits either keep growing or repassivation may occur. According to Virtanen, an alloy with a high pitting corrosion resistance should ideally combine low susceptibility to pit initiation, low pit propagation rate, and fast repassivation [53]. According to Melchers, in case of metals with electron-conducting passive films such as stainless steels, the number of pits usually correlates inversely with their average depth, since the cathodic current consumed by the large passive surface area fosters anodic dissolution inside the pits [54]. The pits may grow at different rates, depending on the number of active pits [54]. According to Abbasi Aghuy et al., changes in electrochemical behavior of metal due to grain refinement as a consequence of changing grain boundary densities may occur [55].

3.3. Crevice Corrosion

Figure 12 shows the "variations of the current" curves as a function of time, for values of imposed potential, for the longitudinal surface of samples #8 and #10, respectively. The crevice potentials were +150 mV (red points) for sample #8 and +350 mV (red points) for sample #10.

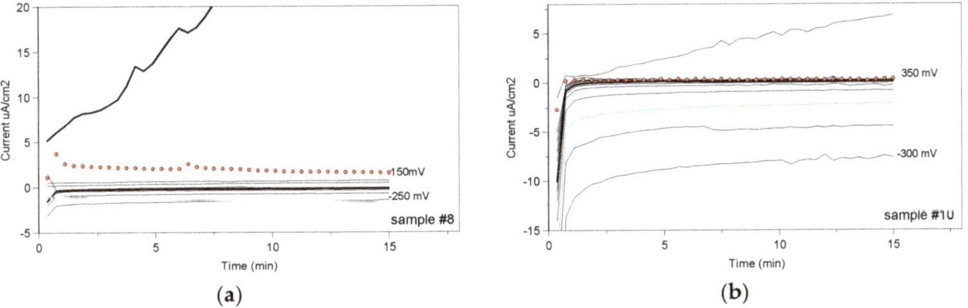

Figure 12. The crevice test potentiostatic curves for the longitudinal surface of sample #8 (**a**) and #10 (**b**).

Figure 13 shows the values of the crevice potentials determined for both longitudinal and transverse surfaces, according to ASTM F746-87 standard [35].

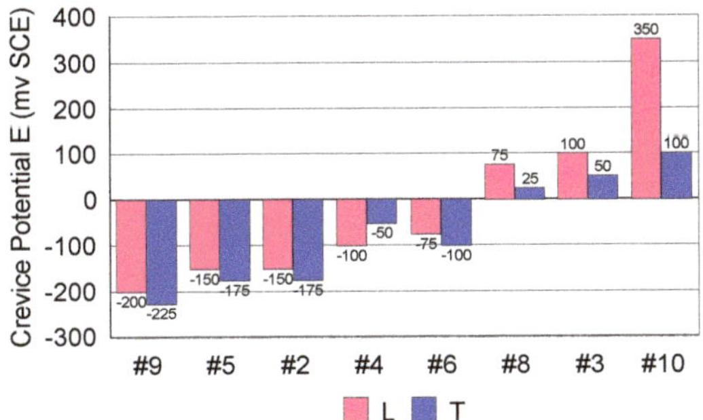

Figure 13. Crevice potential values measured for the transverse and longitudinal surfaces of the steels considered (L: longitudinal, T: transverse).

The values of the crevice potentials measured did not reveal any difference in susceptibility to crevice corrosion between the two surfaces. The only difference was that the transverse surface showed lower values than the longitudinal surface, but these differences remained in the field of experimental errors. In other words, there was no significant difference in the crevice corrosion behavior between the two surfaces.

In case of 316L steels (#2, #3, #4, #6, #8, and #9), the values of the crevice potentials were different, due to the structure type of inclusions and composition in minor chemical elements.

The study of Poyetet et al., involving 18-10 type stainless steels [56], has concluded that the reactivity of the inclusions, in terms of their contribution to the onset of pitting, is a function of their association (Table 7).

Table 7. Types and reactivity of inclusions according to [56].

Inclusion Types and Associations	Number Rating: #Pitted/#Total of Inclusions	Shape Rating: #Pitted/#Total of Inclusions
Sulfide Sulfide-silicate Silicate	2/32 = 6% 22/81 = 27% 0/5 = 0%	Globular 3/20 = 15% Elongated 19/61 = 31%
Sulfide Sulfide-alumina Alumina	1/33 = 3% 9/87 = 10% -	Globular 1/33 = 3% Elongated 8/54 = 15%
Mg-oxide Mg-oxide-sulfide	100%	(rare inclusions)

Mixed oxide-sulfide or silicate-sulfide inclusions are the most susceptible to pitting. The corrosion susceptibility of inclusions might be ranked, in increasing order: sulfides < alumina-sulfides < silicate-sulfides < Mg-oxide-sulfides. By themselves, sulfides do not have a particularly detrimental action on the pitting corrosion resistance of steel, but they become particularly harmful when associated in the form of mixed inclusions. As far as shape is concerned, globular inclusions (present only in the as-cast, undeformed material) seem to be less harmful than inclusions deformed during hot working of the metal [56].

When considering the final values of currents recorded after 15 min for each level and representing the current as a function of potential, a series of "polarization curves", specific to the crevice corrosion process were obtained (Figure 14).

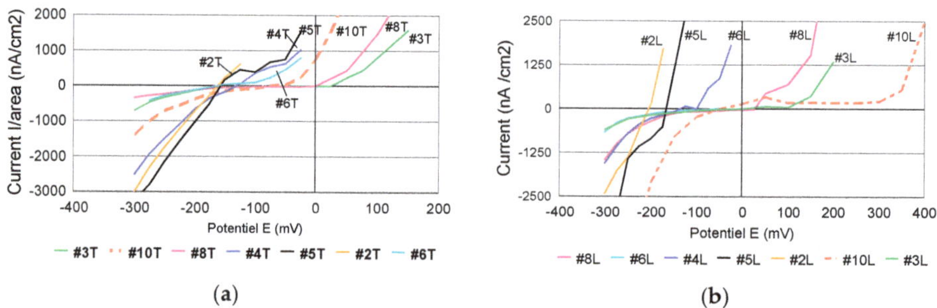

Figure 14. Polarization curves (current value recorded after 15 min vs. preselected potential; (a) transverse and (b) longitudinal surface.

When comparing the crevice corrosion behavior of the two surfaces, no real difference in susceptibility to corrosion was noticed. On the other hand, in accordance to Bryant et al. [57], each steel has a different behavior to crevice corrosion. Some steels do not reveal a "passivation capacity" before reaching the value of crevice initiation potential. In case of 316L, respectively #2, #3, #4, #6 and #8 this difference was noticed. According to Liu et al., in case of 316L stainless steel, widely used as a metallic biomaterial, crevice corrosion has been a serious concern [58]. In case of #10, a difference was expected, due to the better corrosion resistance compared to the 316L family.

In conclusion, the evaluation of the crevice corrosion resistance did not reveal marked differences between the behavior of the transverse surface compared to the longitudinal one. This shows that the pitting and crevice corrosion mechanisms, although having some similarities, are different. In case of certain grades of steel, particularly sensitive to crevice corrosion, sometimes crevice corrosion can interfere with pitting corrosion measurements. Figure 15 shows a situation where crevice corrosion strongly interfered during the measurement of pitting corrosion by the rotating electrode technique. The crevice corrosion developed under a defective collar, making the measurements unusable for the characterization of pitting corrosion. Sometimes, the observation of the corrosion morphology provides information on the metallographic structure of the alloy. The morphology of crevice corrosion on the transverse surface (Figure 15) showed a particular structure, the orientation of the corroded structures suggesting a preferential longitudinal dissolution. This reveals a manifestation of a higher corrosion sensitivity of the transverse direction compared to longitudinal direction.

Figure 15. Scanning electron microscopy (SEM) of the transverse surface, corroded under a defective PTFE collar. The columnar morphology suggests preferential longitudinal dissolution due to the texture of the material.

3.4. Intergranular Corrosion

3.4.1. Chemical Tests

After completing the test, the examination of the interior of the tube showed corrosion signs. In case of each sample, a mass loss of about 40 mg/cm^2 was determined (Table 8). An optical or scanning electron microscopy (SEM) examination was also carried out, for each sample (Figure 16). The metallographic section from the external part of the corroded tube had a structurally disturbed surface area to a depth of about 70 μm (Figure 17).

Figure 16. The interior of a corroded tube (#3, 1.4435/316L steel).

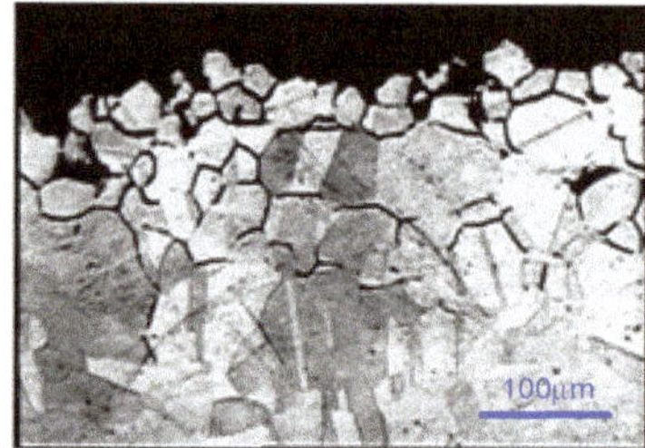

Figure 17. Metallographic section of the tube (#3, 1.4435/316L steel) after Strauss's test.

Table 8. Tube mass loss after Strauss test.

Material	Mass Loss (g)	Surface (cm^2)	Mass Loss (mg/cm^2)
	0.31651	7.38	43
Tubes	0.32816	7.80	42
1.4435/316 L	0.28091	7.02	40
	0.22120	5.84	38
	0.05145	6.04	9

The Strauss test clearly and unequivocally showed that the 316L steel tubes were sensitized to intergranular corrosion. The corrosion rate was higher on the interior compared to the exterior.

The energy-dispersive X-ray spectroscopy (EDX) analysis of the corroded areas showed the presence of elements which did not belong to the alloy: sulfur, chlorine, calcium, sodium, aluminum and potassium. These elements most likely resulted from lubricants formulated as additives or base oil. The sensitization to intergranular corrosion was probably due to the pyrolysis of residual oil present on the tube surface—in other words, poor cleaning during the manufacturing process.

3.4.2. Electrochemical Tests: EPR Method ASTM G108–94 (2015)

Electrochemical Tests were carried out according to the EPR Method ASTM G108–94 (2015) [48]. The grain index was determined according to the ASTM E112-13 method [49] (Figures 18 and 19).

(a)   (b)

Figure 18. (a)#Test 1 and (b) #Test 2. Grain index (ASTM E112-13) G = 11.

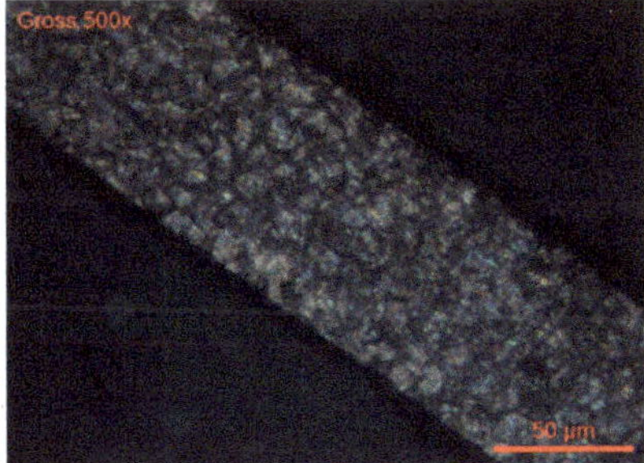

Figure 19. #Brut 1. Grain index (ASTM E112-13) G = 11.

Grain index values are given in Table 9.

Table 9. Grain index according to ASTM E112-13 [49].

Code	Grain Index
#Test 1	11
#Test 2	11
#Brut 1	11

According to the EPR method ASTM G108-94 (2015) [48], after the cyclic polarization scans, the evaluation parameter is the normalized charge (Pa), measured in coulombs/cm^2, calculated with the formula:

$$P_a = Q/X \tag{1}$$

where Q = measured on current integration measuring instrument (coulombs), normalized for both specimen size and grain size X = $A_s[5.1 \times 10^{-3}e^{0.35G}]$, where A_s = specimen area (cm^2), G = grain index at 100× according to ASTM E112-13 [49].

In the derivation of the equation, it was assumed that the Q value was the result of the attack on the specimen surface that was distributed uniformly over the entire grain boundary region of a constant width of $2 \times (5 \times 10^{-5})$ cm. This may not represent the actual physical processes.

The potentiokinetic electrochemical reactivation results are presented in Table 10.

Table 10. Potentiokinetic electrochemical reactivation results.

Code	E_{oc} (mV)	I_r (mA/cm^2)	Q (C/cm^2)	P_a(C/cm^2)
#Test 1	−400	8.21	4.96	20.67
#Test 2	−423	12.92	4.01	16.70
#Brut 1	−387	11.84	0.28	1.15
#Brut 2	−410	8.21	0.33	1.37
#500_1	−410	30.70	1.33	5.54
#500_2	−388	59.22	1.94	8.10
#620_1	−404	173.30	8.17	34.04
#620_2	−410	132.60	8.07	33.63
#620_3	−395	162.20	9.94	41.42
#750_1	−407	34.22	2.14	8.90
#750_2	−407	59.22	4.01	16.70

E_{oc} = Initial open circuit potential, I_r = maximum anodic current density.

Figure 20 shows the potentio kinetic reactivation curves in linear axes.

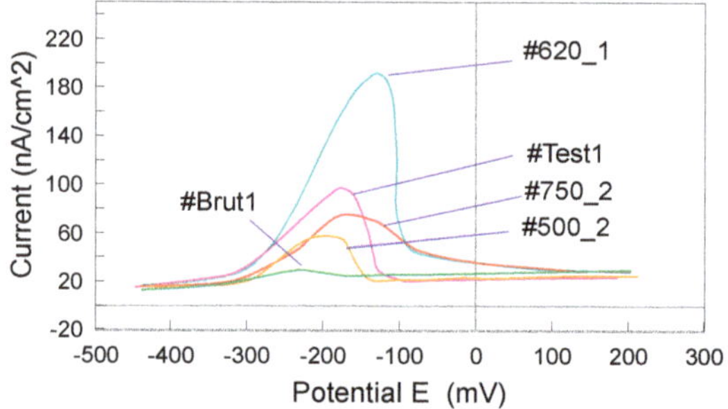

Figure 20. Potentio kinetic reactivation curves recorded for #Brut 1, #Test 1, #500_1, #620_2, and #750_2.

The peak valuesfor I_r, given in Table 10, were specific to the intergranular corrosion degradation of the tubes. The higher the intensity, the greater the degradation. Thus, according to Figure 20, it was noted that the highest sensitization of the tubes was generated by the heat treatment at 620 °C. The minimum sensitization corresponded to the 500 °C heat treatment. The overall results (Table 10) for the normalized charge (Pa) calculated (Figure 21) for all the samples indicated that the heat treatment over 500 °C for 304 steels is not indicated, the risks of inducing an intergranular corrosion process being obvious. Consequently, the 500 °C heat treatment should be used in the manufacturing process.

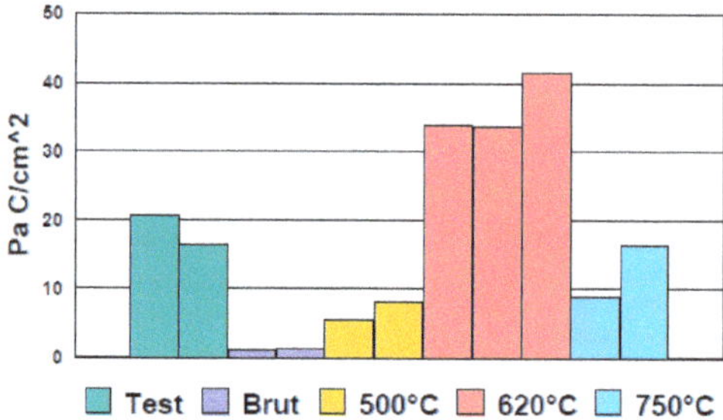

Figure 21. Normalized charge (Pa) measured by EPR.

Type 304 steel was more sensitive to intergranular corrosion compared to other steels. Consequently, in the manufacturing process, great importance must be given to this type of corrosion morphology. The temperature of 620 °C was critical for generating the process and therefore used in the ASTM tests (A262-15 and G108-94) for evaluating intergranular corrosion. The goal is to near the behavior of the tube in raw state (#Brut 1).

The susceptibility to intergranular corrosion of stainless steels is not always due to heat treatment with precipitation of chromium carbides. Under certain conditions, the precipitation of intermetallic compounds of $(Fe, Cr)Mo_2$ or $(Cr, Ni, Fe)_3P_2$ type can occur. According to Stonawská et al., the structural sensitization of 316 L steel is due to the precipitation of secondary phases along the grain boundaries [59]. The studies of Liu et al. regarding 316 L steels [60] and Fujii et al. [61] regarding 304 steel, also supports this statement. According to Liu et al. [62], the chromium-depleted zones near grain boundaries represent the corrosion nucleation sites for austenitic steels. According to Eliaz, since the carbon content in 316 stainless steel was lowered in the 316L and 316LVM grades, sensitization of this steel is less problematic as it used to be [7].

3.5. Galvanic Corrosion

Table 11 presents the corrosion potentials measured in artificial sweat (EN 1811-2011+A1:2015) [50] for the precious alloys and austenitic steels. The measured potential values enable to establish a relative comparison of the alloys nobility in the considered medium and to construct a galvanic series. Higher potential values mean higher corrosion resistance. The alloys series with higher negative potentials (anodic) generally tend to undergo greater corrosion in the event of a galvanic coupling, while other metals (cathodic) will generally undergo a reduced attack. According to Mansfeld and Kenkel, the corrosion potential of each alloy is a criterion in the analysis of galvanic corrosion behavior, but it is still insufficient. The electrical potential values can only indicate a trend and state absolutely nothing about the rate of corrosion and the type of control of the galvanic cell (mixed, cathodic or anodic) [63,64].

Table 11. Galvanic series established in an EN1811-2011+A1:2015 artificial sweat type environment.

A	Precious Alloys	Corrosion Potential E_{corr} (mV)	B	Steels	Corrosion Potential E_{corr} (mV)
952	Pt 950CoNi	175	316L series	1.4441	−21
150	AuPdCu150	70	-	1.4435	−160
109	AuAgNi109	53	-	316L F	−164
141	AuCuNi130	43	-	316L F Cu	−282
142	AuNiCu142	42	304 series	1.4301	−169
112	AuNiCu112	22	303 series	1.4305	−266
374	AuCuZn374	6.4	-	Sandvik 1802	−168
-	-	-	-	1.4104	−234
-	-	-	-	1.4105	−389
-	-	-	904L series	1.4539	−72
-	-	-		12/12	−163

The most frequent cases we encountered are precious metal-austenitic steel assemblies in watch straps. Thus millions of gold-steel links are produced to assemble straps, this is the ideal case for the formation of a galvanic cell, a significant difference in electrical potential being involved. In the case of gold-steel, a difference in electrical potential of around 300 mV can be calculated.

According to Gilbert and Mali, while corrosion per se may not be of great concern, when combined with mechanical effects, restricted crevice-like geometries or any combination thereof, considerably amplified corrosion rates might arise [65].

One of the several available techniques to realize the gold-steel assembly is brazed gold caps. The brazed gold caps reveal a particularity due to brazing. The solder acts as the anode (a small area) and the steel and gold parts are the cathode (large area). Thus, the corrosion process results in the dissolution of the solder (Figures 22 and 23). In such a type of assembly it is particularly important to make the right choice of solder.

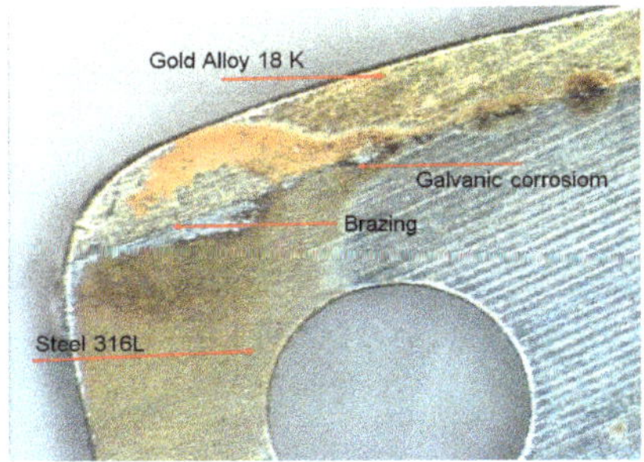

Figure 22. Galvanic corrosion in a gold-steel assembly.

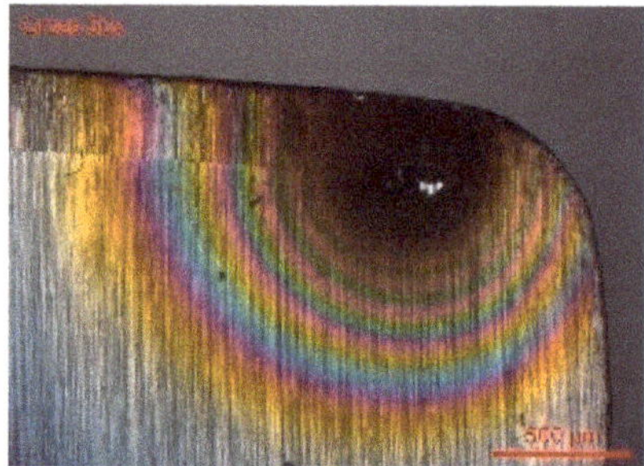

Figure 23. Corrosion of the transverse surface, at the level of the gold-steel interface.

Two aspects of great importance have to be considered:

(a) The cathode–anode relationship. The precious metal surfaces will act as the cathode, and the less noble parts will be the anode. Constructions with large cathode surfaces and small anode surfaces are very dangerous. The galvanic cell will output a strong anodic current which will lead to the rapid degradation of the anodic part by mechanisms of crevice or pitting corrosion;

(b) The nickel release in contact with the skin has to be considered as the current legislation tolerates an amount of 0.5 μg/cm^2·week.

Table 12 presents a series of tests carried out on the same types of gold-stainless steel links. The 14441/316 LM steel originated from five different steelmakers from the EU, USA, Japan and China.

Table 12. Tests results for gold-stainless steel links.

#	Gold	Steel	Nickel Release (μg·cm^{-2}·Week^{-1})	Corroded Parts Rate	Corrosion Rate (%)
#1	5N18	1.4441	0.14	0/6	0%
		316LM	0.25 0.13 0.04 0.22 0.03		
#2	5N18	1.4441	0.44	2/6	33%
		316LM	0.05 0.01 0.10		
#4	5N18	1.4441	0.03	6/9	67%
		316LM	0.06 0.09		
#5	5N18	1.4441	0.09	0/3	0%
		316LM	0.08 0.06	-	-
#6	5N18	1.4539	2.3	0/3	0%
		904L	2.6 2.2		

In case of #6 (gold-steel 904L), the nickel release was greater than 2 μg·cm^{-2}·week^{-1}, despite the absence of visible corrosion. This was due to a different behavior compared to a medical 316L steel. The comparison of the EDX profile between a gold-904L steel (Figure 24) and a gold-316LM steel (Figure 25) revealed a very different nickel profile with the disappearance of the nickel peak in the gold-904L steel solder (Figure 24). In the gold-904L system, the solder was in the anodic position (gold and 904L steel being cathodic), with a very unfavorable surface report. It revealed a selective corrosion morphology powered by a galvanic battery; this would explain the significant nickel release from the gold-904L steel coupling, despite the absence of visible corrosion.

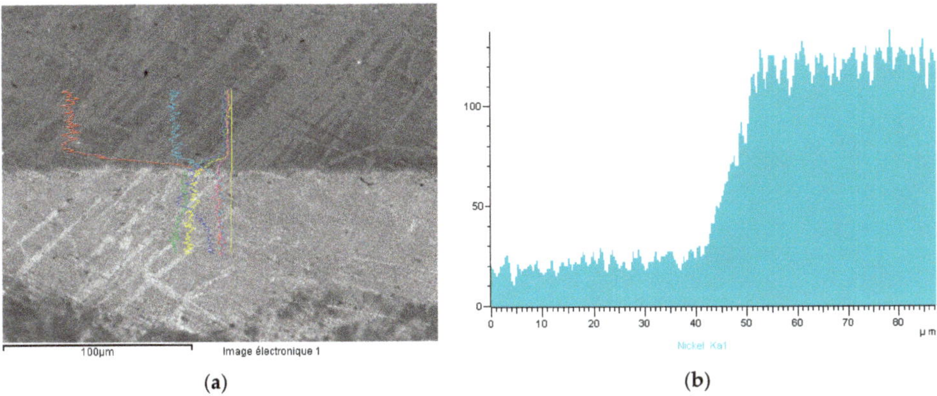

(**a**) (**b**)

Figure 24. Sample #6. (**a**) EDX profiles of the gold-904 L steel solder for iron, nickel, silver, gold, copper and zinc; (**b**) EDX Ni profile.

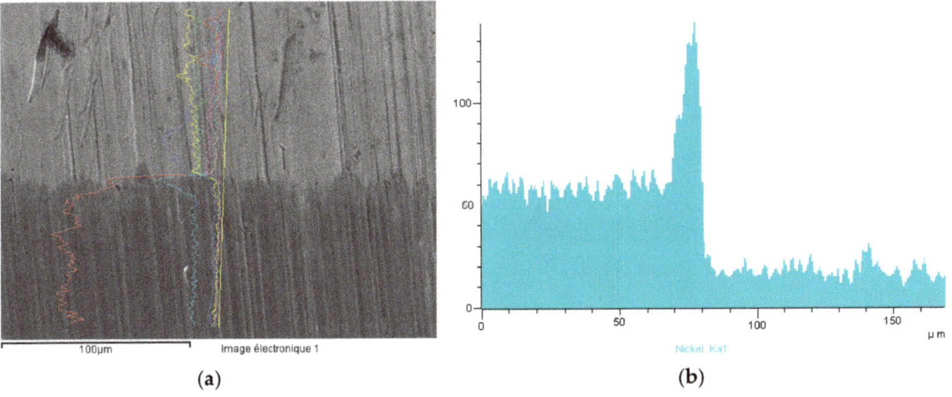

(**a**) (**b**)

Figure 25. Sample #4. (**a**) EDX profiles of the gold-316 LM steel solder for iron, nickel, silver, gold, copper and zinc; (**b**) EDX Ni profile.

SEM examination of sample #4 (5N18) showed that the corrosion was localized and did not develop at the level of the steel, but of the brazing, causing its dissolution (Figure 26). This demonstrates that the solder represents the weak point of the gold-steel assembly.

The steels which are being used for watch straps are of grades 316 and 904L. The other steel grades, such as 304, 304L, 316LS, and 316Ti, are not usable; their rate of Ni release does not respect the limits imposed by the EU directives, or other countries legislations (USA, Japan, China, Korea, Canada). The difficulty consists in eliminating the corrosion process and achieving a rate of nickel release which respects the legislation: max 0.5 $\mu g \cdot cm^{-2} \cdot week^{-1}$.

The use of a Ni-Cr-P solder involves the risk of increasing the nickel amounts by chemical dissolution or corrosion of the solder. In the case of a gold base solder (melting range 750–850 °C) the risk is to initiate corrosion in the steel; with a Ni-Cr-P based solder (melting range 800–950 °C) the risk is to start corrosion in the solder. To find the best compromise, testing the link assemblies is necessary.

Because the quality of 316 LM steel is highly dependent on the supplier, most straps manufacturers use steels they have exclusivity for.

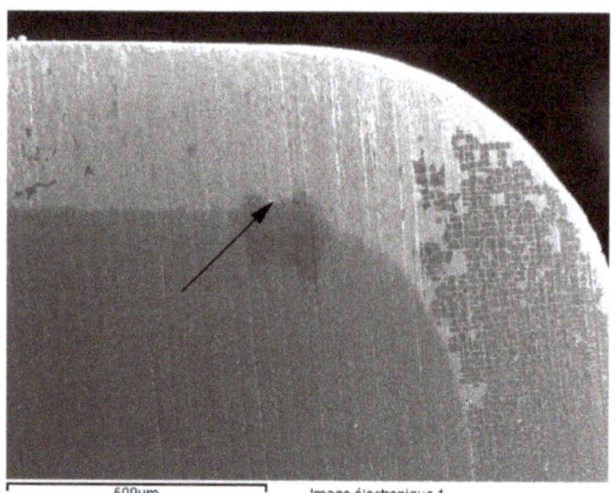 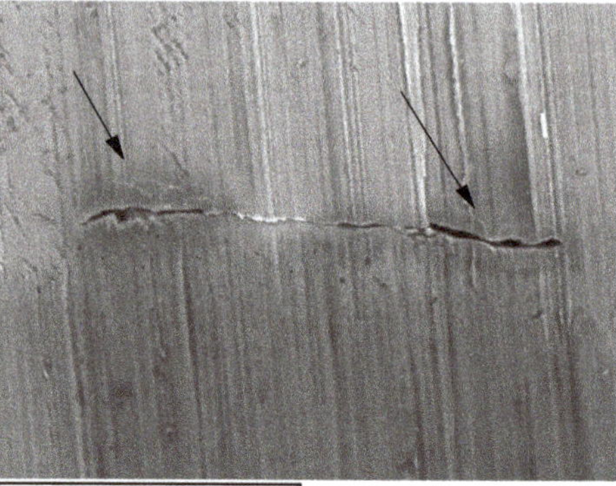

Figure 26. SEM examination of sample #4.

3.6. Nickel Release in Relation with the Manufacturing Process

In our laboratories, a large number of nickel extraction tests in compliance with EN 1811-2011+A1:2015 standard [50] were carried out on the 316 L grades manufactured by five different steelmakers from EU, Japan and USA. There was a significant difference in the quantities of Ni released, compared to the chemical composition of steels, which depends on the steelmaker. The conclusions are presented in Figure 27. It should be pointed out that despite the fact that the grade of steel is in conformity with the classification standards, their behavior was markedly different from one manufacturer to another, due to production parameters, such as the casting volume, alloying additions, and deoxidizing agents.

	PARAMETER	EFFECT	QUANTITY OF Ni RELEASE
RAW MATERIALS	VARIABLE, IN FUNCTION OF THE LOT		STRONG DISPERSION
HEAT TREATMENT	100 % H2		STRONG DECREASE
	100 % N2		SLIGHT DECREASE
SURFACE	ROUGH POLISHED SATINY		SLIGHT INFLUENCE
WORK HARDENING	STRAIN >10 %		INCREASE
STRUCTURE	INCLUSIONS AND SECOND PHASES		INCREASE

Figure 27. Factors influencing the amounts of nickel released during the manufacturing process.

The heat treatments resulted in a reduction of the nickel release rates. The surface state had little influence. On the other hand, the hardening processes strongly influenced the quantities of nickel released. The increase in hardness greatly decreased the corrosion resistance and increased the amount of nickel released. Another factor which strongly influenced the quantities of nickel released was the inclusion state and the existence of secondary phases in the structure of steels. Being aware of these causes, subcontractors demand from the steelmakers a very strict specification respecting in the manufacturing of steels.

4. Conclusions

Seventeen grades of stainless steels were assessed for specific types of corrosion: general, pitting, crevice, intergranular and galvanic. It was noted that there are significant differences between the grades of the austenitic steels studied.

The conclusions are as follows:

- The intensity of the corrosion was dependent on the production parameters, such as the casting volume, alloying additions, and deoxidizing agents;
- The amount of nickel release was dependent on the heat treatment, hardening rate, and other parameters of the manufacturing process;
- The quantity of nickel released is strongly influenced by the inclusion state and the existence of secondary phases;
- There is a clear difference of corrosion between the transverse surface and the longitudinal surface. The longitudinal surface (in the rolling direction) reveals a better resistance to corrosion than the transverse surface.
- The quantities of nickel released are highly dependent on the grade of steel. As a result, manufacturers can use only steels that meet the current legislative requirements;
- Top range watches manufacturers use steels with exclusivity labels, so the chemical compositions, structures, inclusive states, mechanical properties, machinability, polishing are very well defined in their specifications. The rule also applies to medical devices manufacturers;
- Finally, a compromise in choosing a steel over another has to be made, depending on the application and the legal requirements for the final products on a specific market.

Author Contributions: Conceptualization, L.R.; methodology, L.R.; validation, L.R.; formal analysis, L.R.; investigation, L.R.; resources, L.R.; data curation, L.R.; writing—original draft preparation, L.R., L.C.A.; writing—review and editing, L.C.A. All authors have read and agreed to the published version of the manuscript.

Funding: This research received no external funding.

Conflicts of Interest: The authors declare no conflict of interest.

References

1. Eliaz, N. *Degradation of Implant. Materials*; Springer: New York, NY, USA, 2012.
2. Blackwood, D.J. Biomaterials: Past successes and future problems. *Corros. Rev.* **2003**, *21*, 97–124. [CrossRef]
3. Boonruang, C.; Thong-On, A.; Kidkhunthod, P. Effect of nanograin-boundary networks generation on corrosion of carburized martensitic stainless steel. *Sci. Rep.* **2018**, *8*, 2289. [CrossRef]
4. Mudali, U.K.; Sridhar, T.M.; Eliaz, N.; Raj, B. Failures of stainless steel orthopaedic devices—Causesand remedies. *Corros. Rev.* **2003**, *21*, 231–267. [CrossRef]
5. Pound, B.G. Corrosion behavior of metallic materials in biomedical applications. I. Ti and its alloys. *Corros. Rev.* **2014**, *32*, 1–20. [CrossRef]
6. Revie, R.; Uhlig, H.H. *Corrosion and Corrosion Control: An Introduction to Corrosion Science and Engineering*, 4th ed.; John Wiley & Sons: Hoboken, NJ, USA, 2008; p. 84.
7. Eliaz, N. Corrosion of metallic biomaterials: A review. *Materials* **2019**, *12*, 407. [CrossRef] [PubMed]
8. Tiamiyu, A.A.; Eduok, U.; Szpunar, J.A.; Odeshi, A.G. Corrosion behavior of metastable AISI 321 austenitic stainless steel: Investigating the effect of grain size and prior plastic deformation on its degradation pattern in saline media. *Sci. Rep.* **2019**, *9*, 12116. [CrossRef]
9. Gupta, R.K.; Birbilis, N. The influence of nanocrystalline structure and processing route on corrosion of stainless steel: A review. *Corros. Sci.* **2015**, *92*, 1–15. [CrossRef]
10. Ralston, K.D.; Birbilis, N.; Davies, C.H.J. Revealing the relationship between grain size and corrosion rate of metals. *Scr. Mater.* **2010**, *63*, 1201–1204. [CrossRef]
11. Schino, A.D.I.; Materiali, C.S.; Brin, V.B. Effects of grain size on the properties of a low nickel austenitic stainless steel. *J. Mater. Sci.* **2003**, *8*, 4725–4733. [CrossRef]

12. *Key to Steel-Stahlschlüssel*, 25th ed.; Verlag Stahlschlüssel Wegst GmbH: Marbach, Germany, 2019. Available online: http://www.online.stahlschluessel.de/BrowseProducts.aspx (accessed on 30 April 2020).

13. Amaro Vicente, T.; Oliveira, L.A.; Correa, E.O.; Barbosa, R.P.; Macanhan, V.B.P.; Alcântara, N.G. Stress corrosion cracking behaviour of dissimilar welding of AISI 310S austenitic stainless steel to 2304 duplex stainless steel. *Metals* **2018**, *8*, 195. [CrossRef]

14. Lo, K.H.; Shek, C.H.; Lai, J.K.L. Recent developments in stainless steels. *Mater. Sci. Eng. R.* **2009**, *65*, 39–104. [CrossRef]

15. Ha, H.-Y.; Lee, T.-H.; Bae, J.-H.; Chun, D.W. Molybdenum effects on pitting corrosion resistance of FeCrMnMoNC austenitic stainless steels. *Metals* **2018**, *8*, 653. [CrossRef]

16. Ha, H.-Y.; Jang, J.H.; Lee, T.-H.; Won, C.; Lee, C.-H.; Moon, J.; Lee, C.-G. Investigation of the localized corrosion and passive behavior of type 304 stainless steels with 0.2–1.8 wt% B. *Materials* **2018**, *11*, 2097. [CrossRef] [PubMed]

17. Desestret, A.; Charle, J. Les aciersinoxydablesausténico-ferritiques. In *Les AciersInoxydables*; Lacombe, P., Baroux, B., Beranger, G., Colombier, L., Hochmann, J., Eds.; Les Ulis, Les Editions de Physique: Paris, France, 1990; pp. 631–678.

18. Reclaru, L.; Ziegenhagen, R.; Eschler, P.Y.; Blatter, A.; Lemaître, J. Comparative corrosion study of "Ni-free" austenitic stainless steels in view of medical applications. *Acta. Biomater.* **2006**, *2*, 433–444.

19. Baker, M. European standards developed in support of the european union nickel directive. In *Metal Allergy*; Chen, J.K., Thyssen, J.P., Eds.; Springer: Berlin, Germany, 2018; pp. 23–29. [CrossRef]

20. Liden, C. Nickel in jewelry and associated products. *Contact Dermat.* **1992**, *26*, 73–75. [CrossRef]

21. Belsito, D.V. The diagnostic evaluation, treatment, and prevention of allergic contact dermatitis in the new millennium. *J. Allergy Clin. Immunol.* **2000**, *105*, 409–420. [CrossRef]

22. Schafer, T.; Bohler, E.; Ruhdorfer, S.; Weigl, L.; Wessner, D.; Filipiak, B.; Wichmann, H.E.; Ring, J. Epidemiology of contact allergy in adults. *Allergy* **2001**, *56*, 1192–1196. [CrossRef] [PubMed]

23. Publication Office of the European Union. Commission communication in the framework of the implementation of regulation (EC) No 1907/2006 of the european parliament and of the council concerning the registration, evaluation, authorisation and restriction of chemicals (REACH). *Off. J. Eur. Union* **2012**, *C142*, 8.

24. Publication Office of the European Union. European parliament and council directive 94/27/EC of 30 June 1994: Amending for the 12th time directive 76/769/EEC on the approximation of the laws, regulations and administrative provisions of the member States relating to restrictions on the marketing and use of certain dangerous substances and preparations. *Off. J. Eur. Commun.* **1994**, *L188*, 1–2.

25. Heim, K.; Basketter, D. Metal exposure regulations and their effect on allergy prevention. In *Metal Allergy*; Chen, J.K., Thyssen, J.P., Eds.; Springer: Berlin, Germany, 2018; pp. 39–54. [CrossRef]

26. Fontana, M. *Corrosion Engineering*, 5th ed.; McGraw Hill International Edition: New York, NY, USA, 1987; p. 556.

27. ASTM B117-97. *Standard Practice for Operating Salt Spray (Fog) Apparatus*; ASTM International: West Conshohocken, PA, USA, 2010. Available online: www.astm.org (accessed on 30 April 2020).

28. Bech-Nielsen, G. The anodic dissolution of iron–V. Some observations regarding the influence of cold working and annealing on the two anodic reactions of the metal. *Electrochim. Acta* **1974**, *19*, 821–828. [CrossRef]

29. Cigada, A.; Mazza, B.; Pedeferri, P.; Sinigaglia, D. Influence of cold plastic deformation on critical pitting potential of AISI 316 L and 304 L steels in an artificial physiological solution simulating the aggressiveness of the human body. *J. Biomed. Mater. Res.* **1977**, *11*, 503–512. [CrossRef] [PubMed]

30. ISO 3160-2:2015. Watch-Cases and Accessories—Gold Alloy Coverings—Part 2: Determination of Fineness, Thickness, Corrosion Resistance and Adhesion. Available online: https://www.iso.org/standard/66162.html (accessed on 30 April 2020).

31. ASTM G48-11. *Standard Test Methods for Pitting and Crevice Corrosion Resistance of Stainless Steels and Related Alloys by Use of Ferric Chloride Solution*; ASTM International: West Conshohocken, PA, USA, 2015. Available online: www.astm.org (accessed on 30 April 2020).

32. Hoar, T.P.; Mears, D.C. Corrosion-resistant alloys in chloride solutions: Materials for surgical implants. *Proc. R. Soc. Lond.* **1966**, *294*, 486–510.

33. Zabel, D.D.; Brown, S.A.; Merritt, K.; Payer, J.H. AES analysis of stainless steel corroded in saline, in serum and in vivo. *J. Biomed. Mater. Res.* **1988**, *22*, 31–44. [CrossRef] [PubMed]

34. Turnbull, A. The solution composition and electrode potential in pits, crevices and cracks. *Corros. Sci.* **1983**, *23*, 833–870. [CrossRef]

35. ASTM F746-87. *Standard Test Method for Pitting or Crevice Corrosion of Metallic Surgical Implant Materials*; ASTM International: West Conshohocken, PA, USA, 1999. Available online: www.astm.org (accessed on 30 April 2020).

36. ASTM A262-15. *Standard Practices for Detecting Susceptibility to Intergranular Attack in Austenitic Stainless Steels*; ASTM International: West Conshohocken, PA, USA, 2015. Available online: www.astm.org (accessed on 2 May 2020).

37. Streicher, M.A. Intergranular. In *Corrosion Tests and Standards: Application and Interpretation*, 2nd ed.; Baboian, R., Ed.; ASTM International: West Conshohocken, PA, USA, 2004; pp. 244–266.

38. Henthorne, M. Localized Corrosion. *ASTM Tech. Publ.* **1972**, *518*, 108.

39. Cihal, V.; Stefec, R. On the development of the electrochemical potentiokinetic method. *Electrochim. Acta* **2001**, *46*, 3867–3877. [CrossRef]

40. Iacoviello, F.; Di Cocco, V.; D'Agostino, L. Analysis of the intergranular corrosion susceptibility in stainless steel by means of potentiostatic reactivation tests. *Procedia Struct. Integr.* **2017**, *3*, 269–275. [CrossRef]

41. Clarke, W.L.; Romero, V.M.; Danko, J.C. Detection of sensitization in stainless steel using electrochemical techniques. In *Corrosion 77, International Corrosion Forum*; National Association of Corrosion Engineers: San Francisco, CA, USA, 1977; p. 180.

42. Novak, P.; Stefec, R.; Franz, F. Testing the susceptibility of stainless steels to intergranular corrosion by a reactivation method. *Corrosion* **1975**, *31*, 344–347. [CrossRef]

43. Desestret, A.; Froment, M.; Guiraldenq, P. Intergranular corrosion of austenitic stainless steels in the sensitized or solution-annealed condition. In Proceedings of the 23rd Meeting of I.S.E., Stockholm, Sweden, 27 August–2 September 1972; Volume 87.

44. Aydoğdu, G.H.; Aydinol, K. Determination of susceptibility to intergranular corrosion and electrochemical reactivation behaviour of AISI 316L type stainless steel. *Corr. Sci.* **2006**, *48*, 3565–3583. [CrossRef]

45. Cíhal, V.; Lasek, S.; Blahetová, M.; Kalabisová, E.; Krhutová, Z. Trends in the electrochemical polarization potentiodynamic reactivation method-EPR. *Chem. Biochem. Eng. Q.* **2007**, *21*, 47–54.

46. Parvathavarthini, N.; Dayal, R.K.; Seshadri, S.K.; Gnanamoorthy, J.B. Influence of prior deformation on the sensitisation of AISI type 304 stainless steel and applicability of EPR technique. *Br. Corros. J.* **1991**, *26*, 67–76. [CrossRef]

47. Majidi, A.P.; Streicher, M.A. The double loop reactivation method for detecting sensitization in AISI 304 stainless steels. *Corrosion* **1984**, *40*, 584–593. [CrossRef]

48. ASTM G108-94. *Standard Test Method for Electrochemical Reactivation (EPR) for Detecting Sensitization of AISI Type 304 and 304L Stainless Steels*; ASTM International: West Conshohocken, PA, USA, 2015; Available online: www.astm.org (accessed on 2 May 2020).

49. ASTM E112-13. *Standard Test Methods for Determining Average Grain Size*; ASTM International: West Conshohocken, PA, USA, 2013. Available online: www.astm.org (accessed on 6 May 2020).

50. EN 1811-2011+A1:2015, Reference Test Method for Release of Nickel From All Post Assemblies which are Inserted Into Pierced Parts of the Human Body and Articles Intended to Come Into Direct or Prolonged Contact With the Skin, CEN/TC347. 2011. Available online: https://www.en-standard.eu/din-en-1811 (accessed on 10 October 2018).

51. Zanotto, F.; Grassi, V.; Balbo, A.; Monticelli, C.; Zucchi, F. Resistance of thermally aged DSS 2304 against localized corrosion attack. *Metals* **2018**, *8*, 1022. [CrossRef]

52. ASTM G46-94. *Standard Guide for Examination and Evaluation of Pitting Corrosion*; ASTM International: West Conshohocken, PA, USA, 2018. Available online: www.astm.org (accessed on 6 May 2020).

53. Virtanen, S. Degradation of titanium and its alloys. In *Degradation of Implant Materials*; Eliaz, N., Ed.; Springer: New York, NY, USA, 2012; pp. 29–55.

54. Melchers, R.E. A review of trends for corrosion loss and pit depth in longer-term exposures. *Corros. Mater. Degrad.* **2020**, *1*, 4. [CrossRef]

55. Abbasi Aghuy, A.; Zakeri, M.; Moayed, M.H.; Mazinani, M. Effect of grain size on pitting corrosion of 304L austenitic stainless steel. *Corros. Sci.* **2015**, *94*, 368–376. [CrossRef]

56. Poyet, P.; Desestret, A.; Coriou, H.; Grall, L. *Contribution à L'étude de la Corrosion par Piqûres des Aciers Inoxydables 18/10. Influence de Divers Facteurs sur le Processus D'amorçage des Piqûres*; Société française de métallurgie. Journées d'automne: Paris, France, 1973.

57. Bryant, M.; Hu, X.; Farrar, R.; Brummitt, K.; Freeman, R.; Neville, A. Crevicecorrosion of biomedical alloys: A novel method of assessing the effects of bone cement and its chemistry. *J. Biomed. Mater. Res. B Appl. Biomater.* **2013**, *101*, 792–803. [CrossRef]

58. Liu, Y.; Zhu, D.; Pierre, D.; Gilbert, J.L. Fretting initiated crevicecorrosion of 316LVM stainless steel in physiological phosphate buffered saline: Potential and cycles to initiation. *Acta Biomater.* **2019**, *97*, 565–577. [CrossRef]

59. Stonawská, Z.; Svoboda, M.; Sozańska, M.; Krístková, M.; Sojka, J.; Dagbert, C.; Hyspecká, L. Structural analysis and intergranularcorrosion tests of AISI 316L steel. *J. Microsc.* **2006**, *224*, 62–64. [CrossRef]

60. Liu, T.; Xia, S.; Bai, Q.; Zhou, B.; Lu, Y.; Shoji, T. Evaluation of grain boundary network and improvement of intergranular cracking resistance in 316L stainless steel after grain boundary engineering. *Materials* **2019**, *12*, 242. [CrossRef]

61. Fujii, T.; Furumoto, T.; Tohgo, K.; Shimamura, Y. Crystallographic evaluation of susceptibility to intergranular corrosion in austenitic stainless steel with various degrees of sensitization. *Materials* **2020**, *13*, 613. [CrossRef]

62. Liu, G.; Liu, Y.; Cheng, Y.; Li, J.; Jiang, Y. The intergranular corrosion susceptibility of metastable austenitic Cr-Mn-Ni-N-Cu high-strength stainless steel under various heat treatments. *Materials* **2019**, *12*, 1385. [CrossRef] [PubMed]

63. ASTM G82-98. *Standard Guide for Development and Use of a Galvanic Series for Predicting Galvanic Corrosion Performance*; ASTM International: West Conshohocken, PA, USA, 2014. Available online: www.astm.org (accessed on 6 May 2020).

64. Mansfeld, F.; Kenkel, J.V. Laboratory studies of galvanic corrosion of aluminium alloys. In *Galvanic and Pitting Corrosion–Field and Laboratory Studies*; Baboian, R., France, W.D., Eds.; ASTM: West Conshohocken, PA, USA, 1976; pp. 20–47.

65. Gilbert, J.L.; Mali, S. Medical implant corrosion: Electrochemistry at metallic biomaterial surfaces. In *Degradation of Implant Materials*; Eliaz, N., Ed.; Springer: New York, NY, USA, 2012; Chapter 1, pp. 1–28.

© 2020 by the authors. Licensee MDPI, Basel, Switzerland. This article is an open access article distributed under the terms and conditions of the Creative Commons Attribution (CC BY) license (http://creativecommons.org/licenses/by/4.0/).

Article

Effect of Cr on Aqueous and Atmospheric Corrosion of Automotive Carbon Steel

Sang-won Cho [1], Sang-Jin Ko [1], Jin-Seok Yoo [1], Yun-Ha Yoo [2], Yon-Kyun Song [2] and Jung-Gu Kim [1,*]

1 Department of Materials Science and Engineering, Sungkyunkwan University, 2066 Seobu-Ro, Jangan-Gu, Suwon-Si 16419, Korea; jsw2811@gmail.com (S.-w.C.); tkdwls121@skku.edu (S.-J.K.); wlstjr5619@skku.edu (J.-S.Y.)
2 POSCO Global R&D Center, Steel Solution Research Laboratory, 100 Songdogwahak-Ro, Yeonsu-Gu, Incheon 21985, Korea; yunha778@posco.co.kr (Y.-H.Y.); petersong@posco.com (Y.-K.S.)
* Correspondence: kimjg@skku.edu

Abstract: This study investigated the effect of Cr alloying element on the corrosion properties of automotive carbon steel (0.1C, 0.5Si, 2.5Mn, Fe Bal., composition given in wt.%) in aqueous and atmospheric conditions using electrochemical measurement and cyclic corrosion tests. Three steels with 0, 0.3, and 0.5 wt.% Cr were studied by electrochemical impedance spectroscopy. Polarization resistance (R_p) of 0.3 Cr and 0.5 Cr steels was higher than that of 0 Cr steel, and the R_p also increased as the Cr content increased. Therefore, Cr increases the corrosion resistance of automotive carbon steel immersed in a chloride ion (Cl^-)-containing aqueous solution. In the cyclic corrosion test results, Cl^- was concentrated at the metal/rust interface in all of the steels regardless of Cr content. The Cl^- was uniformly concentrated and distributed on the 0 Cr steel, but locally and non-uniformly concentrated on the Cr-added steels. The inner rust layer consisted of β-FeOOH containing Cl^- and Cr-goethite, while the outer rust layer was composed of amorphous iron oxyhydroxide mixed with various types of rust. $FeCl_2$ and $CrCl_3$ are formed from the Cl^- nest developed in the early stage, and the pitting at $CrCl_3$-formed regions are locally accelerated because Cr is strongly hydrolyzed to a very low pH.

Keywords: automotive steel; atmospheric corrosion; electrochemical impedance spectroscopy; cyclic corrosion test; iron oxide

Citation: Cho, S.-w.; Ko, S.-J.; Yoo, J.-S.; Yoo, Y.-H.; Song, Y.-K.; Kim, J.-G. Effect of Cr on Aqueous and Atmospheric Corrosion of Automotive Carbon Steel. *Materials* **2021**, *14*, 2444. https://doi.org/10.3390/ma14092444

Academic Editor: Marián Palcut

Received: 6 April 2021
Accepted: 5 May 2021
Published: 8 May 2021

Publisher's Note: MDPI stays neutral with regard to jurisdictional claims in published maps and institutional affiliations.

Copyright: © 2021 by the authors. Licensee MDPI, Basel, Switzerland. This article is an open access article distributed under the terms and conditions of the Creative Commons Attribution (CC BY) license (https://creativecommons.org/licenses/by/4.0/).

1. Introduction

Steel sheets for automobiles are exposed to various corrosive environments due to climate change. In particular, the increased inundation from heavy rain and the use of salt for snow removal accelerate the corrosion of the automobile steel sheet, which leads to the deterioration of the durability and the collision safety of the vehicle [1]. Furthermore, with the recent rapid development of industry, these corrosive environments are becoming more and more severe. Therefore, it is essential to evaluate the corrosion life of the steel for predicting the durability of automobile parts. Currently, corrosion life is most often evaluated by the salt spray test (SST) and cyclic corrosion test (CCT). The accelerated CCT method for simulating an actual environment is used by many automakers. When an automotive carbon steel (ACS) sheet is evaluated through a CCT, atmospheric corrosion occurs on the test specimen and is greatly affected by environmental factors such as the type of material, humidity, time of wetness (TOW), and temperature [2]. For example, in the case of Cu and Ag, the corrosion rate is most affected by sulfides such as H_2S and SO_2 in the air, whereas in the case of Fe, acid fumes and fine dust are known to be more important. Furthermore, in coastal cities, the corrosion rate changes depending on the chloride concentration in the air. When the salt particles in the air are adsorbed on the metal surface, water may be condensed on the surface even if the relative humidity is low, which facilitates the formation of a water film [3]. As the TOW lengthens, the corrosion rate increases.

Generally, weathering steel is widely used to prevent atmospheric corrosion and contains alloying elements such as Cr, Cu, and Ni. Particularly, Cr is known as an element that improves the corrosion resistance of low alloy steel in various corrosive environments. When the weathering steel is exposed to a corrosive environment, Fe oxides (porous rust layer) are initially formed under attack from oxygen, similarly to common steel. However, over time, a dense rust layer (protective rust) is formed on the steel surface. This rust layer protects the steel surface, inhibiting corrosion and reducing corrosion rates compared to common steel. In common steel, the oxide rust layer penetrates into the substrate, and corrosion of the substrate continues, while in weathering steel, the amorphous layer enriched with Cu, Cr, and Ni inhibits further corrosion progress. Weathering steel forms a thin rust layer with FeOOH as the main component on the steel surface in the early stages. Then, a Cr-enriched layer (Cr-goethite) with very small particles containing Cr is formed on the steel surface [4,5]. The Cr-goethite layer has cation selectivity, preventing the penetration of corrosion substances such as SO_4^{2-} and Cl^- from the outside [6–8]. However, unlike these positive effects, negative effects have also been reported. According to Park et al. [9], in the flue gas desulfurization environment, Cr induces localized corrosion when Cr and Cu are added together in low-carbon steel because Cr segregates into the grain boundary, forming a Cr depletion region.

As described above, many research endeavors have been undertaken with respect to Cr's effect on the corrosion of metallic materials; however, only a few studies on the effect of Cr on the corrosion of ACS have been conducted. In this study, the effect of the Cr alloying element on the aqueous corrosion and atmospheric corrosion of ACS was investigated. The aqueous corrosion properties were analyzed using electrochemical measurements in a chloride (Cl^-)-containing solution, and atmospheric corrosion properties were analyzed via a CCT.

2. Materials and Methods

The specimens used in the electrochemical test and CCT were ACS containing 0, 0.3, and 0.5 wt.% Cr (produced by POSCO, Gwangyang, Korea), as described in Table 1. The microstructure images of the specimens are shown in Figure 1. In Figure 1, all the steels are composed of ferrite and martensite phases. Most martensite was formed along the grain boundary, with a small amount present inside the ferrite matrix. There is no noticeable difference in the three steels except for the grain size. The higher the Cr content, the bigger the grain size.

Table 1. Chemical compositions of the specimens (unit: wt.%).

Steels	Cr	C	Si	Mn	Fe
0 Cr	0.01	0.10	0.52	2.49	Bal.
0.3 Cr	0.32	0.10	0.52	2.49	Bal.
0.5 Cr	0.50	0.10	0.52	2.49	Bal.

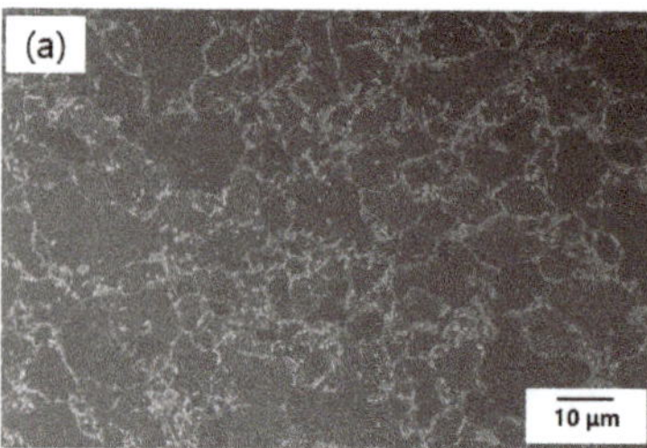

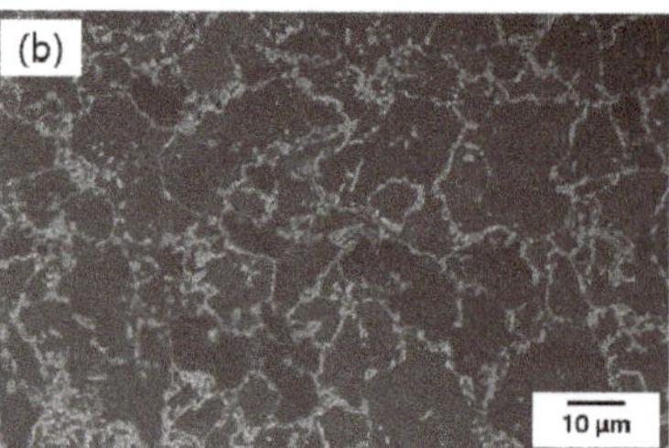

Figure 1. SEM images of the microstructure of (**a**) 0 Cr, (**b**) 0.3 Cr, and (**c**) 0.5 Cr, etched with 2% nital solution.

The specimens were cut to a size of 1.5 cm × 1.5 cm, polished with 600-grit SiC paper, and cleaned with distilled water. Additionally, the solution for electrochemical measurements was 3.5 wt.% NaCl. All of the electrochemical measurements were conducted in a 3-electrode electrochemical cell. The test specimen was used as the working electrode, a carbon rod was used as the counter electrode, and a saturated calomel electrode was used as the reference electrode. Potentiodynamic polarization tests were performed with a potential sweep of 0.166 mV/s according to ASTM G5. To establish a stable potential, the scan was initiated after the specimen was stabilized in the solution [10]. Electrochemical impedance spectroscopy (EIS) tests were performed with an amplitude of 10 mV in the frequency range of 100 kHz to 10 mHz. Electrochemical tests were conducted by a potentiostat (BioLogics, VMP-2, Seyssinet-Pariset, France). After the CCT, the test specimens were mounted with epoxy and analyzed by an optical microscope (OM), and the components of the corrosion product were analyzed by an electron probe micro-analyzer (EPMA; JEOL, JXA-8530F, Fukuoka, Japan), X-ray diffraction (XRD; Rigaku, D/max-2500V/PC, Tokyo, Japan), and transmission electron microscopy (TEM; FEI, Tecnai F20 G2, Hillsboro, OR, USA).

The specific CCT process is shown in Figure 2. The specimens used for the CCT were cut to a size of 3 cm × 7 cm, and exposed on only one side to the corrosive environment. The CCT was performed for 10, 20, and 30 cycles, respectively. The salt solution used for the CCT was 5 wt.% NaCl. The length of a CCT cycle was 24 h, consisting of a wet stage for 21 h and a dry stage at 30% of relative humidity and 50 °C for 3 h.

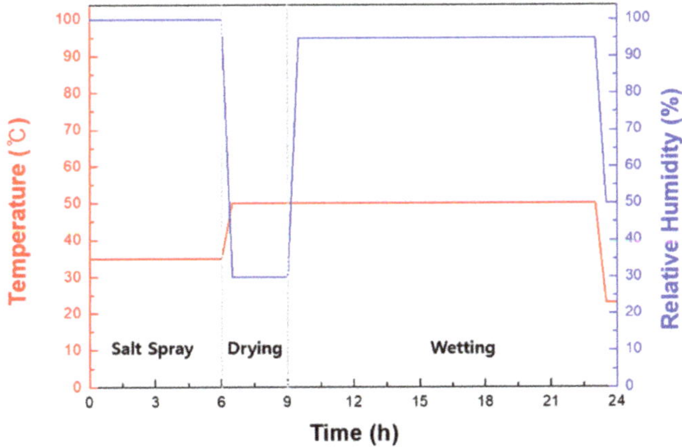

Figure 2. Specific conditions for the cyclic corrosion test.

3. Results and Discussion

3.1. Electrochemical Measurement

Potentiodynamic polarization tests were conducted to analyze the difference in corrosion characteristics depending on the Cr content, and the results are shown in Figure 3 and Table 2. In Figure 3, all of the specimens show an active corrosion behavior that increases with increasing potential in Cl^--containing environments without passivation. Additionally, there was no significant difference in the corrosion potential regardless of the Cr content. The corrosion potential (E_{corr}) and corrosion current density of 0.5 Cr steel were slightly lower than that of 0 Cr and 0.3 Cr steels, but this is an insignificant difference that can be regarded as an experimental error.

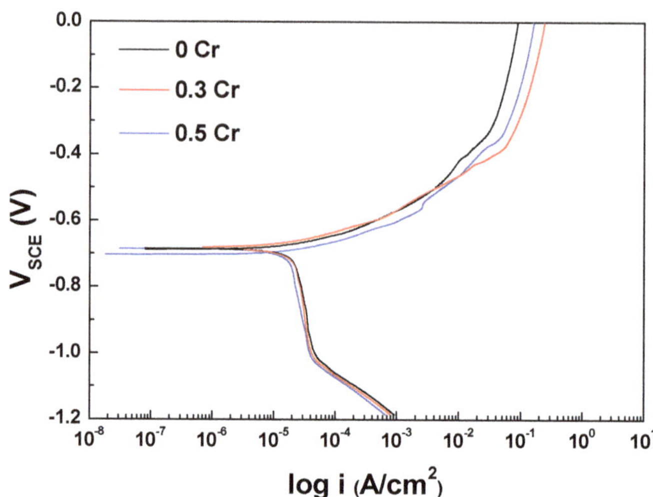

Figure 3. Potentiodynamic polarization curves in 3.5 wt.% NaCl solution.

Table 2. Potentiodynamic polarization test results in 3.5 wt.% NaCl solution.

Parameter	0 Cr	0.3 Cr	0.5 Cr
E_{corr} (mV$_{SCE}$)	-686.5 ± 1.4	-683.6 ± 5.4	-693.2 ± 12.6
I_{corr} (µA/cm^2)	21.3	20.2	17.4

In order to obtain a better understanding of the effect of Cr on the corrosion behavior of ACS under aqueous conditions, EIS measurement was performed in a 3.5 wt.% NaCl solution at room temperature. Figure 4 shows the results of EIS measurement in the form of Nyquist and Bode plots under open circuit potential (OCP) according to various immersion times. The Nyquist plots were not perfect semicircles due to dispersion effects that are often caused by the geometrical inhomogeneity or non-uniform current distribution on the electrode surface [11]. The capacitive loops in the high- and low-frequency regions overlapped. The capacitive loop of the high-frequency region showed the resistance of the film, whereas that of the low-frequency region showed the charge transfer resistance [12–14]. In the 0 Cr steel, the size of the semicircle for 1 h was larger than that for 0 h, and then became smaller with respect to immersion time. This means that a thin and weak oxide layer was formed on the steel surface at the initial stage of the corrosion process, and deteriorated with respect to immersion time due to its instability [15]. Similar to 0 Cr steel, the size of the semicircles for the 0.3 Cr and 0.5 Cr steels also increased immediately after immersion, and then decreased with immersion time. However, the size of the overall capacitive semicircles is ordered as 0.5 Cr > 0.3 Cr > 0 Cr. In general, the size of the capacitive semicircle on the Nyquist plot represents corrosion resistance. This means that the Cr alloying element within 0.5 wt.% improves corrosion resistance in an aqueous environment. In the Bode plots, the impedance at a low frequency and the shoulder width at the phase angle were increased and wider immediately after immersion, and then decreased and narrower with immersion time. This result is consistent with the Nyquist impedance interpretation.

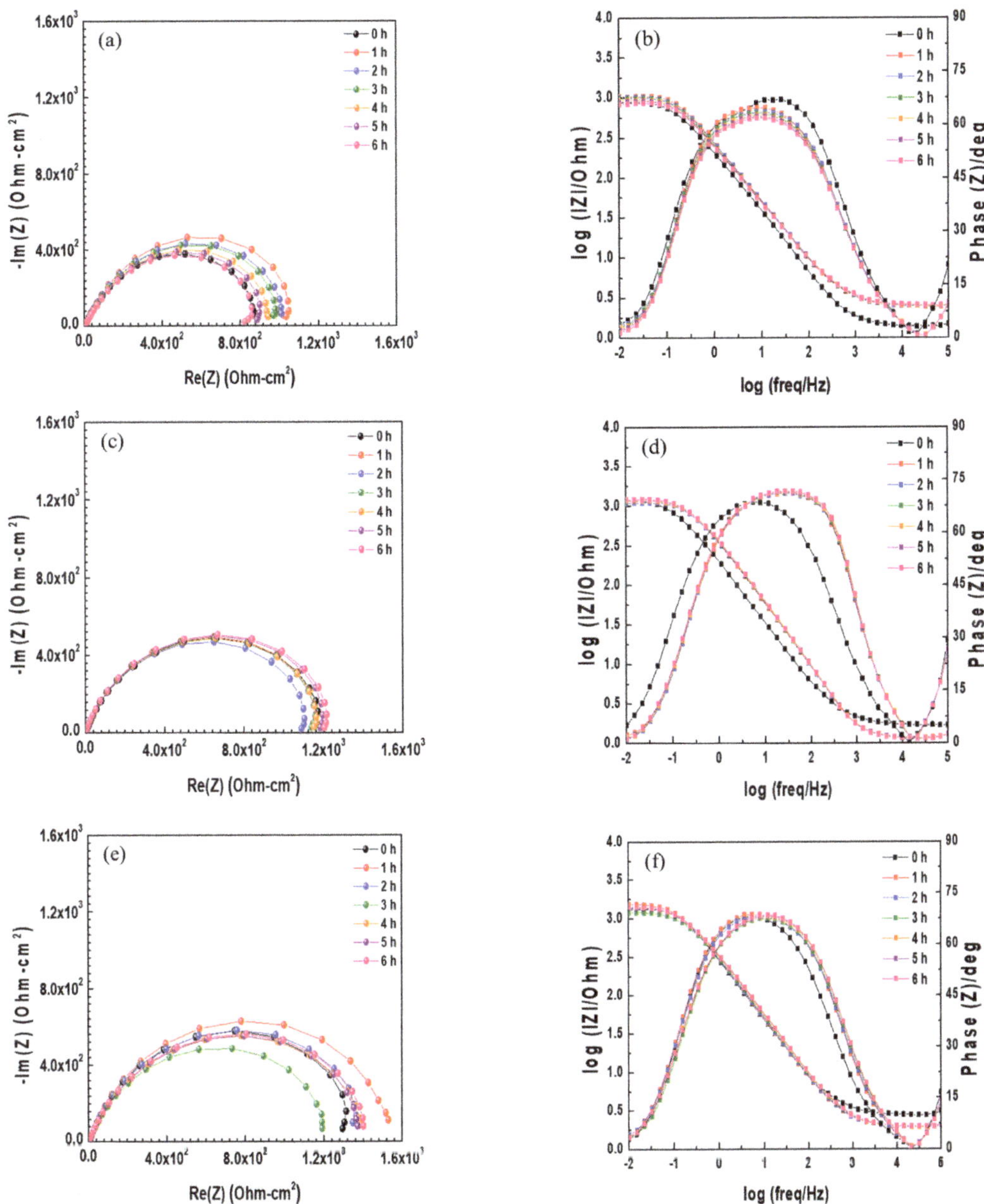

Figure 4. Nyquist and Bode impedance plots of EIS data of (**a,b**) 0 Cr steel, (**c,d**) 0.3 Cr steel, and (**e,f**) 0.5 Cr steel in 3.5 wt.% NaCl solution.

To determine the optimized values for the resistance and capacitance parameters, the equivalent circuit was used as shown in Figure 5. R_s is the test solution resistance, R_{film} is the oxide film resistance, R_{ct} is the charge transfer resistance, and $R_{film} + R_{ct}$ is total resistance or polarization resistance (R_p), which is proportional to the radius of the capacitive loop in the Nyquist plot. The constant phase element (CPE) is the capacitive response of the system. CPE1 is the capacitive response of the oxide film, and CPE2 is the capacitive response of the double layer caused by the dissolution of the metal and the charge separation between the metal/electrolyte interface [16–19]. In the equivalent circuit, CPE is defined as below:

$$Z_{CPE} = Q_0^{-1}(j\omega)^{-n} \tag{1}$$

where Z is the impedance, Q_0 is the coefficient of proportionality, j is the imaginary number, ω is the angular frequency, and n is the empirical CPE exponent ($0 \leq n \leq 1$) measuring the deviation from the behavior of an ideal electric capacity [20,21]. CPE can represent resistance (n = 0), capacitance (n = 1), inductance (n = −1), or Warburg impedance (n = 0.5) in accordance with n [22].

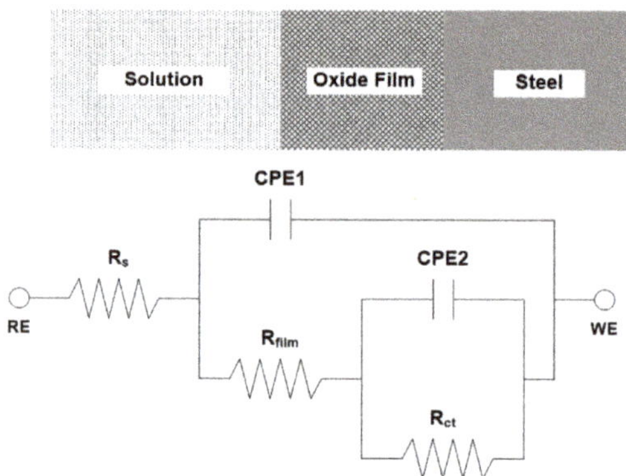

Figure 5. Equivalent circuit for ACS in 3.5 wt.% NaCl solution.

The EIS data were fitted using the ZSimpWin (Princeton Applied Research, Oak Ridge, TN, USA) program and the results are shown in Table 3. The R_p of 0.3 Cr and 0.5 Cr steels was higher than that of 0 Cr steel, and the R_p also increased as the Cr content increased. This indicates that the corrosion resistance is increased as the Cr content increases. This is believed to be due to the bigger grain size of the steel with higher Cr content. Metal with active polarization behavior decreases the corrosion rate with bigger grain size [23]. Therefore, Cr improved the corrosion resistance of the ACS that was immersed in the Cl^--containing aqueous solution.

Table 3. Parameters from electrochemical impedance spectroscopy measurements.

Steel	Immersion Time	R_s ($\Omega\cdot cm^{-2}$)	CPE1 Q_{film} ($\Omega^{-1} cm^{-2}\cdot s^n$)	n_1	R_{film} ($\Omega\cdot cm^{-2}$)	CPE2 Q_{ct} ($\Omega^{-1}\cdot cm^{-2}\cdot s^n$)	n_1	R_{ct} ($\Omega\cdot cm^{-2}$)	R_p ($\Omega\cdot cm^{-2}$)
	0 h	1.424	7.52×10^{-4}	0.8355	29.9	3.97×10^{-4}	0.9671	636.5	665.7
	1 h	2.531	6.11×10^{-4}	0.8146	192.6	1.29×10^{-4}	0.9516	951.6	1144.2
	2 h	2.522	6.36×10^{-4}	0.8037	294	1.48×10^{-4}	0.9883	801.2	1095.2
0 Cr	3 h	2.542	6.65×10^{-4}	0.801	239.9	1.55×10^{-4}	0.9724	831.3	1071.2
	4 h	2.549	6.84×10^{-4}	0.796	267.4	1.61×10^{-4}	0.997	752.1	1019.5
	5 h	2.574	6.72×10^{-4}	0.7959	217.5	1.73×10^{-4}	0.9593	766.3	983.8
	6 h	2.589	6.81×10^{-4}	0.7953	203.7	1.85×10^{-4}	0.9509	733.1	936.8
	0 h	1.773	1.71×10^{-4}	1	3	9.06×10^{-4}	0.7621	1304	1307
	1 h	1.227	2.05×10^{-4}	0.9653	32.3	3.76×10^{-4}	0.7516	1210	1242.3
	2 h	1.22	1.92×10^{-4}	0.9714	29.2	3.79×10^{-4}	0.7541	1151	1180.2
0.3 Cr	3 h	1.212	1.83×10^{-4}	0.9759	26.5	3.90×10^{-4}	0.7497	1215	1241.5
	4 h	1.212	1.69×10^{-4}	0.9843	24.3	4.00×10^{-4}	0.7475	1223	1247.3
	5 h	1.217	1.59×10^{-4}	0.9914	23.7	4.10×10^{-4}	0.7442	1266	1289.7
	6 h	1.215	1.57×10^{-4}	0.993	23.1	4.08×10^{-4}	0.7424	1285	1308.1

Table 3. *Cont.*

Steel	Immersion Time	R_s ($\Omega \cdot cm^{-2}$)	CPE1			R_{film} ($\Omega \cdot cm^{-2}$)	CPE2		R_{ct} ($\Omega \cdot cm^{-2}$)	R_p ($\Omega \cdot cm^{-2}$)
			Q_{film} ($\Omega^{-1} cm^{-2} \cdot s^n$)	n_1			Q_{ct} ($\Omega^{-1} \cdot cm^{-2} \cdot s^n$)	n_1		
	0 h	2.809	4.71×10^{-4}	0.8449	21.6	2.14×10^{-4}	0.8069	1414	1435.6
	1 h	1.986	5.09×10^{-4}	0.8474	59	1.63×10^{-4}	0.7983	1543	1602
	2 h	1.978	5.18×10^{-4}	0.8405	63.5	1.67×10^{-4}	0.7818	1419	1482.5
0.5 Cr	3 h	1.964	3.86×10^{-4}	0.8637	20.6	2.95×10^{-4}	0.7261	1270	1290.6
	4 h	1.963	3.24×10^{-4}	0.8782	14.9	3.39×10^{-4}	0.7095	1489	1503.9
	5 h	2.001	1.18×10^{-4}	0.9798	4.5	5.31×10^{-4}	0.7418	1502	1506.5
	6 h	2.013	9.42×10^{-4}	0.9999	3.6	5.44×10^{-4}	0.748	1542	1545.6

3.2. Cyclic Corrosion Test Results

The specimens after the CCT were cut and the cross-section was observed with an OM and the results are shown in Figure 6. After 10 cycles, the rust of all steels was thin and relatively uniform. However, after 20 cycles, brown and black oxides were formed on the inner layer and outer layer, respectively. Especially after 30 cycles, the amount of corrosion product of 0.3 Cr and 0.5 Cr steels was greater than that of 0 Cr steel, and corroded in a more localized way. Furthermore, since the thickness of the oxide layer is proportional to the amount of corrosion of the base metal, a thick oxide layer was locally formed on the 0.3 Cr and 0.5 Cr steels.

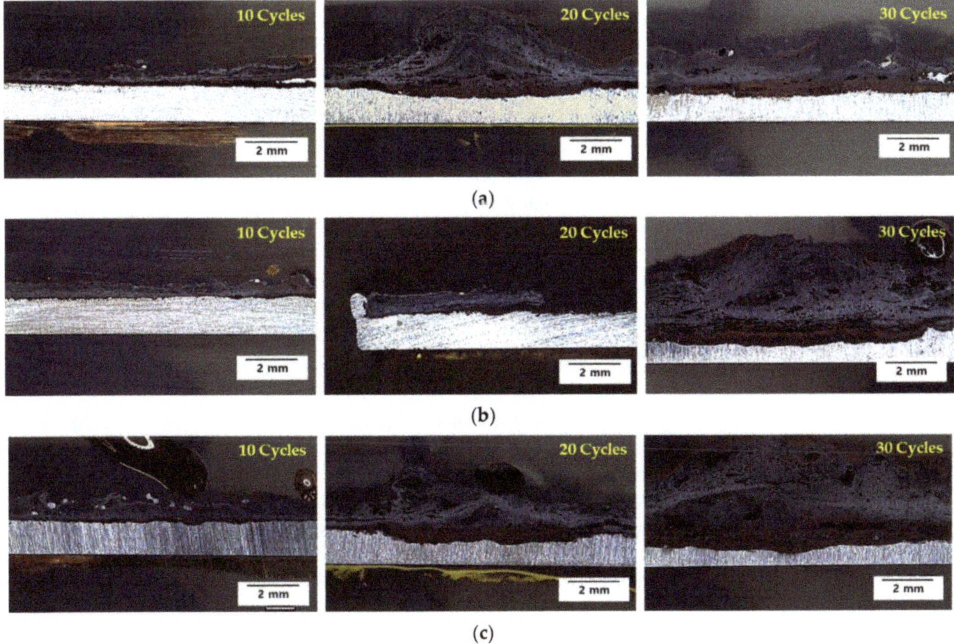

Figure 6. Cross-section OM images of (**a**) 0 Cr steel, (**b**) 0.3 Cr steel, and (**c**) 0.5 Cr steel after CCT.

To determine the localized corrosion tendency, the pitting factor (PF) with a concept similar to that given in ASTM G46 was used. A PF value of 1 means perfect uniform corrosion, and a higher PF means an increased localized corrosion tendency. The PF for each CCT cycle was derived by the following equation, and the variation of the PF according to CCT cycle is shown in Figure 7.

$$PF = \frac{p}{d} \tag{2}$$

where p is the maximum penetration depth, and d is the average penetration depth.

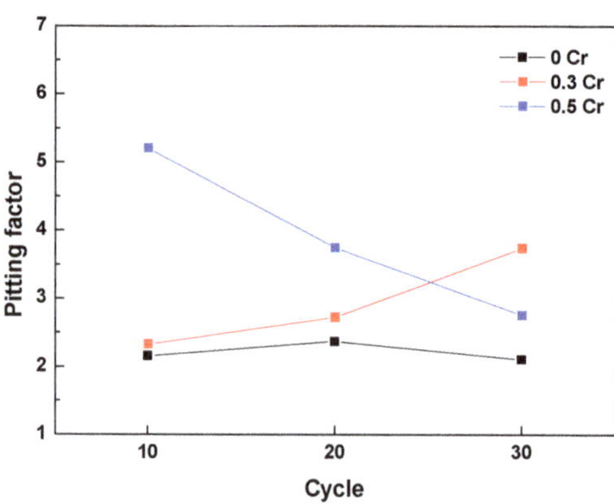

Figure 7. Variation of pitting factor according to CCT cycle.

In the case of 0 Cr steel, the PF was approximately 2 regardless of the cycle, while the PF of 0.3 Cr and 0.5 Cr steels changed depending on the cycle. In all cycles, the PF of the Cr-added steels was higher than that of the 0 Cr steel, but the PF was not proportional to Cr content. This indicates that the Cr alloying element can accelerate localized corrosion, and the presence or absence of Cr greatly affects the localized corrosion, not the Cr content.

The cross-section of the specimen after 10 and 30 cycles was analyzed to determine the chemical composition using EPMA, and the results are shown in Figure 8. The rust layer of 0 Cr steel was composed entirely of porous iron oxide (e.g., γ-FeOOH, γ-Fe$_2$O$_3$, Fe$_3$O$_4$). In addition, Cl$^-$ was accumulated at the metal/rust interface and on the inner layer with uniform concentration and distribution. The Cr-added steels had a very dense and uniform Cr-enriched region in the inner rust layer, while the outer rust layer was composed of porous iron oxide, like 0 Cr. Cl$^-$ was detected underneath the Cr-enriched layer and at the metal/rust interface, but unlike 0 Cr steel, it was localized and non-uniformly concentrated. The rust layer of the 30-cycle steel was exfoliated from the metal, and the Cl$^-$ concentration in the inner rust layer was increased significantly compared to 10 cycles. Therefore, it is considered that the corrosion is accelerated because the protective oxide layer loses its protective property after the rust layer exfoliates.

To summarize the above results, Cl$^-$ was concentrated at the metal/rust interface in all of the specimens regardless of Cr content. Generally, since the localized corrosion in an atmospheric environment is caused by Cl$^-$ enrichment [4,24], localized corrosion with a PF of approximately 2 or higher occurred in all of the steels, as shown in Figure 7. However, 0.3 Cr and 0.5 Cr steels had higher PFs than 0 Cr steel because Cl$^-$ was localized and non-uniformly concentrated as compared with 0 Cr steel.

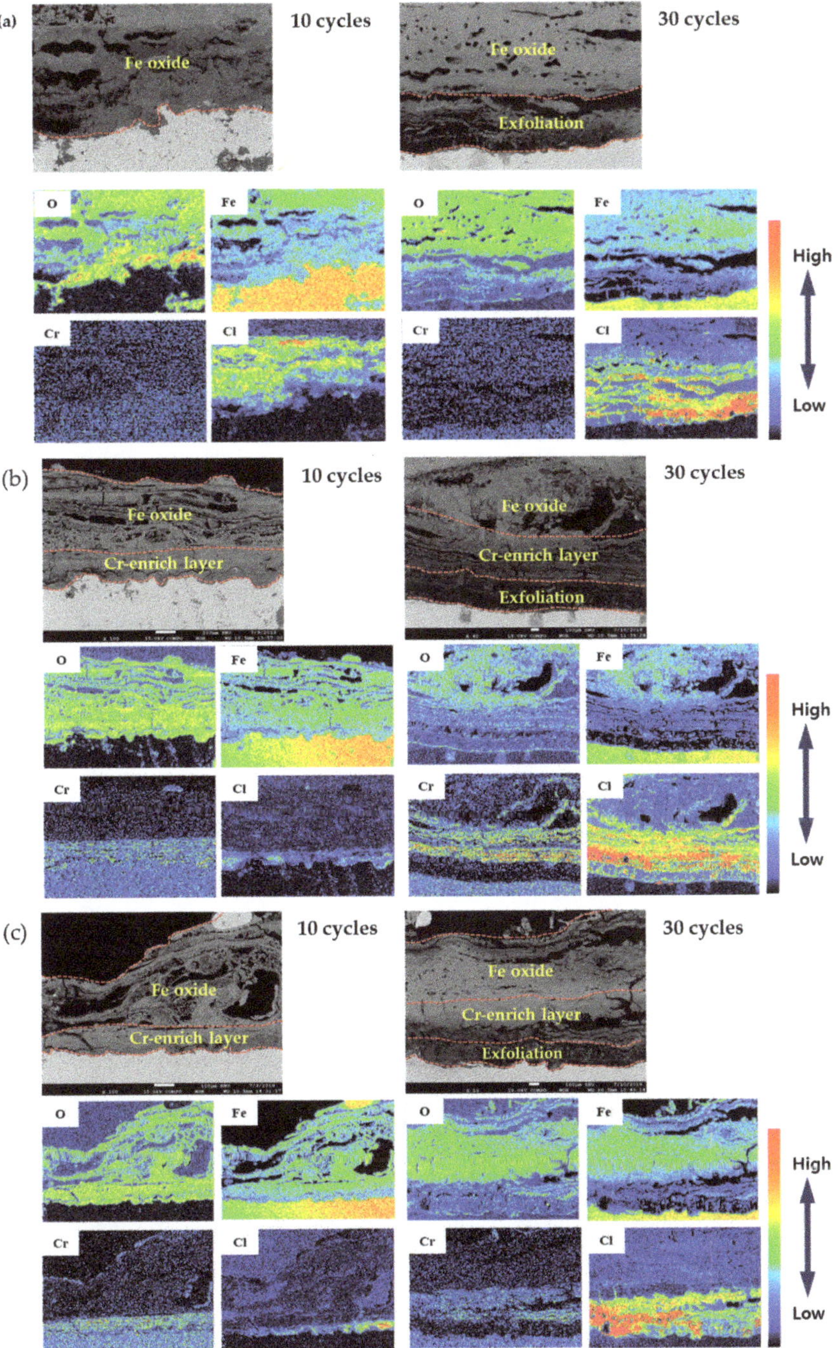

Figure 8. EPMA analysis of (**a**) 0 Cr steel, (**b**) 0.3 Cr steel, and (**c**) 0.5 Cr steel after CCT.

3.3. Rust Constituent Analysis

The constituent of the rust formed after the CCT was analyzed by XRD, EDS, and TEM, and the results are shown in Figures 9 and 10. As shown in Figure 9, the phase of rust formed by the CCT was almost the same for all steels. The rusts were composed of various oxides and hydroxides such as α-FeOOH (goethite), β-FeOOH (akaganeite), γ-FeOOH (lepidocrocite), γ-Fe$_2$O$_3$ (maghemite), and Fe$_3$O$_4$ (magnetite). Among them, akaganeite always contains Cl$^-$ in the lattice because it is stabilized by the Cl$^-$ entering the lattice structure. In addition, akaganeite is formed only during the dry stage in an atmospheric environment, and Cl$^-$ in the akaganeite is dissolved in water to promote corrosion during the wet stage. That is, the akaganeite acts as a Cl$^-$ reservoir. As a result, a large amount of akaganeite is formed inside the pit generated by atmospheric corrosion [3,6,24–26]. Therefore, akaganeite was formed in the inner rust layer, and Cl$^-$ was observed at the metal/rust interface, as shown in Figure 8.

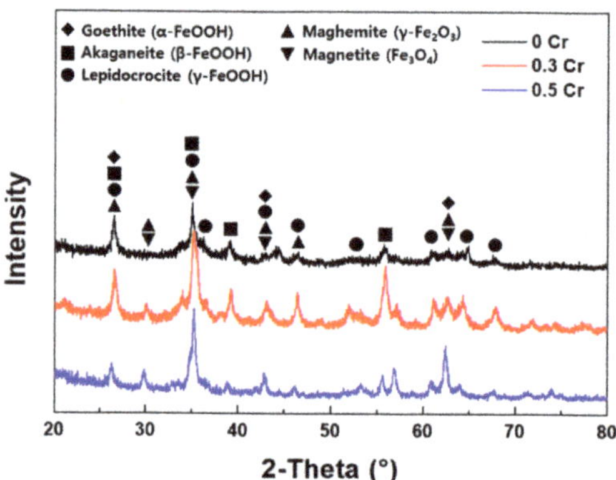

Figure 9. XRD analysis of the specimen surfaces after 20 cycles of CCT.

The EDS results, TEM images, and diffraction patterns of the inner and outer rust formed on the 0.5 Cr steel were analyzed, and the results are shown in Figure 10. Chlorine was observed in the inner rust particle as an acicular single crystal with a size of about 100 nm, as shown in Figure 10a. This is the major characteristic of akaganeite [27]. In Figure 10b, Cr and Cl$^-$ were observed together in the inner rust particle. The particle was polycrystalline and was a spherical agglomeration with a size of several nanometers. As the spherical-shaped rust is the main feature of goethite [28], the particle is Cr-containing nanoscale goethite (Cr-goethite). Since dissolved or enriched Cr suppresses the growth of goethite crystals [7], the size of the Cr-goethite particles is very small. Cr-goethite is so small in size that it acts as a protective film that is densely formed in the inner rust layer. Furthermore, Cr-goethite has cation selectivity so it can inhibit the penetration of aggressive anions such as Cl$^-$ and SO$_4{}^{2-}$ and improve corrosion resistance [7,25,29–31]. In short, Cr-goethite was formed in the inner rust layer of Cr-added steels, which blocked the inflow of additional Cl$^-$ from the outside and consequently improved the corrosion resistance. As shown in Figure 10c, Cl$^-$ and Cr were not detected in the outer rust particle. Therefore, the outer rust layer is composed of various rusts such as lepidocrocite, maghemite, and magnetite detected from the XRD analysis results. In summary, the inner rust layer consists of akaganeite containing Cl$^-$ and Cr-goethite, while the outer rust layer is composed of amorphous iron oxyhydroxide mixed with various types of rust.

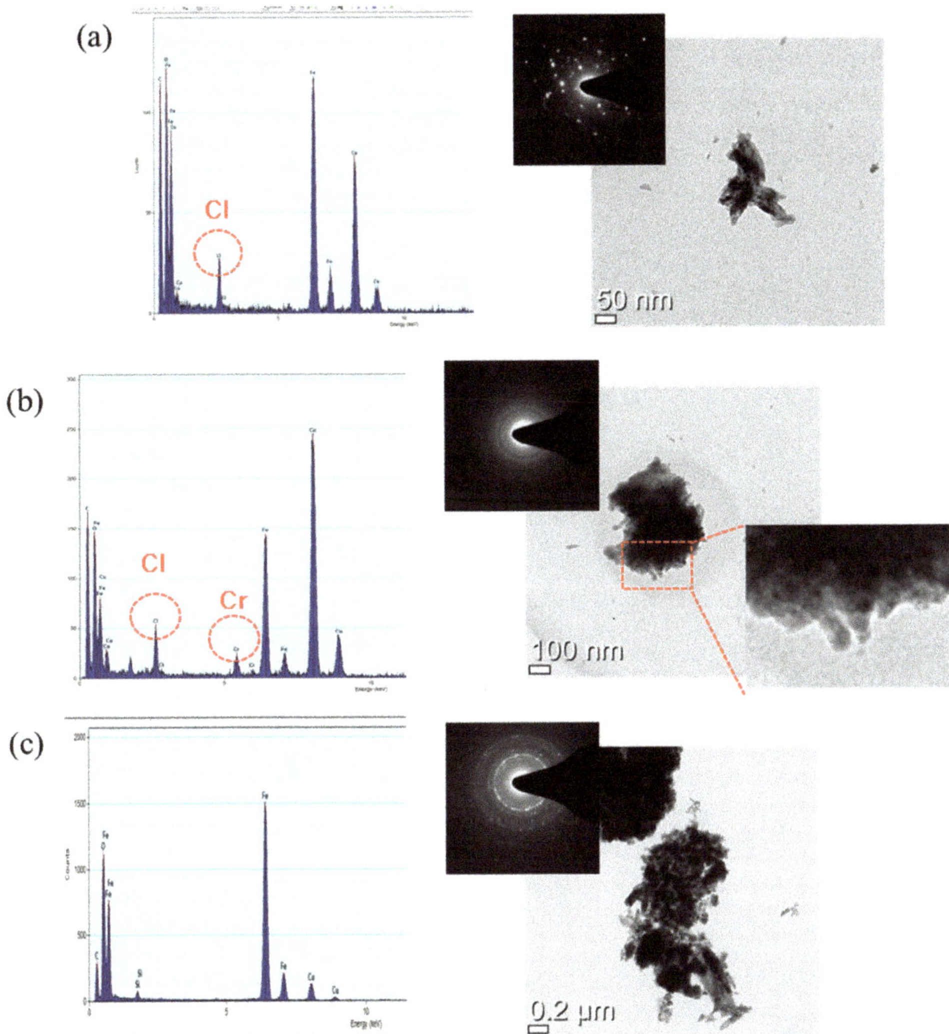

Figure 10. TEM images and the diffraction patterns of (**a,b**) inner rust and (**c**) outer rust formed on the 0.5 Cr steel.

3.4. Localized Corrosion Mechanism of Cr-Added Steel under Wet/Dry Conditions

The Cr alloying element accelerates localized corrosion under Cl-containing wet/dry conditions unlike the immersion condition. The mechanism of localized corrosion of Cr-added steel under wet/dry conditions is as follows, and a schematic diagram is shown in Figure 11.

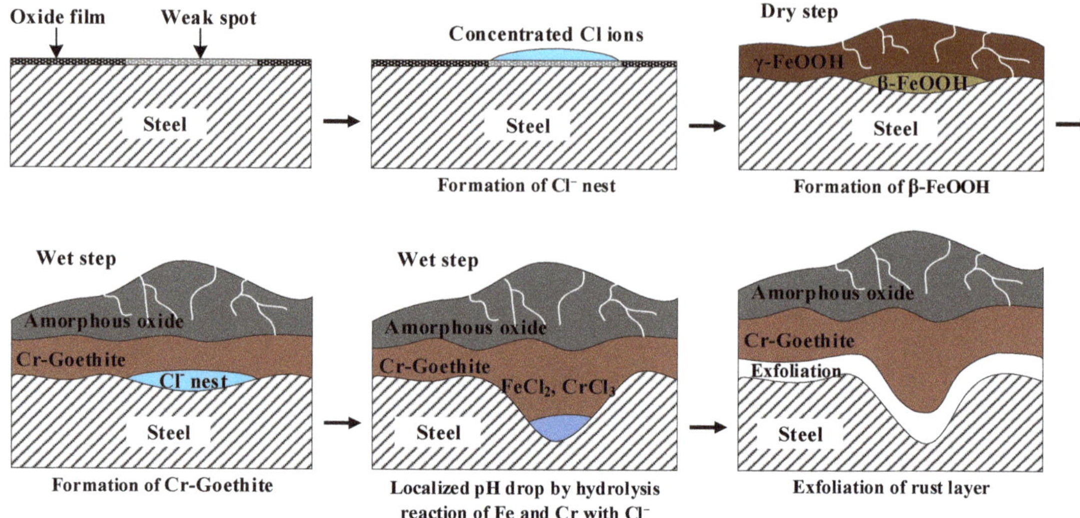

Figure 11. Schematic diagram of the mechanism of localized corrosion of Cr-added steel under wet/dry conditions.

Corrosion of steel begins in areas where the inherent oxide film is weak. During the wet stage, Cl^- ions existing in the aqueous adsorption layer move to these weak areas, and a Cl-concentrated region (nest) is formed [32]. The Cl^- is adsorbed on the steel surface, and then atmospheric corrosion initiates. Thereafter, Fe^{2+} reacts with H_2O to form $Fe(OH)_2$, and with salt or Cl^- in the air to form $FeCl_2$.

$$Fe \rightarrow Fe^{2+} + 2e^- \tag{3}$$

$$Fe^{2+} + 2H_2O \rightarrow Fe(OH)_2 + 2H^+ \tag{4}$$

$$Fe^{2+} + 2Cl^- \rightarrow FeCl_2 \tag{5}$$

During the dry stage, $Fe(OH)_2$ is transformed into lepidocrocite, and $FeCl_2$ formed in the Cl-concentrated region is transformed into akaganeite. After that, lepidocrocite and akaganeite are reduced to amorphous oxide or magnetite in the wet stage. Next, magnetite is re-oxidized into lepidocrocite.

$$2\gamma\text{-FeOOH} + Fe^{2+} \rightarrow Fe_3O_4 + 2H^+ \tag{6}$$

$$Fe_3O_4 + 3/2O_2 + H_2O \rightarrow 3\gamma\text{-FeOOH} \tag{7}$$

In the atmospheric rusting process, lepidocrocite on the steel surface transforms into amorphous ferric hydroxide, then it converts to goethite. Cl^- may facilitate this reaction and promote goethite formation [33–35].

$$\gamma\text{-FeOOH} \rightarrow FeO_x(OH)_{3-2x} \text{ (amorphous ferric oxyhydroxide)} \rightarrow \alpha\text{-FeOOH} \tag{8}$$

$$\gamma\text{-FeOOH} \rightarrow FeO_x(OH)_{2-2x}Cl \rightarrow \alpha\text{-FeOOH} + HCl \tag{9}$$

Cr^{3+} ions dissolved in the early stages of the corrosion process are more easily deposited as hydroxides near the steel surface compared to the Fe^{2+} ions since the solubility of Fe^{2+} ions is higher than that of Cr^{3+}. Additionally, the Cr^{3+} ions act as nuclei for the growth of Cr-goethite. Finally, an ultrafine Cr-goethite layer is formed in the inner rust layer when the wet/dry process is repeated [6]. Since Cr-goethite has cation selectivity, it suppresses the penetration of aggressive anions and improves corrosion resistance. Then, after the formation of Cr-goethite, the inflow of extra Cl^- from the outside is blocked, so that Cl^- is locally accumulated underneath the Cr-enriched layer.

During the wet stage, Cl^- ions in the akaganeite formed in the inner rust layer are dissolved and eluted in water, resulting in the formation of $FeCl_2$ and $CrCl_3$. The hydrolysis reactions of these Fe and Cr salts occur.

$$FeCl_2 + 2H_2O \rightarrow Fe(OH)_2 + 2HCl \tag{10}$$

$$CrCl_3 + 3H_2O \rightarrow Cr(OH)_3 + 3HCl \tag{11}$$

The pH of the steel surface is reduced by the hydrolysis reaction of Fe and Cr [36,37]. Since Cr tends to strongly hydrolyze up to pH 1.4, it enhances the susceptibility to localized corrosion compared to the hydrolysis reaction of Fe [38]. That is, $FeCl_2$ and $CrCl_3$ are formed from the Cl^- nest developed in the early stage, and the $CrCl_3$-formed regions are locally accelerated. Even if Cr is not added, Cl^- may cause localized corrosion. However, when Cr is added, localized corrosion is more accelerated since Cr is strongly hydrolyzed to a very low pH.

Finally, when the wet/dry cycle is continuously repeated, the rust layer is exfoliated and loses its protective property. Then, Cl^- easily penetrates into the gap between the separated rust layer and substrate, which accelerates the localized corrosion of ACS.

4. Conclusions

In this study, the effect of Cr alloying element on the corrosion properties of ACS in aqueous and atmospheric conditions was investigated using electrochemical measurements and a CCT. The conclusions based on the investigations are as follows:

- In the electrochemical measurement results, the Cr alloying element improves the corrosion resistance of the ACS that was immersed in the Cl-containing aqueous solution.
- Cl is concentrated at the metal/rust interface in all of the specimens regardless of Cr content after the CCT. The Cl is uniformly concentrated and distributed on the 0 Cr steel, whereas Cl is localized and non-uniformly concentrated on the Cr-added steels. The PF of the Cr-added steels is higher than that of the 0 Cr steel during the CCT.
- The inner rust layer consists of Cl-containing akaganeite and Cr goethite, while the outer rust layer is composed of amorphous iron oxyhydroxide mixed with various types of rust.
- $FeCl_2$ and $CrCl_3$ are formed from the Cl nest developed in the early stage, and the pitting at $CrCl_3$-formed regions is locally accelerated because Cr is strongly hydrolyzed to a very low pH.

Author Contributions: Conceptualization, S.-w.C. and Y.-H.Y.; Data curation, S.-w.C.; Formal analysis, S.-w.C.; Funding acquisition, J.-G.K.; Investigation, S.-w.C., S.-J.K., and J.-G.K.; Methodology, J.-G.K.; Project administration, Y.-H.Y., Y.-K.S., and J.-G.K.; Resources, Y.-H.Y. and Y.-K.S.; Software, S.-J.K. and J.-S.Y.; Validation, Y.-H.Y. and J.-G.K.; Visualization, S.-w.C.; Writing—original draft, S.-w.C.; Writing—review and editing, S.-w.C., S.-J.K., J.-S.Y., Y.-H.Y., Y.-K.S., and J.-G.K. All authors have read and agreed to the published version of the manuscript.

Funding: This research was funded by POSCO, grant number 2018Z098.

Institutional Review Board Statement: Not applicable.

Informed Consent Statement: Not applicable.

Conflicts of Interest: The authors declare no conflict of interest.

References

1. Lee, C.; Kang, B.; Choi, B.-H.; Lee, J.; Lee, K. Observation and characterization of squeak noises of polymeric materials for automotive interior parts under field-degradation. *Trans. KSAE* **2017**, *25*, 257–265. [CrossRef]
2. Adikari, A.; Munasinghe, R.D.S.; Jayatileke, S. Prediction of atmospheric corrosion—A Review. *Engineer* **2014**, *47*, 75–83. [CrossRef]

3. Alcántara, J.; Chico, B.; Simancas, J.; Díaz, I.; Morcillo, M. Marine atmospheric corrosion of carbon steel: A review. *Materials* **2017**, *10*, 406. [CrossRef]

4. Kamimura, T.; Stratmann, M. The influence of chromium on the atmospheric corrosion of steel. *Corros. Sci.* **2001**, *43*, 429–447. [CrossRef]

5. Asami, K.; Kikuchi, M. Characterization of rust layers on weathering steels air-exposed for a long period. *Mater. Trans.* **2002**, *43*, 2818–2825. [CrossRef]

6. Yamashita, M.; Shimizu, T.; Konishi, H.; Mizuki, J.; Uchida, H. Structure and protective performance of atmospheric corrosion product of Fe–Cr alloy film analyzed by Mössbauer spectroscopy and with synchrotron radiation X-rays. *Corros. Sci.* **2003**, *45*, 381–394. [CrossRef]

7. Yamashita, M.; Miyuki, H.; Matsuda, Y.; Nagano, H.; Misawa, T. The long term growth of the protective rust layer formed on weathering steel by atmospheric corrosion during a quarter of a century. *Corros. Sci.* **1994**, *36*, 283–299. [CrossRef]

8. Zhao, Q.-H.; Liu, W.; Zhu, Y.-C.; Zhang, B.-L.; Li, S.-Z.; Lu, M.-X. Effect of small content of chromium on wet–dry acid corrosion behavior of low alloy steel. *Acta Metall. Sin-Engl.* **2017**, *30*, 164–175. [CrossRef]

9. Park, S.-A.; Kim, J.-G.; Lee, B.-H.; Yoon, J.-B. Development of sulfuric and hydrochloric acid dew-point corrosion-resistant steels: 1. Effect of alloying elements on the corrosion resistance of low-alloy steels. *Korean J. Met. Mater.* **2014**, *52*, 837–855.

10. Kim, S.-H.; Lee, J.-H.; Kim, J.-G.; Kim, W.-C. Effect of the crevice former on the corrosion behavior of 316L stainless steel in chloride-containing synthetic tap water. *Met. Mater. Int.* **2018**, *24*, 516–524. [CrossRef]

11. Cheng, Q.; Chen, Z. The cause analysis of the incomplete semi-circle observed in high frequency region of EIS obtained from TEL-covered pure copper. *Int. J. Electrochem. Sci* **2013**, *8*, 8282–8290.

12. Keddam, M.; Mottos, O.R.; Takenouti, H. Reaction model for iron dissolution studied by electrode impedance: I. Experimental results and reaction model. *J. Electrochem. Soc.* **1981**, *128*, 257–266. [CrossRef]

13. Liu, W.; Dou, J.; Lu, S.; Zhang, P.; Zhao, Q. Effect of silty sand in formation water on CO_2 corrosion behavior of carbon steel. *Appl. Surf. Sci.* **2016**, *367*, 438–448. [CrossRef]

14. Zeng, L.; Zhang, G.; Guo, X. Erosion–corrosion at different locations of X65 carbon steel elbow. *Corros. Sci.* **2014**, *85*, 318–330. [CrossRef]

15. Srinivasan, A.; Blawert, C.; Huang, Y.; Mendis, C.; Kainer, K.; Hort, N. Corrosion behavior of Mg–Gd–Zn based alloys in aqueous NaCl solution. *J. Magnes. Alloy* **2014**, *2*, 245–256. [CrossRef]

16. Kim, M.; Jang, S.; Woo, S.; Kim, J.; Kim, Y. Corrosion resistance of ferritic stainless steel in exhaust condensed water containing aluminum cations. *Corrosion* **2015**, *71*, 285–291. [CrossRef]

17. Cho, S.; An, J.-H.; Lee, S.-H.; Kim, J.-G. Effect of pH on the passive film characteristics of lean duplex stainless steel in chloride-containing synthetic tap water. *Int. J. Electrochem. Sci* **2020**, *15*, 4406–4420. [CrossRef]

18. Chen, Y.; Hong, T.; Gopal, M.; Jepson, W. EIS studies of a corrosion inhibitor behavior under multiphase flow conditions. *Corros. Sci.* **2000**, *42*, 979–990. [CrossRef]

19. Hamdy, A.S.; El-Shenawy, E.; El-Bitar, T. Electrochemical impedance spectroscopy study of the corrosion behavior of some niobium bearing stainless steels in 3.5% NaCl. *Int. J. Electrochem. Sci.* **2006**, *1*, 171–180.

20. Bentiss, F.; Lebrini, M.; Vezin, H.; Chai, F.; Traisnel, M.; Lagrené, M. Enhanced corrosion resistance of carbon steel in normal sulfuric acid medium by some macrocyclic polyether compounds containing a 1,3,4-thiadiazole moiety: AC impedance and computational studies. *Corros. Sci.* **2009**, *51*, 2165–2173. [CrossRef]

21. Lopez, D.A.; Simison, S.; De Sanchez, S. The influence of steel microstructure on CO_2 corrosion. EIS studies on the inhibition efficiency of benzimidazole. *Electrochim. Acta* **2003**, *48*, 845–854. [CrossRef]

22. Mansfeld, F. Recording and analysis of AC impedance data for corrosion studies. *Corrosion* **1981**, *37*, 301–307. [CrossRef]

23. Ralston, K.D.; Birbilis, N.; Davies, C.H.J. Revealing the relationship between grain size and corrosion rate of metals. *Scripta Mater.* **2010**, *63.12*, 1201–1204. [CrossRef]

24. Xiao, H.; Ye, W.; Song, X.; Ma, Y.; Li, Y. Evolution of akaganeite in rust layers formed on steel submitted to wet/dry cyclic tests. *Materials* **2017**, *10*, 1262–1275. [CrossRef] [PubMed]

25. Asami, K.; Kikuchi, M. In-depth distribution of rusts on a plain carbon steel and weathering steels exposed to coastal–industrial atmosphere for 17 years. *Corros. Sci.* **2003**, *45*, 2671–2688. [CrossRef]

26. Alcántara, J.; Chico, B.; Díaz, I.; De la Fuente, D.; Morcillo, M. Airborne chloride deposit and its effect on marine atmospheric corrosion of mild steel. *Corros. Sci.* **2015**, *97*, 74–88. [CrossRef]

27. Senthilnathan, A.; Dissanayake, D.; Chandrakumara, G.; Mantilaka, M.; Rajapakse, R.; Pitawala, H.; Nalin de Silva, K. Akaganeite nanorices deposited muscovite mica surfaces as sunlight active green photocatalyst. *R. Soc. Open Sci.* **2019**, *6*, 1–12. [CrossRef] [PubMed]

28. Verma, S.; Baig, R.N.; Nadagouda, M.N.; Varma, R.S. Oxidative CH activation of amines using protuberant lychee-like goethite. *Sci. Rep.* **2018**, *8*, 1–7. [CrossRef] [PubMed]

29. Kimura, M.; Kihira, H. Nanoscopic mechanism of protective-rusts formation on weathering steel surfaces. *SHINNITTETSU GIHO* **2004**, *91*, 77–81.

30. Xu, Q.; Gao, K.; Lv, W.; Pang, X. Effects of alloyed Cr and Cu on the corrosion behavior of low-alloy steel in a simulated groundwater solution. *Corros. Sci.* **2016**, *102*, 114–124. [CrossRef]

31. Melchers, R.E. Effect of small compositional changes on marine immersion corrosion of low alloy steels. *Corros. Sci.* **2004**, *46*, 1669–1691. [CrossRef]
32. Henriksen, J. The distribution of NaCl on Fe during atmospheric corrosion. *Corros. Sci.* **1969**, *9*, 573–576. [CrossRef]
33. Ma, Y.; Li, Y.; Wang, F. Corrosion of low carbon steel in atmospheric environments of different chloride content. *Corros. Sci.* **2009**, *51*, 997–1006. [CrossRef]
34. Ma, Y.; Li, Y.; Wang, F. The effect of β-FeOOH on the corrosion behavior of low carbon steel exposed in tropic marine environment. *Mater. Chem. Phys.* **2008**, *112*, 844–852. [CrossRef]
35. Misawa, T.; Hashimoto, K.; Shimodaira, S. The mechanism of formation of iron oxide and oxyhydroxides in aqueous solutions at room temperature. *Corros. Sci.* **1974**, *14*, 131–149. [CrossRef]
36. Yang, W.; Ni, R.-C.; Hua, H.-Z.; Pourbaix, A. The behavior of chromium and molybdenum in the propagation process of localized corrosion of steels. *Corros. Sci.* **1984**, *24*, 691–707. [CrossRef]
37. SA, P.; DP, L. Alloying effect of chromium on the corrosion behavior of low-alloy steels. *Meter. Trans.* **2013**, *54*, 1770–1778.
38. Jones, D.A. *Principles and Prevention of Corrosion*, 2nd ed.; Prentice Hall, Inc.: Upper Saddle River, NJ, USA, 1996; p. 217.

 *materials*

Article

Effect of Grain Size on the Corrosion Behavior of Fe-3wt.%Si-1wt.%Al Electrical Steels in Pure Water Saturated with CO$_2$

Gaetano Palumbo [1,*], Dawid Dunikowski [1], Roma Wirecka [2,3], Tomasz Mazur [2], Urszula Lelek-Borkowska [1], Kinga Wawer [4] and Jacek Banaś [1]

[1] Faculty of Foundry Engineering, Department of Chemistry and Corrosion of Metals, AGH University of Science and Technology, Mickiewicza St. 30, 30-059 Krakow, Poland; dundaw@agh.edu.pl (D.D.); lelek@agh.edu.pl (U.L.-B.); jbs@agh.edu.pl (J.B.)
[2] Academic Centre for Materials and Nanotechnology, AGH University of Science and Technology, Mickiewicza St. 30, 30-059 Kraków, Poland; roma.wirecka@gmail.com (R.W.); tmazur@agh.edu.pl (T.M.)
[3] Department of Condensed Matter Physics, Faculty of Physics and Applied Computer Science, AGH University of Science and Technology, Mickiewicza St. 30, 30-059 Krakow, Poland
[4] Łukasiewicz Research Network–Institute of Aviation, Al. Krakowska 110/114, 02-256 Warsaw, Poland; Kinga.Wawer@ilot.lukasiewicz.gov.pl
* Correspondence: gpalumbo@agh.edu.pl; Tel.: +48-12-888-27-63

Citation: Palumbo, G.; Dunikowski, D.; Wirecka, R.; Mazur, T.; Lelek-Borkowska, U.; Wawer, K.; Banaś, J. Effect of Grain Size on the Corrosion Behavior of Fe-3wt.%Si-1wt.%Al Electrical Steels in Pure Water Saturated with CO$_2$. *Materials* **2021**, *14*, 5084. https://doi.org/10.3390/ma14175084

Academic Editor: Marián Palcut

Received: 27 July 2021
Accepted: 1 September 2021
Published: 5 September 2021

Publisher's Note: MDPI stays neutral with regard to jurisdictional claims in published maps and institutional affiliations.

Copyright: © 2021 by the authors. Licensee MDPI, Basel, Switzerland. This article is an open access article distributed under the terms and conditions of the Creative Commons Attribution (CC BY) license (https://creativecommons.org/licenses/by/4.0/).

Abstract: The corrosion behavior of two silicon steels with the same chemical composition but different grains sizes (i.e., average grain area of 115.6 and 4265.9 μm^2) was investigated by metallographic microscope, gravimetric, electrochemical and surface analysis techniques. The gravimetric and electrochemical results showed that the corrosion rate increased with decreasing the grain size. The scanning electron microscopy/energy dispersive x-ray spectroscopy and X-ray photoelectron spectroscopyanalyses revealed formation of a more homogeneous and compact corrosion product layer on the coarse-grained steel compared to fine-grained material. The Volta potential analysis, carried out on both steels, revealed formation of micro-galvanic sites at the grain boundaries and triple junctions. The results indicated that the decrease in corrosion resistance in the fine-grained steel could be attributed to the higher density of grain boundaries (e.g., a higher number of active sites and defects) brought by the refinement. The higher density of active sites at grain boundaries promote the metal dissolution of the and decreased the stability of the corrosion product layerformed on the metal surface.

Keywords: silicon steel; electrical steel; grain size; electrochemical corrosion; carbon dioxide; AM-KPFM Volta potential measurements

1. Introduction

Electrical steels, also referred to as silicon steels (i.e., Si is the major additive element), are used as soft magnetic materials for construction of stators and rotors due to their magnetic properties and low cost [1,2]. The magnetic properties of the electrical steel are influenced by different parameters, such as sheet thickness, chemical composition and microstructure, and particularly by the grain size. Previous studies reported that large grain sizes are desired to improve the soft magnetic properties of the electrical steel [3,4]. Lee et al. [4] analyzed the magnetic properties of electrical steel as function of the grain size and found that samples with finer grains exhibited approximately 15% higher core loss W with little effect on the magnetic flux density B, compared to samples with larger grains. It should be noted that electrical steels with a low core loss and high magnetic flux density are preferable for electrical machinery cores from the magnetic properties point of view [4].

Stators and rotors are composed of hundreds of thin electrical steel sheets and used in the cores of electromagnetic devices [2]. Although this equipment is designed to avoid the

249

introduction of any liquids, they usually operate in severe working environments, such as high pressure, high temperature, and presence of aggressive gases (e.g., CO_2, H_2S), which can easily compromise their mechanical integrity. With time, the steam condenses into droplets of liquid inside this equipment. The CO_2 corrosion in the oil and gas industry is one of the greatest challenges [5,6]. The gaseous CO_2 dissolves in the condensed water, forming carbonic acid, which successively dissociates into bicarbonate and carbonate anions [5,6]. The combination of liquid water and CO_2 creates aggressive conditions, which may lead to severe corrosion attack, leading to their performance degradation and hence, compromising the functionality of the plant over time. The grain size plays an important role in the design of electrical steel. From the magnetic properties point of view, large grains are more benefical. Grain size has also a strong effect on the mechanical and corrosion properties of the steel [7–11]. The relationship between the grain size and mechanical properties of the steel is well defined by the Hall-Petch relationship. However, the correlation between the grain size and its corrosion behavior is still an open field for investigation. Onyeji et al. [8] studied the corrosion behavior of two X65 steels with the same chemical composition but different grain sizes in aerated and deaerated brine solutions. The authors reported that the steel with coarser grains showed a higher corrosion resistance in both solutions. Li et al. [12] observed that the corrosion resistance of nanocrystallized low-carbon steels in 0.05 M H_2SO_4 + 0.05 M Na_2SO_4 aqueous solution increased with decreasing the grain size. Palumbo et al. [13] observed that an increase in grain refinement leads to an increase in the volume fraction of intercrystalline areas such as grain boundaries and triple junctions. Many authors argued that the grain boundaries and triple junctions have higher energies compared to the bulk and, as such, are more chemically active with respect to the adjacent matrix [7,8,11,12,14–17]. Therefore, the grain refinement enhances the reactivity of the surface, which may cause a preferential dissolution of the grains [7,8,11,12,14–16,18]. However, it is worth mentioning that there is not an unanimous consensus regarding the effect of the grain size on the corrosion resistance of ferrous alloys. Some studies showed that the environment plays a crucial role. Wang et al. [7] reported that the grain refinement decreased the corrosion resistance of the low alloy steel in a 3.5 wt.% NaCl solution, but the same steel showed an improvement in corrosion resistance in a 0.1 M $NaHCO_3$ solution. Similar behavior was observed by Zeiger et al. [16]. The authors found that the grain refinement led to a decrease in the corrosion resistance of the steel in a Na_2SO_4 solution with pH = 1, but the corrosion resistance increased in a Na_2SO_4 solution with pH = 6. The little consensus reported in the literature is related to the difficulty of isolating the effect of the grain size from other microstructural changes introduced during the grain refinement processes such as, for example, rolling or plastic deformation. Consequently, a case-by-case study is needed to understand the corrosion effect of the grain size of a given metal in a given environment.

The objective of this work is to study the corrosion behavior of two electrical steel sheets with similar chemical composition, but different grain sizes in pure water saturated with CO_2. To this end, the study was carried out using weight loss and electrochemical measurements. The scanning electron microscopy-energy dispersive x-ray spectroscopy (SEM-EDS) and x-ray photoelectron spectroscopy (XPS) analyses were employed to characterize the corrosion product layer and to support the gravimetric and electrochemical results. Moreover, to highlight the micro-galvanic activities occurring at the grain boundaries and triple junctions on the metal surface, Volta potential measurements were performed.

2. Experimental Procedures

2.1. Materials

The study was performed on two types of silicon steels labeled 200 and 300. The samples were supplied by Łukasiewicz Research Network according to IEC 60404 8 5 standard. Both samples in as-received conditions were covered with a phosphate protective coating. Each time prior to a test, the coating was removed by grinding the surface with SiC abrasive paper up to 1200 grit and finishing the surface with levigated alumina, cleaned ultrasonically in absolute ethanol and dried before immersion in the tested solution.

The coating-free sample surface was analyzed on a spark spectrometer to identify the chemical composition of samples (Table 1).

Table 1. Chemical composition of the tested samples.

Element	Sample	
wt.%	200	300
C	0.011 ± 0.01	0.007 ± 0.01
Si	2.93 ± 0.17	3.24 ± 0.11
Mn	0.22 ± 0.03	0.18 ± 0.02
P	0.03 ± 0.01	0.026 ± 0.03
S	0.004 ± 0.001	0.003 ± 0.001
Al	0.91 ± 0.13	1.05 ± 0.08
Fe	Bal.	Bal.

2.2. Metallographic Analysis

The metallographic analysis was carried out by etching the samples with a nital solution (e.g., 4% HNO_3 solution) and then degreased with absolute ethanol and dried. Grain size measurements were performed according to the ASTM E112 [19] method using a LEICA DM4000 M LED microscope (Leica Microsystems, Wetzlar, Germany).

2.3. Gravimetric Measurement

Gravimetric measurements were carried out by immersing the samples for 24 h in pure water saturated with carbon dioxide at 25 °C. The solution was thermostated in a water bath. After the immersion time had elapsed, the specimens were removed and ultrasonically washed in ethanol, dried, and reweighed. The weight loss was determined using an analytical balance with the accuracy of ±0.1 mg. In each case, the experiment was conducted three times. The corrosion rate (*CR*) in $mm \cdot y^{-1}$ was obtained from the following equation [6]:

$$CR \ (mm \ y^{-1}) = \frac{87.6 \ \Delta m}{d \ A \ t} \tag{1}$$

where Δm is the weight loss, A is the surface of the sample (cm^2), d is the density ($7.87 \ g \ cm^{-3}$), and t is the immersion time (h).

2.4. Electrochemical Measurement

The electrochemical measurements were carried out with a Gamry 600 potentiostat (Gamry Instruments, Warminster, PA, USA). The electrochemical cell consisted of a working electrode, a saturated calomel reference electrode (SCE) and a platinum counter electrode. The experiments were performed at 25 °C in pure water saturated with carbon dioxide. The electrolyte conductivity was 190 $\mu S \ cm^{-1}$ and the pH was 4.12. The tests were carried out using electrochemical impedance spectroscopy (EIS) and potentiodynamic polarization (PDP). The EIS measurements were performed at an amplitude of 40 mV in the frequency range from 1 kHz to 10 mHz. Both measurements were carried out at intervals of 3, 6, 12, 18, and 24 h. The PDP measurements were performed by sweeping the potential from -1.0 to -0.4 V vs. SCE with a scan rate of 1 mV s^{-1} after holding the specimen at open circuit potential for 24 h in the tested solutions. The corrosion rate (*CR*) was calculated according to the ASTM G102 [20] using the following equation:

$$CR \ (mm \ y^{-1}) = \frac{3.27 \times 10^{-3} \ i_{corr} \ E_W}{d} \tag{2}$$

where i_{corr} is the corrosion current density, E_w is the equivalent weight of the metal, and d is the density of the metal. 3.27×10^{-3} is the conversion factor.

2.5. Surface Analysis

The morphological analyses were carried out on specimens exposed in the tested solution for 24 h, rinsed with deionized water and dried. The surface analysis was performed using SEM-EDS and XPS. The SEM-EDS investigation was performed using a JEOL scanning electron microscope aquiped with a IXRF EDS detector (JEOL, Inc., Peabody, MA, USA). The XPS analysis was performed with a PHI 5000 VersaProbe II spectrometer (ULVAC-PHI, Inc., Kanagawa, Japan) with an Al Kα monochromatic X-ray beam as described elsewhere [6].

The amplitude modulation Kelvin probe force microscopy (AM-KPFM) Volta potential ($\Delta\Psi$) analysis was carried out with a Dimension Icon XR (Bruker, Santa Barbara, CA, USA) working in the tapping mode, using platinum-iridium coated, electrically conductive SCM-PIT-V2 probes with cantilevers with a nominal spring constant of 3.0 N/m. AM-KPFM mode was used with 500 mV bias voltage and 100 nm lift height.

3. Results and Discussion

3.1. Metallographic Measurement

Figure 1 shows the microstructures and the grain size distribution of the tested steels, with the microstructure parameters summarized in Table 2. It is clear from the data that both samples show a similar chemical composition, with very little variation between different elements, but different microstructures. The 300 steel shows a coarse-grained microstructure with an average grain size of *circa* 4265.9 μm^2, whereas the 200 steel shows a fine-grained microstructure with an average grain size of *circa* 115.6 μm^2.

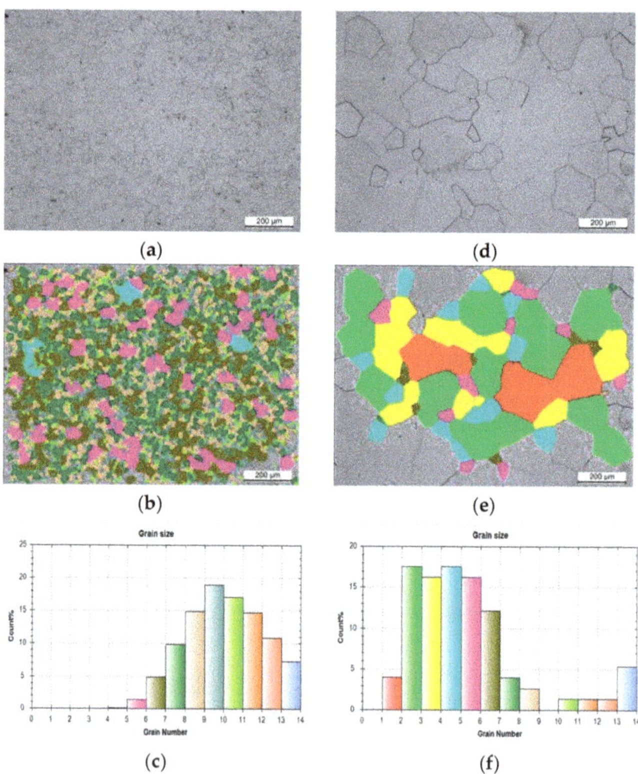

Figure 1. Microstructure and the histogram showing the grain size number distribution of 200 (**a–c**) and 300 steels (**d–f**).

Table 2. Grain size analysis of the steels.

200			300		
Average grain size number (G)	Average diameter (μm)	Average grain area (μm²)	Average grain size number (G)	Average diameter (μm)	Average grain area (μm²)
10.1 ± 0.23	10.8 ± 0.87	115.6 ± 20.9	4.9 ± 0.32	65.3 ± 7.49	4265.9 ± 499.7

3.2. Gravimetric Measurements

Table 3 shows the corrosion rates of the samples obtained from the weight loss measurements after 24 h of immersion in the tested solution. It follows from the data that the *CR* of the sample with fine grains is approximately two times higher compared to the sample with coarse grains. The analysis of the chemical composition (Table 1) shows that both specimens have a similar composition, with very little variation in concentration of the alloying elements. It can be inferred that the small differences observed in microalloying do not affect the corrosion rate of the metal and that the difference in the corrosion rate is therefore related to the microstructure. Similar results were also reported in [8,11,12,14–16,18]. Onyeji et al. [8] reported that the corrosion behavior of two X65 steels with identical chemical composition but different microstructures varied. The authors observed that the coarse-grained steel showed higher corrosion resistance compared to the steel with a fine grained microstructure. The authors suggested that this behavior was attributed to the enhanced reactivity of the surface after the grain refinement, which causes a preferential dissolution of the grains.

Table 3. Weight loss and corrosion rate of the steels after 24 h of immersion time.

Sample	Weight Loss (mg)	Corrosion Rate (mm y^{-1})
200	3.57 ± 0.25	0.25 ± 0.02
300	0.79 ± 0.04	0.12 ± 0.01

3.3. Electrochemical Measurements

Figures 2 and 3 show the EIS plots carried out under different immersion times at open circuit potential for 200 and 300 steels, respectively. As can be seen from the Nyquist and Bode plots (Figure 2a,c and Figure 3a,c), the conductivity of the solution was very low. This result is understandable since the tested electrolyte consisted of pure water saturated with CO_2 with a conductivity of approximately 190 μS cm^{-1}. However, as time increased, the conductivity of the solution increased, likely due to the release of ions into the bulk solution from the metal surface. To compare the corrosion behavior of both samples, the IR drop was manually compensated. The corrected plots (Figure 2b,d,f, and Figure 3b,d,f) were then fitted with the equivalent circuit displayed in Figure 4 Due to the imperfection of the metal surface, the double-layer capacitance (C_{dl}) was simulated using a constant phase element (CPE) [14]. The impedance of the CPE is described by the following equation [5,21]:

$$Z_{CPE} = \frac{1}{Q(j\omega)^n} \tag{3}$$

where Q stands for CPE constant, n is the exponent, j is the imaginary unit, and ω is the angular frequency at which Z reaches its maximum value.

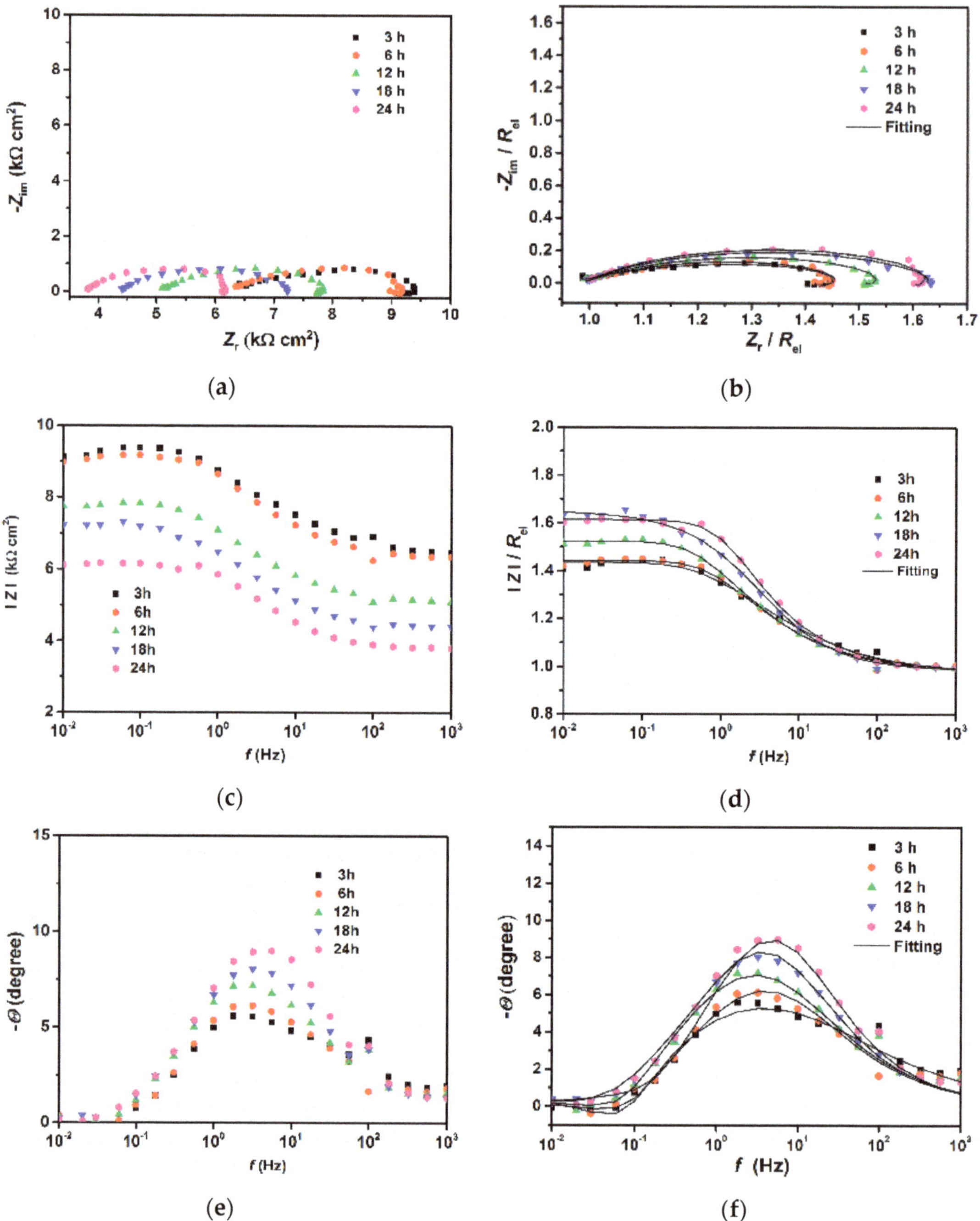

Figure 2. EIS plots obtained after different immersion times for the 200 steel before (**a,c,e**) and after (**b,d,f**) the IR drop correction.

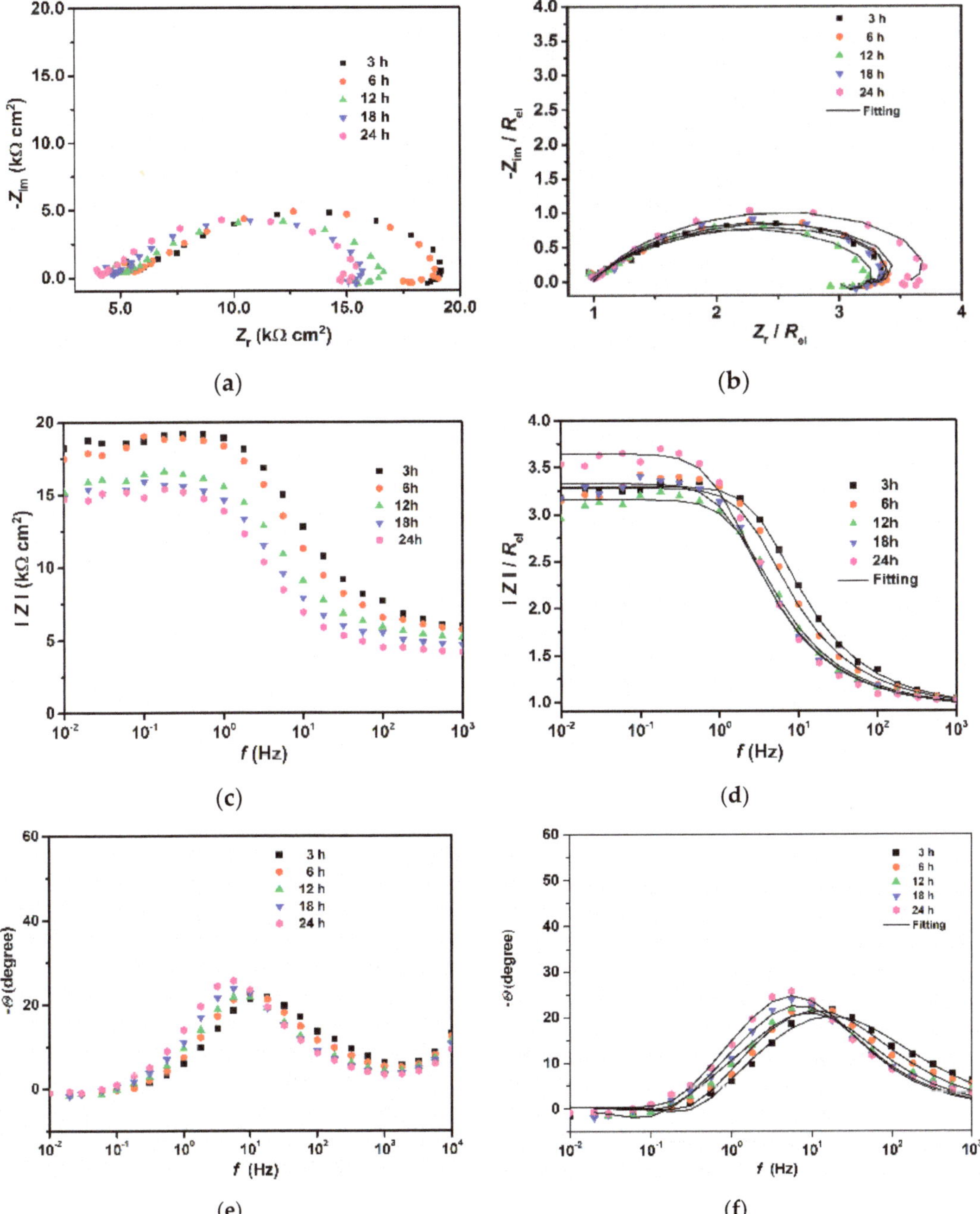

Figure 3. EIS plots obtained after different immersion times for the 300 steel before (**a–c**) and after (**d–f**) the IR drop correction.

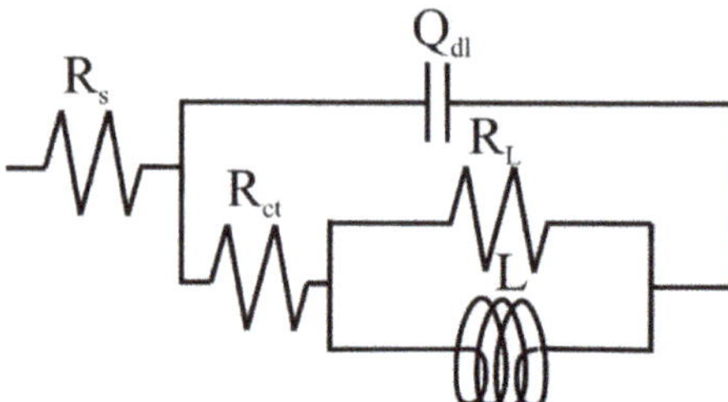

Figure 4. Equivalent circuit used to fit the EIS plots. Here, R_s is the electrolyte resistance, Q_{dl} is the constant phase element representing the double charge layer capacitance and R_{ct} is the charge transfer resistance. L and R_L represent the inductance and inductance resistance, respectively.

As can be seen from figures (Figure 2b,d,f and Figure 3b,d,f), the fitted results were similar to those obtained experimentally and the values of χ^2 were very low (Table 4), indicating that the equivalent circuit, employed to simulate the system under investigation, was the most appropriate one. It follows from the Nyquist diagrams (Figures 2c and 3c) that the shape of the curve did not change with the immersion time and exhibited a depressed semicircle in the whole frequency range due to the inherent charge transfer processes controlling the corrosion reactions. An inductive loop is also visible at low frequencies, likely due to the relaxation time of the intermediate adsorbed species.

Table 4. Electrochemical impedance parameters of the two steels in the tested solution after the IR drop correction.

Sample	Time (h)	R_s (Ω cm^2)	Q_1 (mΩ^{-1} s^n cm^{-2})	n	R_{ct} (Ω cm^2)	L (H cm^2)	R_L (Ω cm^2)	χ^2 (10^{-4})
	3	0.96	0.48 ± 0.06	0.57 ± 0.02	0.47 ± 0.03	0.27 ± 0.03	0.14 ± 0.011	1.05
	6	0.98	0.38 ± 0.03	0.59 ± 0.04	0.45 ± 0.05	0.15 ± 0.02	0.01 ± 0.005	0.99
200	12	0.98	0.36 ± 0.03	0.58 ± 0.05	0.54 ± 0.06	0.19 ± 0.04	0.01 ± 0.003	0.38
	18	0.97	0.26 ± 0.02	0.63 ± 0.03	0.65 ± 0.03	0.12 ± 0.02	0.04 ± 0.001	0.46
	24	0.99	0.19 ± 0.01	0.67 ± 0.05	0.64 ± 0.03	0.07 ± 0.01	0.07 ± 0.002	1.6
	3	0.95	0.04 ± 0.001	0.61 ± 0.04	2.32 ± 0.16	0.24 ± 0.01	0.70 ± 0.06	2.02
	6	0.98	0.04 ± 0.001	0.65 ± 0.03	2.35 ± 0.21	0.25 ± 0.02	0.69 ± 0.02	2.06
300	12	0.89	0.04 ± 0.003	0.69 ± 0.04	2.03 ± 0.22	0.64 ± 0.01	0.49 ± 0.02	1.99
	18	0.98	0.05 ± 0.005	0.71 ± 0.03	2.26 ± 0.15	0.36 ± 0.02	0.52 ± 0.03	1.41
	24	0.97	0.05 ± 0.003	0.72 ± 0.02	2.61 ± 0.28	0.36 ± 0.01	0.51 ± 0.05	0.94

The CO_2 gas dissolves in the solution forming carbonic acid, which successively dissociates into bicarbonate and carbonate anions, according to the following reactions:

$$CO_2 + H_2O_{(l)} \leftrightarrow H_2CO_3 \tag{4}$$

$$H_2CO_3 \leftrightarrow H^+ + HCO_3^- \tag{5}$$

$$HCO_3^- \leftrightarrow H^+ + 2CO_3^{2-} \tag{6}$$

In the presence of CO_2, the process is controlled by the three cathodic reactions [6,22,23]:

$$2H_2CO_3 + 2e^- \rightarrow H_2 + 2HCO_3^- \tag{7}$$

$$2HCO_3^- + 2e^- \rightarrow H_2 + 2CO_3^{2-} \tag{8}$$

$$2H^+ + 2e^- \rightarrow H_2 \tag{9}$$

The anodic reaction in a CO_2-saturated solution can be summarized by the multi-step dissolution of carbon steel [24]:

$$Fe + H_2O \rightarrow (FeOH)_{ads} + H^+ + e^- \tag{10}$$

$$(FeOH)_{ads} \rightarrow FeOH^{+} + e^{-} \tag{11}$$

$$FeOH^{+} + H^{+} \rightarrow Fe^{2+} + H_2O \tag{12}$$

The inductive loop, observed at low frequencies, is likely due to the adsorption of $(FeOH)_{ads}$ on the metal surface [24].

After 24 h of immersion the 300 steel showed a higher capacitive semicircle compared to the 200 steel (Figure 5) The EIS findings are in agreement with the gravimetric results. Since the corrosion resistance of a given metal is a function of the size of the capacitive loop, it follows from the figure that the 300 steel, with a wider Nyquist curve capacitive loop, shows a higher corrosion resistance than the 200 steel. The SEM-EDS and XPS analyses (Section 3.4) indicate that the higher corrosion resistance of the 300 steel was likely ascribed to formation of a more stable and/or compact protective layer comprised of Al_2O_3, SiO_2, Fe_2O_3 and traces of FeCO3. Since both materials had a similar chemical composition this behavior can be ascribed to different microstructure. Di Schino et al. also observed a similar result [9,10]. They studied the effects of the grain size on the corrosion behavior of refined austenitic stainless steel in a 5% H_2SO_4 boiling solution after 10 h of immersion. The authors reported that the corrosion rate decreased with increasing the grain size. They suggested that an increase of the grain boundary surface area due to grain refining, caused the passive film to become less stable, due to the defects concentrated in the grain boundaries. In their study, the layer formed on the coarse-grained steel was more stable and could thus provide more protection to the steel substrate by a blocking effect, thereby reducing the diffusion of the aggressive substances from the bulk solution to the metal surface.

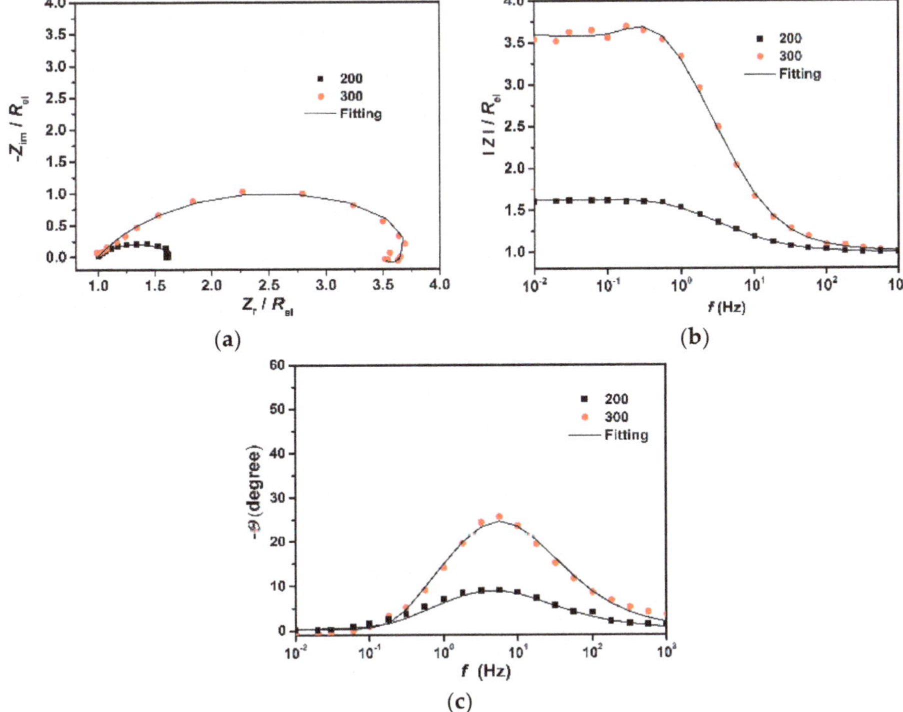

Figure 5. EIS plots comparing the two steels after 24 h of immersion and after the IR drop correction in the tested solution. (**a**) Nyquist, (**b**) Bode and (**c**) Phase angle.

Figure 6 and Table 5 show the potentiodynamic polarization measurements and corrosion kinetic parameters observed after 24 h of immersion in the tested solution, respectively. It is evident from the data that the corrosion current density of the steel with coarse grains was significantly lower compared to steel with fine grains. The corrosion rate of the metal with coarse grains was found to be 0.15 mm y^{-1} against 0.28 mm y^{-1} for the metal with fine grains, in agreement with the results observed with the gravimetric measurements. Furthermore, both the anodic and cathodic branches of the polarization curves were shifted towards the lower current densities for the steel with coarser grains. Li et al. [12] also reported a similar result. The authors suggested that the higher anodic current density observed for the steel with finer grains was related to the high energies that the atoms have at the grain boundaries. These atoms are the first to take part in the reaction. An increase in grain refinement leads to an increase in the volume fraction of intercrystalline areas such as grain boundaries and triple junctions [13]. Therefore, as the volume fraction of the grain boundary increases, the amount of the active atoms on the steel surface also increases, which in turn leads to an increase in the anodic current density. The result suggests that the stable layer, formed on the coarse grain size metal surface, hinders both the rate of the cathodic reaction (Equations (7)–(9)) and anodic dissolution (Equations (10)–(12)), by either covering part of the metal surface or blocking the active corrosion sites on the steel surface [5,6].

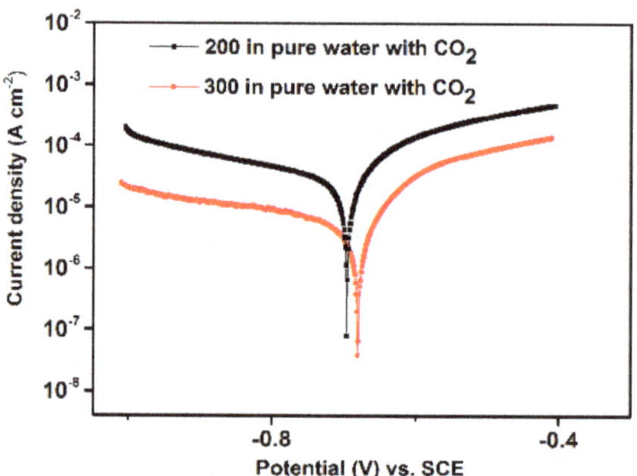

Figure 6. Potentiodynamic polarization curves obtained after 24 h of immersion in the tested solution.

Table 5. Potentiodynamic polarization parameters in the tested solution obtained after 24 h of immersion.

Sample	E_{corr} (V vs. SCE)	$-\beta_c$ (V dec^{-1})	i_{corr} (μA cm^{-2})	CR (mm y^{-1})
200	-0.696 ± 0.015	0.433 ± 0.087	24.44 ± 1.87	0.28 ± 0.02
300	-0.680 ± 0.011	0.588 ± 0.051	12.98 ± 2.57	0.15 ± 0.03

To confirm the correctness of the results obtained in pure water, the electrochemical experiments were also carried out in a 3.5 wt.% NaCl aqueous solution saturated with CO_2. The EIS and PDP results are displayed in Figures 7 and 8 and Tables 6 and 7, respectively. The results are in agreement with ones observed in pure water saturated with CO_2, which shows that the coarse-grained steel still displays a better corrosion resistance compared to the fine grained steel. In particular, it can be seen from the potentiodynamic experiments (Figure 8) that the 300 steel exhibits a pseudo-passive region, only, extends to a small range of potential (i.e., -612 to -586 mV vs. SCE). This result confirms that the coarse-grained steel tends to form a more stable corrosion product layer on the surface, which led to an increase in its corrosion resistance.

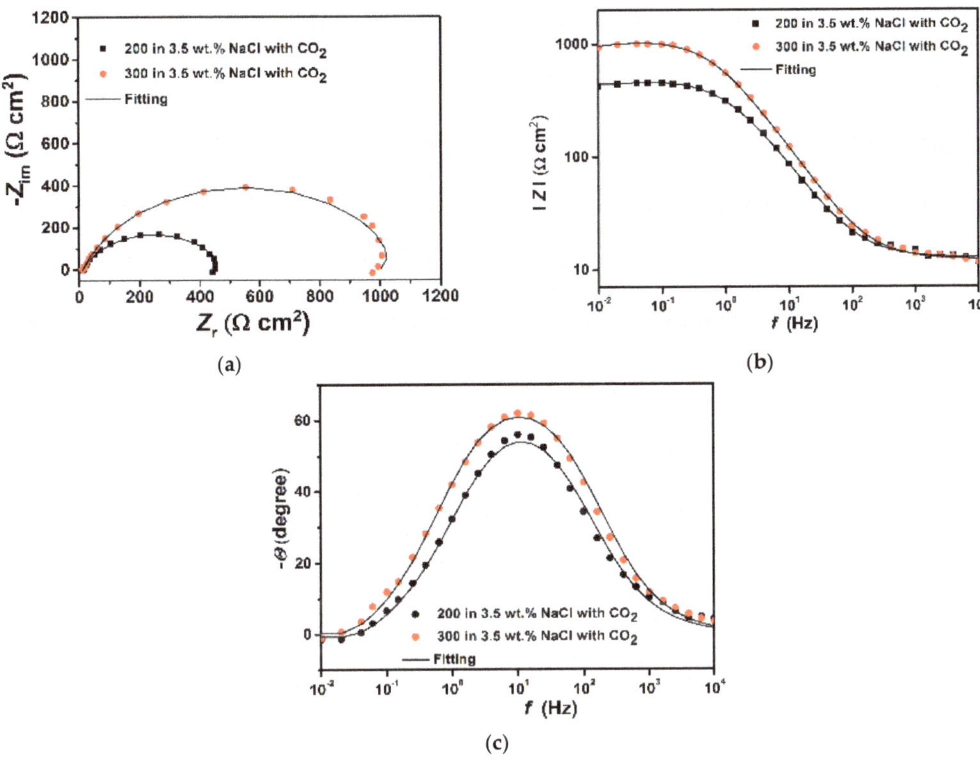

Figure 7. EIS plots comparing the two different samples after 24 h of immersion in a 3.5 wt.% NaCl saturated with CO_2. (**a**) Nyquist, (**b**) Bode, and (**c**) Phase angle.

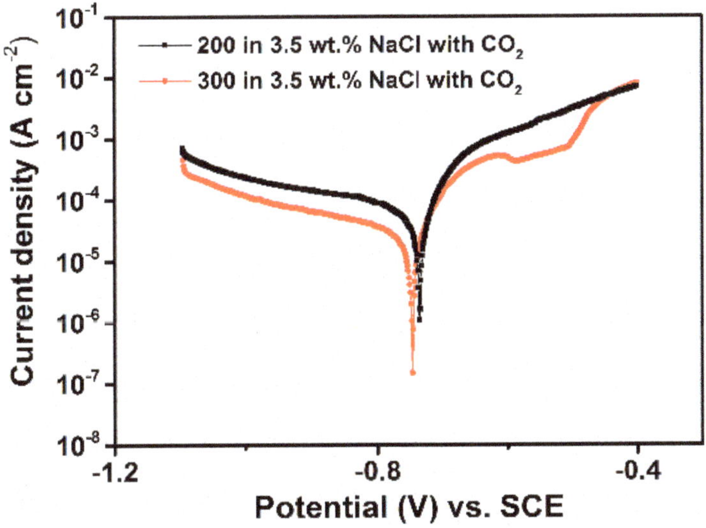

Figure 8. Potentiodynamic polarization curves obtained after 24 h of immersion in a 3.5 wt.% NaCl aqueous solution saturated with CO_2.

Table 6. Electrochemical impedance parameters of the steels after 24 h of immersion in a 3.5 wt.% NaCl aqueous solution saturated with CO_2.

Sample	Time (h)	R_s (Ω cm^2)	Q_1 (mΩ^{-1} s^n cm^{-2})	n	R_{ct} (Ω cm^2)	L (H cm^2)	R_L (Ω cm^2)	χ^2 (10^{-4})
200	24	10.04	0.46 ± 0.03	0.78 ± 0.03	421.10 ± 28.53	50.11 ± 9.07	49.22 ± 8.59	8.27
300	24	10.26	0.30 ± 0.06	0.80 ± 0.05	968.30 ± 40.89	60.10 ± 10.03	31.81 ± 6.89	5.69

Table 7. Potentiodynamic polarization parameters obtained after 24 h of immersion in a 3.5 wt.% NaCl aqueous solution saturated with CO_2.

Sample	E_{corr} (V vs. SCE)	$-\beta_c$ (V dec^{-1})	i_{corr} (μA cm^{-2})	CR (mm y^{-1})
200	-0.735 ± 0.019	0.517 ± 0.044	70.42 ± 2.34	0.82 ± 0.03
300	-0.746 ± 0.016	0.471 ± 0.067	29.43 ± 1.37	0.34 ± 0.02

3.4. Morphological Analysis

Figure 9 shows the surface morphology after 24 h of immersion in pure water saturated with CO_2. It is clear from figures that the surface morphology of the tested samples differs significantly. The severity of the corrosion process revealed the microstructure of the 200 steel (Figure 9a). By contrast, the surface of the 300 steel appears much smoother. The EDS analysis listed in Table 8 confirmed formation of a protective corrosion product layer, mainly composed of aluminum, oxygen, and other alloying elements. However, the data shows that the surface of the 300 steel was covered by a more homogeneous layer with the concentration of the abovementioned elements uniformly distributed over the entire surface. On the other hand, the surface of the 200 steel shows the presence of darker areas (e.g., red square 2 in Figure 9a), where the concentration of Al and O was higher compared to lighter areas (e.g., red square 1 from Figure 9a) and much similar to that found on the surface of the 300 steel. These results confirmed the observations reported in this study and are in agreement with the literature [9–16]. The results suggest that the better corrosion resistance performance observed of the coarse-grained steel was ascribed to its ability to form a more homogeneous and stable layer over the entire surface, which shields the material from the aggressive solution. Moreover, the morphological analysis also shows the presence of carbon on both surfaces. The presence of carbon was likely due to the precipitation of $FeCO_3$. Iron carbonate can form when the concentration of Fe^{2+} and CO_3^{2-} ions exceeds its solubility product (i.e., super-saturation):

$$Fe^{2+} + CO_3^{2-} \rightarrow FeCO_3 \tag{13}$$

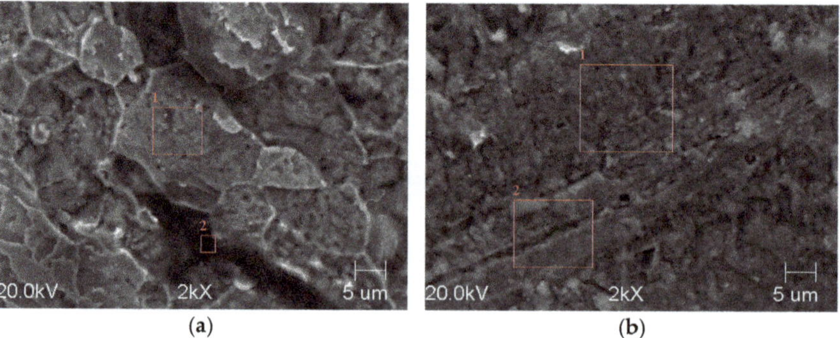

(a)　　　　　　(b)

Figure 9. SEM analysis on the sample (**a**) 200 and (**b**) 300, after 24 h of immersion. (The red square area corresponds to the area of the EDS analysis).

Table 8. EDS analysis of the tested samples.

| Element | 200 | | 300 | |
(wt.%)	1	2	1	2
C	11.97 ± 2.95	23.04 ± 2.17	21.78 ± 0.08	20.83 ± 1.35
O	7.11 ± 1.62	14.75 ± 1.04	10.71 ± 1.44	14.38 ± 1.03
Al	0.90 ± 0.08	3.54 ± 0.53	4.96 ± 0.55	7.10 ± 0.47
Si	2.60 ± 0.14	3.39 ± 0.30	3.19 ± 0.10	3.50 ± 0.05
P	0.09 ± 0.01	0.24 ± 0.02	0.10 ± 0.01	0.11 ± 0.01
S	0.08 ± 0.03	0.11 ± 0.02	0.11 ± 0.01	0.06 ± 0.01
Mn	0.13 ± 0.03	0.06 ± 0.01	0.07 ± 0.01	0.13 ± 0.01
Fe	77.12 ± 4.58	54.87 ± 1.42	59.12 ± 2.37	50.88 ± 5.55

The precipitation of $FeCO_3$ depends not only on the concentration of Fe^{2+} and CO_3^{2-} ions, but is also affected by other factors such as temperature, CO_2 partial pressure, and pH. Among the abovementioned factors, the pH of the solution can be regarded as one of the most influential factors [25]. Dugstad [25] reported that an increase in pH of the solution significantly reduced the concentration of Fe^{2+} ions required to exceed the $FeCO_3$ solubility product and therefore promoteing its precipitation. In this study, the pH increased as the immersion time increased, going from 4.12, at the beginning of the experiment, to 5.61 after 24 h of immersion. As such, making the precipitation of $FeCO_3$ more probable. Moreover, Dugstad [25] also reported that the concentration of Fe^{2+} ions is higher at the surface/solution interface compared to the bulk solution. Consequently, the concentration of Fe^{2+} ions required to promote the formation of $FeCO_3$ on the metal surface is lower and thus, increasing the likelihood of having $FeCO_3$ on surface. However, only small traces of $FeCO_3$ could be found as confirmed by the XPS analysis (Figure 10). Previous studies reported that $FeCO_3$ begins to decompose at temperatures below $100\ °C$ according to the following reaction [6,23,26]:

$$FeCO_3 \rightarrow FeO + CO_2 \tag{14}$$

In the presence of CO_2 or water vapor, FeO transforms into Fe_3O_4 [23,26].

$$3FeO + CO_2 \rightarrow Fe_3O_4 + CO \tag{15}$$

$$3FeO + H_2O \rightarrow Fe_3O_4 + H_2 \tag{16}$$

In the presence of oxygen, FeO and Fe_3O_4 transform into Fe_2O_3 [23,26].

$$4FeO + O_2 \rightarrow 2Fe_2O_3 \tag{17}$$

$$\text{In the air}: 4Fe_3O_4 + O_2 \rightarrow 6Fe_2O_3 \tag{18}$$

The XPS was employed to characterize the composition of the thin corrosion layer formed after 24 h of immersion at $25\ °C$ (Figure 10). The high-resolution spectra of aluminum (Al2p), silicon (Si2p), oxygen (O1s), carbon (C1s), and iron (Fe2p) are presented in Figure 10 and the binding energies and the corresponding quantification (%) of each peak are presented in Table 9. The deconvolution of the Al2p spectrum (Figure 10b,c) shows two peaks located at 74.7 and ~77.5 eV that can be attributed to Al_2O_3 and anhydrous Al_2O_3, respectively [27]. The high-resolution XPS spectrum of Si2p (Figure 10d,e) shows two main peaks at around 102 and 102.5 eV corresponding to Si-O/Si-O-C bond (Silicon oxide/Silicon oxycarbide). The silicon oxycarbide phase might have been formed on the surface as part of the protective layer when the material was exposed to the atmosphere. The O1s spectrum (Figure 10f,g) is fitted into three distinct peaks namely, 530, 531.6, and 533 eV. The peak observed at 530 eV was ascribed to O^{2-} and could be related to oxygen atoms bonded to the metal, (i.e., Al_2O_3, Fe_2O_3, and SiO_2 oxides) [6], whereas the peaks at 531.6 and 533 eV are associated with single bonded and double-bonded oxygen in $FeCO_3$ [6,28]. A fourth peak was observed at 354.6 eV for the 300 steel and can be attributed to the COO^- of $FeCO_3$. The C1s spectrum (Figure 10h,i) corroborated with the data observed for the O1s spectrum, which shows three peaks at 284.8, ~286.3, and ~288.5 eV. The 284.6 eV peak may correspond to secondary carbon [28], whereas the peaks at ~286.3 are ~288.5 eV are attributed to the C–O and C=O bonds of $FeCO_3$ [6]. As in the case of the O1S spectrum, for the 300 steel, a fourth peak was observed at 289.8 eV, which is characteristic of $FeCO_3$ [29]. The deconvoluted $Fe2p_{3/2}$ peaks (Figure 10j,k) could be attributed to α-Fe_2O_3 or/and γ-Fe_2O_3 oxides [6,30], likely due to the partial decomposition of iron carbonate (i.e., Equations (14)–(18)).

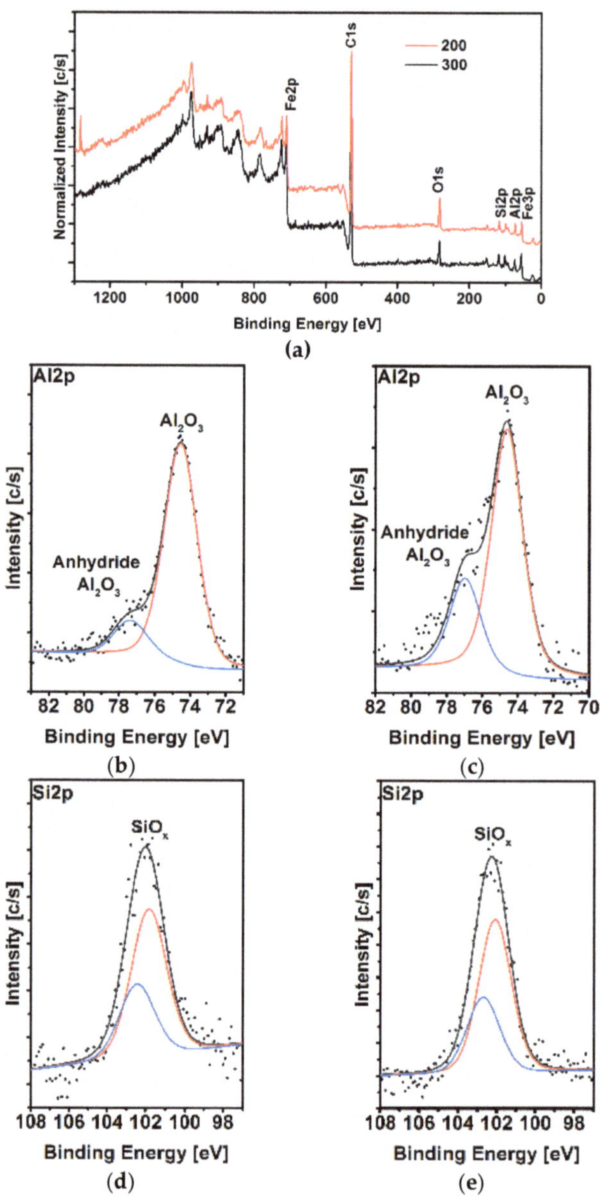

Figure 10. *Cont.*

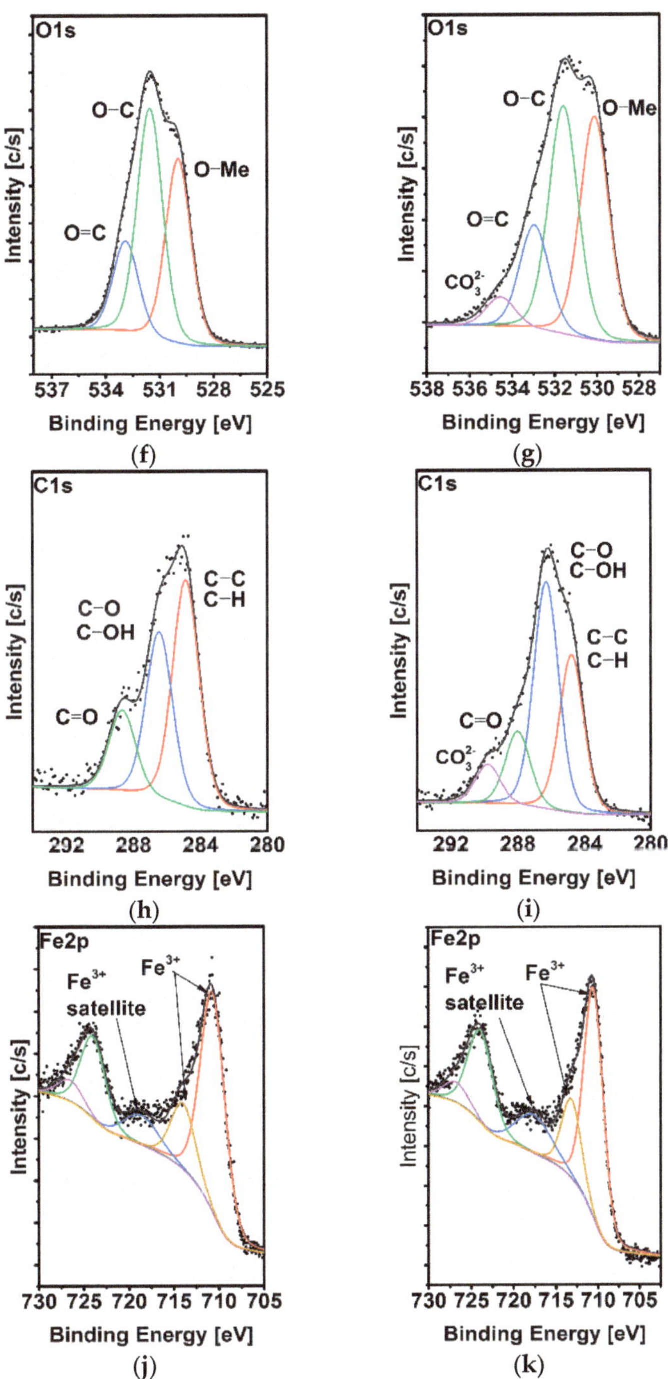

Figure 10. XPS analysis carried out after 24 h of immersion. Survey (**a**), 200 steel (**b,d,f,h,j**) and 300 steel (**c,e,g,i,k**).

Table 9. XPS analysis of sample surfaces after 24 h of immersion.

Peak Assignment	200		300	
	Binding Energy (eV)	%Area	Binding Energy (eV)	%Area
C1s	-	23.0	-	14.5
C–C, C–H	284.8	47.5	284.8	52.4
C–O, C–OH	286.5	35.2	286.3	24.5
C=O	288.7	17.3	288.5	14.2
CO_3^{2-}	-	-	289.8	8.9
Al2p	-	12.0	-	14.5
Al_2O_3	74.7	85.8	74.7	72.6
AlOOH	77.5	14.2	77.3	27.4
Si2p	-	4.6	-	7.2
SiO_x	101.8	66.7	102.1	66.7
splitting	102.5	33.3	102.7	33.3
O1s	-	53.7	-	54.0
O–Me	530	36.6	530.1	38.2
O–C	531.6	45.3	531.6	38.9
O=C	532.9	18.1	533.0	17.8
CO_3^{2-}	-	-	534.6	5.1
Fe2p	-	5.7	-	8.3
Fe^{3+}	710.5	44.5	710.7	51.0
Fe^{3+}	713.2	15.2	714.0	15.3
satellite	717.6	13.2	718.6	9.7
-	723.9	22.2	724.0	19.1
-	726.7	4.9	726.5	4.9

The XPS analysis is in agreement with the EDS analysis (Table 8), indicating that the corrosion products formed on both samples were mainly composed of Al_2O_3 and SiO_2, with traces of $FeCO_3/Fe_2O_3$.

To study the micro-galvanic activities occurring at the grain boundaries and triple junction on the metal surface, AM-KPFM measurements were carried out. AM-KPFM is a powerful technique for assessing the Volta potential ($\Delta\Psi$) of a metal surface. The $\Delta\Psi$ is a characteristic property of the metal surface and can provide an insight into the local electrochemical activities on the metal surface [31]. Figure 11 the topography and the corresponding Volta potential maps of the two tested steels. The Volta potential maps show darker color representing the anodic regions, whereas the lighter color representing cathodic regions. The $\Delta\Psi$ mapping clearly shows potential differences at grain boundaries and triple junctions, indicating higher electrochemical activity in these regions. The grain boundaries and triple junctions are characterized in the maps as lighter zones, thus representing the cathodic areas and displaying a relative $\Delta\Psi$ difference of *circa* +30 mV with respect to the adjacent matrix. The results confirmed that these regions are more active compared to the adjacent matrix, which makes them more susceptible to corrosion attack during the exposure to electrolytes. Therefore, the grain refinement enhances the reactivity of the surface, which could promote a preferential dissolution of the grains [7,8,11,12,14–16,18].

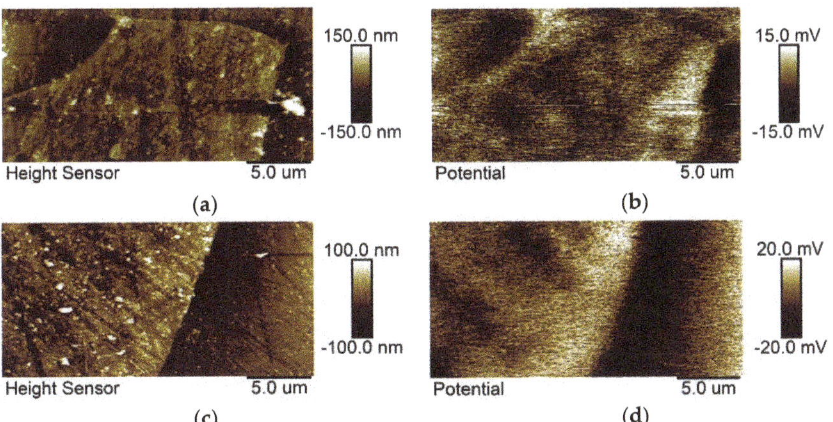

Figure 11. AM-KPFM analysis of the steels with the topography and the corresponding Volta potential map, respectively. Steel 200 (**a**,**b**); 300 (**c**,**d**).

4. Conclusions

The effect of the grain size on the corrosion behavior of two electric steels in pure water saturated by CO_2 can be summarized as follows:

- The gravimetric results indicated that the grain refinement decreases the corrosion resistance of the steel.
- The EIS measurements showed that the coarse-grained sample displayed a higher capacitive loop compared to fine-grained steel. This result is related to the formation of a thicker and more stable protective corrosion product layer.
- The potentiodynamic measurements showed that the corrosion current density of the coarse-grained steel was much smaller compared fine-grained steel. Both the anodic and cathodic current densities were found to be lower for the coarse-grained steel.
- The SEM-EDS and XPS analyses confirmed presence of a thicker and more homogenous protective layer on the coarse-grained steel, consisting mainly of Al_2O_3, SiO_2, and traces of $FeCO_3$.
- The Volta potential measurements showed potential differences between the grains and grain boundaries, indicating a higher electrochemical activity in these regions, which would couse a preferential dissolution of grains.

Author Contributions: G.P. conceived and designed this study. G.P. performed the electrochemical experiments, analyzed the experimental data, wrote and edited the manuscript; D.D. performed the gravimetric and SEM-EDS experiments; R.W. performed the XPS analysis; T.M. performed the AM-KPFM analysis; U.L.-B. prepared the samples; K.W. supplied materials, supervision, and funding acquisition; J.B. supervision and funding acquisition. All authors have read and agreed to the published version of the manuscript.

Funding: Part of this work was supported by AGH University of Science and Technology, Faculty of Foundry Engineering, Department of Chemistry and Corrosion of Metals, project no. 16.16.170.654. R.W. has been partly supported by the EU Project POWR.03.02.00-00-I004/16.

Institutional Review Board Statement: Not applicable.

Informed Consent Statement: Not applicable.

Data Availability Statement: The data presented in this study are available on request from the corresponding author.

Conflicts of Interest: The authors declare no conflict of interest. The funders had no role in the design of the study; in the collection, analyses, or interpretation of data; in the writing of the manuscript, or in the decision to publish the results.

References

1. Xu, Y.; Jiao, H.; Qiu, W.; Misra, R.D.K.; Li, J. Effect of cold rolling process on microstructure, texture and properties of strip cast Fe-2.6% Si Steel. *Materials* **2018**, *11*, 1161. [CrossRef] [PubMed]
2. Xia, C.; Wang, H.; Wu, Y.; Wang, H. Joining of the laminated electrical steels in motor manufacturing: A Review. *Materials* **2020**, *13*, 4583. [CrossRef] [PubMed]
3. Petryshynets, I.; Kováč, F.; Petrov, B.; Falat, L.; Puchý, V. Improving the magnetic properties of non-oriented electrical steels by secondary recrystallization using dynamic heating conditions. *Materials* **2019**, *12*, 1914. [CrossRef] [PubMed]
4. Lee, K.; Park, S.; Huh, M.; Kim, J.; Engler, O. Effect of texture and grain size on magnetic flux density and core loss in non-oriented electrical steel containing 3.15% Si. *J. Magn. Magn. Mater.* **2014**, *354*, 324–332. [CrossRef]
5. Palumbo, G.; Górny, M.; Banaś, J. Corrosion inhibition of pipeline carbon steel (N80) in CO_2-saturated chloride (0.5 M of KCl) solution using gum arabic as a possible environmentally friendly corrosion inhibitor for shale gas industry. *J. Mater. Eng. Perform.* **2019**, *28*, 6458–6470. [CrossRef]
6. Palumbo, G.; Kollbek, K.; Wirecka, R.; Bernasik, A.; Górny, M. Effect of CO_2 partial pressure on the corrosion inhibition of N80 carbon steel by gum arabic in a CO_2-water saline environment for shale oil and gas industry. *Materials* **2020**, *13*, 4245. [CrossRef] [PubMed]
7. Wang, P.; Ma, L.; Cheng, X.; Li, X. Effect of grain size and crystallographic orientation on the corrosion behaviors of low alloy steel. *J. Alloys Compd.* **2021**, *857*, 158258. [CrossRef]
8. Onyeji, L.; Kale, G. Preliminary investigation of the corrosion behavior of proprietary micro-alloyed steels in aerated and deaerated brine solutions. *J. Mater. Eng. Perform.* **2017**, *26*, 5741–5752. [CrossRef]
9. Di Schino, A.; Barteri, M.; Kenny, J.M. Grain size dependence of mechanical, corrosion and tribological properties of high nitrogen stainless steels. *J. Mater. Sci.* **2003**, *38*, 3257–3262. [CrossRef]
10. Di Schino, A.; Kenny, J.M. Effects of the grain size on the corrosion behavior of refined AISI 304 austenitic stainless steels. *J. Mater. Sci. Lett.* **2002**, *21*, 1631–1634. [CrossRef]
11. Ralston, K.D.; Birbilis, N. Effect of grain size on corrosion: A review. *Corrosion* **2010**, *66*, 075005. [CrossRef]
12. Li, Y.; Wang, F.; Liu, G. Grain size effect on the electrochemical corrosion behavior of surface nanocrystallized low-carbon steel. *Corrosion* **2004**, *60*, 891–896. [CrossRef]
13. Palumbo, G.; Thorpe, S.; Aust, K. On the contribution of triple junctions to the structure and properties of nanocrystalline materials. *Scr. Metall. Mater.* **1990**, *24*, 1347–1350. [CrossRef]
14. Soleimani, M.; Mirzadeh, H.; Dehghanian, C. Effect of grain size on the corrosion resistance of low carbon steel. *Mater. Res. Express* **2019**, *7*, 016522. [CrossRef]
15. Chen, Y.T.; Zhang, K.G. Influence of grain size on corrosion resistance of a HSLA steel. *Adv. Mater. Res.* **2012**, *557–559*, 143–146. [CrossRef]
16. Zeiger, W.; Schneider, M.; Scharnweber, D.; Worch, H. Corrosion behaviour of a nanocrystalline FeA18 alloy. *Nanostructured Mater.* **1995**, *6*, 1013–1016. [CrossRef]
17. Wang, S.G.; Shen, C.B.; Long, K.; Zhang, T.; Wang, F.H.; Zhang, Z.D. The electrochemical corrosion of bulk nanocrystalline ingot iron in acidic sulfate solution. *J. Phys. Chem. B* **2006**, *110*, 377–382. [CrossRef] [PubMed]
18. Seikh, A.H. Influence of heat treatment on the corrosion of microalloyed steel in sodium chloride solution. *J. Chem.* **2013**, *2013*, 587514. [CrossRef]
19. ASTM International. *ASTM-E112. Standard Test Methods for Determining Average Grain Size*; ASTM-E112: West Conshohocken, PA, USA, 2013.
20. ASTM International. *ASTM-G102. Standard Practice for Calculation of Corrosion Rates and Related Information from Electrochemical Measurements*; ASTM-G102: West Conshohocken, PA, USA, 1994.
21. Palumbo, G.; Berent, K.; Proniewicz, E.; Banaś, J. Guar gum as an eco-friendly corrosion inhibitor for pure aluminium in 1-M HCl solution. *Materials* **2019**, *12*, 2620. [CrossRef]
22. Mustafa, A.H.; Ari-Wahjoedi, B.; Ismail, M.C. Inhibition of CO_2 corrosion of X52 steel by imidazoline-based inhibitor in high pressure CO_2-water environment. *J. Mater. Eng. Perform.* **2012**, *22*, 1748–1755. [CrossRef]
23. Islam, M.A.; Farhat, Z.N. Characterization of the corrosion layer on pipeline steel in sweet environment. *J. Mater. Eng. Perform.* **2015**, *24*, 3142–3158. [CrossRef]
24. Zhang, G.; Cheng, Y. Electrochemical characterization and computational fluid dynamics simulation of flow-accelerated corrosion of X65 steel in a CO_2-saturated oilfield formation water. *Corros. Sci.* **2010**, *52*, 2716–2724. [CrossRef]
25. Dugstad, A. Fundamental aspects of CO_2 metal loss corrosion—Part 1: Mechanism. In *Corrosion 2006*; NACE International: San Diego, CA, USA, 2006; p. 18.
26. Heuer, J.; Stubbins, J. An XPS characterization of $FeCO_3$ films from CO_2 corrosion. *Corros. Sci.* **1999**, *41*, 1231–1243. [CrossRef]
27. Barrera, A.; Tzompantzi, F.; Campa-Molina, J.; Casillas, J.E.; Pérez-Hernández, R.; Ulloa-Godinez, S.; Velásquez, C.; Arenas-Alatorre, J. Photocatalytic activity of Ag/Al_2O_3-Gd_2O_3 photocatalysts prepared by the sol–gel method in the degradation of 4-chlorophenol. *RSC Adv.* **2018**, *8*, 3108–3119. [CrossRef]
28. Dong, B.; Zeng, D.; Yu, Z.; Cai, L.; Shi, S.; Yu, H.; Zhao, H.; Tian, G. Corrosion mechanism and applicability assessment of N80 and 9Cr steels in CO_2 auxiliary steam drive. *J. Mater. Eng. Perform.* **2019**, *28*, 1030–1039. [CrossRef]

29. Garcia, S.; Rosenbauer, R.; Palandri, J.; Maroto-Valer, M. Sequestration of non-pure carbon dioxide streams in iron oxyhydroxide-containing saline repositories. *Int. J. Greenh. Gas Control.* **2012**, *7*, 89–97. [CrossRef]

30. Singh, A.; Ansari, K.R.; Quraishi, M.A.; Lgaz, H. Effect of electron donating functional groups on corrosion inhibition of J55 steel in a sweet corrosive environment: Experimental, density functional theory, and molecular dynamic simulation. *Materials* **2018**, *12*, 17. [CrossRef]

31. Örnek, C.; Engelberg, D. SKPFM measured Volta potential correlated with strain localisation in microstructure to understand corrosion susceptibility of cold-rolled grade 2205 duplex stainless steel. *Corros. Sci.* **2015**, *99*, 164–171. [CrossRef]

Article

Effect of Sub-Zero Treatments and Tempering on Corrosion Behaviour of Vanadis 6 Tool Steel

Peter Jurči [1,*], Aneta Bartkowska [2], Mária Hudáková [1], Mária Dománková [1], Mária Čaplovičová [3] and Dariusz Bartkowski [4]

[1] Institute of Materials Science, Faculty of Materials Science and Technology in Trnava, Slovak University of Technology, J. Bottu 25, 917 24 Trnava, Slovakia; maria.hudakova@stuba.sk (M.H.); maria.domankova@stuba.sk (M.D.)

[2] Institute of Materials Science and Engineering, Faculty of Materials Engineering and Technical Physics, Poznan University of Technology, ul. Jana Pawła II 24, 61-138 Poznan, Poland; aneta.bartkowska@put.poznan.pl

[3] Laboratories of STU Centre for Nanodiagnostics, Slovak University of Technology in Bratislava, Vazovova 5, 812 43 Bratislava, Slovakia; maria.caplovicova@stuba.sk

[4] Institute of Materials Technology, Faculty of Mechanical Engineering, Poznan University of Technology, ul. Piotrowo 3, 61-138 Poznan, Poland; dariusz.bartkowski@put.poznan.pl

* Correspondence: p.jurci@seznam.cz

Abstract: Sub-zero treatment of Vanadis 6 steel resulted in a considerable reduction of retained austenite amount, refinement of martensite, enhancement of population density of carbides, and modification of precipitation behaviour. Tempering of sub-zero-treated steel led to a decrease in population density of carbides, to a further reduction of retained austenite, and to precipitation of M_3C carbides, while M_7C_3 carbides precipitated only in the case of conventionally quenched steel. Complementary effects of these microstructural variations resulted in more noble behaviour of sub-zero-treated steel compared to the conventionally room-quenched one, and to clear inhibition of the corrosion rate at the same time.

Keywords: ledeburitic steel; sub-zero treatment; microstructure; potentiodynamic polarisation; corrosion resistance

Citation: Jurči, P.; Bartkowska, A.; Hudáková, M.; Dománková, M.; Čaplovičová, M.; Bartkowski, D. Effect of Sub-Zero Treatments and Tempering on Corrosion Behaviour of Vanadis 6 Tool Steel. *Materials* 2021, *14*, 3759. https://doi.org/10.3390/ma14133759

Academic Editor: Thomas Fiedler

Received: 5 June 2021
Accepted: 28 June 2021
Published: 5 July 2021

Publisher's Note: MDPI stays neutral with regard to jurisdictional claims in published maps and institutional affiliations.

Copyright: © 2021 by the authors. Licensee MDPI, Basel, Switzerland. This article is an open access article distributed under the terms and conditions of the Creative Commons Attribution (CC BY) license (https://creativecommons.org/licenses/by/4.0/).

1. Introduction

Chromium-vanadium (Cr-V) ledeburitic steels are widely used as materials for the manufacturing of the powder compaction dies, seamless tube pilgering, press tools, paper cutters, extruders, and metal cutting tools, because of the attractive combination of the hardness, wear resistance, and toughness. Generally, these materials obtain their properties through appropriate heat treatment. The standard heat treatment of Cr-V ledeburitic tool steels is comprised of austenitizing, quenching, and double (or triple) tempering, which results in a hardness of around 60 HRC, high compressive strength, excellent wear resistance, and relatively good toughness when properly treated.

Sub-zero treatment (SZT) was introduced to the industries in the late 1950s. This kind of treatment is defined as a process which is carried out at temperatures from 0 to $-269\ °C$. The main benefits of the SZT have been reported as elevated hardness [1–4], additional wear resistance [2,3,5–11], and better dimensional stability [12]. In some cases, the toughness and fracture toughness also manifested certain, but very limited, improvement [6,13].

The above-mentioned ameliorations originate from several sources. The latest investigations arrived at general conclusions, finding that the application of SZT results in the following four main microstructural changes [1,9,11,14–19]:

i. Sub-zero-treated materials contain significantly reduced retained austenite (γ_R) amounts, as a result of isothermal and time-dependent martensitic transformation, which takes place during the SZT.

269

ii. The martensite of SZT steel is refined compared to that developed by conventional room-temperature quenching.

iii. Sub-zero treatment leads to acceleration of precipitation of transient carbides during either the re-heating to room temperature or short-term storage at room temperature.

iv. The last consequence of the SZT is the considerably enhanced number and population density of small globular carbides (SGCs). The formation of the SGCs at cryogenic temperature is time-dependent and follows the reduction of the retained austenite amount well.

The corrosion resistance is not commonly considered as a key property for tool materials. This is due to the fact that the dominant number of technological applications is conducted in corrosion-friendly environments, or, if not, there exist surfacing techniques (such as physical vapour deposition, for instance) through which a sufficiently high corrosion resistance of tools can be ensured. Despite this, it might be desirable to keep at least acceptable corrosion behaviour of tools made of Cr and Cr-V ledeburitic tool steels in some applications, where, for instance, surface techniques fail for several reasons. These applications may comprise industrial branches, such as mining, earth-handling, milling, powder compaction, mineral processing, etc. Here, high processing reliability and sufficient tools' durability require materials with not only excellent wear performance but also with at least acceptable corrosion resistance.

However, the characterisation of the effect of SZT on the corrosion behaviour of Cr and Cr-V ledeburitic steels is almost completely lacking. There are only a few relevant studies published on this topic, and the obtained results manifest clear inconsistency. Amini et al. [20] reported on the worsened corrosion behaviour of 1.2080 grade (AISI D3) steel due to the SZT in liquid nitrogen for 24–48 h. According to their consideration, the worsening of corrosion resistance can be ascribed to the increased carbide percentage, which decreases the number of solutionised chromium atoms in the martensite, as well as to the increased martensite/carbide interfaces (galvanic cell areas). Hill et al. [21], on the contrary, recorded an improvement of corrosion resistance of sub-zero-treated ($-196\,°C/15$ min) X190CrVMo 20-4 ledeburitic steel when tested in 0.5 molar sulphuric acid solution.

The present study attempts to overcome the limitation of the lack of data on the corrosion behaviour of sub-zero-treated ledeburitic tool steels. It describes the effects of different sub-zero treatment temperatures (-75, -140, -196, and $-269\,°C$) and tempering regimes on the corrosion behaviour of Cr-V ledeburitic steel Vanadis 6. Conventionally treated steel has been used as a reference. In addition, the relationships between the microstructures, heat treatment parameters, and corrosion behaviour are presented and thoroughly discussed.

2. Materials and Experimental Methods

2.1. Materials and Processing

The tool steel Vanadis 6, with chemical composition as shown in Table 1, was used for the examinations.

Table 1. Chemical composition of the experimental steel.

Element (Mass %)	C	Si	Mn	Cr	V	Mo	Fe
Content	2.1	1.0	0.4	6.8	5.4	1.5	balance

Specimens of the studied steel with a size of 20 mm × 20 mm × 7 mm were gradually heated (step (1) in Figure 1) up to the austenitizing temperature of 1050 °C in a vacuum, held there for 30 min (2), and quenched by nitrogen gas to room temperature (3). Then, the specimens were divided to five batches. The first one was immediately subjected to tempering treatment (7, 8). This treatment route is called conventional heat treatment, CHT. Immediately after quenching, the other four sets were moved to the cryogenic system, where they were cooled down at a cooling rate of 1 °C/min to a pre-determined

SZT temperature (4). The SZTs were carried out at temperatures of −75, −140, −196, or −269 °C, and for 17 h each (5). After that, the material was re-heated to room temperature, at a heating rate of 1 °C/min (6). Immediately after re-heating, sub-zero-treated specimens were subjected to tempering. Tempering was carried out at temperatures of 170, 330, 450, or 530 °C, 2 times each, for 2 h (7, 8).

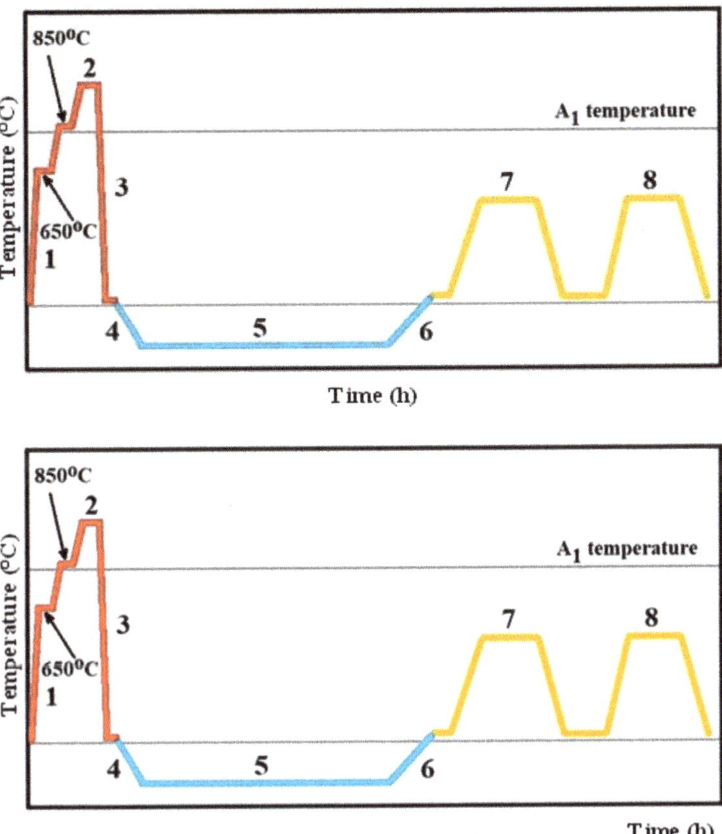

Figure 1. A schematic of the heat treatment schedules used.

2.2. Experimental Methods

The specimens for the microstructural examinations were mounted in conductive resin, ground on silicon carbide papers up to 1600 grit, and then progressively polished with 9, 3, and 1 μm diamond paste. The etchant used was 3% ethanol solution of picric acid. The secondary electron (SE) micrographs were acquired with a JEOL JSM 7600 F (Jeol Ltd., Tokyo, Japan) high-resolution field emission scanning electron microscope (SEM), operating at an acceleration voltage of 15 kV. The microstructural examinations were coupled with semi-quantitative analysis of chemical composition of phases by using energy-dispersive spectroscopy (EDS). The basis of the carbide particles' classification and the determination of their quantitative characteristics were elaborated on earlier, and are demonstrated in recent papers [14,22]. In brief, the eutectic carbides (ECs) are vanadium-based MC-particles, hence, they differ from the secondary M_7C_3-carbides (SCs) in terms of their chemistry. Therefore, EDS analysis was used to differentiate between these two carbide types. In order to differentiate between the small globular carbides (SGCs) and other particles, a classification according to their size was used, and the particles smaller than 500 nm were then denoted as SGCs. For the determination of the population densities of ECs, SCs, and SGCs, 25 randomly acquired SEM micrographs, at a magnification of 3000×, were used. Then, the mean values and standard deviations were calculated from the obtained experimental data.

Semi-quantitative chemical composition of phases of sub-micron size (metallic elements only) and the microstructure of both the martensite and the austenite were examined by transmission electron microscopy (TEM). Thin foils for TEM were prepared by sectioning specimens, with a special saw, to obtain pieces with a thickness of 0.1 mm, which were then ground mechanically on silicon carbide paper (1200 grit) to a thickness of approximately 0.02 mm. Final thinning was carried out using an electro-polisher Struers-Tenupol 5. A TEM JEM ARM 200 cF (Jeol Ltd., Tokyo, Japan) equipped with an energy-dispersive spectroscopy (EDS) detector was used for acquisition of micrographs as well as for the determination of phase compositions.

The phases in differently heat-treated specimens were identified from the X-ray diffraction (XRD) profiles by using a Phillips PW 1710 diffractometer. A filtered $Co_{\alpha 1,2}$ characteristic radiation, obtained at the voltage of 40 kV and current of 40 mA, has been used for acquisition of diffraction line profiles, within a range of 37–130° of two-theta angles. The analysis was coupled with Rietveld refinement of the X-ray line profiles. The retained austenite amount was determined following the ASTM E975-13 standard [23].

Corrosion resistance studies were carried out by the potentiodynamic polarisation measurements (TAFEL). The TAFEL measurements were completed by using the potentiostat-galvanostat ATLAS 0531 EU (Atlas-Sollich, Rębiechowo, Poland) and IA ATLAS SOLLICH (Atlas-Sollich, Rębiechowo, Poland). A platinum electrode was used as an auxiliary electrode, while a calomel electrode was used as the reference electrode and tested specimens as the working electrodes. The data were recorded by AtlasCorr (Atlas-Sollich, version 3.19) and AtlasLab (Atlas-Sollich, version 2.24) computer software. After corrosion tests, the tested specimens were examined by using SEM coupled with EDS.

Before testing, all the specimens were ground using metallographic emery papers with a grit size up to 2000, and finally, were polished using 3 µm diamond slurry. Just prior to measurements, the specimens were degreased in ethanol and warm air-dried. The 3.5 (in mass %) NaCl water solution was prepared using distilled water and high-purity reagent grade sodium chloride. Then, the solution was subjected to Ar gas bubbling for 30 min in order to its deaerate. For the corrosion tests, the temperature of the solution was kept constant at 22 °C.

The potentiodynamic polarisation measurements were carried out within the range of potentials of −1.5 to 1.5 V, and at a scan rate of 1 mV/s.

The corrosion rate for each specimen was estimated according to the ASTM G 102-89 standard [24]. The calculation was based on the validity of the Faraday's Law, and it followed Equation (1):

$$C_R = K_1 \cdot \frac{i_{cor}}{\rho} \cdot E_W \qquad (1)$$

where C_R is the corrosion rate (mm/year), I_{cor} is the current density (µA/cm^2), ρ is the specific density of the alloy (g/cm^3), $K_1 = 3.27 \cdot 10^{-3}$, (mm × g/µA × cm × year), and E_W is the equivalent weight of the experimental steel.

Equivalent weight, E_W, represents the mass of metal, in grams, that will be oxidised by the passage of one Faraday of electric charge. The value of E_W of the experimental material was calculated upon its known chemical composition, atomic weight of the major elements, W, the most common valence of a particular element n, and as a reciprocal value of the sum of the electron equivalents of all the major elements, Q, following Equations (2) and (3):

$$E_W = \frac{1}{\sum \frac{n_i \cdot f_i}{W_i}} \qquad (2)$$

$$Q = \sum \frac{n_i \cdot f_i}{W_i} \tag{3}$$

where f_i is the mass percentage of the ith element in the alloy, W_i is the atomic weight of the ith element in the alloy, and n_i is the most common valence of the ith element of the alloy. All the input data for the calculations are collected in Table 2. Additionally, it should be noted that an assumption of corrosion uniformity was adopted in the present study, in order to simplify the considerations.

Table 2. The mass percentage, f, atomic weight, W, most common valance, n, electron equivalents of the main elements in the Vanadis 6 steel, Q, and calculated electron equivalent of the steel, Q_{total}, as well as its equivalent weight, E_w.

Element	f (Mass %)	W (g)	n (-)	$Q = (n \times f)/W$
C	2.1	12.0107	4	0.699376389
Si	1.0	28.0855	4	0.142422246
Mn	0.4	54.9380	2	0.014561857
Cr	6.8	51.9961	3	0.392337118
Mo	1.5	95.9400	6	0.093808630
V	5.4	50.9415	5	0.530019729
Fe	82.8	55.8450	3	4.448025786
Q_{total}				6.320551755
E_W				15.821403555

In order to analyse the differences in nobility between the carbides and matrix, the Kelvin probe force microscopy (KPFM) has been adopted. This technique enables to distinguish between local potentials of different phases, at the sub-micron level. Hence, it is suitable for analysing fine-grained PM ledeburitic steels. The analyses were performed at The University of Manchester, Department of Materials Corrosion, by using a Multimode 8 instrument (Bruker), in the amplitude modulated mode. The imaging parameters were the following: 50×50 µm scans, potential maps obtained from a 50 nm lift height, using a 0.3 Hz scan rate, 512 points per line, 256 lines, and height images obtained using 100 nm peak force amplitude.

3. Results

3.1. Microstructure

SEM images (Figure 2) show the microstructures of the Vanadis 6 steel obtained by the conventional quenching and by different sub-zero treatments. The steel after conventional room temperature quenching is composed of martensitic matrix with the presence of a relatively large retained austenite amount, and of three carbide types (Figure 2a). The carbides are, according to the classification reported in [14]: eutectic carbides (ECs), secondary carbides (SCs), and small globular carbides (SGCs). The character of the matrix microstructure does not change significantly with the application of sub-zero treatment, at the magnification used for the SEM observations (Figure 2b–e). The formations of retained austenite become almost invisible after SZT, due to the significant reduction of γ_R amount due to this kind of treatment. Alternatively, sub-zero-treated steel differs from that after CHT in terms of the number and population density of carbide particles (compare the SEM micrograph in Figure 2a with micrographs in Figure 2b–e). This concerns mainly the SGCs: the population density of these particles is much higher after SZT, while the SZT does not alter the population densities of ECs and SCs, as demonstrated recently [4,13,19].

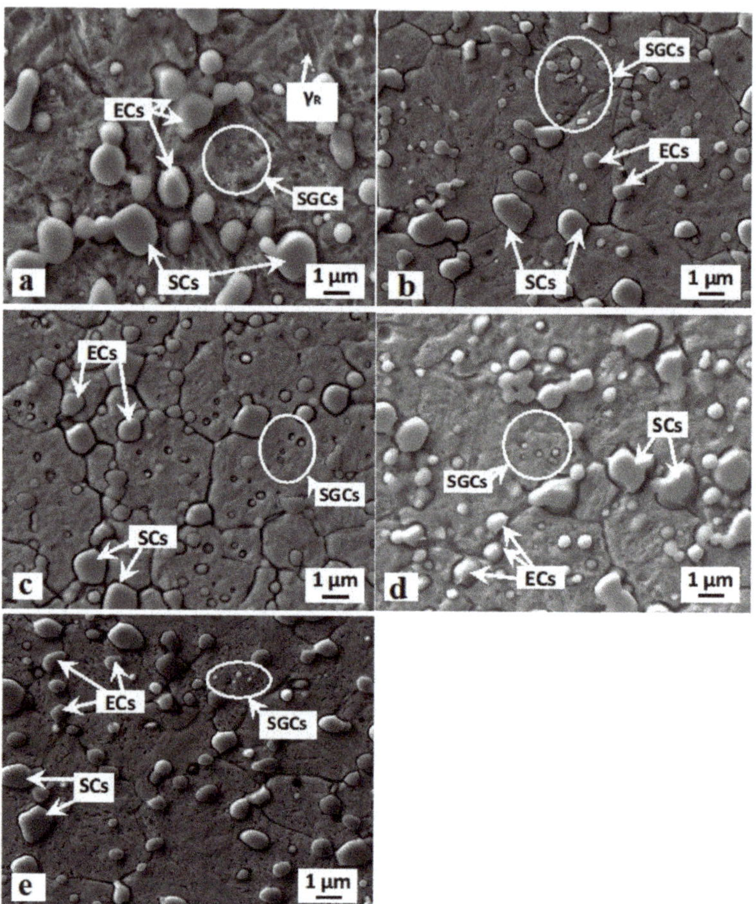

Figure 2. Scanning electron microscopy (SEM) micrographs showing the microstructure of Vanadis 6 ledeburitic steel after conventional room temperature quenching (**a**) and after SZT at −75 °C (**b**), −140 °C (**c**), −196 °C (**d**), and −269 °C (**e**).

Figure 3 summarizes the population densities of SGCs, which were obtained by different combinations of SZT and tempering. It is shown that the tempering after CHT does not influence the population density of these particles, while the tempering after SZT makes their population density lower than that obtained by SZT without tempering. Despite this fact, the population density of SGCs is much higher than what can be obtained by CHT and tempering. It is also shown that the application of SZT at −140 °C acts in the most effective way in the enhancement of the SGCs population density, and the population densities obtained by other sub-zero treatments are much lower at equal tempering temperatures.

Figure 4 shows the TEM images of specimens after CHT and after CHT with subsequent SZT at −196 °C. CHT produces a needle-like martensitic microstructure with a relatively large amount of retained austenite. The width of martensitic needles is typically in the range of 100–200 nm (Figure 4a). The retained austenite formations are located at the interfaces of martensitic domains. Their width range is much wider than that of the martensite: some of the austenite formations have a width of a few tens of nanometres, while others are much greater, as illustrated in Figure 4b. The presence of γ_R is confirmed by diffraction patterns in Figure 4c. The martensite produced by SZT manifests considerable refinement compared to that developed by CHT. The typical width of martensitic domains ranges between 50 and 150 nm (Figure 4d). It is also shown here that the size of the martensitic domains manifests a great level of variability—there are some coarser domains visible in the micrograph, while in some sites, the domains are much smaller. A much higher amount of dislocations are generated within the martensite, as a result of plastic deformation that takes place during the isothermal hold at cryotemperatures. The retained austenite amount is significantly reduced by application of SZT. Additionally, the size of γ_R formations is much smaller compared to the state after CHT (Figure 4e). The presence of retained austenite at the interfaces of martensitic domains is confirmed by diffraction patterns in Figure 4f.

Figure 5 shows X-ray diffraction spectra obtained on the conventionally quenched specimen and on specimens with SZT at −140 and −269 °C. All the spectra contain the peaks of the martensite, retained austenite, and major carbides. In the diffraction profile of the CHT specimen, there are the peaks of $(111)\gamma$, $(200)\gamma$, $(220)\gamma$, and $(311)\gamma$, all well-visible. On the other hand, the last two peaks disappear almost completely from the XRD spectra of SZT steel, suggesting a significant reduction of retained austenite amounts after this kind of treatment.

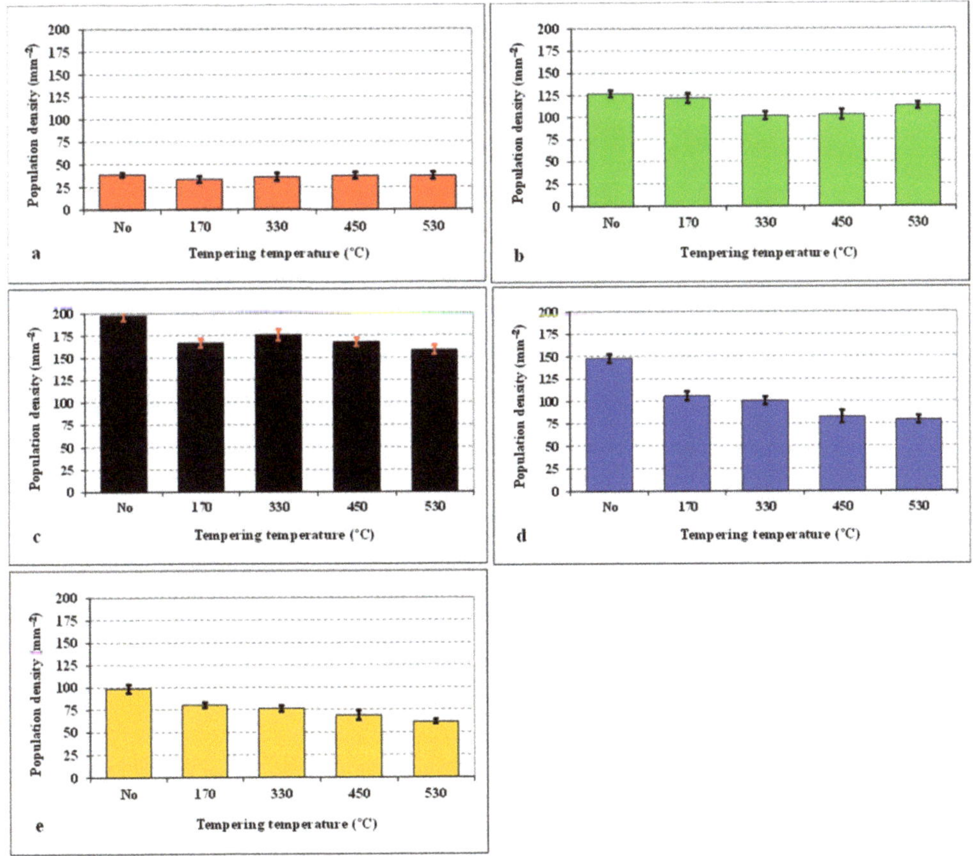

Figure 3. Population density of small globular carbides (SGCs) for differently sub-zero treated (SZT) specimens as a function of tempering temperature: (**a**) conventional heat treatment (CHT), (**b**) SZT −75 °C, (**c**) SZT −140 °C, (**d**) SZT −196 °C, (**e**) SZT −269 °C.

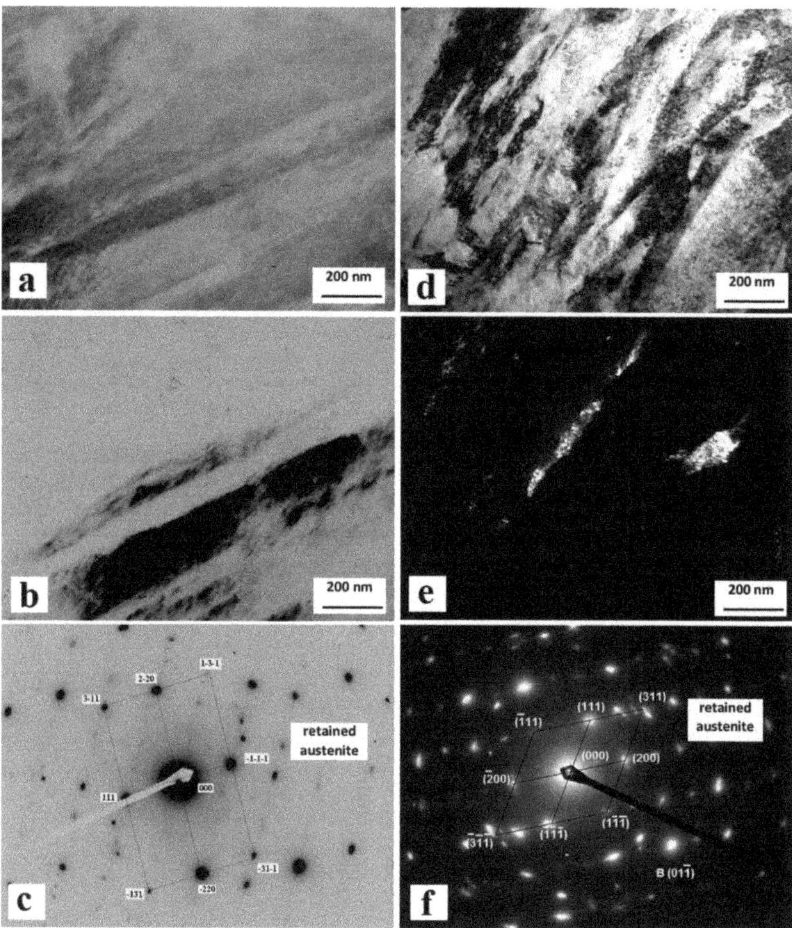

Figure 4. Transmission electron microscopy (TEM) micrographs of the CHT specimen (**a–c**) and the specimen that was subjected to SZT at −196 °C (**d–f**). (**a**) Bright-field image showing martensitic needle microstructure with retained austenite at needle interfaces, (**b**) corresponding dark-field image showing the retained austenite, (**c**) diffraction patterns of the retained austenite, (**d**) bright-field image showing martensitic microstructure with a small amount of retained austenite at the interfaces of martensitic domains, (**e**) corresponding dark-field image, (**f**) diffraction patterns of the retained austenite.

Variations in the retained austenite amounts for CHT steel, and for steel samples subjected to different SZTs, as a function of tempering temperature are summarised in Figure 6. It is shown that the application of SZT reduces the γ_R amount to one tenth—one fourth compared to the state after room-temperature quenching. Tempering at low temperatures does not influence the γ_R amounts. Alternatively, tempering in the secondary hardening temperature range evokes almost complete retained austenite decomposition in the case of CHT steel. The application of SZT, in addition, accelerates the decomposition of γ_R. Therefore, the amounts of metastable retained austenite lie below the detection limit of XRD after tempering at 450 °C and higher, in the case of the steel SZT in either liquid nitrogen or in liquid helium.

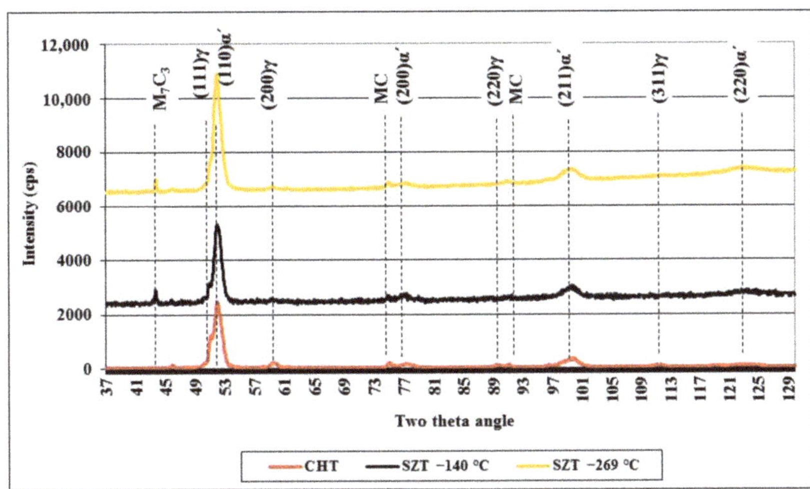

Figure 5. X-ray diffraction spectra of CHT steel and steel after SZT at −140 and −269 °C.

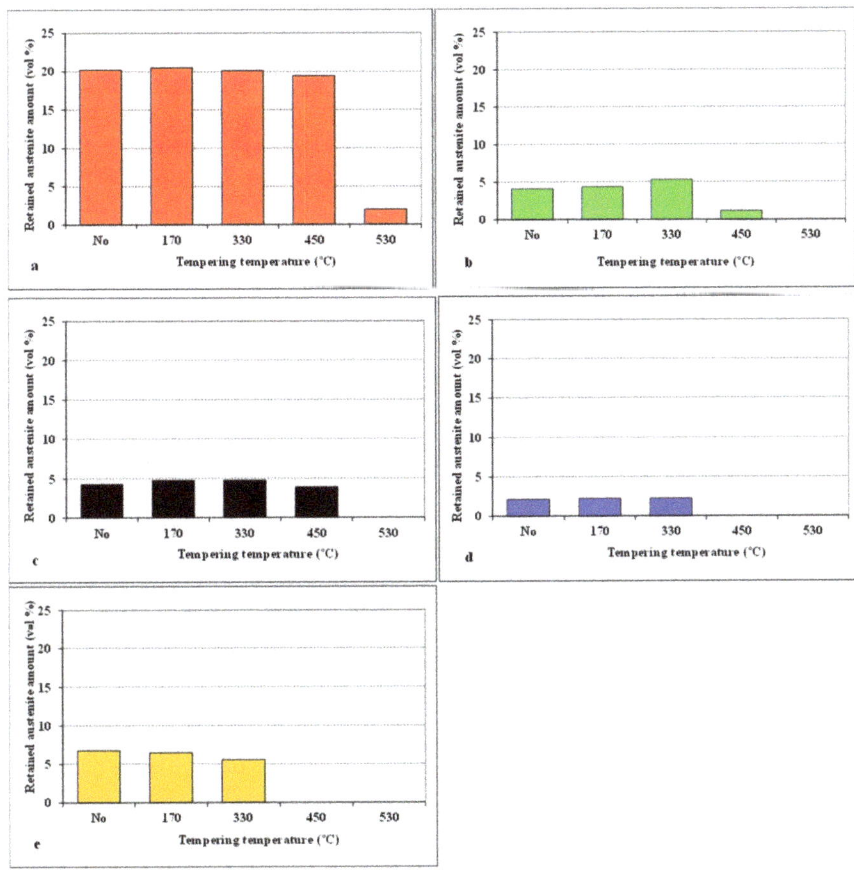

Figure 6. Retained austenite amounts for different SZT specimens, as a function of tempering temperature: (**a**) CHT, (**b**) SZT −75 °C, (**c**) SZT −140 °C, (**d**) SZT −196 °C, (**e**) SZT −269 °C.

A detailed description of the precipitation behaviour of differently sub-zero-treated Vanadis 6 steel exceeds the scope of the present paper. This topic has been extensively studied recently, and the main results can be summarised as follows: Application of sub-zero treatments evokes acceleration of precipitation of transient M_3C carbide at low tempering temperatures [19,25]. In CHT steel, the M_7C_3 phase precipitates during tempering within the secondary hardening temperature range, while the precipitation of M_7C_3 is inhibited in the case of SZT steel, and only growth of M_3C particles takes place.

3.2. Corrosion Behaviour

Figure 7 summarizes the potentiodynamic curves of CHT specimen and specimens after different SZTs, in un-tempered state. The specimens with SZT at −140 °C manifest the highest corrosion potential (E_{corr}) (the most anodic), and the E_{corr} decreases in the order: steel SZT at −269 °C, SZT at −75 °C, SZT at −196 °C, and CHT steel. This suggests that the SZT lowers the steel tendency towards oxidation, and the most convenient SZT temperature for lowering the oxidation tendency was found to be −140 °C. Additionally, the pitting potentials, E_{pit}, are shifted to the more anodic values (see also Table 3), indicating slightly better resistance of SZT steel to the stable pitting corrosion. As follows from the values of corrosion current, I_{corr} (Table 3), the CHT specimens manifest the highest corrosion rate, and the corrosion rate (dissolution) decreases in the order: SZT at −75 °C, SZT at −269 °C, SZT at −196 °C, and SZT at −140 °C. These results indicate that the application of SZT improves the corrosion resistance of the Vanadis 6 steel in the prior-to-tempering material state.

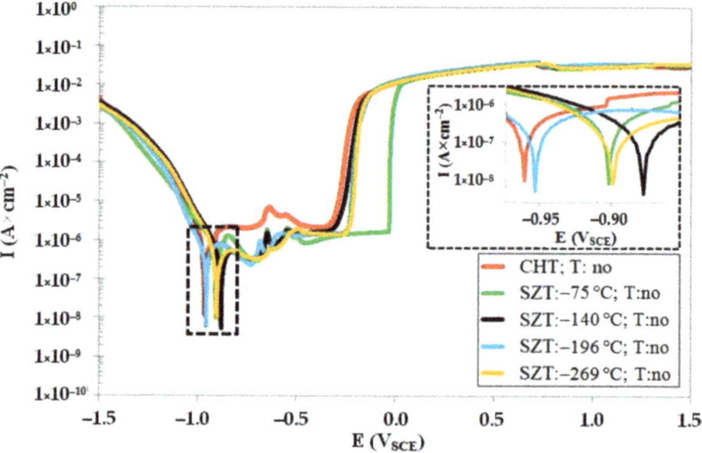

Figure 7. Potentiodynamic polarisation curves of CHT and different SZT specimens in un-tempered state.

The potentiodynamic curves of CHT steel in un-tempered state as well as in the states after different tempering treatments are shown in Figure 8. The un-tempered specimen manifests the highest corrosion potential (E_{corr}), and the E_{corr} decreases with the increasing tempering temperature. This indicates that the tempering treatment deteriorates the resistance of conventional room-temperature-quenched Vanadis 6 steel against oxidation in 3.5% water solution of NaCl. The variations in the pitting potentials, E_{pit} (see also Table 4), do not manifest a clear tendency with respect to the level of tempering temperature. The specimens treated at either low (170 °C) or high (530 °C) tempering temperatures manifest more anodic behaviour as compared with un-tempered steel. Conversely, the pitting potentials, E_{pit}, of specimens tempered at intermediate temperatures indicate higher susceptibility to the stable pitting corrosion. The corrosion current, I_{corr} (see also Table 4), does not manifest significant changes after tempering at 170 °C. However, it raises rapidly after tempering at 330 °C and higher, suggesting that the corrosion rate of the material increases dramatically. In other words, tempering treatment induces considerable worsening of the corrosion resistance of CHT Vanadis 6 steel.

Table 3. Corrosion current, I_{corr}, corrosion potential, E_{corr}, and pitting potential, E_{pit}, values acquired from corrosion tests in 3.5 mass % NaCl water solution, for CHT and different SZT specimens in the prior-to-tempering state.

Heat Treatment	I_{corr} (A·cm^{-2})	E_{corr} (V)	E_{pit} (V)
CHT, un-tempered	1.54×10^{-7}	-0.964	-0.264
SZT $-75\,^{\circ}$C, un-tempered	9.82×10^{-8}	-0.903	-0.026
SZT $-140\,^{\circ}$C, un-tempered	1.02×10^{-8}	-0.878	-0.221
SZT $-196\,^{\circ}$C, un-tempered	4.41×10^{-8}	-0.956	-0.188
SZT $-269\,^{\circ}$C, un-tempered	4.82×10^{-8}	-0.901	-0.179

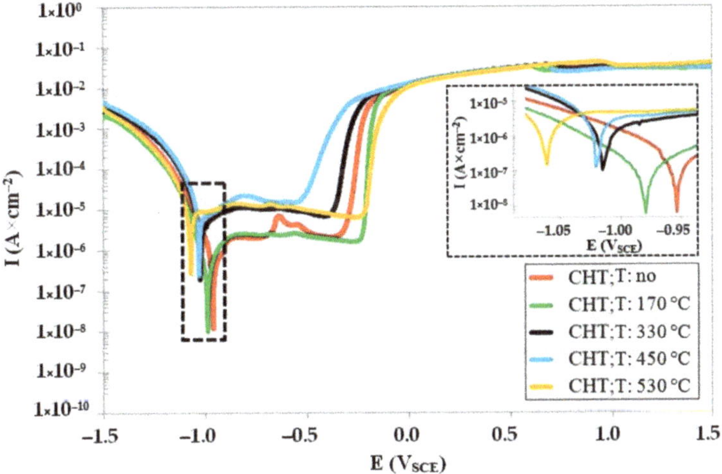

Figure 8. Potentiodynamic polarisation curves of CHT Vanadis 6 steel in un-tempered state and in the states after different tempering treatments.

Table 4. Corrosion current, I_{corr}, corrosion potential, E_{corr}, and pitting potential, E_{pit}, values acquired from corrosion tests in 3.5 mass % NaCl water solution, for CHT and differently tempered specimens.

Heat Treatment	I_{corr} (A·cm^{-2})	E_{corr} (V)	E_{pit} (V)
CHT, un-tempered	1.54×10^{-7}	-0.964	-0.264
CHT, tempered at 170 °C	1.21×10^{-7}	-0.990	-0.189
CHT, tempered at 330 °C	1.74×10^{-6}	1.031	-0.283
CHT, tempered at 450 °C	2.32×10^{-6}	-1.035	-0.395
CHT, tempered at 530 °C	2.49×10^{-6}	-1.072	-0.142

Potentiodynamic curves of the steel that was subjected to sub-zero treatment at $-140\,^{\circ}$C, un-tempered and tempered at different temperatures, are shown in Figure 9. The E_{corr} decreases with tempering (in a similar way to CHT steel, see Figure 8). However, it is also obvious that the values of E_{corr} (for the same tempering regimes) are shifted to higher potentials, suggesting that the SZT improves the resistance of the Vanadis 6 steel towards oxidation, not only in un-tempered state, but also after tempering treatments. The pitting potentials, E_{pit} (see also Table 5), are practically the same for differently tempered specimens, indicating almost no effect of tempering on the steel susceptibility to the pitting corrosion with SZT at $-140\ ^{\circ}\mathrm{C}$. The I_{corr} follows a very similar tendency to what was recorded for CHT steel. However, it is obvious (see also Table 5) that the values of I_{corr} are lower than what was recorded for the CHT steel tempered at the same temperatures. It should be noted that similar variations in corrosion characteristics were recorded for other SZTs. One can thus surmise that the SZT makes an overall amelioration of corrosion behaviour of the Vanadis 6 steel when tested in the 3.5 mass % NaCl water solution.

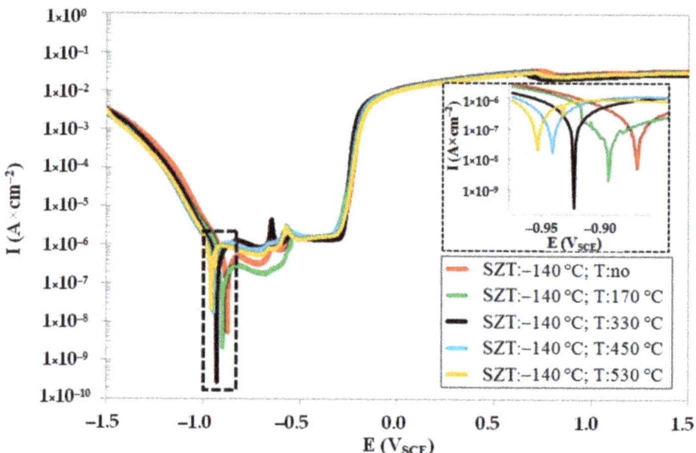

Figure 9. Potentiodynamic polarisation curves of Vanadis 6 steel subjected to SZT at $-140\ ^{\circ}\mathrm{C}$, in un-tempered state and in the states after different tempering treatments.

Table 5. Corrosion current, I_{corr}, corrosion potential, E_{corr}, and pitting potential, E_{pit}, values acquired from corrosion tests in 3.5 mass % NaCl water solution, for specimens after SZT at $-140\ ^{\circ}\mathrm{C}$ and different tempering regimes.

Heat Treatment	I_{corr} (A·cm^{-2})	E_{corr} (V)	E_{pit} (V)
SZT $-140\ ^{\circ}$C, un-tempered	1.02×10^{-8}	-0.878	-0.221
SZT $-140\ ^{\circ}$C, tempered at $170\ ^{\circ}$C	5.17×10^{-8}	-0.902	-0.219
SZT $-140\ ^{\circ}$C, tempered at $330\ ^{\circ}$C	1.13×10^{-7}	-0.930	-0.232
SZT $-140\ ^{\circ}$C, tempered at $450\ ^{\circ}$C	1.88×10^{-7}	-0.948	-0.230
SZT $-140\ ^{\circ}$C, tempered at $530\ ^{\circ}$C	2.52×10^{-7}	-0.961	-0.198

Figure 10 shows the potentiodynamic polarisation curves for the CHT steel specimen and steel specimens after SZT at -75, -140, -196, and $-269\ ^{\circ}\mathrm{C}$, after tempering at $530\ ^{\circ}\mathrm{C}$. As shown, the corrosion potential of SZT specimens is higher (more anodic) as compared to the material state after CHT. Further, it is evident that the SZTs at -75, -140, and $-269\ ^{\circ}\mathrm{C}$ produced higher E_{corr} than the treatment at $-196\ ^{\circ}\mathrm{C}$. The pitting potentials, E_{pit} (see also Table 6), on the other hand, are shifted to the more cathodic values. This suggests that the application of SZTs provides the examined steel with higher susceptibility to the stable formation of pitting, in the state after tempering at $530\ ^{\circ}\mathrm{C}$. The variations of corrosion current well-follow the changes in E_{corr}: the samples after CHT have the highest dissolution rate (the highest I_{corr}), and the I_{corr} decreases in the order of SZT $-196\ ^{\circ}\mathrm{C}$, SZT $-75\ ^{\circ}\mathrm{C}$, SZT $-140\ ^{\circ}\mathrm{C}$, and SZT $-269\ ^{\circ}\mathrm{C}$. Here, a very important finding is that the tendency of the corrosion behaviour improvement, due to the SZT, is partly maintained after tempering at temperatures normally used for the secondary hardening.

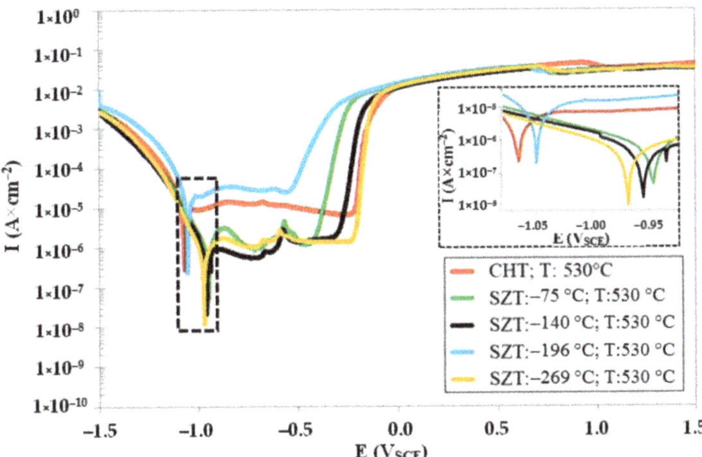

Figure 10. Potentiodynamic polarisation curves of CHT Vanadis 6 steel, and Vanadis 6 subjected to different SZTs, after tempering at 530 °C.

Table 6. Corrosion current, I_{corr}, corrosion potential, E_{corr}, and pitting potential, E_{pit}, values acquired from corrosion tests in 3.5 mass % NaCl water solution, for CHT and different SZT specimens, after tempering at 530 °C.

Heat Treatment	I_{corr} (A·cm^{-2})	E_{corr} (V)	E_{pit} (V)
CHT, tempered at 530 °C	2.49×10^{-6}	-1.072	-0.142
SZT -75 °C, tempered at 530 °C	4.61×10^{-7}	-0.952	-0.262
SZT -140 °C, tempered at 530 °C	2.52×10^{-7}	-0.961	-0.198
SZT -196 °C, tempered at 530 °C	1.27×10^{-6}	-1.065	-0.372
SZT -269 °C, tempered at 530 °C	1.70×10^{-7}	-0.975	-0.145

Figure 11 provides an overview of the dependence of the corrosion rate on the tempering temperature for CHT specimens and for the specimens that were subjected to different SZTs. There are two main tendencies apparently shown. The first one is that the tempering treatment accelerates the corrosion rate (CR), and the higher the tempering temperature, the more accelerated the CR. The second general tendency is that SZTs retard the CR, and that the SZTs at -140 and -269 °C act most effectively in this way.

The surfaces after the potentiodynamic measurements of differently heat-treated specimens are presented in Figure 12. It is shown that the carbide/matrix boundaries are preferentially attacked by corrosion. This mainly concerns the boundaries between coarser eutectic and secondary carbides, as they differ from the matrix in terms of their chemistry. The M_7C_3 carbides (secondary carbides, SCs), for instance, contain 37.2 ± 0.8 mass % of Cr, 46.4 ± 0.5 mass % of Fe, and 12.7 ± 0.3 mass % of V (there are only metallic elements considered). The eutectic MC carbides are formed mainly by vanadium (73.6 ± 0.9 mass %), but they also contain limited amounts of chromium (9.4 ± 0.5 mass %) and iron (0.8 ± 0.1 mass %). On the other hand, the chromium content in the matrix is around 5.5 mass % only, and that of vanadium is correspondingly much lower (up to 1%). The steel also contains small globular carbides, after SZT in particular, see Figure 3. SEM micrographs in Figure 12c,d clearly delineate that the boundaries between SGCs and matrix are less intensively attacked by corrosion; alternatively, they remain well-embedded in the matrix and assist to lower the corrosion rate at the sites where they are present in sufficiently high amounts.

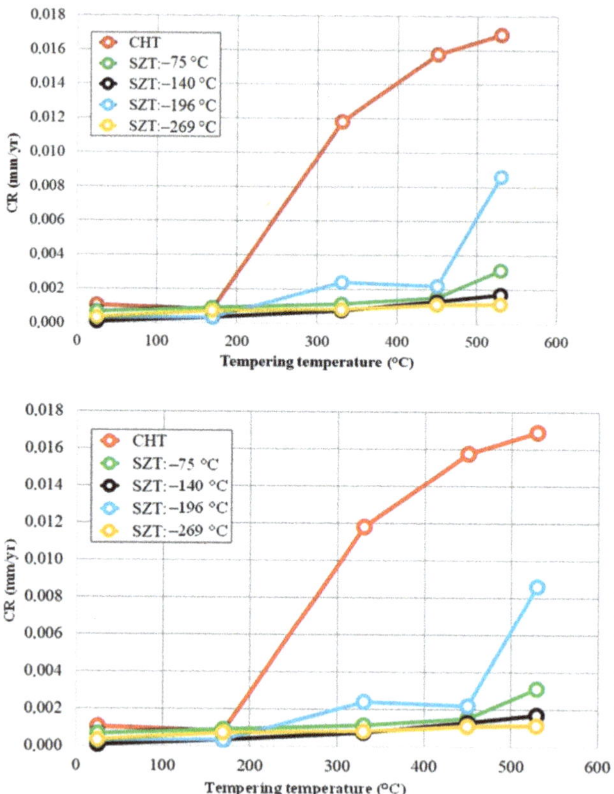

Figure 11. Corrosion rate in dependence on tempering temperature for CHT specimens and for specimens after application of different SZTs.

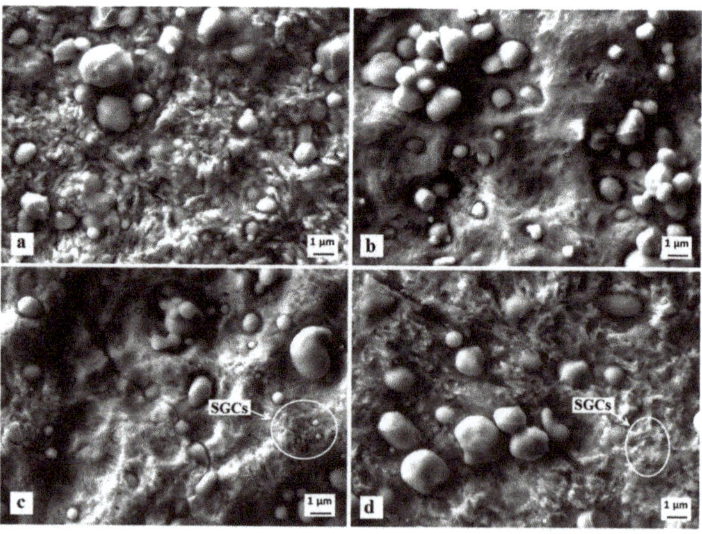

Figure 12. SEM micrographs showing the surfaces after the potentiodynamic polarisation measurements, CHT (**a**), CHT + tempering at 530 °C (**b**), SZT at −140 °C (**c**), SZT at −140 °C + tempering at 530 °C (**d**).

The SEM image in Figure 13 presents the corrosion-attacked specimen that was subjected to the SZT at −140 °C. The same features as in Figure 12 are visible. This mainly concerns the behaviour of different carbides. These carbides are highlighted in EDS maps of chromium (SCs) and vanadium (ECs). The other EDS maps (chlorine, oxygen, sodium) clearly demonstrate that the corrosion-exposed surface is covered by products of this process. However, it is also seen that the corrosion products' layer is not uniform. While the matrix is fully covered, the carbides are attacked by the corrosion environment to a much lower extent.

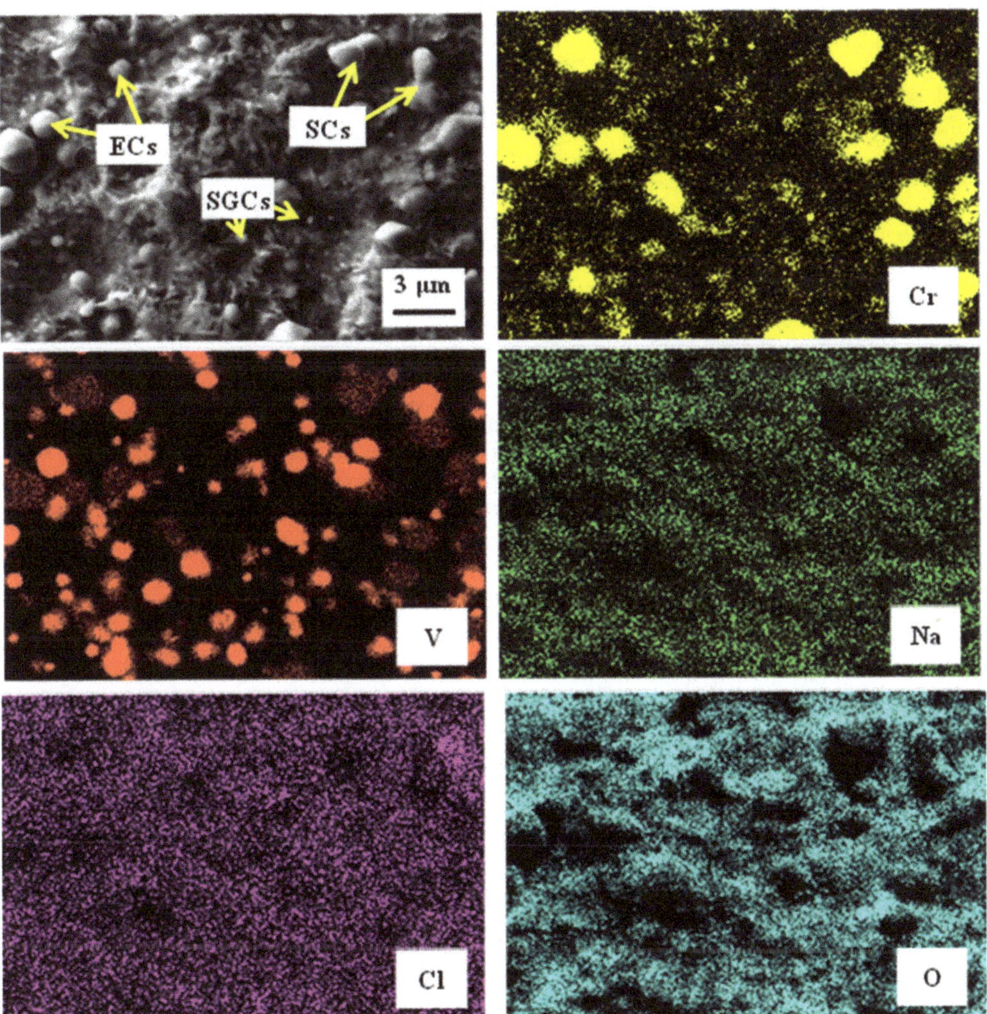

Figure 13. SEM micrograph showing the surface of the specimen that was subjected to the SZT at −140 °C, after the potentiodynamic polarisation measurements, and corresponding EDS maps of Cr, V, Na, Cl, and O.

Figure 14 shows the topography and work function mapping of the phases on the clean surface of the examined steel specimen that was sub-zero-treated at −140 °C. The white spots on the topography image, in Figure 14a, correspond to the carbide particles. The corresponding work function map in Figure 14b undoubtedly confirms that these sites have higher potential than the matrix. The difference is almost fully consistent through the whole measured area, and is about 50–60 mV. In other words, these measurements indicate a more noble behaviour of carbides than the mixture of the martensite and retained austenite (matrix microstructure).

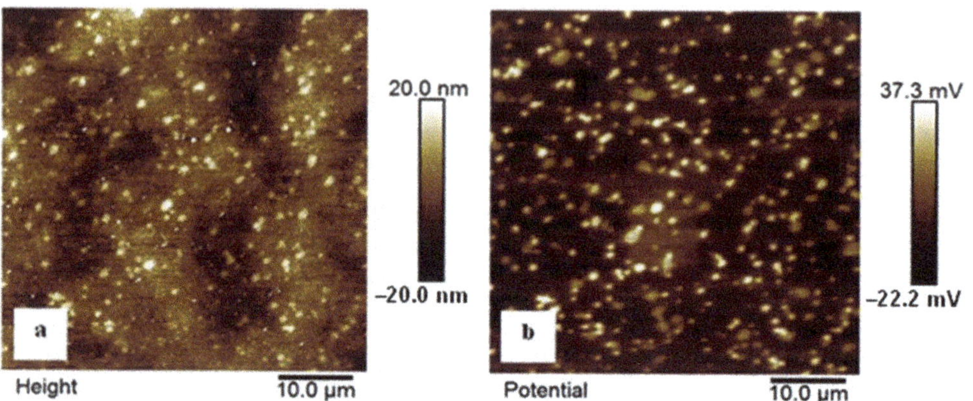

Figure 14. KPFM images of the examined steel sample that was subjected to the SZT at −140 °C. (**a**) Topography and (**b**) work function mapping. Image size 50 × 50 µm^2.

4. Discussion

The obtained results infer that the corrosion resistance of the Vanadis 6 steel is generally improved by the application of sub-zero treatments. Additionally, it was demonstrated that ameliorations in corrosion behaviour are maintained after tempering treatment, even though the tempering generally deteriorates the corrosion behaviour. The mentioned variations in corrosion behaviour are the topic of the following discussion.

It has been summarised recently [26] that the application of SZTs evokes a significant reduction of retained austenite amount, refines the martensite, induces an acceleration of the precipitation rate of transient carbides, and produces an enhanced number and population density of small globular carbides.

4.1. Retained Austenite Amount and Characteristics

The first factor that makes a clear difference in microstructures of SZTs Vanadis 6 steel and the CHT material is the retained austenite amount. Figures 4 and 6 provide clear evidence on the reduction of this phase in SZT steel, by 70–90%, depending on the temperature of SZT. The mentioned results appear counterintuitive with regard to the corrosion behaviour of the steel at the first glance because it has been experimentally proven that the austenite manifests a more noble behaviour than the ferrite (or martensite) [27–29]. The more noble behaviour of the austenite was attributed to a weaker internal stress state and to lower amounts of defects in the gamma phase compared to either ferrite or martensite [30].

However, it was demonstrated recently that sub-zero treatment evokes an introduction of high compressive stresses into the retained austenite of different steels [4,31], due to the combined effect of different thermal expansions of the austenite and the martensite, and to the volumetric effect of the martensitic transformation.

Beneficial effects of compressive residual stresses in the γ_R on the corrosion resistance can be assumed based upon the obtained results of recent investigations. Peyre et al. [32], for instance, reported increased corrosion performance of the AISI 316L austenitic stainless steel, as a result of compressive stresses that were introduced into the surface by either laser- or shot-peening. Takakuwa and Soyama [33] investigated the effect of various surface finish techniques (and thereby different stress level) on the corrosion resistance of the same steel grade in 5 mass % aqueous solution of H_2SO_4, and they arrived at very similar findings. Moreover, they claimed that the principal explanation of improved corrosion behaviour may be based on the fact that the reduction of interatomic spacing due to the compressive stress on the surface facilitates the growth and maintenance of the passivation film. It is also interesting to note that the introduction of compressive stresses enhances the corrosion behaviour not only for ferrous alloys, but, for instance, also for aluminium alloys [34]. Therefore, one can conclude that the high state of compression in the retained austenite contributes to the overall improvement of corrosion resistance of SZT Vanadis 6 steel, even though the γ_R amount was significantly reduced by this kind of treatment.

4.2. Microstructure of the Martensite

Refinement of martensitic domains is the second main consequence of SZT. At the beginning, it should be mentioned that the martensitic transformation that takes place in SZT can be divided into two components: (i) the diffusion-less component that takes place during continuous cooling down from the austenitizing temperature, and (ii) the isothermal component that occurs during the hold at the cryotemperatures. As reported recently [26], the refinement of martensite concerns only the second component, due to spatial limitation effects in the growth of the martensite as well as the result of slow plastic deformation of virgin martensite.

The effect of martensitic domain size on the corrosion behaviour of complex-phase ledeburitic tool steels can only be roughly judged. First, the size of martensite can only be estimated by viewing the TEM micrographs. More exact quantification of this parameter fails, often due to the very small transparent area on thin foils. Additionally, it is clear that the area fraction of refined martensite has a maximum of 17–18%, depending on the extent of the retained austenite reduction due to the particular SZT regime, as illustrated in Figure 7. However, it is logical that the martensitic domains' boundary density is increased when the domains become smaller, meaning that more domains' boundaries are present per unit volume. The boundaries of martensitic domains (laths, plates, or needles) will have higher energy levels, as compared to the bulk of the domains, and thereby the finer martensite is expected to be more corrosion-active as compared to the coarse one. Here, it should also be underlined that the refinement of domains concerns only a minor part of the martensite (the isothermally formed one, as mentioned above), and the unfavourable effect of the refinement of martensitic domains on the corrosion performance of SZT steel is hereby significantly reduced.

A clear difference between the martensite produced via CHT and that developed by SZT was seen at the lattice tetragonality level. It is commonly accepted that SZTs reduce the tetragonality of the martensitic lattice of different steels, such as AISI D2 [35] or Vanadis 6 [25]. A logical interpretation is that carbon clusters are formed at dislocations in the martensite during the hold at the cryotemperatures, and that SZT induces an acceleration of precipitation of transient nano-sized cementite particles [14,19]. The carbon atoms in clusters as well as those in precipitates can essentially not contribute to the tetragonality of the martensitic lattice. The carbon content in the martensite of CHT steel was estimated, considering the austenitizing temperature of 1050 °C, carbon contents in major carbides MC and M_7C_3 [36], and the level of their dissolution in the austenite at given T_A [22], to be at around 1.3 wt %. One can expect that the solutionised amount of carbon atoms in the martensite of the steel after SZTs would be correspondingly lower.

Even though a lot of research has been done about the understanding of metallurgical aspects of corrosion behaviour of different steels, only little attention has been paid to the effect of different carbon contents in the martensite, at medium chromium contents, on the corrosion resistance of ferrous materials. There are only two studies devoted to this topic. In the first one, Gulbrandsen et al. [37] reported that the corrosion rate decreased slightly as the carbon content in the martensite rose from 0.095 to 0.12 mass %. In the second study, de Waard et al. [38] established that the addition of up to 2 mass % Cr decreases the effect of carbon content in the martensite on the corrosion resistance to an insignificant level.

However, the Cr content in the matrix of Vanadis 6 steel is at around 5.5 mass % after austenitizing at 1050 °C and quenching, and the results obtained by Gulbrandsen et al. [37] and by de Waard et al. [38] are hardly comparable with the current ones from this point of view. Nevertheless, one can assume a much stronger effect of Cr on the corrosion performance of steels than that caused by carbon (considering the results of de Waard et al. [38], for instance); hence, one can expect almost "no effect" of reduced amounts of carbon atoms solutionised in martensite on the corrosion resistance at 5.5 mass % Cr.

4.3. Enhanced Number of Small Globular Carbides

The role of enhanced number and population density of SGCs in the corrosion behaviour of examined steel seems to be a controversial issue. It has been reported in many scientific papers that the presence of carbides, inclusions, or precipitates has a detrimental effect on the corrosion resistance, since there are microelectrochemical cells formed at the carbide/matrix interfaces [39,40]. This was reaffirmed by many authors for ledeburitic steels containing lamellar eutectic mixtures [41], high chromium white-cast irons [42–46], and for Fe-C alloys containing lamellar pearlite [46].

It is obvious from Figures 2 and 3 that SZT increases the amount and population density of small globular carbides. On the other hand, the amounts and population densities of eutectic (ECs) and secondary carbides (SCs) are not affected by SZTs [14]. In Figure 12, it is shown that the areas around the coarser ECs and SCs manifest more distinct corrosion attacks compared to the areas around the SGCs. In addition, it appears that the areas with higher amounts of these small particles undergo corrosion to a lesser extent than the matrix with no presence of SGCs.

Potentiodynamic curves in Figures 7 and 10, and the data in Table 3, provide clear information on the shift of corrosion potential of SZT specimens to higher (more anodic) values, and show that the dissolution rate (corrosion current, I_{corr}) decreases with the application of SZTs.

For the explanation of "unexpected" ameliorations of corrosion behaviour of SZT steel, it should first be noted that the carbides in experimental materials used in [41–47] were formed either by the eutectic solidification or by the eutectoid decomposition of the austenite, i.e., at high temperatures where diffusion is possible. Hence, an extensive partitioning of carbon and alloying elements between carbides and solid solutions occurred, which resulted in considerable differences in chemistry between these phases. As a consequence, the galvanic corrosion occurred on the materials' surfaces due to the difference in corrosion potentials between the carbides and matrix. In the corrosion process, the carbides have a much nobler corrosion potential than the matrix (solid solutions), and hence act as cathodes in galvanic corrosion cells [45]. This is the case of ECs and SCs in the current experimental work. As mentioned above, however, these two carbide types are not influenced by the SZTs, and hence their contributions to the corrosion behaviour can be expected to be invariant to the heat treatment route used. Alternatively, it has been experimentally proven that the SGCs are formed during the hold of the steel at the cryotemperature [1,4,14], where the partitioning of carbon and alloying elements is very limited as there is only little atomic movement at such a low temperature. These particles are a by-product of a more complete martensitic transformation [48]. Additionally, it was indicated that the temperature of −140 °C provides the best balance between the plastic deformation rate of virgin martensite during the isothermal hold of the steel at the cryotemperature and the transformation rate of retained austenite (also takes place at the cryotemperature). Therefore, it is also logical that the presence of SGCs has the most beneficial effect on corrosion behaviour in the case of the steel treated at −140 °C.

The TEM micrograph in Figure 15a shows different carbides, i.e., the ECs, SCs, and SGCs, in martensitic matrix. In corresponding EDS maps of chromium (Figure 15b) and vanadium (Figure 15c), and also in Table 7, it is shown that the ECs (marked by number 2, and other dark particles) contain much more vanadium than the matrix (marked with number 6). Additionally, it is shown that the SCs differ from the matrix by significantly enhanced chromium content (particle with number 1 as well as two carbides on the right side of the image). On the other hand, the particles numbered 3 and 5 do not manifest any significant partitioning of alloying elements, suggesting that they were formed under diffusion-less conditions.

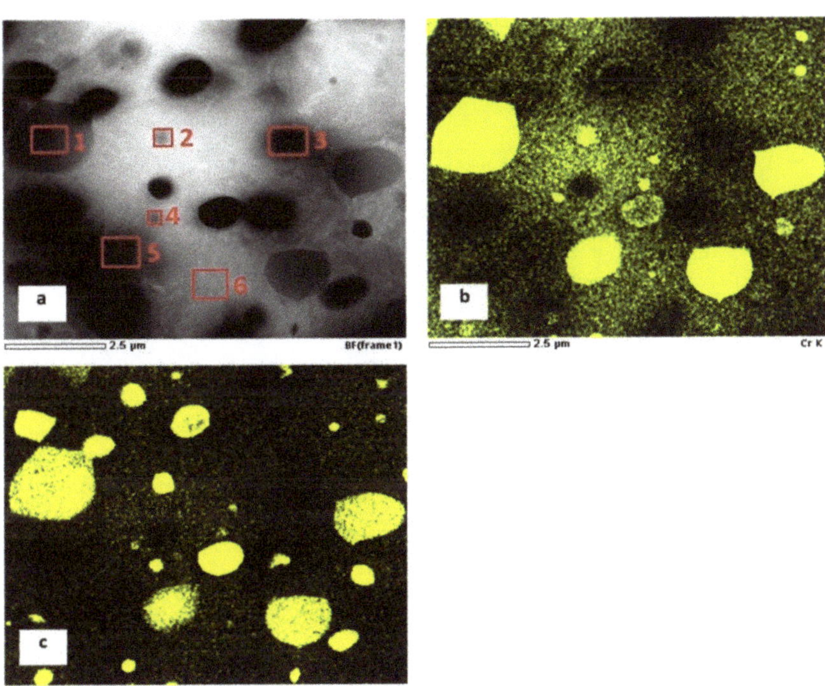

Figure 15. TEM micrograph showing the carbides in martensitic matrix of the specimen after quenching followed by SZT at −140 °C (**a**), EDS map of Cr (**b**), EDS map of V (**c**). The sites of semi-quantitative EDS measurements are labelled and numbered in the TEM image (**a**).

Table 7. Recorded values of EDS measurements from sites in Figure 16a.

Site No.	Chemical Composition (mass %)				
	Si	V	Cr	Fe	Mo
1	-	12.8	37.8	47.1	2.3
2	-	73.4	17.7	8.9	-
3	0.5	0.5	5.4	92.4	1.2
4	0.2	16.2	7.4	74.8	1.4
5	0.5	0.7	5.7	91.6	1.5
6	0.5	0.6	5.6	92.4	0.9

Therefore, enhanced amount and population density of carbides may not inevitably lead to increasing the overall area ratio of anode (carbides) to cathode (matrix). In addition, an opposite effect can occur, where increased carbides/matrix surface area ratio may contribute to the retardation of corrosion since a more stable protective film on the surface of these carbides can be formed. Experimental investigations of the effect of cementite on the corrosion resistance of carbon steel provided a good example of a much nobler response of cementite on corrosion attacks and confirmed improved corrosion behaviour of the material when coated with Fe_3C [49].

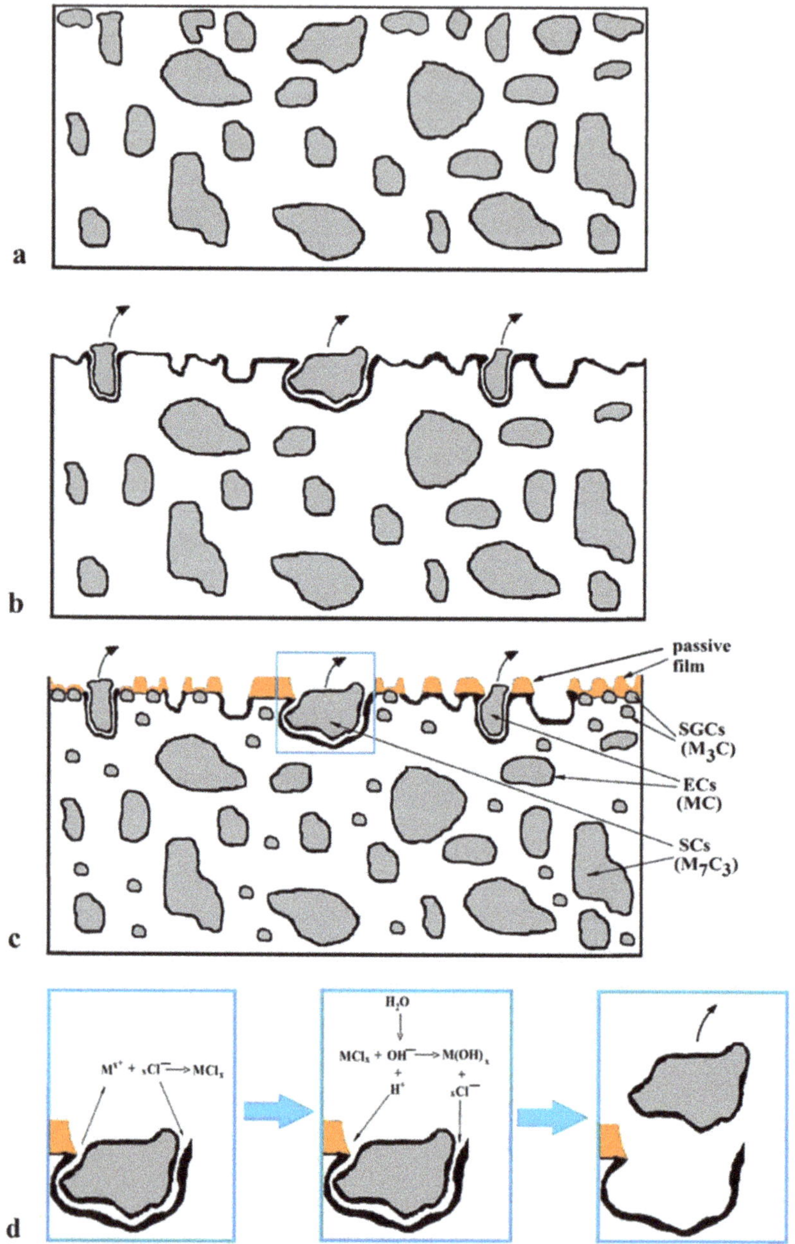

Figure 16. A schematic of the corrosion attack of the Vanadis 6 steel: before testing (**a**), after testing, CHT steel (**b**), SZT steel—overview (**c**), detail from (**d**).

4.4. Precipitation of Carbides

The last difference between the material state after CHT and that after SZTs is an accelerated precipitation rate of nano-sized carbides. According to recent studies [28,50], the precipitation of the Cr-rich $M_{23}C_6$ carbides during tempering is considered to induce Cr-depleted zones around them, and thereby retards the formation of a protective passive film on the steels' surfaces. On the other hand, the precipitation of M_3C carbides with similar Cr content as the matrix has a less detrimental effect on the growth of protective films [28].

For the Vanadis 6 steel, it has been demonstrated recently that the SZT accelerates the precipitation rate of transient cementitic carbides at low tempering temperatures, but these treatments suppress the precipitation of stable M_7C_3 phase during tempering in the secondary hardening temperature range [19,25].

Figures 8 and 9 show that the corrosion potential, E_{corr}, decreases with increasing tempering temperature for CHT steel, as well as for the steel which was subjected to the SZT at $-140\,°C$. It should be mentioned here that the potentiodynamic measurements of specimens subjected to other regimes of SZT (-75, -196, and $-269\,°C$) provided similar qualitative results. Additionally, it is shown (Table 4) that the corrosion current, I_{corr}, increases rapidly with the tempering, which is clearly reflected in the corrosion rate of different SZT specimens (Figure 11). The mentioned changes in corrosion behaviour characteristics can be ascribed to the precipitation of different carbides and the corresponding changes in the matrix. Only cementitic particles were found in the experimental steel after tempering within the low-tempering temperature range [19,25]. The precipitation of M_3C does not evoke the Cr depletion of the matrix as the M_3C contain only very low chromium amount. The only factor that increases the corrosion may be the higher number of activated sites by forming large amounts of M_3C/matrix boundaries. Increased tempering temperature leads to precipitation of M_7C_3 particles in the case of CHT specimens, which reduces the number of solutionised Cr atoms in the microstructure and thereby considerably deteriorates the corrosion characteristics. Instead, the precipitation of M_7C_3 carbides was not evidenced after SZTs, and the only consequence of the tempering treatment is the increase in the number of M_3C particles and their coarsening [19]. Hence, the corrosion characteristics of SZT Vanadis 6 steel are less negatively influenced by high-temperature tempering.

Based on the obtained results, the possible corrosion mechanism of the Vanadis 6 steel in 3.5% NaCl water solution could be delineated. As mentioned above, the steel contains ECs (vanadium-rich, MC), SCs (chromium rich, M_7C_3), and certain but very limited amounts of SGCs (Figure 16a). During the corrosion tests, both the ECs/matrix and SCs/matrix interface types are extensively attacked by the corrosion environment, and the carbides are extracted from the surface, which enhances further corrosion (Figure 16b). Additionally, the matrix is considerably attacked by corrosion in this case, as Figure 12a illustrates, and the specimen surface manifests significantly enhanced roughness.

Conversely, the examined steel contains considerably enhanced population density of SGCs after an application of SZTs. The SGCs/matrix interfaces are attacked less extensively by the corrosion environment (Figures 12c and 16c). Moreover, the area percentage of carbides increases at the same time by the application of SZTs. The carbides manifest more noble behaviour than the matrix (Figure 14), and these particles are less covered by the corrosion products (Figure 13). The resulting effect is that the corrosion rate of the SZT specimens is lowered (Figure 11), implying that the application of SZT generally improves the corrosion resistance of the Vanadis 6 steel in 3.5% NaCl water solution.

5. Conclusions

The microstructural changes in Cr-V ledeburitic steel Vanadis 6 with different sub-zero treatments and tempering were investigated using SEM, TEM, and X-ray diffraction. The corresponding changes in corrosion resistance in a 3.5% water solution of NaCl were studied by potentiodynamic polarisation tests, calculation of the corrosion rate, and by analysis of attacked surfaces by SEM.

The following conclusions can be expressed from the present study:

1. The austenitizing at 1050 °C for 30 min followed by nitrogen gas quenching produced the microstructure composed of needle-like martensite, retained austenite, and undissolved carbides.

2. Application of sub-zero treatments considerably reduced the retained austenite amount, refined the martensite, enhanced the number and population density of small globular carbides, and modified the precipitation behaviour of nano-sized carbides. The retained austenite amount decreased with the increasing tempering temperature, and it was reduced to values below the detection limit of X-ray diffraction after tempering at 530 °C. Additionally, the number and population density of small globular carbides decreased with tempering in sub-zero-treated steel.

3. In the un-tempered state, the application of sub-zero treatments increased the corrosion potential of the material and inhibited the dissolution. Additionally, sub-zero treatments made the passive films on the surface more stable. These phenomena were most pronounced in the case of the steel treated at -140 °C.

4. Tempering treatment evoked a shift of the corrosion potential to lower values, and at the same time, the corrosion current (dissolution rate) to higher values. For equal tempering temperatures, however, sub-zero-treated steel manifested better corrosion behaviour.

5. The coarse carbide/matrix interfaces were preferentially attacked by corrosion, which resulted in the extraction of these particles, and thereby enhanced corrosion. However, the corrosion attack of the small globular carbides/matrix interfaces was very limited, which significantly contributed to the improvement of the corrosion behaviour of sub-zero-treated material.

Author Contributions: Conceptualization, P.J.; methodology, P.J., M.D., A.B., M.Č., software, A.B., D.B.; validation, P.J., A.B.; formal analysis, P.J.; investigation, P.J., M.D., A.B., M.Č.; resources, M.H.; data curation, P.J., D.B.; writing—original draft preparation, P.J.; writing—review and editing, P.J., A.B.; visualization, P.J., M.H., M.D., D.B.; supervision, P.J.; project administration, M.H.; funding acquisition, P.J. All authors have read and agreed to the published version of the manuscript.

Funding: The authors would like to acknowledge that the article is an outcome implementation of the following two projects: scientific project VEGA 1/0112/20 and APRODIMET, ITMS: 26220120048, supported by the Research & Development Operational Programme funded by the European Regional Development Fund.

Institutional Review Board Statement: Not applicable.

Informed Consent Statement: Not applicable.

Data Availability Statement: The data presented in this study are available on request from the corresponding author. The data are not publicly available due to the fact the this is an ongoing research.

Acknowledgments: The authors would like to acknowledge that the article is an outcome implementation of the following two projects: scientific project VEGA 1/0112/20 and APRODIMET, ITMS: 26220120048, supported by the Research & Development Operational Programme funded by the European Regional Development Fund. Special thanks are expressed to Suzanne Morsch from The University of Manchester, for realizing the Kelvin probe force microscopy.

Conflicts of Interest: The authors declare no conflict of interest.

References

1. Jurči, P. Sub-zero treatment of cold work tool steels—Metallurgical background and the effect on microstructure and properties. *HTM J. Heat Treat. Mater.* **2017**, *72*, 62–68. [CrossRef]
2. Akhbarizadeh, A.; Amini, K.; Javadpour, S. Effects of applying an external magnetic field during the deep cryogenic heat treatment on the corrosion resistance and wear behaviour of 1.2080 tool steel. *Mater. Des.* **2012**, *41*, 114–123. [CrossRef]
3. Akhbarizadeh, A.; Javadpour, S.; Amini, K. Investigating the effect of electric current flow on the wear behaviour of 1.2080 tool steel during the deep cryogenic heat treatment. *Mater. Des.* **2013**, *45*, 103–109. [CrossRef]
4. Jurči, P.; Kusý, M.; Ptačinová, J.; Kuracina, V.; Priknerová, P. Long-term sub-zero treatment of P/M Vanadis 6 ledeburitic tool steel–a preliminary study. *Manuf. Technol.* **2015**, *15*, 41–47. [CrossRef]
5. Stratton, P.F. Optimising nano-carbide precipitation in tool steels. *Mater. Sci. Eng. A* **2007**, *449–451*, 809–812. [CrossRef]
6. Sobotová, J.; Jurči, P.; Dlouhý, I. The effect of sub-zero treatment on microstructure, fracture toughness, and wear resistance of Vanadis 6 tool steel. *Mater. Sci. Eng. A* **2016**, *652*, 192–204. [CrossRef]

7. Das, D.; Dutta, A.K.; Ray, K.K. On the enhancement of wear resistance of tool steels by cryogenic treatment. *Philos. Mag. Lett.* **2008**, *88*, 801–811. [CrossRef]

8. Amini, K.; Akhbarizadeh, A.; Javadpour, S. Investigating the effect of the quench environment on the final microstructure and wear behaviour of 1.2080 tool steel after deep cryogenic heat treatment. *Mater. Des.* **2013**, *45*, 316–322. [CrossRef]

9. Akhbarizadeh, A.; Shafyei, A.; Golozar, M.A. Effects of cryogenic treatment on wear behaviour of D6 tool steel. *Mater. Des.* **2009**, *30*, 3259–3264. [CrossRef]

10. Das, D.; Dutta, A.K.; Ray, K.K. Sub-zero treatments of AISI D2 steel: Part II. Wear behaviour. *Mater. Sci. Eng. A* **2010**, *527*, 2194–2206. [CrossRef]

11. Das, D.; Ray, K.K.; Dutta, A.K. Influence of temperature of sub-zero treatments on the wear behaviour of die steel. *Wear* **2009**, *267*, 1361–1370. [CrossRef]

12. Surberg, C.H.; Stratton, P.; Lingenhoele, K. The effect of some heat treatment parameters on the dimensional stability of AISI D2. *Cryogenics* **2008**, *48*, 42–47. [CrossRef]

13. Ptačinová, J.; Sedlická, V.; Hudáková, M.; Dlouhý, I.; Jurči, P. Microstructure Toughness relationships in sub-zero treated and tempered Vanadis 6 steel compared to conventional treatment. *Mater. Sci. Eng. A* **2017**, *702*, 241–258. [CrossRef]

14. Jurči, P.; Dománková, M.; Čaplovič, L.; Ptačinová, J.; Sobotová, J.; Salabová, P.; Prikner, O.; Šuštaršič, B.; Jenko, D. Microstructure and hardness of sub-zero treated and no tempered P/M Vanadis 6 ledeburitic tool steel. *Vacuum* **2015**, *111*, 92–101. [CrossRef]

15. Tyshchenko, A.I.; Theisen, W.; Oppenkowski, A.; Siebert, S.; Razumov, O.N.; Skoblik, A.P.; Sirosh, V.A.; Petrov, J.N.; Gavriljuk, V.G. Low-temperature martensitic transformation and deep cryogenic treatment of a tool steel. *Mater. Sci. Eng. A* **2010**, *527*, 7027–7039. [CrossRef]

16. Das, D.; Dutta, A.K.; Ray, K.K. Sub-zero treatments of AISI D2 steel: Part I. Microstructure and hardness. *Mater. Sci. Eng. A* **2010**, *527*, 2182–2193. [CrossRef]

17. Das, D.; Ray, K.K. Structure-property correlation of sub-zero treated AISI D2 steel. *Mater. Sci. Eng. A* **2012**, *541*, 45–60. [CrossRef]

18. Meng, F.; Tagashira, K.; Azuma, R.; Sohma, H. Role of eta-carbide precipitation's in the wear resistance improvements of Fe-12Cr-Mo-V-1.4C tool steel by cryogenic treatment. *ISIJ Int.* **1994**, *34*, 205–210. [CrossRef]

19. Jurči, P.; Dománková, M.; Hudáková, M.; Ptačinová, J.; Pašák, M.; Palček, P. Characterization of microstructure and tempering response of conventionally quenched, short- and long-time sub-zero treated PM Vanadis 6 ledeburitic tool steel. *Mater. Charact.* **2017**, *134*, 398–415. [CrossRef]

20. Amini, K.; Akhbarizadeh, A.; Javadpour, S. Effect of Carbide Distribution on Corrosion Behavior of the Deep Cryogenically Treated 1.2080 Steel. *J. Mater. Eng. Perform.* **2016**, *25*, 365–373. [CrossRef]

21. Hill, H.; Huth, S.; Weber, S.; Theisen, W. Corrosion properties of a plastic mould steel with special focus on the processing route. *Mater. Corros.* **2011**, *62*, 436–443. [CrossRef]

22. Bílek, P.; Sobotová, J.; Jurči, P. Evaluation of the microstructural changes in Cr-V ledeburitic tool steel depending on the austenitization temperature. *Mater. Technol.* **2011**, *45*, 489–493.

23. ASTM E975-13. Standard Practice for X-Ray Determination of Retained Austenite in Steel with Near Random Crystallographic Orientation. In *ASTM Book of Standards*; ASTM: West Conshohocken, PA, USA, 2004; Volume 3.01.

24. ASTM G 102. Standard Practice for Calculation of Corrosion Rates and Related Information from Electrochemical Measurements. In *ASTM Book of Standards*; ASTM: West Conshohocken, PA, USA, 1994.

25. Jurči, P.; Dománková, M.; Ptačinová, J.; Pašák, M.; Kusý, M.; Priknerová, P. Investigation of the Microstructural Changes and Hardness Variations of Sub-Zero Treated Cr-V Ledeburitic Tool Steel Due to the Tempering Treatment. *J. Mater. Eng. Perform.* **2018**, *27*, 1514–1529. [CrossRef]

26. Jurči, P.; Ptačinová, J.; Sahul, M.; Dománková, M.; Dlouhý, I. Metallurgical principles of microstructure formation in sub-zero treated cold-work tool steels—A review. *Mater. Tech.* **2018**, *106*, 104–112. [CrossRef]

27. Lee, J.S.; Fushimi, K.; Nakanishi, T.; Hasegawa, Y.; Park, Y.S. Corrosion behaviour of ferrite and austenite phases on super duplex stainless steel in a modified green-death solution. *Corros. Sci.* **2014**, *89*, 111–117. [CrossRef]

28. Cheng, X.; Wang, Y.; Li, X.; Dong, C. Interaction between austenite-ferrite phases on passive performance of 2205 duplex stainless steel. *J. Mater. Sci. Technol.* **2018**, *34*, 2140–2148. [CrossRef]

29. Lu, S.Y.; Yao, K.F.; Chen, Y.B.; Wang, M.H.; Chen, N.; Gea, X.Y. Effect of quenching and partitioning on the microstructure evolution and electrochemical properties of a martensitic stainless steel. *Corros. Sci.* **2016**, *103*, 95–104. [CrossRef]

30. Dobelaar, J.A.L.; Herman, E.C.M.; De Wit, J.F.W. The influence of the microstructure on the corrosion behaviour of Fe-25Cr. *Corros. Sci.* **1992**, *33*, 779–790. [CrossRef]

31. Villa, M.; Pantleon, K.; Somers, M.A.J. Evolution of compressive strains in retained austenite during sub-zero Celsius martensite formation and tempering. *Acta Mater.* **2014**, *65*, 383–392. [CrossRef]

32. Peyre, P.; Scherpereel, X.; Berthe, L.; Carboni, C.; Fabbro, R.; Béranger, G.; Lemaitre, C. Surface modifications induced in 316L steel by laser peening and shot-peening. Influence on pitting corrosion resistance. *Mater. Sci. Eng. A* **2000**, *280*, 294–302. [CrossRef]

33. Takakuwa, O.; Soyama, H. Effect of Residual Stress on the Corrosion Behavior of Austenitic Stainless Steel. *Adv. Chem. Eng. Sci.* **2015**, *5*, 62–71. [CrossRef]

34. Liu, X.; Frankel, G.S. Effects of compressive stress on localized corrosion in AA2024-T3. *Corros. Sci.* **2006**, *48*, 3309–3329. [CrossRef]

35. Gavriljuk, V.G.; Theisen, W.; Sirosh, V.V.; Polshin, E.V.; Kortmann, A.; Mogilny, G.S.; Petrov, Y.N.; Tarusin, Y.V. Low-temperature martensitic transformation in tool steels in relation to their deep cryogenic treatment. *Acta Mater.* **2013**, *61*, 1705–1715. [CrossRef]

36. Fredriksson, H.; Hillert, M.; Nica, M. The decomposition of the M_2C carbide in high speed steel. *Scand. J. Metall.* **1979**, *8*, 115–122.
37. Gulbrandsen, E.; Nyborg, R.; Loland, T.; Nisancioglu, K. *Effect of Steel Microstructure and Composition on Inhibition of CO_2 Corrosion*; Corrosion 2000, Paper Nr. 23; NACE International: Houston, TX, USA, 2000.
38. De Waard, C.; Lotz, U.; Dugstad, A. *Influence of Liquid Flow Velocity on CO_2 Corrosion. A Semi-Empirical Model*; Corrosion 1995, Paper Nr. 128; NACE International: Houston, TX, USA, 1995.
39. Ansari, T.Q.; Luo, J.L.; Shi, S.Q. Multi-Phase-Field Model of Intergranular Corrosion Kinetics in Sensitized Metallic Materials. *J. Electrochem. Soc.* **2020**, *167*, 061508. [CrossRef]
40. Hong, Y.Y.; Wang, X.Z.; Cadien, K.; Luo, J. Transient Potential Induced Anodic Dissolution of 316L Stainless Steel in Sulfuric Acid Solution. *J. Electrochem. Soc.* **2019**, *166*, C3355–C3363. [CrossRef]
41. Rogal, L.; Dutkiewicz, J.; Szklarz, Z.; Krawiec, H.; Kot, M.; Zimowski, S. Mechanical properties and corrosion resistance of steel X210CrW12 after semi-solid processing and heat treatment. *Mater. Charact.* **2014**, *88*, 100–110. [CrossRef]
42. Kawalec, M.; Krawiec, H. Corrosion Resistance of High-Alloyed White Cast Iron. *Arch. Metall. Mater.* **2015**, *60*, 301–303. [CrossRef]
43. Abd El-Aziz, K.; Zohdy, K.; Saber, D.; Sallam, H.E.M. Wear and Corrosion Behavior of High-Cr White Cast Iron Alloys in Different Corrosive Media. *J. Bio Tribo Corros.* **2015**, *1*, 25. [CrossRef]
44. Tang, X.H.; Chung, R.; Li, D.Y.; Hinckley, B.; Dolman, K. Variations in microstructure of high chromium cast irons and resultant changes in resistance to wear, corrosion and corrosive wear. *Wear* **2009**, *267*, 116–121. [CrossRef]
45. Tang, X.H.; Chung, R.; Pang, C.J.; Li, D.Y.; Hinckley, B.; Dolman, K. Microstructure of high (45 wt.%) chromium cast irons and their resistances to wear and corrosion. *Wear* **2011**, *271*, 1426–1431. [CrossRef]
46. Wiengmoon, A.; Pearce, J.T.H.; Chairuangsri, T. Relationship between microstructure, hardness and corrosion resistance in 20 wt.%Cr, 27 wt.%Cr and 36 wt.%Cr high chromium cast irons. *Mater. Chem. Phys.* **2011**, *125*, 739–748. [CrossRef]
47. Ferhat, M.; Benchettara, A.; Amara, S.E.; Naijar, D. Corrosion behaviour of Fe-C alloys in a Sulfuric Medium. *J. Mater. Environ. Sci.* **2014**, *5*, 1059–1068.
48. Ďurica, J.; Ptačinová, J.; Dománková, M.; Čaplovič, L.; Čaplovičová, M.; Hrušovská, L.; Malovcová, V.; Jurči, P. Changes in microstructure of ledeburitic tool steel due to vacuum austenitizing and quenching, sub-zero treatments at −140 °C and tempering. *Vacuum* **2019**, *170*, 108977. [CrossRef]
49. Wu, J.; Wang, B.; Zhang, Y.; Liu, R.; Xia, Y.; Li, G.; Xue, W. Enhanced wear and corrosion resistance of plasma electrolytic carburized layer on T8 carbon steel. *Mater. Chem. Phys.* **2016**, *171*, 50–56. [CrossRef]
50. Bonagani, S.K.; Bathula, V.; Kaina, V. Influence of tempering treatment on microstructure and pitting corrosion of 13wt.% Cr martensitic stainless steel. *Corros. Sci.* **2018**, *131*, 340–354. [CrossRef]

Article

Effect of CO$_2$ Partial Pressure on the Corrosion Inhibition of N80 Carbon Steel by Gum Arabic in a CO$_2$-Water Saline Environment for Shale Oil and Gas Industry

Gaetano Palumbo [1,*]**, Kamila Kollbek** [2]**, Roma Wirecka** [2,3]**, Andrzej Bernasik** [3] **and Marcin Górny** [4]

[1] Department of Chemistry and Corrosion of Metals, Faculty of Foundry Engineering, AGH University of Science and Technology, 30-059 Krakow, Poland

[2] Academic Centre for Materials and Nanotechnology, AGH University of Science and Technology, Mickiewicza St. 30, 30-059 Kraków, Poland; kamila.kollbek@agh.edu.pl (K.K.); roma.wirecka@fis.agh.edu.pl (R.W.)

[3] Department of Condensed Matter Physics, Faculty of Physics and Applied Computer Science, AGH University of Science and Technology, Mickiewicza St. 30, 30-059 Krakow, Poland; bernasik@agh.edu.pl

[4] Department of Cast Alloys and Composites Engineering, Faculty of Foundry Engineering, AGH University of Science and Technology, 30-059 Krakow, Poland; mgorny@agh.edu.pl

* Correspondence: gpalumbo@agh.edu.pl; Tel.: +48-12-888-27-63

Received: 25 August 2020; Accepted: 21 September 2020; Published: 23 September 2020

Abstract: The effect of CO$_2$ partial pressure on the corrosion inhibition efficiency of gum arabic (GA) on the N80 carbon steel pipeline in a CO$_2$-water saline environment was studied by using gravimetric and electrochemical measurements at different CO$_2$ partial pressures (e.g., P$_{CO_2}$ = 1, 20 and 40 bar) and temperatures (e.g., 25 and 60 °C). The results showed that the inhibitor efficiency increased with an increase in inhibitor concentration and CO$_2$ partial pressure. The corrosion inhibition efficiency was found to be 84.53% and 75.41% after 24 and 168 h of immersion at P$_{CO_2}$ = 40 bar, respectively. The surface was further evaluated by scanning electron microscopy (SEM), energy dispersive spectroscopy (EDS), grazing incidence X-ray diffraction (GIXRD), and X-ray photoelectron spectroscopy (XPS) measurements. The SEM-EDS and GIXRD measurements reveal that the surface of the metal was found to be strongly affected by the presence of the inhibitor and CO$_2$ partial pressure. In the presence of GA, the protective layer on the metal surface becomes more compact with increasing the CO$_2$ partial pressure. The XPS measurements provided direct evidence of the adsorption of GA molecules on the carbon steel surface and corroborated the gravimetric results.

Keywords: high-pressure CO$_2$ corrosion; corrosion inhibition; gum arabic; carbon steel N80

---

1. Introduction

Shale oil and gas are "unconventional" resources of natural oil and gas trapped in fine-grained sedimentary rocks called shale. The rapid expansion of shale oil and gas exploration and the development of a new technology (i.e., hydraulic fracturing (HF) techniques), has seen the popularity of these natural resources to grow over the years. However, after years of exploitation, the oil and gas production in the reservoir declines to result in a major economic challenge for the oil companies. The injection of CO$_2$ at high pressure into the wellbore is an effective method to increase the oil fields lifetime [1–4]. This process is usually referred to as carbon dioxide flooding enhanced oil recovery (CO$_2$-EOR). However, CO$_2$ gas dissolves in the fluid to form the weak carbonic acid, which in turn dissociates into bicarbonate and carbonate anions [4,5]. The presence of this weak acid can lead to severe corrosion attacks on the steel structures [4–6].

Another common problem encountered in the extraction of these natural resources is the use of aggressive fluids with high concentrations of chloride ions (e.g., fracturing fluid) [7]. In the HF process, the fluid usually injected into the wellbore is a neutral water-based chloride solution (up to 4% of potassium chloride) with different additives (i.e., inhibitors of scaling, thickening agents, corrosion inhibitors, etc.) [8]. The literature reports that the presence of a high concentration of chloride ions in a CO_2-containing fluid can exponentially accelerate the dissolution of the steel [6,9].

Carbon and low-alloys steel are often used in the construction of the pipeline in the shale oil and gas industry infrastructures, mainly due to its durability, ductility, high strength, and low cost [7,8,10]. However, due to these harsh operating conditions encountered during the exploitation of these natural resources, the steel is prone to corrode. One practical and relatively cheap method for controlling sweet corrosion in the shale oil and gas industry is the use of corrosion inhibitors. Corrosion inhibitors are substances that added to the solution greatly reduce the dissolution of the metal by forming a protective layer on its surface. The literature reports that over the last decades the use of corrosion inhibitors as a means to mitigate CO_2 corrosion that occurs inside the carbon steel pipelines has received a wide interest. Nitrogen-based compounds such as pyridine derivatives [11] imidazolines [12], benzimidazole derivatives [13], and amines [14] were found to be effective corrosion inhibitors against CO_2 corrosion. However, most of these compounds are reported to be toxic and their synthesis can be very expensive [15,16]. These drawbacks and the increase in environmental awareness have led many researchers to focus on the use of more naturally occurring substances as corrosion inhibitors. Plant extracts substances, such as berberine extract [17], *Momordica charantia* [18], *Gingko biloba* [19] were successfully tested as green corrosion inhibitors in CO_2-saturated saline solutions.

The last trend of research has also seen the use of many naturally occurring polymers as green corrosion inhibitors in various corrosive environments [10,20–22]. They are abundant in nature, environmentally sustainable, and have an appreciable solubility. Additionally, polymers, unlike small molecules, with their multiple adsorption sites for bonding on the metal surface, are expected to show a higher corrosion inhibition efficiency, compared to their monomer counterpart.

Umoren et al. [15] studied the corrosion inhibition effect of two naturally occurring polymers such as carboxymethyl cellulose and chitosan for API 5 L X60 steel in a CO_2 saline solution at P_{CO_2} = 1 bar. The results showed that both inhibitors reduced the corrosion rate of the metal due to the formation of a protective layer on its surface. Singh et al. [23] studied the corrosion inhibition effect of a modified natural polysaccharide (e.g., guar gum + methylmethacrylate) in a 3.5 wt% NaCl solution saturated with CO_2 (e.g., P_{CO_2} = 1 bar) at 50 °C. The authors found that this modified polysaccharide acted like a good corrosion inhibitor for P110 steel with maximum inhibition efficiency found to be 90%. However, most of these studies were carried out at atmospheric pressure (e.g., P_{CO_2} = 1 bar). The CO_2-EOR process can significantly increase the dissolution of the tube. As reported by many studies, the severity of the CO_2 corrosion attack increases with an increase in CO_2 partial pressure due to the increase in acidity of the fluid [4–6]. Therefore, understanding how the CO_2 partial pressure can influence the inhibitory action of certain corrosion inhibitors in CO_2 saline environments is important and can help to minimize the material and economic losses.

Mustafa et al. [4] studied the effect of the CO_2 partial pressure (e.g., 10, 40, and 60 bar) on the corrosion inhibition of an imidazoline-based inhibitor for X52 steel exposed to CO_2 water saline solution at 60 °C. The authors reported that the inhibitor efficiency of the tested inhibitor was observed to be strongly affected by the concentration of inhibitor and CO_2 partial pressure. Ansari et al. [16] studied the influence of a modified chitosan corrosion inhibitor on J55 carbon steel in a 3.5 wt% NaCl solution saturated with CO_2 at 60 bar and 65 °C, reporting a corrosion inhibition efficiency of 95%. Yet, all inhibitors tested so far are labeled either as toxic or are expensive to synthesize.

Gum arabic (GA) is a natural polymer obtained from the Acacia trees of the Leguminosae family [22] and it has been reported to successfully inhibit the corrosion of the steel in different environments [7,21,22,24–28]. Furthermore, GA is often used in the fracturing fluid as a thickening agent to increase the viscosity of the fluid [29]. Therefore, due to the encouraging results presented by these studies and the continuous research of affordable and eco-friendly corrosion inhibitors, this work was undertaken to study the efficacy of GA as an eco-friendly corrosion inhibitor to mitigate high-pressure CO_2 corrosion for carbon steel pipeline in a saline solution. This paper also aims to show that GA not only can be used as a thickening agent in the make-up of the fracturing fluid, but it could also be used as an active component in corrosion inhibitor in the shale gas industry. To this end, the study was performed in an autoclave in the presence and absence of different concentrations of GA, different CO_2 partial pressures, and different temperatures using weight loss and electrochemical measurements. SEM-EDS, GIXRD, and XPS measurements were also employed to characterize the corrosion product layer and to support the gravimetric and electrochemical results.

2. Experimental Procedure

2.1. Materials

The study was carried out on carbon steel (N80) with composition of (weight %): C 0.39%, Mn 1.80%, Si 0.26%, Cu 0.26%, V 0.19%, Cr 0.04%, Ni 0.04%, Al 0.03%, Mo 0.003%, Co 0.002%, Sn 0.004%, S 0.001%, P 0.001% and the remainder Fe. Figure 1 shows that the microstructure of the N80 carbon steel pipeline is composed of perlite and ferrite (α-Fe) phases, where the latter phase accounting for circa 41% of the total. The samples used in this study were machined from pipeline carbon steel, ground with silicon carbide abrasive paper up to 1200 grit, then were ultrasonically washed with distilled water, dried with absolute alcohol.

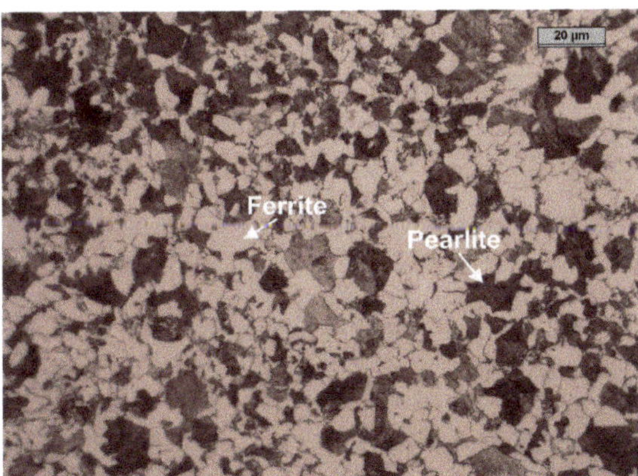

Figure 1. Optical micrographs of the N80 carbon steel microstructures.

All experiments were carried out in 3% of potassium chloride (KCl). Potassium salt was used in this study instead of the more common NaCl, because in the fracturing fluid, the potassium (K^+) ions formed a semi-permeable membrane on the shale rock and therefore, preventing the water from entering the shale.

The tested solution was prepared from reagent grade material potassium chloride (Sigma-Aldrich) and pure deionized water with an electrical resistivity of 0.055 µS/cm at T = 25 °C. The tested inhibitor was purchased from Sigma-Aldrich (Warsaw, Poland). The concentrations of inhibitor solution prepared and used for the study ranged from 0.6 up to 2.0 g L^{-1}.

2.2. Gravimetric Measurements

The gravimetric experiments were carried out in a 1.2 L high-pressure autoclave (PARR instrument) at different CO_2 partial pressures (1, 20, and 40 bar). The coupons were suspended in a 1.0 L solution in the presence and absence of different concentrations of the inhibitor (i.e., from 0.6 up to 2.0 g L^{-1}) at 25 and 60 °C. Before each experiment, the tested solution was deaerated with CO_2 for 2 h under atmospheric pressure and then CO_2 was purged for another 2 h at the tested pressure after the introduction of the samples. After saturation, the pH and conductivity of the tested solution were 4.5 and 60.30 mS cm^{-1} at 1 bar and 25 °C, respectively. To ensure homogeneous mixing, a Teflon-coated blade agitator was used (e.g., 200 rpm). The weight loss was determined by retrieving the coupons after 24 h of immersion by means of an analytical balance with an accuracy of ±0.1 mg. To assess the effect of time, the samples were immersed for 168 h in the presence and absence of 1.0 g L^{-1} of GA at different CO_2 partial pressures (1, 20, and 40 bar). The corrosion products were removed according to the ASTM G1-90 [30], then the specimens were ultrasonically washed with distilled water, dried with absolute alcohol, and reweighed. In each case, the experiment was conducted thrice and the corrosion rate (*CR*) in mm y^{-1} was obtained from the following equation:

$$CR \ (\text{mm y}^{-1}) \ = \ \frac{87.6 \ \Delta m}{dAt} \tag{1}$$

where, Δm is the weight loss calculated form the difference between the initial (W_i) and the final (W_f) weight (mg). d is the density (7.87 g cm^{-3}), A is the surface of the sample (cm^{-2}) and t is the immersion time (h). The inhibition efficiency (*IE%*) was determined using the following equation [20,22,26]:

$$IE\% \ = \ \frac{CR - CR^{\text{inh}}}{CR} \times 100 \tag{2}$$

where CR^{inh} and CR are the corrosion rates of the steel with and without the inhibitor, respectively.

2.3. Electrochemical Experiments

The electrochemical experiments were carried out in a 1.2 L high-pressure autoclave (PARR instrument) at different CO_2 partial pressures (1, 20, and 40 bar) with a conventional three-electrode system. N80 carbon steel specimen was used as a working electrode, a platinum foil as a counter electrode (CE), and a high-pressure 0.1 M KCl Ag/AgCl probe was used as a reference electrode. To ensure homogeneous mixing, a Teflon-coated blade agitator was used (200 rpm). The electrochemical impedance spectroscopy (EIS) and potentiodynamic polarization (PDP) measurements were carried in a Gamry reference 600 potentiostat/galvanostat electrochemical system after the sample was exposed for 24 h in the tested solution, with and without the presence of 1.0 g L^{-1} of GA. The EIS tests were performed over the frequency range of 100 kHz to 10 mHz and amplitude of 10 mV at open circuit potential. The data were then fitted by means of Echem Analyst 5.21 software using the opportune equivalent circuit. The *IE%* was calculated from the polarization resistances (R_p) determined from the fitting process using the following equation [21,26]:

$$IE\% \ = \ \frac{R_p^{\text{inh}} - R_p}{R_p^{\text{inh}}} \times 100 \tag{3}$$

where $R_p{}^{\text{inh}}$ and R_p are the values of the polarization resistances in the presence and absence of the inhibitor, respectively. The PDP measurements were carried out at a potential of ±−0.3 V from the OCP and a scan rate of 1 mV s^{-1}. The potentiodynamic parameters were determined by means of Echem Analyst 5.21 software. The values of *IE%* were calculated from the measured i_{corr} values using the relationship [21,26]:

$$IE\% \ = \ \frac{i_{\text{corr}} - i_{\text{corr}}^{inh}}{i_{\text{corr}}} \times 100 \tag{4}$$

where i_{corr} and i_{corr}^{inh} represent the values of the corrosion current densities without and with inhibitor, respectively.

2.4. Surface Analysis

The surface of the samples, prepared as described above, were analyzed in the presence and absence of 1.0 g L^{-1} of GA. After the immersion, the samples were removed and rinsed with deionized water and dried. The surface analysis was carried out by means of different techniques such as a scanning electron microscopy combined with an energy dispersive spectroscopy, grazing incidence X-ray diffraction (GIXRD), and an X-ray photoelectron spectroscopy (XPS). The SEM measurements were carried out by using a JEOL scanning electron microscope. The GIXRD analysis was carried out to further determine the composition of the corrosion products film. Grazing incident X-ray diffraction (GIXRD) with an incident angle of 3° was applied to study samples phase composition. A Panalytical Empyrean X-ray diffractometer in the parallel beam geometry (Goebel mirror in the incident beam optics and parallel plate collimator in the secondary beam optics) with Co lamp (Kα = 1.7902 Å) was used to perform measurements. The samples were scanned with a 0.02° step in the range of 20°–70° at room temperature. The XPS analysis was carried out in a PHI 5000 VersaProbe II spectrometer with an Al Kα monochromatic X-ray beam. The X-ray source was operated at 25 W and 15 kV beam voltages. Dual-beam charge compensation with 7 eV Ar$^+$ ions and 1 eV electrons was used to maintain a constant sample surface potential regardless of the sample conductivity. The pass energy of the hemispherical analyzer for the iron (Fe 2p) spectra was fixed at 23.5 eV and for other elements at 46.95 eV. The spectra were charge corrected to the mainline of the carbon C 1 s spectrum set to 284.8 eV.

3. Results and Discussion

3.1. Effect of Pressure and Temperature

3.1.1. Gravimetric Experiments

Figure 2 and Table S1 show the corrosion rate and the variation of the inhibition efficiency obtained at different concentrations of GA and CO_2 partial pressures. It follows from the data that *CR* increases with an increase of CO_2 partial pressure, going from 1.28 to 10.95 mm y^{-1} at P$_{CO_2}$ = 1 bar and P$_{CO_2}$ = 40 bar at 25 °C, respectively.

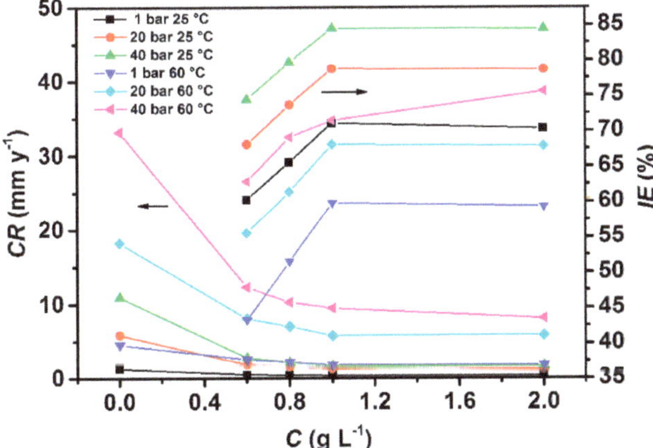

Figure 2. Corrosion inhibitor efficiency at different concentrations of gum arabic (GA) and CO_2 partial pressures after 24 h of immersion.

The solubility of CO_2 in water increases sharply with increasing the pressure of the system [31]. The high corrosion rate observed at higher CO_2 partial pressures can be explained with the increase of the acidity of the solution. In fact, in the presence of CO_2, the weak carbonic acid is formed, which in turn dissociates in HCO_3^- and in CO_3^{2-}, according to the following reactions:

$$CO_2 + H_2O \leftrightarrow H_2CO_3 \tag{5}$$

$$H_2CO_3 \leftrightarrow H^+ + HCO_3^- \tag{6}$$

$$HCO_3^- \leftrightarrow H^+ + CO_3^{2-} \tag{7}$$

The corrosion process in a CO_2 containing solution is controlled by the anodic reaction (Equation (8)) and the three cathodic reactions (Equation (9)–(11)) [4,5]:

$$Fe \rightarrow Fe^{2+} + 2e^- \tag{8}$$

$$2H_2CO_3 + 2e^- \rightarrow H_2 + 2HCO_3^- \tag{9}$$

$$2HCO_3^- + 2e^- \rightarrow H_2 + 2CO_3^{2-} \tag{10}$$

$$2H^+ + 2e^- \rightarrow H_2 \tag{11}$$

The pH of the solution plays an important role in determining the corrosion rate of carbon steel in a CO_2 environment. As the CO_2 partial pressure increases, its solubility also increases, resulting in an increase of the carbonic acid concentration in the solution (Equation (5)). Nesic' predicted that the concentrations of H_2CO_3 in the solution would increase of about 40 times with changing the pressure from P_{CO_2} = 1 bar to P_{CO_2} = 40 bar [31]. Increasing the concentration of carbonic acid leads to an increase in the rate of reduction of carbonic acid and bicarbonate ions (Equations (9) and (10)), and ultimately the anodic dissolution of the steel (Equation (8)) as reported by several studies [4,5,31].

After the addition of the inhibitor, it can be seen that the corrosion rate of the metal is greatly reduced going from 1.28 to 0.37 mm y^{-1}, with a maximum corrosion inhibition efficiency found to be 71.09% at P_{CO_2} = 1 bar, after 24 h of immersion. The data shows that in contrast to the uninhibited solution, an increase in CO_2 partial pressure has a favorable effect on the corrosion rate of the metal in the presence of the inhibitor. It follows from Figure 2 that *IE*, which varies inversely with *CR*, significantly increased after the addition of GA and with CO_2 partial pressure, with a maximum corrosion inhibition efficiency of 78.77% and 84.53% at P_{CO_2} = 20 bar and P_{CO_2} = 40 bar, after 24 h of immersion, respectively [4].

The literature reports that GA [7,21], and in general polysaccharides-like inhibitors [15,20], is mainly adsorbed on the metal surface in acidic condition by weak electrostatic interaction between the protonated inhibitor molecules and the chloride ions adsorbed on the metal surface. In a weak acid solution GA molecules are in equilibrium with their protonated molecules according to the following reaction (see also Section 3.5.1) [7]:

$$GA + xH^+ \leftrightarrow [GAH_x]_{(sol)}^{x+} \tag{12}$$

where $[GAH_x]_{(sol)}^{x+}$ is the protonated inhibitor in the solution. As mentioned before, an increase in CO_2 partial pressure leads to an increase in the acidity of the solution [32]. The higher value of *IE* observed at higher CO_2 partial pressures can be ascribed to the higher concentration of H^+ ions present in the solution, which in turn leads to an increase in the number of protonated inhibitor molecules that can be adsorbed on the metal surface. Moreover, Figure 2 also reveals that *IE* varies with the concentration of the inhibitor until the system reached a state (e.g., 1.0 g L^{-1} of GA), in which it can be said that the inhibitor molecules are in equilibrium with their protonated counterpart. For further increase in GA concentration, *IE* remains almost stable. The results clearly demonstrate that GA has greatly reduced the *CR* of the metal in the tested environment, and the high corrosion inhibition activity of GA was influenced by both its concentration and CO_2 partial pressure. The lower values of *CR* observed in the

presence of the inhibitor can be ascribed to its adsorption on the metal surface, covering the metal surface and thereby, blocking the active corrosion sites on its surface [4,7,28]. The gravimetric results are also supported by the SEM analysis presented from Figures 7–9, where it can be seen that the surface coverage increases and the protective layer becomes more compact in the presence of GA and with increasing CO_2 partial pressure.

As the temperature rises, *IE* slightly decreased. This decrease may be due to the combination of two different reasons. For instance, the solubility of CO_2 decreases with increasing the temperature of the solution [31], which can lead to a less acid environment. The pH of the solution increases slightly and therefore shifting the equilibrium reaction Equation (12) to the left. At higher pH, the concentration of H^+ ions in the solution is smaller, which would result in the formation of fewer protonated inhibitor molecules available for the absorption process. Another possible reason may be due to the fact that these types of inhibitors get absorbed via electrostatic interactions (e.g., van der Waal forces) onto the surface of the metal, and it is known that this types of interaction generally grow weaker with an increase in temperature due to larger thermal motion [3,20]. Consequently, an increase in temperature will increase the metal surface kinetic energy, which has a detrimental effect on the adsorption process and encourages desorption processes [15,20].

Table S2 lists the inhibition efficiency of various corrosion inhibitors used to mitigate sweet corrosion obtained at different immersion times and temperatures. It is worth mentioning that most of these inhibitors are labeled either as toxic or are expensive to synthesize. Umoren et al. [15] reported the corrosion inhibition efficiency of a commercial inhibitor to be 87 and 88% at 25 and 60 °C, respectively after 24 h of immersion. The table shows that GA, compared to other studied corrosion inhibitors, and the commercial corrosion inhibitor, can be considered a good environmentally friendly corrosion inhibitor for carbon steel in a CO_2-saturated saline solution. Moreover, since GA is already used as a thickening agent in the make-up of the fracturing fluid, can also work as an active component in corrosion inhibitor in the shale gas industry.

3.1.2. Electrochemical Experiments

The electrochemical experiments such as electrochemical impedance spectroscopy (EIS) and potentiodynamic polarization (PDP) were also employed as a means to support the gravimetric findings. These experiments were carried out at 1.0 g L^{-1} of GA at different CO_2 partial pressures after 24 of immersion. 1.0 g L^{-1} is the concentration in which the tested inhibitor exhibited a maximum in the concentration-efficiency curve.

The EIS measurements were used to evaluate the resistance of the protective layer from the electrochemical angle and are presented in Figure 3. It can be seen from the Bode (Figure 3b) and phase angle plots (Figure 3c) that the system is characterized by two-time constants at low (LF) and high frequencies (HF). The presence of these two-time constants suggests that the electrochemical reaction process of the N80 carbon steel in a CO_2 saturated saline solution is affected by two state variables i.e., the corrosion products layer and/or the protective adsorptive layer, and electric double-layer, as also reported by Dong et al. [33]. For this reason, the EIS plots presented in Figure 3 were fitted with the help of the equivalent circuit (EC) presented in Figure 3d, consisting of the following elements: R_s is the electrolyte resistance. CPE_1 and R_1 are the constant phase element and the resistance of the layer formed on the metals surface, respectively. CPE_{dl} and R_{ct} are the constant phase element representing the double-charge layer capacitance and the charge transfer resistance, respectively. The EIS parameters are listed in Table 1 and from the small values of χ^2 (i.e., the goodness of fit) it can be said that the EC used to fit the system under investigation was the most appropriate one.

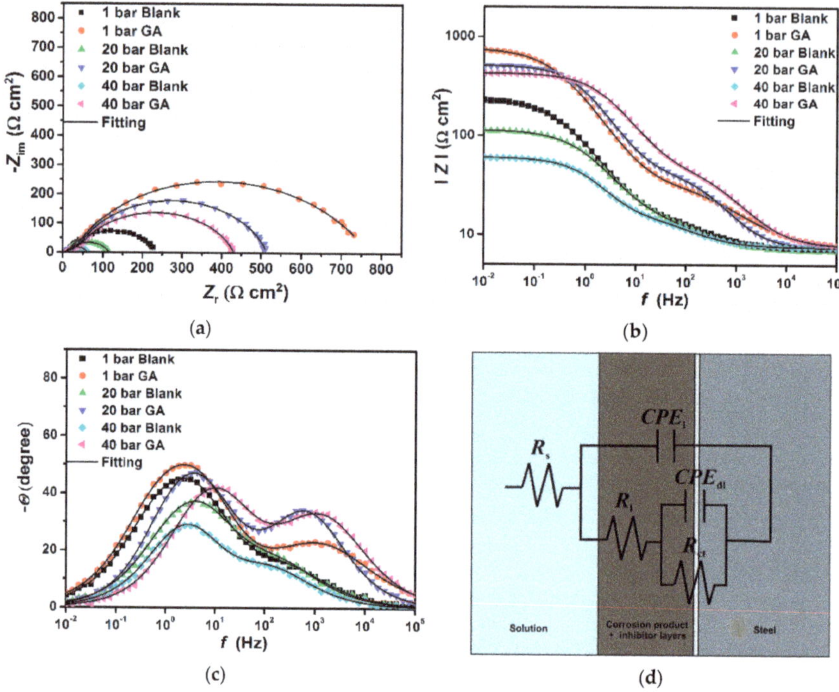

Figure 3. EIS plot recorded in the presence and absence of 1 g L^{-1} of GA after 24 h of immersion at different CO_2-partial pressures. (**a**) Nyquist; (**b**) Bode; (**c**) phase angle; (**d**) equivalent circuit

Table 1. Electrochemical impedance parameters with and without the presence of 1.0 g L^{-1} concentrations of GA after 24 h of immersion.

C (g L^{-1})	R_s (Ω cm^2)	CPE_f Y_f (mΩ^{-1} s^n cm^{-2})	n_f	R_f (Ω cm^2)	CPE_{dl} Y_{dl} (mΩ^{-1} s^n cm^{-2})	n_{dl}	R_{ct} (Ω cm^2)	$R_p = R_f + R_{ct}$ (Ω cm^2)	χ^2 ($\times 10^{-3}$)	IE (%)
1 bar Blank	7.13	1.24	0.66	11.59	1.68	0.79	217.90	229.49	1.12	-
1 bar Ga	7.39	0.34	0.61	30.29	0.68	0.81	730.60	760.89	1.50	69.83
20 bar Blank	6.92	1.28	0.68	12.18	1.85	0.76	95.07	107.25	1.34	-
20 bar GA	7.32	0.10	0.78	38.89	0.55	0.82	469.90	507.79	1.28	78.68
40 bar Blank	7.49	0.15	0.70	9.44	3.99	0.78	43.87	53.31	1.11	-
40 bar GA	7.56	0.01	0.71	50.11	0.28	0.81	374.50	424.61	1.99	87.44

The presence of a time constant at HF is reported in several studies [34,35] and it is often observed in a Fe/water system. This time constant may be due to the capacity of a porous thin layer formed onto the metal surface. In this study, and without the inhibitor, the presence of this time constant at HF is due to the formation of a thin layer of Fe_3C onto the metal surface. As mentioned before, the microstructure of the tested carbon steel is composed of circa 41% of a ferritic phase and the remaining of a perlitic phase (Figure 1). The ferritic phase is more active than the Fe_3C contained in the perlitic phase [7], in this case, the former phase will act as an anode and the latter one as a cathode. This will generate a micro-galvanic effect, which will eventually lead to the formation of a thin layer of Fe_3C onto the metal surface. However, it follows from the data that both the values of R_f and R_{ct} greatly increased in the presence of the inhibitor, which indicated that the GA molecules were adsorbed onto the metal surface leading to the formation of a protective layer that covers the surface, as confirmed also from the morphological analysis (e.g., SEM-EDS and XPS). Moreover, the difference between these two values obtained in the absence and the presence of GA increased even more with increasing CO_2 partial pressure, suggesting that this protective layer becomes more stable and compact, with the corrosion inhibition efficiency going from 69.83% up to 87.44% at $P_{CO_2} = 1$ bar and $P_{CO_2} = 40$ bar, respectively. The increase in *IE* observed with an increase in CO_2 partial pressure agrees with the results obtained with the gravimetric measurements and is in agreement with the ones reported in the literature [4]. It is evident that the addition of GA had a remarkable effect on the corrosion process of the metal and that its inhibition not only depends on the concentration of GA but also from CO_2 partial pressure. The results show that the coverage and thickness of the formed protective layer increased with CO_2 partial pressure, acting both as a barrier against the charge and the mass transfer processes that occur onto the metal surface owing to the corrosive attack of the aggressive electrolyte.

Figure 4 and Table 2 show the potentiodynamic polarization measurements and the corrosion kinetic parameters obtained from the polarization plots in the presence and absence of GA at different CO_2 partial pressures, respectively. As can be seen from Figure 4, the anodic polarization curve of the blank solution does not show the typical Tafel behavior consequently, the corrosion current densities were calculated from the extrapolation of the cathodic Tafel region.

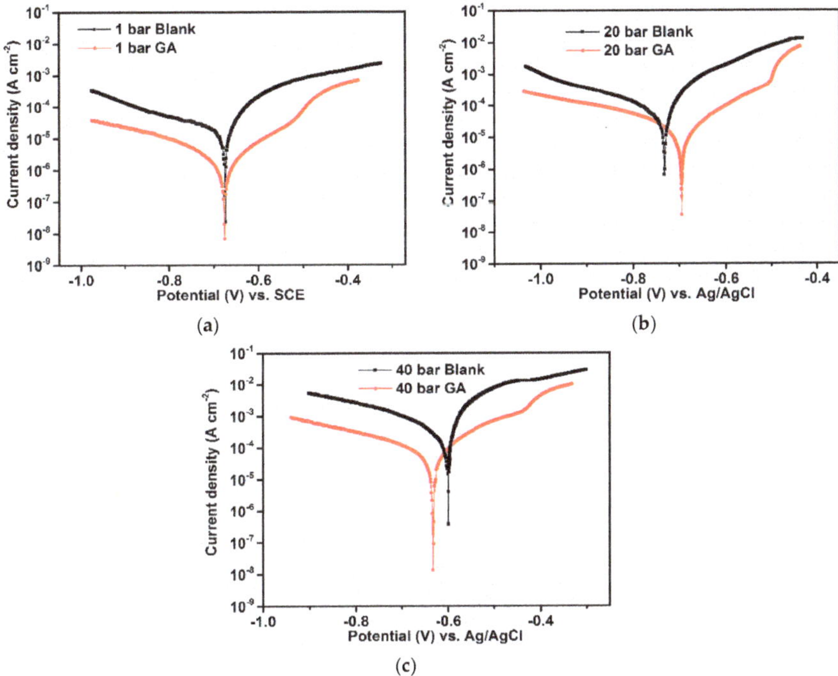

Figure 4. Potentiodynamic polarization parameters obtained in the absence and presence of 1.0 g L^{-1} of GA at different CO_2-partial pressures, after 24 h of immersion. (**a**) $P_{CO_2} = 1$ bar, (**b**) $P_{CO_2} = 20$ bar and (**c**) $P_{CO_2} = 40$ bar.

Table 2. Potentiodynamic polarization parameters obtained after 24 h of immersion without and with 1.0 g L^{-1} of GA.

C (g L^{-1})	E_{corr} (V)	i_{corr} (µA cm^{-2})	β_c (V dec^{-1})	IE (%)
1 bar Blank	−0.673	17.98	0.286	-
1 bar GA	−0.676	5.54	0.334	69.23
20 bar Blank	−0.696	99.90	0.311	-
20 bar GA	−0.736	24.09	0.391	75.88
40 bar Blank	−0.600	647.05	0.316	-
40 bar GA	−0.634	84.90	0.312	86.76

The data shows that in absence of GA, the corrosion current density of the steel increased with an increase in CO_2 partial pressure, which is linked with the increased acidity of the solution, in agreement with the gravimetric experiments. However, it is evident from the data that the corrosion current density of the steel was prominently reduced after the addition of GA to the solution. Furthermore, both the cathodic and anodic curves of the polarization curves were shifted towards lower current densities after the addition of GA. The result suggests that the inhibitor impeded both the rate of the anodic dissolution (Equation (8)) and the cathodic reactions (Equations (9)–(11)), by either covering part of the metal surface and/or blocking the active corrosion sites on the steel surface. The dominant cathodic reaction depends on the pH value of the solution. At lower pH (e.g., less than 4) the reduction of H^+ ions would be the dominant cathodic reaction (Equation (11)). At pH > 4 the dominant cathodic reaction will be the reduction of HCO_3^- ions and H_2CO_3 (Equations (9) and (10)). At higher values of CO_2 partial pressure, GA suppresses the Equation (11) (e.g., the pH of the solution is circa 3.5 at P_{CO_2} = 40 bar), through the formation of H-bonding between the hydroxyl groups of the inhibitor units and the H^+ ions, adsorbed onto the steel surface, as discussed in more detail in Section 3.5.2.

Moreover, after the addition of GA, the E_{corr} can be seen to shift with no definite trend toward both the anodic and cathodic regions. This result suggested that GA behaves as a mixed type inhibitor as also reported by other studies for this inhibitor [7,21,27,28].

3.2. Effect of Time

The effect of immersion time on the corrosion inhibition efficiency of the tested inhibitor was also assessed in this paper. Figure 5 and Table S3 show the corrosion rate and the corrosion inhibition efficiency after 168 h of immersion in the presence and absence of 1.0 g L^{-1} of GA at different CO_2 partial pressures at 25 °C. It follows from the table that GA still shows a very high *IE* even after a longer immersion time. However, it should be noted that *IE* slightly decreases after 168 h of immersion, compared to the one observed after 24 h of immersion.

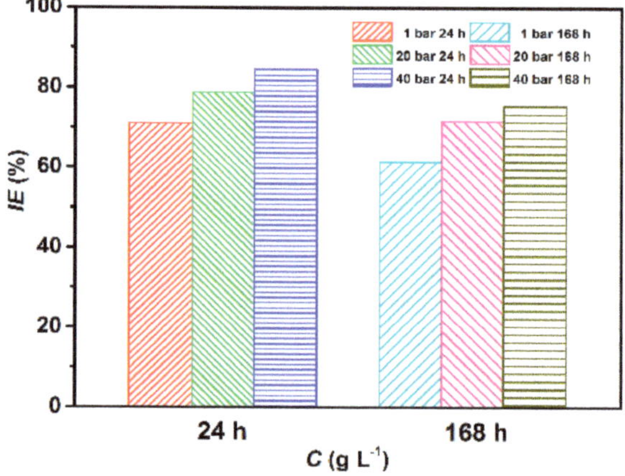

Figure 5. Corrosion inhibitor efficiency obtained at different CO_2 partial pressures after 24 and 168 h of immersion at 25 °C.

This behavior has also been reported by several studies [36,37]. The decrease in *IE* may be due to the desorption of the inhibitor from the metal surface, which makes the protective layer unstable. In this study, the desorption of GA is likely ascribed to its deprotonation due to the consumption of CO_2 from the tested solution because of the electrochemical reactions occurring into the system [5,38]. This leads to a decrease in the acidity of the solution and shifting the Equation (12) towards the deprotonation of the inhibitor.

These results confirm that GA is effectively able to protect the steel surface from sweet corrosion at high CO_2 partial pressures even after a prolonged immersion time, reflecting a strong molecular adsorption of GA on the metal surface and the formation of a stable protective layer.

3.3. Adsorption Study and Standard Adsorption Free Energy

The corrosion inhibition adsorption process of the tested inhibitor on the N80 carbon steel surface was carried out by several adsorption isotherms, such as Temkin's, Frumkin's, Langmuir's, and El-Awady's adsorption isotherms. The Temkin's adsorption isotherm was found to give the best description of the adsorption behavior of the studied inhibitor. The Temkin's adsorption isotherm is defined by the following equations:

$$\theta = \frac{-2.303 \; \log K_{ads}}{2a} - \frac{2.303 \; \log C}{2a} \tag{13}$$

where θ is the surface coverage ($\theta = IE\%/100$), K_{ads} the adsorption-desorption equilibrium constant, C is the inhibitor concentration, a is the molecules interaction parameter. Positive values of a imply attractive forces between the inhibitor molecules, while negative values indicate repulsive forces between them.

K_{ads} is related to the free energy of adsorption by the following equation:

$$\Delta G^{\circ}_{ads} = -RT \; Ln(K_{ads}) \tag{14}$$

where R is the gas constant ($8.314 \; J \; K^{-1} \; mol^{-1}$), T is the absolute temperature (K). The plot of surface coverage (θ) as a function of the logarithm of the inhibitor concentration at different CO_2 partial pressures is shown in Figure 6.

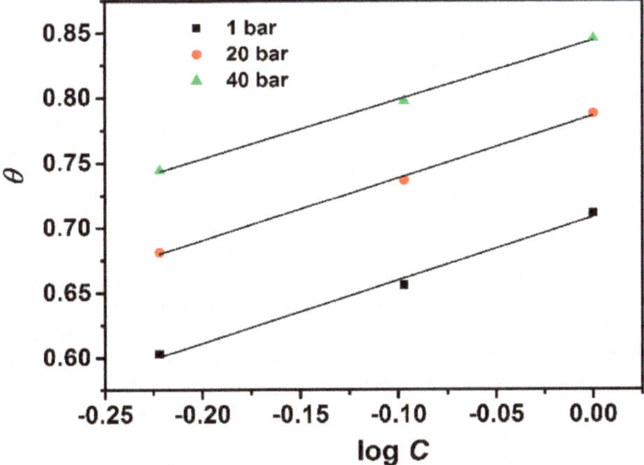

Figure 6. Temkin's adsorption isotherm for carbon steel (N80) pipeline steel in CO_2-saturated chloride at different pressures.

The plot of θ vs. log C yields a straight line and the regression coefficient ranges from 0.985 to 0.996. The calculated values of adsorption parameters ΔG°_{ads}, a and K at different CO_2 partial pressures are presented in Table 3 and the following notes can be written: (i) The values of ΔG°_{ads} are negative for all three pressures, indicating that the adsorption of GA on the steel surface in the tested solution is a spontaneous process [7,21,22,28]. Furthermore, the value of ΔG°_{ads} ranges between -10.64 to -8.37 kJ mol^{-1} indicating that the adsorption of GA on the steel occurs through a physical adsorption process [7,21,22,28]; (ii) The values of "a" are negative for all three pressures, indicating that repulsion forces exist between the adsorbed inhibitor molecules in the adsorption layer, as also reported by other studies for the same tested inhibitor [7,22]; (iii) The values of K_{ads} increases with an increase in CO_2 partial pressure. It should be noted that K_{ads} denotes the strength between adsorbate and adsorbent. It can be inferred that a large value of K_{ads} implies a more efficient adsorption process and thus, a better corrosion inhibition efficiency [21,22]. The results suggest that the adsorption of GA increases with an increase of the environment pressure, leading to a greater surface coverage and consequently, a better protection performance.

Table 3. Parameters of the Temkin's adsorption isotherm calculated from weight loss measurements after 24 h of immersion time.

Pressure (bar)	R^2	Slope	Intercept	a	K_{ads}	ΔG_{ads} (kJ mol^{-1})
1	0.985	0.483	0.708	−2.38	29.10	−8.37
20	0.995	0.478	0.786	−2.41	44.09	−9.39
40	0.996	0.453	0.844	−2.50	72.97	−10.64

3.4. Surface Analysis

The surface morphology of the samples exposed for 24 h at different CO_2 partial pressures in the absence and presence of 1.0 g L^{-1} of GA are presented in Figures 7–9. For instance, it can be seen that the surface morphology of the samples exposed to the blank and inhibited solution differs significantly. For the blank solution, at P_{CO_2} = 1 bar, the microstructure of the sample is clearly visible (Figure 7a). The metal surface appears corroded resulting from the selective dissolution of the ferritic phase over the cementite contained in the perlitic phase. By contrast, Figure 7c,d show that after the addition of the inhibitor the metal surface becomes much smoother. It is clear from the image that the metal surface was partially covered by a protective layer, although some areas of the surface still show signs of corrosion attacks.

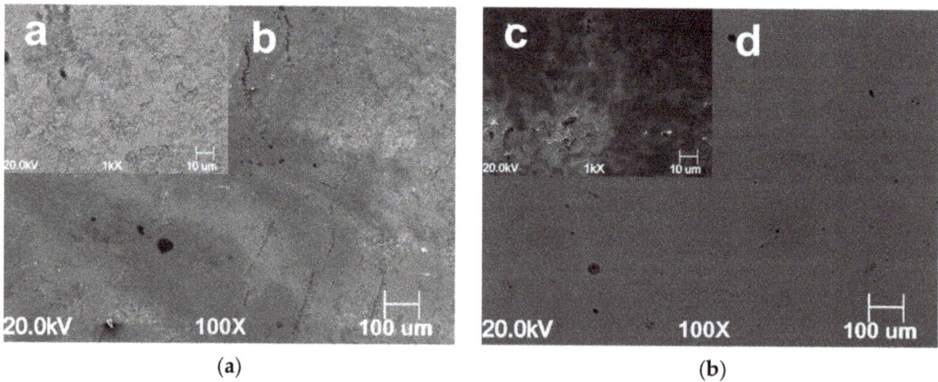

Figure 7. SEM images of the N80 carbon steel surface morphology after 24 h of immersion in the uninhibited ((**a**) **a** lower and **b** higher magnification) and inhibited ((**b**) **c** lower and **d** higher magnification) solution at 25 °C and P_{CO_2} = 1 bar CO_2.

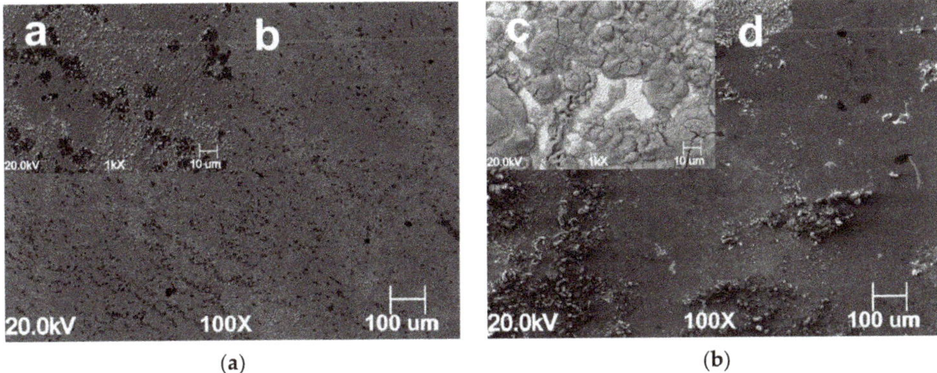

Figure 8. SEM images of the N80 carbon steel surface morphology after 24 h of immersion in the uninhibited ((**a**) **a** lower and **b** higher magnification) and inhibited ((**b**) **c** lower and **d** higher magnification) solution at 25 °C and P_{CO_2} = 20 bar CO_2.

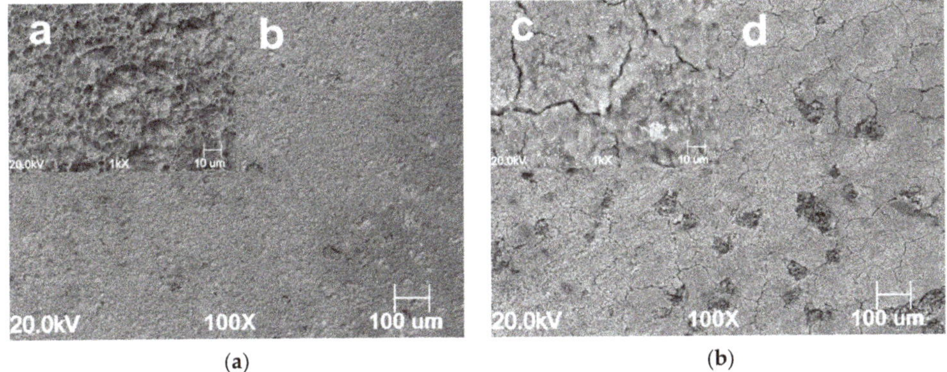

Figure 9. SEM images of the N80 carbon steel surface morphology after 24 h of immersion in the uninhibited ((**a**) **a** lower and **b** higher magnification) and inhibited ((**b**) **c** lower and **d** higher magnification) solution at 25 °C and P_{CO_2} = 40 bar CO_2.

The severity of the corrosion attack increases with an increase in CO_2 partial pressure in the blank solution, as shown in Figure 8a,b and Figure 9a,b respectively carried out at P_{CO_2} = 20 bar and P_{CO_2} = 40 bar. However, it can be seen that in the presence of GA, an increase in CO_2 partial pressure led to a gradual increase in the surface coverage on the metal surface, as a result of an increase of GA molecules adsorbed onto the metal surface (Figure 8c,d and Figure 9c,d). At higher CO_2 partial pressure (e.g., P_{CO_2} = 40 bar, Figure 9) the protective action of the inhibitor is even more evident. The images show that for the uninhibited solution, the surface of the metal appears severely corroded, while the one obtained in the presence of the inhibitor shows the formation of a uniform protective layer over its entire metal surface. The results indicate that in the presence of GA and with a gradual increase in CO_2 partial pressure, the protective layer gradually becomes more compact and thicker [4]. As discussed in Section 3.1, the solubility of CO_2 increases with its partial pressure, and as a result of this, the concentration of H^+ ions into the solution also increases, hence the number of the inhibitor molecules that can be protonated and adsorbed onto the metal surface also increases according to the Equation (12), leading to a substantial reduction of the corrosion rate of the metal.

The morphology of the metal surface was also analyzed with the help of an energy-dispersive spectroscopy with the result listed in Table 4. In the absence of GA, the metal surface was characterized by a corrosion product layer mainly consisting of carbon, iron, and a small amount of oxygen elements, indicating that this corrosion layer is mainly composed of Fe_3C. These results are in agreement with that previously observed in the literature [5,7,39,40]. Other researchers reported that at a temperature below 40 °C, the corrosion product layer is generally composed of Fe_3C, and only little traces of $FeCO_3$ were observed on the metal surface [4,7,39,40], as also confirmed by the GIXRD measurements shown in Figure 10. The presence of Fe_3C on the metal surface is due to the anodic dissolution of the ferrite phase over the cementite in the perlitic phase, which leads to an accumulation of the cementite on the metal surface.

Table 4. Weight percentage of the elements calculated from EDS analyses.

Element	Weight%			
	C	**O**	**Fe**	**Total**
Polished	0.70	-	99.30	100
Blank (1 bar)	1.18	-	98.82	100
1.0 g L^{-1} (1 bar)	4.06	3.51	92.43	100
Blank (20 bar)	7.28	0.83	91.89	100
1.0 g L^{-1} (20 bar)	8.00	21.98	70.02	100
Blank (40 bar)	4.99	2.05	92.96	100
1.0 g L^{-1} (40 bar)	9.90	16.21	73.89	100

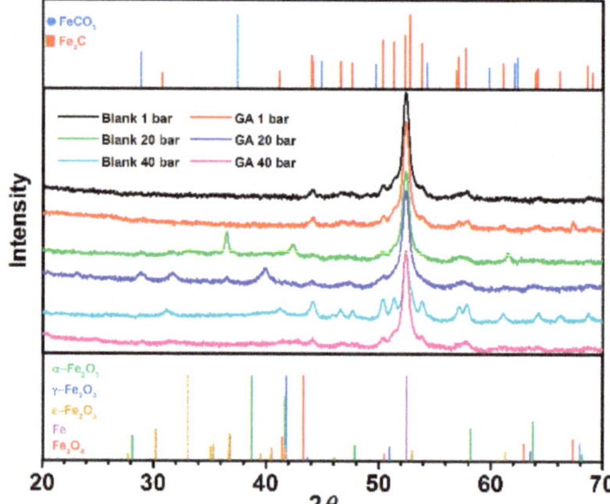

Figure 10. XRD spectra of corrosion product film formed on the metal surface after been exposed for 24 h without and with the presence of 1.0 g L^{-1} of GA at different CO_2 partial pressures at 25 °C.

It is worth mentioning that in the presence of GA the content of carbon and oxygen was found to be higher than those observed for the blank solution. It should be noted that carbon and oxygen are also the main constituents of the tested inhibitor and therefore, their higher concentration on the protective layer formed in the presence of the inhibitor can be attributed to its adsorption onto the metal surface, as also reported by other studies [4,7,20]. Moreover, it can be seen from the table that the percentage of Fe decreased in the presence of GA, likely due to the overlying effect of the inhibitor layer.

The GIXRD analysis for the samples corroded in an inhibited and uninhibited solution at P_{CO_2} = 40 bar and at 25 °C (Figure 10) shows the presence of cementite on the metal surface, although in the presence of GA the intensity of these peaks is much weaker. This result can be explained as follows: Fe_3C accumulates on the metal surface after the dissolution of the ferritic phase. However, in the presence of the inhibitor, it only accumulates in small amounts on the bare metal surface at the early stage of the experiment, since the dissolution of the ferritic phase is quickly suppressed by the absorption of the inhibitor on the surface of the metal.

Figure 11a,b show the surface morphology for specimens corroded in the blank and inhibited solution carried out at 60 °C and P_{CO_2} = 40 bar, after immersion the samples for 24 h in the tested solution, without and with the presence of GA, respectively. The corrosion product layer appears to be different for the inhibited solution compared to one observed in the presence of GA. Figure 11a shows the presence of a porous corrosion product layer formed onto the metal surface corroded in a free-inhibitor solution, pores which create paths for the solution to penetrate it and thereby leading to the dissolution of the underlying metal. On the other hand, the surface of the metal corroded in the presence of GA (Figure 11b) shows the formation of a more compact layer, which forms a better protective barrier and thereby greatly reducing the corrosion rate of the metal.

Figure 11. SEM images of the N80 carbon steel morphology after 24 h of immersion in the tested solution at P_{CO_2} = 40 bar and at 60 °C, without (**a**) and with (**b**) the presence of GA.

EDS analysis reports high content of carbon, oxygen, and iron elements in both layers (C:11.11%, O:6.02% and C:13.69%, O:11.0%, in the blank and inhibited solution, respectively). The GIXRD measurements presented in Figure 12 show the characteristic XRD diffraction patterns associated with $FeCO_3$. By contrast, the intensity of the iron carbonate peaks observed in the presence of GA is almost negligible. These results suggest that the layer observed for the uninhibited solution is mainly composed of Fe_3C and $FeCO_3$, while in the presence of GA is mainly composed of Fe_3C with little traces of $FeCO_3$ [4]. Similar behavior was also reported by Ding et al. [41] related to the study of the effect of an imidazoline-type inhibitor against CO_2 corrosion of mild steel. The authors suggested that the formation of the corrosion inhibitor layer was able to suppress the formation of the iron carbonate. The precipitation of $FeCO_3$ depends on the concentration of the Fe^{2+} and CO_3^{2-} ions, pH, and temperature. When the concentrations of Fe^{2+} and CO_3^{2-} ions exceed the solubility limit, $FeCO_3$ will precipitate on the surface [9,40,42]. At higher temperatures, its solubility decreases, and therefore the likelihood of its precipitation will be also higher. In a free-inhibitor solution, the dissolution of the ferrite phase may lead to an increase in the concentration of Fe^{2+} ions in the bulk solution and thereby favoring the precipitation of $FeCO_3$ onto the surface of the metal. Conversely, in the presence of the inhibitor, the protective layer formed onto the surface of the metal slows down the corrosion processes, and thereby reducing the concentrations of Fe^{2+} ions available for the formation of $FeCO_3$.

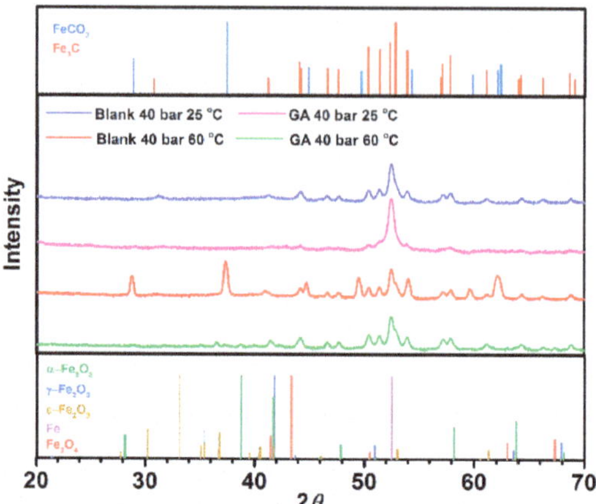

Figure 12. XRD spectra of corrosion product film formed on the metal surface after been exposed for 24 h without and with the presence of 1.0 g L^{-1} of GA at P$_{CO_2}$ = 40 bar and at 60 °C.

Figure 13a–d show the SEM analysis of the metal surface after 168 h of immersion in the absence and presence of 1.0 g L^{-1} GA at P$_{CO_2}$ = 40 bar, respectively. It is apparent from the figures that a thick porous layer covers both surface samples; although it seems that in the presence of the GA, this layer appears denser, thus providing a higher level of protection. To analyze the condition of the metal surface, these porous layers were removed with the help of Clark's solution. It can be seen that both surfaces show clear signs of corrosion attacks (Figure 13b; however, it is also clear from the figures that in the presence of the inhibitor (Figure 13d) the surface of the metal appears to be less damaged and smoother, with the ground scratches still visible on the surface. This result was also confirmed by the atomic force microscopy experiments performed by Azzaoui et al. [28] concerning the use of GA as a corrosion inhibitor in a 1 M HCl solution. The authors reported that in the uninhibited solution the surface of the metal was found to be more corroded with an average roughness of 1.3 μm, while in the presence of GA the average roughness was reduced to 500 nm. The authors justified this behavior due to the formation of a more compact protective layer on the metal surface that strongly reduced the diffusion of the aggressive substances to the metal, and thereby reducing the corrosion rate of the metal.

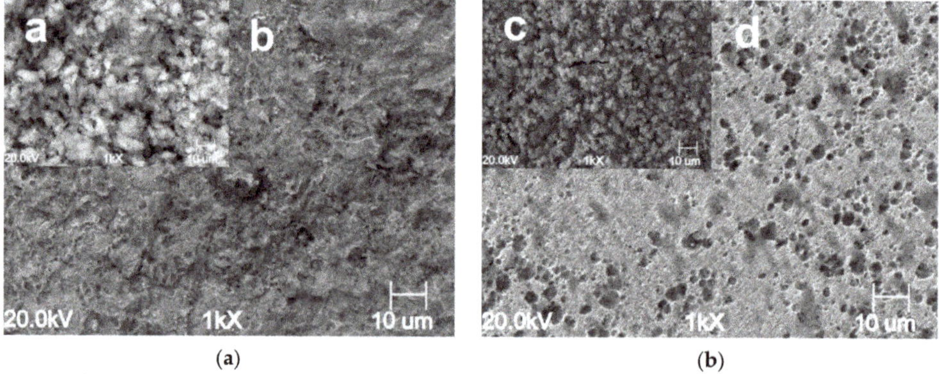

(a)   (b)

Figure 13. SEM images of the N80 carbon steel morphology after 168 h of immersion in the tested solution in the presence of 1.0 g L^{-1} of GA at P$_{CO_2}$ = 40 bar. Without (**a,b**) and with the inhibitor (**c,d**) at 25 °C.

The SEM-EDS and GIXRD result confirm that GA provides adequate protection to the metal surface from sweet corrosion even at high CO_2 partial pressures and after long immersion times. The results are in agreement with the findings obtained with the weight loss measurements, confirming the high inhibition efficiency value observed after a long immersion time.

X-ray photoelectron spectroscopy analysis was employed as a means to confirm the adsorption of the tested inhibitor on the carbon steel surface. The analysis was carried out on the native inhibitor and the steel surface after 24 h of immersion in the tested solution in the presence of 1.0 g L^{-1} of GA at P_{CO_2} = 40 bar and at 25 °C. The XPS results presented in Figure 14a showed evidence of the presence of O, C, N, and Fe on the carbon steel surface, where the O and C contents displayed the highest amount, while the signal of N was detected with small intensity. The high-resolution peaks core levels were analyzed through a deconvolution fitting of the complex spectra. The binding energies and the corresponding quantification (%) of each peak component are presented in Table S4.

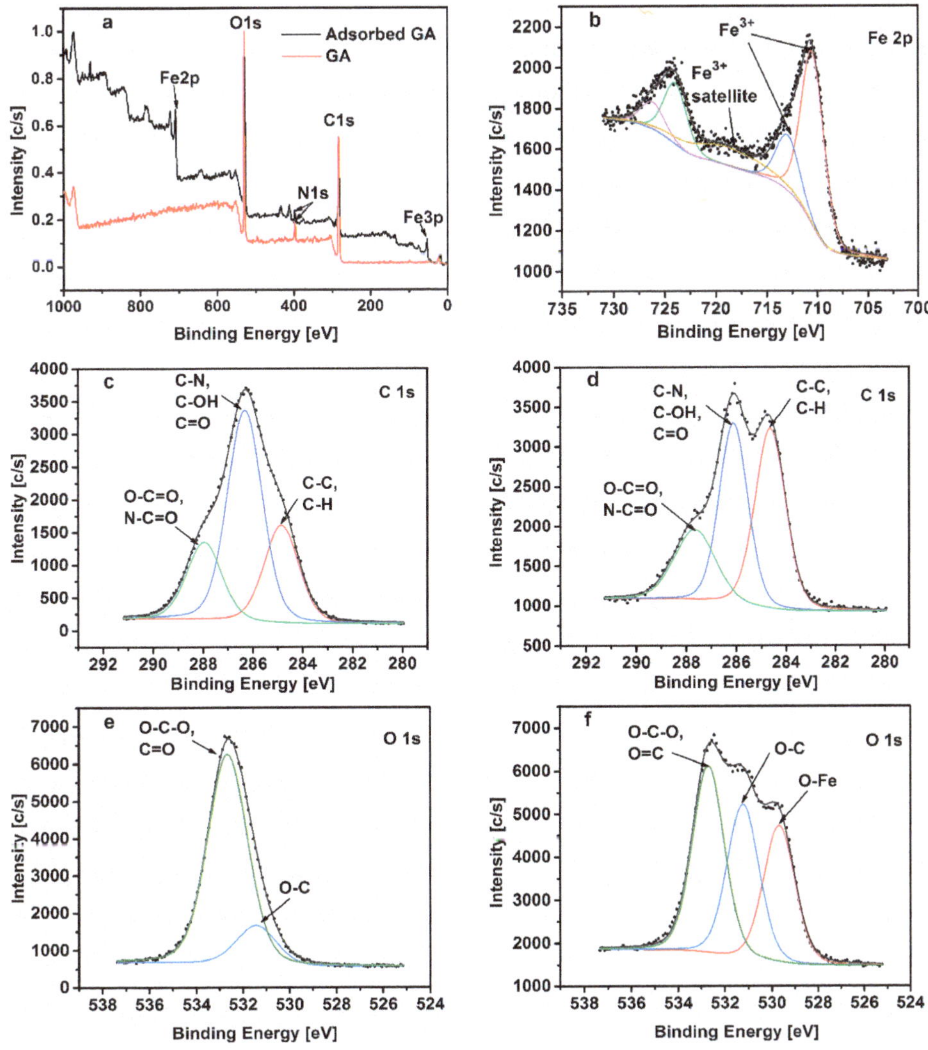

Figure 14. *Cont.*

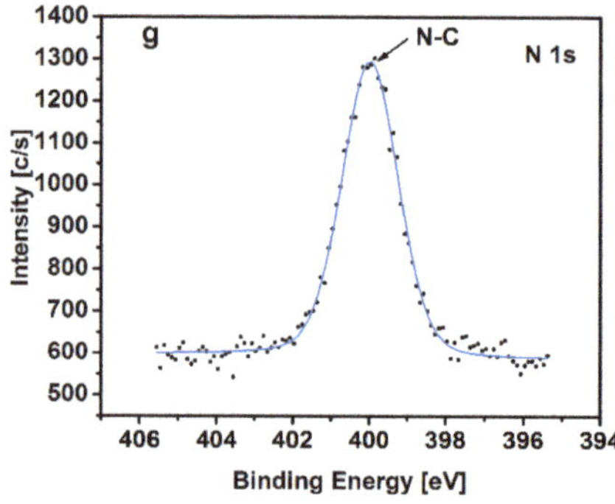
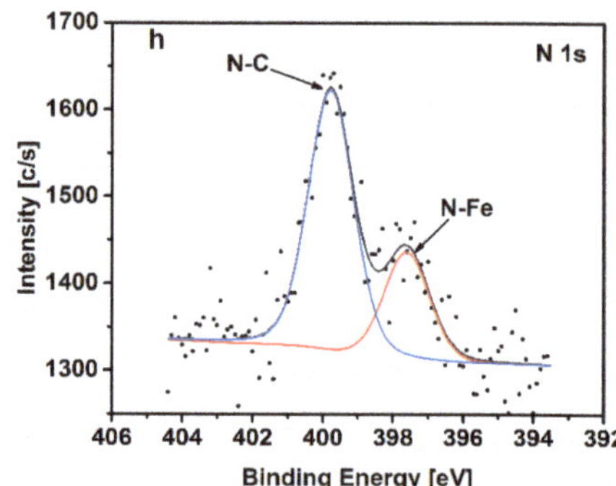

Figure 14. XPS spectra of the native gum Arabic: (**a,c,e,g**). XPS spectra of the film formed on the N80 carbon steel after 24 h exposure in CO_2 at P_{CO_2} = 40 bar in the presence of 1.0 g L^{-1} of GA at 25 °C: (**b,d,f,h**).

The deconvoluted $Fe2p_{3/2}$ peaks (Figure 14b) at 710.5 and 713.0 eV could be associated with the α-Fe_2O_3 or/and γ- Fe_2O_3 [13]. The presence of these species is likely due to the partial decomposition of iron carbonate. The literature reported that $FeCO_3$ begins decomposing at temperatures below 100 °C according to the following reaction [5,42]:

$$FeCO_3 \rightarrow FeO + CO_2 \tag{15}$$

In the presence of CO_2 or water vapor, FeO transforms into Fe_3O_4 [5,42].

$$3FeO + CO_2 \rightarrow Fe_3O_4 + CO \tag{16}$$

$$3FeO + H_2O \rightarrow Fe_3O_4 + H_2 \tag{17}$$

However, in the presence of oxygen, FeO and Fe_3O_4 transform into Fe_2O_3 [5,42].

$$4FeO + O_2 \rightarrow 2Fe_2O_3 \tag{18}$$

$$\text{In the air}: \ 4Fe_3O_4 + O_2 \rightarrow 6Fe_2O_3 \tag{19}$$

The C1s spectra of the native gum arabic and the adsorbed one (Figure 14c,d, respectively) show three main peaks. The C1s peak with binding energy at 284.8 eV could be attributed to the C–C/C–H bonds [26,28]. The C1s peak at 286.2 eV could be attributed to the C–OH/C=O bonds related to the different groups of GA [26,43]. This peak may also be assigned to the carbon atom bonded to nitrogen in C–N bond [13,44] and could be related to the glycoprotein and/or to the arabinogalactan-protein fractions of the inhibitor (Figure 15b,c, respectively). The last C1s peak with a binding energy of 287.7 eV could be associated with the presence of carbonyl type groups O–C=O/N–C=O that result from the protonation of the GA molecule in the acid environment [28].

It is worth mentioning that no peaks assigned to Fe_3C were found with the XPS analysis in contrast to the results reported from the GIXRD analysis, where the characteristic peaks assigned to this compound can be seen in the presence of GA (Figure 10). Fe_3C cannot be detected since the average depth of analysis for an XPS measurement is approximately 5 nm however, the cementite formed on the metal surface at the early stage of the experiment is covered by a thicker layer of inhibitor (Figure 9c,d).

The deconvoluted O1s spectra of the native and adsorbed inhibitor are displayed in Figure 14e,f, respectively. The peaks at 531.2 and 532.7 eV could be attributed to the single bonded oxygen in C–O and the double bonded oxygen C=O and/or to the single bonded oxygen in O–C–O respectively [4,13,26,28]. The latter peak may correspond to the carbonyl type groups and/or to the glycosidic C(1)-O-C(4)/C(1)-O-C(6) linkages of the GA molecules (Figure 15a), as well as, in the case of the sample exposed to the tested solution, to $FeCO_3$ formed on the metals surface, respectively [4,26,28]. Moreover, some authors reported that the peak at 231.2 eV could also be attributed to the oxygen of the hydroxyl groups (–OH) [5,43], likely due to the hydroxyl groups of the tested polysaccharide. The O1s spectrum of the adsorbed inhibitor (Figure 14f) displays an extra peak at 529.7 eV corresponding to O^{2-} related to the oxygen atoms bonded with Fe^{3+} in the Fe_2O_3 oxide [4,43,44]. The O1s results are in good agreement with the findings of the Fe2p spectrum.

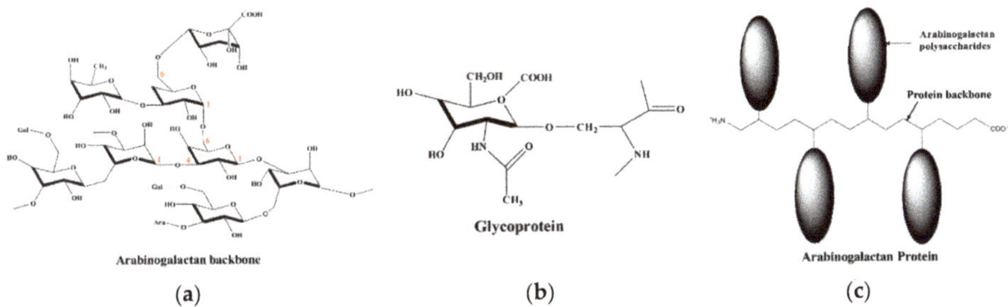

Figure 15. Structure of gum arabic: (**a**) arabinogalactan; (**b**) glycoprotein; (**c**) arabinogalactan-protein.

The presence of N1s peak in the survey for the adsorbed GA on the carbon steel surface (Figure 14a) provides evidence that gum arabic was effectively adsorbed on the tested substrate surface since the N80 carbon steel substrate does not contain nitrogen in its chemical composition. The N1s spectra of the native and adsorbed inhibitor are presented in Figure 14g,h. Both images show the presence of a peak at 400 and 399.8 eV attributed to the nitrogen atom bonded with the carbon atom, for the native and adsorbed inhibitor. However, as it can be seen that the high-resolution N1s spectrum of the tested substrate sample after the addition of GA depicts an extra peak at 397.6 eV. This extra peak can be ascribed to the coordinated nitrogen atom of the amino group with the metal surface (N–Fe bond) [44]. Other authors also suggested that this peak could be attributed to the bond between the nitrogen of the amino groups and the oxide layer on the metal surface (FeO_x) [45].

3.5. Mechanism of Inhibition

Given all the observed results, it can be inferred that the GA was effectively adsorbed on the metal surface, providing good protection to the metal surface against sweet corrosion. However, the complex chemical structure of this inhibitor makes it difficult to determine the exact adsorption mechanism involved. Gum arabic is a heterogeneous mixture of different compounds consisting of three main fractions: 80% of arabinogalactan (AG), 10.4% of arabinogalactan-protein (AGP) and 1.2% glycoprotein (GP) (Figure 15). Each of these fractions contains a range of different molecular weight components and different protein contents. Therefore, some of these compounds can be physically and others chemically adsorbed. Nevertheless, based on the results reported in this study, it can be assumed that the following three types of adsorption mechanisms or likely a combination of them may take place in the inhibiting phenomena involving GA on the steel surface.

3.5.1. Adsorption via Electrostatic Interaction

The functional groups such as hydroxyl, carboxyl, and amino present in the GA molecules, by virtue of the presence of lone pair of electrons, can be easily protonated in acid solutions such that the newly formed polycations are in equilibrium with their neutral counterpart according to the Equation (12). The high corrosion inhibition activity showed by GA is likely due to a synergistic electrostatic interaction between the protonated GA molecules with the adsorbed chloride ions, as shown in Figure 16a. As reported by several studies [7,20–22,28] chloride ions are strongly adsorbed on the positively charged metal surface, thereby creating an excess of electrons so that the metal will be negatively charged. These adsorbed chloride ions can act as an intermediate bridge between the surface and the protonated inhibitor molecules and therefore, assisting the adsorption of GA on the metal surface. This type of adsorption mechanism is likely the one that accounts for the most inhibition action of the inhibitor. In fact, the results presented in this manuscript have demonstrated clearly that the corrosion inhibition action of GA was strongly influenced by both the concentration of the inhibitor, CO_2 partial pressure, and temperature. A change in one of these two factors has a great effect on the equilibrium reaction (Equation (12)), shifting the equilibrium towards the protonated or the deprotonated form of the inhibitor. A shift to the right implies an increase in the number of protonated molecules of GA available to interact with the chloride ions adsorbed on the surface and thus, an increase in *IE* of the system.

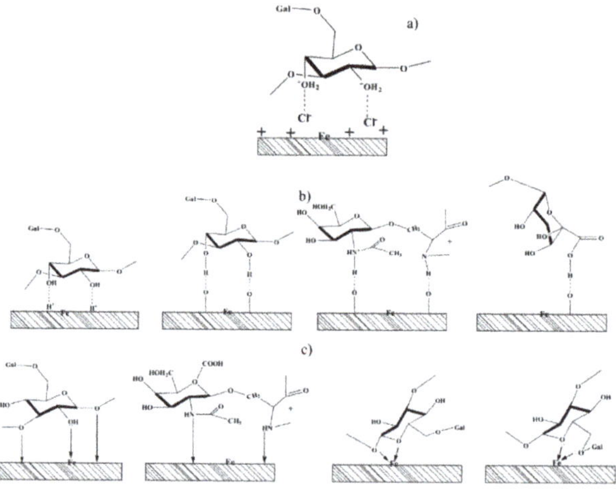

Figure 16. Schematic representation of the corrosion inhibition mechanism of the N80 carbon steel by GA. (**a**) electrostatic; (**b**) H-bond formation; (**c**) chemical adsorption.

3.5.2. Adsorption via Hydrogen Bond Formation Interaction

At higher CO_2 partial pressure (i.e., 40 bar) the pH of the solution is around 3 [32], and among the three possible occurring cathodic reactions (Equations (9)–(11)), the reduction of hydrogen ions to hydrogen gas is the dominant cathodic reaction. It is generally accepted that this reaction can be described using three steps [46]. The first step is the electrochemical adsorption of the H^+ ions (Equation (20)) followed by either the electrochemical desorption (Equation (21)) or the chemical desorption (Equation (22)).

$$H^+_{(aq)} + e^- \rightarrow H_{ads} \tag{20}$$

$$H_{ads} + H^+_{(aq)} \rightarrow H_{2(g)} \tag{21}$$

$$H_{ads} + H_{ads} \rightarrow H_{2(g)} \tag{22}$$

The potentiodynamic measurements presented in Figure 5 showed that the cathodic current density of the system was greatly reduced after the addition of GA in the solution, suggesting that GA was able to suppress the hydrogen evolution reaction (Equation (11)) to some extent. Similar results were also confirmed by other authors [21,26–28]. This assumption was also confirmed by FT-IR and Raman measurements performed on GA [7,26] and other gum-like [20,36,47] compounds. The results showed that the characteristic peak assigned to the hydroxyl groups of the carbohydrate units narrowed down and/or shifted after its adsorption on the metal surface. The authors agreed that this change in shape was likely due to a possible interaction of the hydroxyl groups of the GA molecules with the H adsorbed on the cathodic sites of the metal surface via H-bond formation (Figure 16b). Therefore, the high value of IE observed in this study at different CO_2 partial pressure can be also ascribed to the ability of GA to suppress one of these reactions (Equations (20)–(22)) via H-bonds formation, thus suppressing Equation (11) and consequently the dissolution of the steel (Equation (8)).

The adsorption of GA may also be promoted by the presence of the oxide layer on the metal surface via hydrogen bonding (Figure 16b). Studies concerning the adsorption of GA on oxide nanoparticles (i.e., iron oxide nanoparticles [48] and zinc or aluminum oxide nanoparticles [49]) reported that GA showed a strong affinity toward these oxide nanoparticles. The authors suggested that the adsorption of GA on these oxide nanoparticles surface might be due to the formation of hydrogen bonds between the functional groups of the GA molecules (e.g., hydroxyl, carboxylate, and amino) with the oxidized surface. The XPS analysis presented in this study showed that the metal surface after 24 h of exposure is covered by different oxide species such as Fe_2O_3 and/or Fe_3O_4, (e.g., Equations (15)–(19)). Therefore, the adsorption of GA assisted by the presence of oxide species formed on the metal surface via H-bonds formation is an adsorption mechanism that must be also taken into account.

3.5.3. Chemical Adsorption

The heteroatoms (i.e., O, N) present on the GA molecules by virtue of the presence of lone pair of electrons may promote the adsorption of the inhibitor via the formation of coordinate bonds with the iron from the metal surface and/or with iron from the oxide species formed on the surface [45] (Figure 16c). The XPS measurements observed in this study showed a peak at 397.6 eV likely ascribed to the coordinated nitrogen atom of the amino group with the Fe (N–Fe bond) [44]. This result suggests that although the inhibitor is mainly physically adsorbed on the surface of the metal, a small contribution of the chemical adsorption process cannot be ignored.

4. Conclusions

The corrosion inhibition effect of gum Arabic on the corrosion of carbon steel (N80) exposed in a high-pressure CO_2-saline environment has been studied and the following conclusion can be drawn:

- The weight loss results showed that the thickening agent gum arabic was found to be an efficient corrosion inhibitor for carbon steel in a high-pressure CO_2-saline environment. The Inhibition efficiency increased with an increase in inhibitor concentration and CO_2 partial pressure with the maximum value of IE found to be 84.53% at P_{CO_2} = 40 bar after 24 h of immersion. Moreover, the weight loss results also showed that GA was effectively able to protect the steel surface from sweet corrosion at high CO_2 partial pressures (i.e., 40 bar) even after a prolonged immersion time (i.e., 168 h) with a corrosion inhibition efficiency found to be 74.41%.
- The adsorption of GA on the carbon steel surface follows the Temkin's adsorption isotherm model. The negative free energy of adsorption $\Delta G°_{ads}$ indicates a strong and spontaneous adsorption of GA on the carbon steel surface. Furthermore, the value of $\Delta G°_{ads}$ indicates that the GA adsorbs mainly via physical adsorption on the metal surface.
- The SEM analysis revealed that in the presence of GA the protective layer on the metal surface becomes more compact and dense with an increase in CO_2 partial pressure. Also, the SEM analysis revealed that after 168 h of immersion, in the presence of GA, the metal surface appeared to be less damaged and smother.
- The XPS results confirmed the formation of a protective layer containing GA molecules and iron oxides on the metal surface.

Supplementary Materials: The following are available online at http://www.mdpi.com/1996-1944/13/19/4245/s1, Table S1: Corrosion rate and inhibition efficiency obtained from weight loss measurements for the N80 carbon steel at various concentrations of GA and CO_2 partial pressures after 24 h of immersion time, Table S2: Comparison of reported inhibition efficiency of some other corrosion inhibitors used in a CO_2 saturated saline solution (3.5 wt.% NaCl), Table S3: Corrosion rate and inhibition efficiency obtained from weight loss measurements for the carbon steel (N80) carried out at 1.0 g L−1 of GA and CO_2 partial pressures after 168 h of immersion time at 25 °C, Table S4: XPS analysis of sample steel surface after 24 h of immersion in test solution at P_{CO_2} = 40 bar and at 25 °C in the presence of 1.0 g L^{-1} of GA.

Author Contributions: G.P. conceived, designed, and performed the measurements, analyzed the experimental data, wrote and edited the manuscript; K.K. performed the XRD analysis; R.W. and A.B. performed the XPS. analysis; M.G. performed the SEM-EDS analysis. All authors have read and agreed to the published version of the manuscript.

Funding: This research received no external funding.

Acknowledgments: RW has been partly supported by the EU Project POWR.03.02.00-00-I004/16.

Conflicts of Interest: The authors declare no conflict of interest.

References

1. Sheng, J.J. Enhanced oil recovery in shale reservoirs by gas injection. *J. Nat. Gas Sci. Eng.* **2015**, *22*, 252–259. [CrossRef]
2. Bai, H.; Wang, Y.; Ma, Y.; Zhang, Q.; Zhang, N. Effect of CO_2 Partial Pressure on the Corrosion Behavior of J55 Carbon Steel in 30% Crude Oil/Brine Mixture. *Materials* **2018**, *11*, 1765. [CrossRef] [PubMed]
3. Bai, H.; Wang, Y.; Ma, Y.; Ren, P.; Zhang, N. Pitting Corrosion and Microstructure of J55 Carbon Steel Exposed to CO_2/Crude Oil/Brine Solution under 2–15 MPa at 30–80 °C. *Materials* **2018**, *11*, 2374. [CrossRef] [PubMed]
4. Mustafa, A.H.; Ari-Wahjoedi, B.; Ismail, M.C. Inhibition of CO_2 Corrosion of X52 Steel by Imidazoline-Based Inhibitor in High Pressure CO_2-Water Environment. *J. Mater. Eng. Perform.* **2012**, *22*, 1748–1755. [CrossRef]
5. Islam, A.; Farhat, Z.N. Characterization of the Corrosion Layer on Pipeline Steel in Sweet Environment. *J. Mater. Eng. Perform.* **2015**, *24*, 3142–3158. [CrossRef]
6. Aristia, G.; Hoa, L.Q.; Bäßler, R. Corrosion of Carbon Steel in Artificial Geothermal Brine: Influence of Carbon Dioxide at 70 °C and 150 °C. *Materials* **2019**, *12*, 3801. [CrossRef]
7. Palumbo, G.; Górny, M.; Banaś, J. Corrosion Inhibition of Pipeline Carbon Steel (N80) in CO_2-Saturated Chloride (0.5 M of KCl) Solution Using Gum Arabic as a Possible Environmentally Friendly Corrosion Inhibitor for Shale Gas Industry. *J. Mater. Eng. Perform.* **2019**, *28*, 6458–6470. [CrossRef]
8. Palumbo, G.; Banas, J.; Bałkowiec, A.; Mizera, J.; Lelek-Borkowska, U. Electrochemical study of the corrosion behaviour of carbon steel in fracturing fluid. *J. Solid State Electrochem.* **2014**, *18*, 2933–2945. [CrossRef]
9. Linter, B.; Burstein, G. Reactions of pipeline steels in carbon dioxide solutions. *Corros. Sci.* **1999**, *41*, 117–139. [CrossRef]
10. Palumbo, G.; Banaś, J. Inhibition effect of guar gum on the corrosion behaviour of carbon steel (K-55) in fracturing fluid. *Solid State Phenom.* **2015**, *227*, 59–62. [CrossRef]
11. Tang, J.; Hu, Y.; Han, Z.; Wang, H.; Zhu, Y.; Wang, Y.; Nie, Z.; Wang, Y. Experimental and Theoretical Study on the Synergistic Inhibition Effect of Pyridine Derivatives and Sulfur-Containing Compounds on the Corrosion of Carbon Steel in CO_2-Saturated 3.5 wt.% NaCl Solution. *Molecules* **2018**, *23*, 3270. [CrossRef]
12. Ortega-Toledo, D.; Gonzalez-Rodriguez, J.G.; Casales, M.; Martinez, L.; Martinez-Villafañe, A. Co2 corrosion inhibition of X-120 pipeline steel by a modified imidazoline under flow conditions. *Corros. Sci.* **2011**, *53*, 3780–3787. [CrossRef]
13. Singh, A.; Ansari, K.R.; Quraishi, M.A.; Lgaz, H. Effect of Electron Donating Functional Groups on Corrosion Inhibition of J55 Steel in a Sweet Corrosive Environment: Experimental, Density Functional Theory, and Molecular Dynamic Simulation. *Materials* **2018**, *12*, 17. [CrossRef] [PubMed]
14. Ghareba, S.; Omanovic, S. The effect of electrolyte flow on the performance of 12-aminododecanoic acid as a carbon steel corrosion inhibitor in CO_2-saturated hydrochloric acid. *Corros. Sci.* **2011**, *53*, 3805–3812. [CrossRef]
15. Umoren, S.; Alahmary, A.A.; Gasem, Z.M.; Solomon, M. Evaluation of chitosan and carboxymethyl cellulose as ecofriendly corrosion inhibitors for steel. *Int. J. Boil. Macromol.* **2018**, *117*, 1017–1028. [CrossRef] [PubMed]
16. Ansari, K.; Chauhan, D.S.; Quraishi, M.; Mazumder, M.A.; Singh, A. Chitosan Schiff base: An environmentally benign biological macromolecule as a new corrosion inhibitor for oil & gas industries. *Int. J. Boil. Macromol.* **2020**, *144*, 305–315. [CrossRef]
17. Lin, Y.; Singh, A.; Ebenso, E.E.; Quraishi, M.A.; Zhou, Y.; Huang, Y. Use of HPHT Autoclave to Determine Corrosion Inhibition by Berberine extract on Carbon Steels in 3.5% NaCl Solution Saturated with CO_2. *Int. J. Electrochem. Sci.* **2015**, *10*, 194–208.
18. Singh, A.; Lin, Y.; Liu, W.; Ebenso, E.E.; Pan, J. Extract of Momordica charantia (Karela) Seeds as Corrosion Inhibitor for P110SS Steel in CO_2 Saturated 3.5% NaCl Solution. *Int. J. Electrochem. Sci.* **2013**, *8*, 12884–12893.

19. Singh, A.; Lin, Y.; Ebenso, E.E.; Liu, W.; Pan, J.; Huang, B. Gingko biloba fruit extract as an eco-friendly corrosion inhibitor for J55 steel in CO_2 saturated 3.5% NaCl solution. *J. Ind. Eng. Chem.* **2015**, *24*, 219–228. [CrossRef]

20. Palumbo, G.; Berent, K.; Proniewicz, E.; Banaś, J. Guar Gum as an Eco-Friendly Corrosion Inhibitor for Pure Aluminium in 1-M HCl Solution. *Materials* **2019**, *12*, 2620. [CrossRef]

21. Bentrah, H.; Rahali, Y.; Chala, A. Gum Arabic as an eco-friendly inhibitor for API 5L X42 pipeline steel in HCl medium. *Corros. Sci.* **2014**, *82*, 426–431. [CrossRef]

22. Umoren, S. Inhibition of aluminium and mild steel corrosion in acidic medium using Gum Arabic. *Cellulose* **2008**, *15*, 751–761. [CrossRef]

23. Singh, A.; Ansari, K.; Quraishi, M. Inhibition effect of natural polysaccharide composite on hydrogen evolution and P110 steel corrosion in 3.5 wt% NaCl solution saturated with CO_2: Combination of experimental and surface analysis. *Int. J. Hydrogen Energy* **2020**, *45*, 25398–25408. [CrossRef]

24. Umoren, S.; Ogbobe, O.; Igwe, I.; Ebenso, E. Inhibition of mild steel corrosion in acidic medium using synthetic and naturally occurring polymers and synergistic halide additives. *Corros. Sci.* **2008**, *50*, 1998–2006. [CrossRef]

25. Mobin, M.; Alam Khan, M. Investigation on the Adsorption and Corrosion Inhibition Behavior of Gum Acacia and Synergistic Surfactants Additives on Mild Steel in 0.1 MH_2SO_4. *J. Dispers. Sci. Technol.* **2013**, *34*, 1496–1506. [CrossRef]

26. Abu-Dalo, M.A.; Othman, A.A.; Al-Rawashdeh, N.A.F. Exudate gum from acacia trees as green corrosion inhibitor for mild steel in acidic media. *Int. J. Electrochem. Sci.* **2012**, *7*, 9303–9324.

27. Shen, C.; Alvarez, V.; Koenig, J.D.B.; Luo, J.-L. Gum Arabic as corrosion inhibitor in the oil industry: Experimental and theoretical studies. *Corros. Eng. Sci. Technol.* **2019**, *54*, 444–454. [CrossRef]

28. Azzaoui, K.; Mejdoubi, E.; Jodeh, S.; Lamhamdi, A.; Rodríguez-Castellón, E., Algarra, M.; Zarrouk, A.; Errich, A.; Salghi, R.; Lgaz, H. Eco friendly green inhibitor Gum Arabic (GA) for the corrosion control of mild steel in hydrochloric acid medium. *Corros. Sci.* **2017**, *129*, 70–81. [CrossRef]

29. Spellman, F.R. *Environmental Impacts of Hydraulic Fracturing*; Informa UK Limited: London, UK, 2012.

30. *ASTM-G1-90, Standard Practice for Preparing, Cleaning, and Evaluation Corrosion Test Specimens*; ASTM International: West Conshohocken, PA, USA, 1999.

31. Choi, Y.-S.; Nešić, S. Determining the corrosive potential of CO_2 transport pipeline in high pCO_2–water environments. *Int. J. Greenh. Gas. Control* **2011**, *5*, 788–797. [CrossRef]

32. Li, X.; Peng, C.; Crawshaw, J.; Maitland, G.; Trusler, J.M. The pH of CO_2-saturated aqueous NaCl and NaHCO3 solutions at temperatures between 308 K and 373 K at pressures up to 15 MPa. *Fluid Phase Equilibria* **2018**, *458*, 253–263. [CrossRef]

33. Dong, B.; Liu, W.; Zhang, Y.; Banthukul, W.; Zhao, Y.; Zhang, T.; Fan, Y.; Li, X.; Wei, L.; Yonggang, Z.; et al. Comparison of the characteristics of corrosion scales covering 3Cr steel and X60 steel in CO_2-H2S coexistence environment. *J. Nat. Gas Sci. Eng.* **2020**, *80*, 103371. [CrossRef]

34. Bousselmi, L.; Fiaud, C.; Tribollet, B.; Triki, E. Impedance spectroscopic study of a steel electrode in condition of scaling and corrosion. *Electrochim. Acta* **1999**, *44*, 4357–4363. [CrossRef]

35. Bousselmi, L.; Fiaud, C.; Tribollet, B.; Triki, E. The characterisation of the coated layer at the interface carbon steel-natural salt water by impedance spectroscopy. *Corros. Sci.* **1997**, *39*, 1711–1724. [CrossRef]

36. Roy, P.; Karfa, P.; Adhikari, U.; Sukul, D. Corrosion inhibition of mild steel in acidic medium by polyacrylamide grafted Guar gum with various grafting percentage: Effect of intramolecular synergism. *Corros. Sci.* **2014**, *88*, 246–253. [CrossRef]

37. Saha, S.K.; Dutta, A.; Sukul, D.; Ghosh, P.; Banerjee, P. Adsorption and corrosion inhibition effect of Schiff base molecules on the mild steel surface in 1 M HCl medium: A combined experimental and theoretical approach. *Phys. Chem. Chem. Phys.* **2015**, *17*, 5679–5690. [CrossRef]

38. Outirite, M.; Lagrenée, M.; Lebrini, M.; Traisnel, M.; Jama, C.; Vezin, H.; Bentiss, F. Ac impedance, X-ray photoelectron spectroscopy and density functional theory studies of 3,5-bis(n-pyridyl)-1,2,4-oxadiazoles as efficient corrosion inhibitors for carbon steel surface in hydrochloric acid solution. *Electrochim. Acta* **2010**, *55*, 1670–1681. [CrossRef]

39. Paolinelli, L.; Perez, T.; Simison, S. The effect of pre-corrosion and steel microstructure on inhibitor performance in CO_2 corrosion. *Corros. Sci.* **2008**, *50*, 2456–2464. [CrossRef]

40. Mora-Mendoza, J.; Turgoose, S. Fe3C influence on the corrosion rate of mild steel in aqueous CO_2 systems under turbulent flow conditions. *Corros. Sci.* **2002**, *44*, 1223–1246. [CrossRef]

41. Ding, Y.; Brown, B.; Young, D.; Singer, M. Effectiveness of an Imidazoline-Type Inhibitor Against CO_2 Corrosion of Mild Steel at Elevated Temperatures (120 °C–150 °C). In Proceedings of the CORROSION 2018, Phoenix, AZ, USA, 15–19 April 2018; p. 22.

42. Heuer, J.; Stubbins, J. An XPS characterization of $FeCO_3$ films from CO_2 corrosion. *Corros. Sci.* **1999**, *41*, 1231–1243. [CrossRef]

43. Boumhara, K.; Tabyaoui, M.; Jama, C.; Bentiss, F. Artemisia Mesatlantica essential oil as green inhibitor for carbon steel corrosion in 1M HCl solution: Electrochemical and XPS investigations. *J. Ind. Eng. Chem.* **2015**, *29*, 146–155. [CrossRef]

44. Bouanis, M.; Tourabi, M.; Nyassi, A.; Zarrouk, A.; Jama, C.; Bentiss, F. Corrosion inhibition performance of 2,5-bis(4-dimethylaminophenyl)-1,3,4-oxadiazole for carbon steel in HCl solution: Gravimetric, electrochemical and XPS studies. *Appl. Surf. Sci.* **2016**, *389*, 952–966. [CrossRef]

45. Hashim, N.Z.N.; Anouar, E.H.; Kassim, K.; Zaki, H.M.; Alharthi, A.I.; Embong, Z. XPS and DFT investigations of corrosion inhibition of substituted benzylidene Schiff bases on mild steel in hydrochloric acid. *Appl. Surf. Sci.* **2019**, *476*, 861–877. [CrossRef]

46. Barker, R.; Burkle, D.; Charpentier, T.; Thompson, H.; Neville, A. A review of iron carbonate ($FeCO_3$) formation in the oil and gas industry. *Corros. Sci.* **2018**, *142*, 312–341. [CrossRef]

47. Messali, M.; Lgaz, H.; Dassanayake, R.; Salghi, R.; Jodeh, S.; Abidi, N.; Hamed, O. Guar gum as efficient non-toxic inhibitor of carbon steel corrosion in phosphoric acid medium: Electrochemical, surface, DFT and MD simulations studies. *J. Mol. Struct.* **2017**, *1145*, 43–54. [CrossRef]

48. Williams, D.N.; Gold, K.A.; Holoman, T.R.P.; Ehrman, S.H.; Wilson, O.C. Surface Modification of Magnetic Nanoparticles Using Gum Arabic. *J. Nanopart. Res.* **2006**, *8*, 749–753. [CrossRef]

49. Leong, Y.; Seah, U.; Chu, S.; Ong, B. Effects of Gum Arabic macromolecules on surface forces in oxide dispersions. *Colloids Surf. A Physicochem. Eng. Asp.* **2001**, *182*, 263–268. [CrossRef]

 © 2020 by the authors. Licensee MDPI, Basel, Switzerland. This article is an open access article distributed under the terms and conditions of the Creative Commons Attribution (CC BY) license (http://creativecommons.org/licenses/by/4.0/).

Article

Effect of Phase Transformation on Stress Corrosion Behavior of Additively Manufactured Austenitic Stainless Steel Produced by Directed Energy Deposition

Tomer Ron *, Ohad Dolev, Avi Leon, Amnon Shirizly and Eli Aghion7

Department of Materials Engineering, Ben-Gurion University of the Negev,
Beer-Sheva 8410501, Israel; dolev.ohad@gmail.com (O.D.); avileon12@gmail.com (A.L.);
a.shirizly@gmail.com (A.S.); egyon@bgu.ac.il (E.A.)
* Correspondence: toron@post.bgu.ac.il

Abstract: The present study aims to evaluate the stress corrosion behavior of additively manufactured austenitic stainless steel produced by the wire arc additive manufacturing (WAAM) process. This was examined in comparison with its counterpart, wrought alloy, by electrochemical analysis in terms of potentiodynamic polarization and impedance spectroscopy and by slow strain rate testing (SSRT) in a corrosive environment. The microstructure assessment was performed using optical and scanning electron microscopy along with X-ray diffraction analysis. The obtained results indicated that in spite of the inherent differences in microstructure and mechanical properties between the additively manufactured austenitic stainless steel and its counterpart wrought alloy, their electrochemical performance and stress corrosion susceptibility were similar. The corrosion attack in the additively manufactured alloy was mainly concentrated at the interface between the austenitic matrix and the secondary ferritic phase. In the case of the counterpart wrought alloy with a single austenitic phase, the corrosion attack was manifested by uniform pitting evenly scattered at the external surface. Both alloys showed ductile failure in the form of "cap and cone" fractures in post-SSRT experiments in corrosive environment.

Keywords: additive manufacturing; direct energy deposition; wire arc additive manufacturing; 316L stainless steel; stress corrosion

Citation: Ron, T.; Dolev, O.; Leon, A.; Shirizly, A.; Aghion, E. Effect of Phase Transformation on Stress Corrosion Behavior of Additively Manufactured Austenitic Stainless Steel Produced by Directed Energy Deposition. *Materials* **2021**, *14*, 55. https://dx.doi.org/10.3390/ma14010055

Received: 23 November 2020
Accepted: 21 December 2020
Published: 24 December 2020

Publisher's Note: MDPI stays neutral with regard to jurisdictional claims in published maps and institutional affiliations.

Copyright: © 2020 by the authors. Licensee MDPI, Basel, Switzerland. This article is an open access article distributed under the terms and conditions of the Creative Commons Attribution (CC BY) license (https://creativecommons.org/licenses/by/4.0/).

1. Introduction

Additive manufacturing (AM) has been considered to be a promising technology for producing a variety of complex components in a relatively short time [1–6]. Traditional AM processing of metals mainly focuses on powder bed fusion (PBF) methods such as selective laser melting (SLM) and electron beam melting (EBM) [7–10]. However, PBF technologies are relatively expensive due to the high cost of raw material, high energy consumption and relatively low deposition rate. In addition, the size of the printed component is limited and depends on the printing cell dimension. The inherent disadvantages of PBF technologies highlight the need to use more affordable AM methods such as the wire arc additive manufacturing (WAAM) process. Comparatively, proven PBF technologies can produce a deposition rate of 0.1 kg/h, while the deposition rate of WAAM is about 10 kg/h [11–13]. In addition, the use of relatively inexpensive wires as raw materials and an electric arc as the energy source can reduce the cost of the printing process by 80% compared to PBF [14,15]. Furthermore, the dimensions of components produced by WAAM are almost unlimited [16]. It should be pointed out that WAAM process can be also implemented using computer numerical control (CNC) systems [17]. The almost unlimited dimensions of WAAM is due to the fact that the printing process can be performed by an external robot that is free to move in all directions [18–20]. However, there are some limitations related to the WAAM process compared to PBF technology. This includes relatively increased surface roughness,

317

and limited capabilities to produce complex structures. Currently, most of the research activities related to the WAAM process have focused on optimizing the printing parameters and residual stresses status [21–25], with very limited attention paid to the corrosion performance of the obtained components. This study mainly aims at evaluating the effect of phase transformation on stress corrosion behavior of additively manufactured austenitic stainless steel in the form of 316L alloy produced by the WAAM process. For reference consideration, the obtained stress corrosion behavior was compared to its counterpart wrought alloy AISI 316L. The general corrosion performance was evaluated in terms of potentiodynamic polarization and electrochemical impedance spectroscopy (EIS) analyses all in 3.5% NaCl solution.

2. Materials and Methods

The tested specimens were machined from a hollow cylindrical component produced by the WAAM process with a conventional metal inert gas (MIG) welding system, using austenitic 316L stainless steel welding wire, as shown in Figure 1 [20] along with a tension specimen. No additional heat treatment was applied and all tested were carried out in as-build conditions. The welding wire diameter was 1.14 mm and the dimensions of the cylindrical component were: 120 mm height, 55 mm mid-wall radius and 15 mm wall thickness. The wire deposition was implemented using a Cloos Rotrol V7.13 robot (CLOOS, Haiger, Germany) that was integrated with a welding head system. The welding pathway was controlled by a computer-aided design (CAD) model. The welding parameters in terms of the deposition process included: wire feed rate of 6.1 m/min, electrical current of 210 A, voltage of 23.9–24.1 V, pulse frequency of 120 Hz, and robot deposition speed of 14 cm/min. The deposition process was carried out under a protective gas atmosphere composed of 98% Argon and 2% Oxygen. The printing parameters were selected based on regular welding conditions used by MIG-welding of steels.

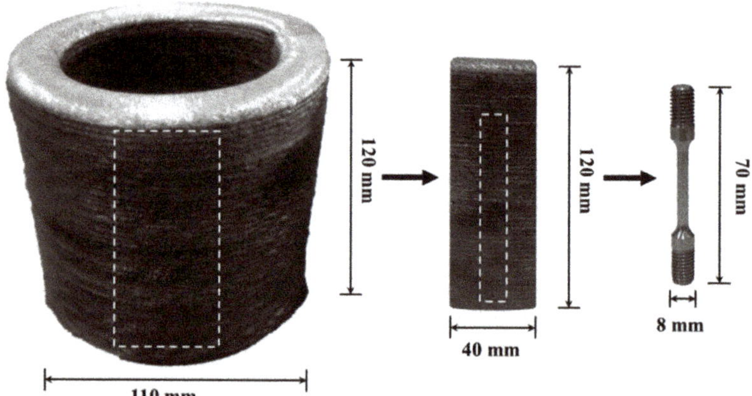

Figure 1. General appearance of the cylindrical component along with a tension specimen obtained by the WAAM process.

Microstructure analysis was carried out by scanning electron microscopy (SEM-JEOL 5600, JEOL Ltd., Tokyo, Japan) [26] equipped with an EDS sensor (Thermo Fisher Scientific, Waltham, MA, USA) for phase chemical composition detection. The metallographic preparation included polishing up to 0.04 μm and subsequent etching with HNO_3 (70%) 20 mL, HCl (32%) 45 mL and ethanol 25 mL for 1.5 min. The presence of a secondary phase in the form of ferrite (BCC) was evaluated by X-ray diffraction analysis using a RIGAKU-2100H X-ray diffractometer (Rigaku Corporation, Tokyo, Japan) with CuKα [27]. The diffraction parameters were 40 KV/30 mA and a scanning rate of 2°/min.

The corrosion performance was examined in terms of electrochemical analysis by potentiodynamic polarization and EIS using a Bio-Logic SP-200 potentiostat (BioLogic Science Instruments, Seyssinet-Pariset, France) equipped with EC-Lab software v11.18. Both analyses were implemented using a standard three-electrode cell with saturated calomel (SCE) as a reference electrode and Platinum as an auxiliary electrode. The potentiodynamic polarization scanning rate was 0.5 mV/s, and the EIS examination was performed between 10 kHz and 0.015 Hz at a 10 mV amplitude signal. The preparation procedure of the samples for the electrochemical testing included cleaning in an ultrasonic bath for 5 min, washing with alcohol and drying in air. The stress corrosion behavior in terms of SSRT was examined according to ASTM G129-00 standard [28], using Cormet C-76 apparatus (Cormet Testing Systems, Vantaa, Finland) [29,30]. The dimensions of the SSRT test samples were: gauge length 25.4 mm and cylindrical cross section 11.4 mm^2. The SSRT strain rates were: 2.5×10^{-7}, 2.5×10^{-6} and 2.5×10^{-5} s^{-1}. All the corrosion tests were carried out in 3.5% NaCl solution at ambient temperature (25 °C).

3. Results

The chemical composition of the welding wire, printed alloy and counterpart AISI 316L stainless steel alloy are shown in Table 1. This reveals that the composition of the printed alloy was in line with the composition of the welding wire and quite similar to that of the counterpart AISI alloy in terms of main alloying elements and carbon content.

Table 1. Chemical composition (wt.%) of welding wire, printed stainless steel part and counterpart AISI 316L.

Material	C	Mn	Si	S	P	Cr	Ni	Cu	Mo	Co	N	Fe
Welding Wire	0.013	1.97	0.51	0.0010	0.018	19.22	11.50	0.15	2.42	0.07	0.092	Bal.
Printed 316L	0.024	1.85	0.44	0.0007	0.020	19.21	11.62	0.09	2.48	0.26	0.052	Bal.
AISI 316L	0.021	1.59	0.40	0.02	0.037	16.58	10.07	0.44	2.04	0.19	0.078	Bal.

X-ray diffraction analysis of the printed alloy and its counterpart AISI 316L are shown in Figure 2. This reveals that the printed alloy was composed from an austenitic matrix (γ-Fe) and a secondary ferritic phase (δ-Fe), as expected from a regular 316L printed alloy [31] while the stock material (wire) has one austenitic phase. In parallel, the counterpart AISI 316L was composed from only an austenitic phase. In addition, significant differences were obtained between the XY-plane (building direction 0°) and XZ-plane (building direction 90°) of the printed alloy in terms of peak intensity. This can be attributed to the epitaxial characteristics of the AM process that display a preferred orientation of the solidification process. The calculated lattice parameter of the γ-Fe and δ-Fe related to the printed alloy were 2.88 and 3.59 Å, respectively, which basically comes in line with the parameters found in the literature: 2.86 Å (PDF 006–096) and 3.59 Å (PDF 33–0397). In the case of the counterpart AISI 316L, the calculated lattice parameter was 3.59 Å, as expected. Furthermore, it should be pointed out that, in contrast to the microstructure of 316L obtained by the WAAM process as presented by X. Chen et al. [25,32], no σ-phase was observed in this study.

Typical microstructure of printed 316L stainless steel in the XY- and XZ-planes is shown in Figure 3 while the microstructure of its counterpart AISI 316L alloy is introduced in Figure 4. The microstructure of the printed alloy was composed from an austenitic matrix and a secondary ferritic phase at the grain boundaries. The ferritic dendrites in the XY-plane present an anisotropic morphology, while the XZ-plane introduces an epitaxial solidification characteristic. In parallel, the counterpart AISI 316L was composed of a single austenitic phase, as expected from conventional 316L stainless steel [33]. The two-phase microstructure of the printed alloy was formed due to the high solidification rate of the AM process. This can also be explained in terms of the Fe-Cr-Ni phase diagram [34] operating under high cooling rate conditions. In addition, the inherent re-heating effect of the AM process produces a non-equilibrium microstructure that displays the presence of a primary δ ferrite phase [25].

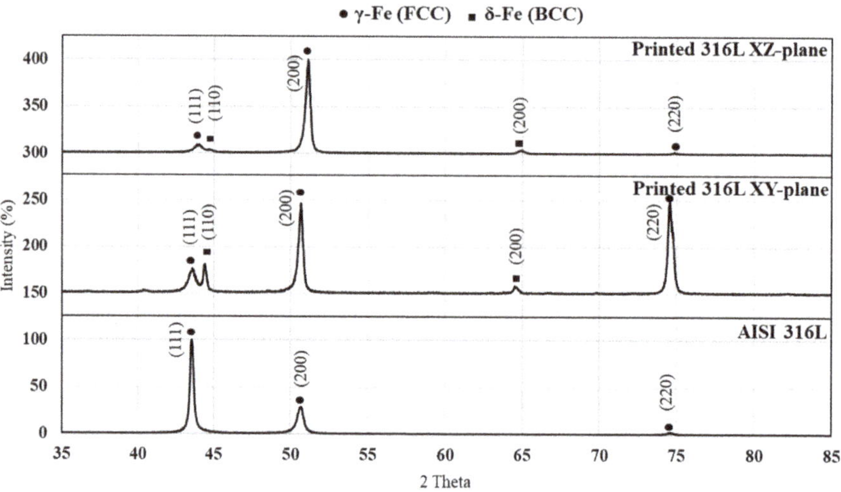

Figure 2. X-ray diffraction analysis of printed alloy and AISI 316L alloy.

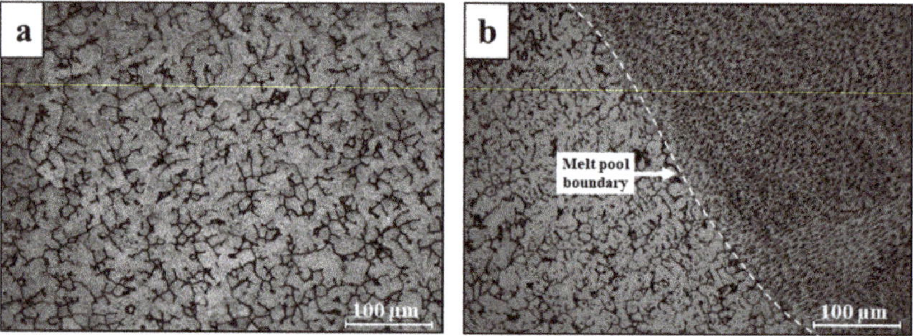

Figure 3. Typical microstructures of printed 316L stainless steel obtained by optical microscopy (**a**) XY-plane, (**b**) XZ-plane.

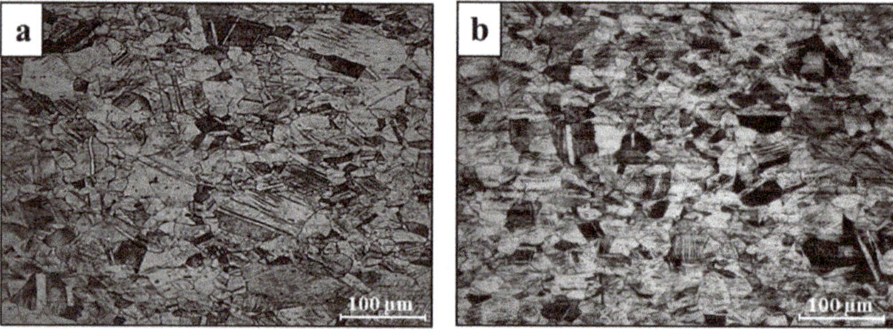

Figure 4. Typical microstructures of AISI 316L stainless steel (**a**) longitudinal cross section, (**b**) transvers cross section.

The general macrostructure of the printed alloy in 3D is shown in Figure 5, along with the corresponding close-up microstructures and spot chemical analysis. This clearly reveals that the microstructure in the XY-plane was quite uniform compared to the non-uniform structure in the XZ-plane that relates to the preferred orientation of the solidification course. In addition, the melt pool boundaries shown in Figure 5d,e clearly illustrate the epitaxial nature of solidification in the XZ-plane. The spot chemical analyses at points 1 and 2 (Table 2) disclose the typical compositions of austenite and ferrite phases, respectively. These compositions reflect the relatively increased amount of Ni and reduced content of Cr in the austenitic phase and vice versa in the ferritic phase.

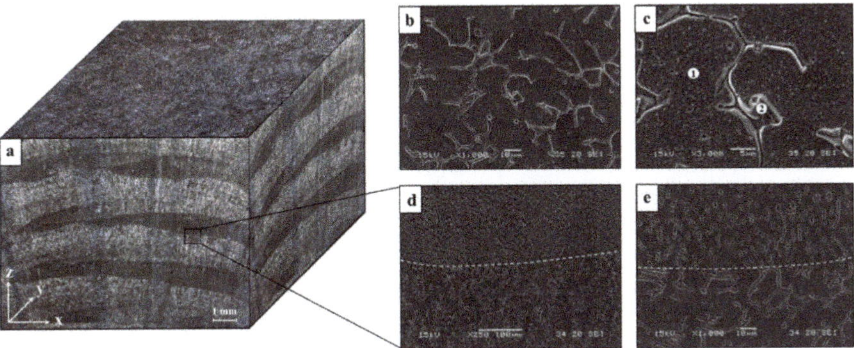

Figure 5. Macrostructure and corresponding microstructures of printed 316L alloy obtained by stereo microscope and SEM (**a**) general macrostructure, (**b,c**) microstructure in the XY-plane, (**d,e**) microstructure in the XZ-plane.

Table 2. Spot chemical analysis of printed 316L at points 1 and 2 (Figure 5c) by EDS in wt.%.

Test Point	C	Mn	Si	Cr	Ni	Mo	Fe
Point 1	0.39	2.3	0.41	17.34	12.08	2.02	Bal.
Point 2	0.25	2.12	0.39	24.86	5.43	4.17	Bal.

The mechanical properties of printed 316L and its counterpart AISI 316L in terms of tensile strength, yield strength, elongation and hardness are shown in Table 3 along with their typical stress-strain curves shown in Figure 6. This reveals that the strength of the printed alloy was relatively reduced, while its ductility was increased compared to the counterpart AISI 316L.

Table 3. The mechanical properties—UTS (Ultimate tensile strength), YP (yield point) elongation, hardness and density of printed and AISI 316L stainless steel.

Material	UTS (MPa)	YP (MPa)	Elongation (%)	Hardness (HV)	Density (gr/cm^3)
Printed 316L	552 ± 11	364 ± 17	42 ± 1	196 ± 5	7.6 ± 0.3
AISI 316L	752 ± 3	695 ± 3	37 ± 1	275 ± 9	7.8 ± 0.2

The corrosion resistance of the printed and counterpart AISI 316L in terms of potentiodynamic polarization analysis is shown in Figure 7. Although the polarization curve of the printed alloy was relatively shifted to higher corrosion currents, which reflects reduced corrosion resistance [35,36], its break potential (E_{b1}) was relatively higher compared to the AISI 316L (E_{b2}), which reflects an improved passivation process. Altogether, the corrosion rate of the printed alloy in terms of Tafel extrapolation was excellent, and quite similar to the counterpart alloy as shown in Table 4 (0.005 vs. 0.001 mmpy, respectively). In addition, as expected from metals having an active-passive transition, both alloys showed localized corrosion attack. In the case of the printed alloy the localized corrosion attack was concentrated at the boundaries between the austenite and ferrite phases (Figure 8a,b) [37].

The corrosion attack in the counterpart AISI 316L was in the form of pitting corrosion (Figure 8c,d) that was evenly scattered on the external surface. The counterpart alloy presented typical pitting morphology for AISI 316L, as well as typical corrosion potential, breakdown potential and corrosion current [38].

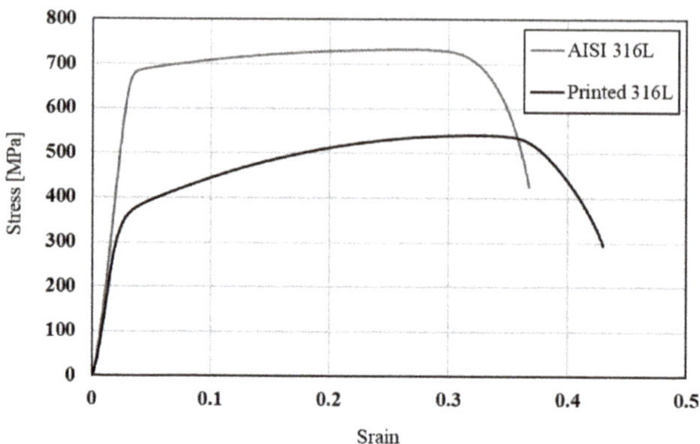

Figure 6. Typical stress-strain curves of printed 316L alloy and counterpart AISI 316L.

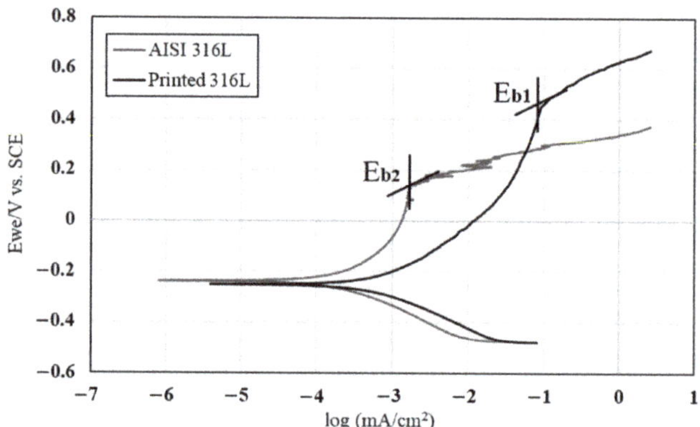

Figure 7. Potentiodynamic polarization analysis of printed 316L alloy and counterpart AISI 316L in 3.5% NaCl solution.

Table 4. Corrosion rate of tested alloys as obtained by Tafel extrapolation from potentiodynamic polarization curves.

Material	E_{corr} (v)	I_{corr} ($\mu A/cm^2$)	Corrosion Rate (mmpy)	E_{break} (v)
Printed 316L	-0.25 ± 0.02	0.48 ± 0.12	0.005 ± 0.001	0.47 ± 0.03
AISI 316L	-0.21 ± 0.02	0.09 ± 0.004	0.001 ± 0.0003	0.18 ± 0.004

The corrosion performance of printed and AISI 316L obtained by EIS analysis are shown in Figure 9. The Nyquist diagrams of both alloys (Figure 9a) in terms of curve radius, which represents the surface corrosion resistance, were quite similar. This similarity was also maintained by the Bode magnitude diagram (Figure 9b) that introduces the solution resistance. The related electrical equivalent circuit and corresponding fitting parameters (R1-solution resistance, R2 and Q1-capacitor) [39,40] are introduced in Figure 10 and Table 5, respectively. Altogether, the EIS analysis clearly indicates that the corrosion resistance of printed and counterpart alloys was quite similar.

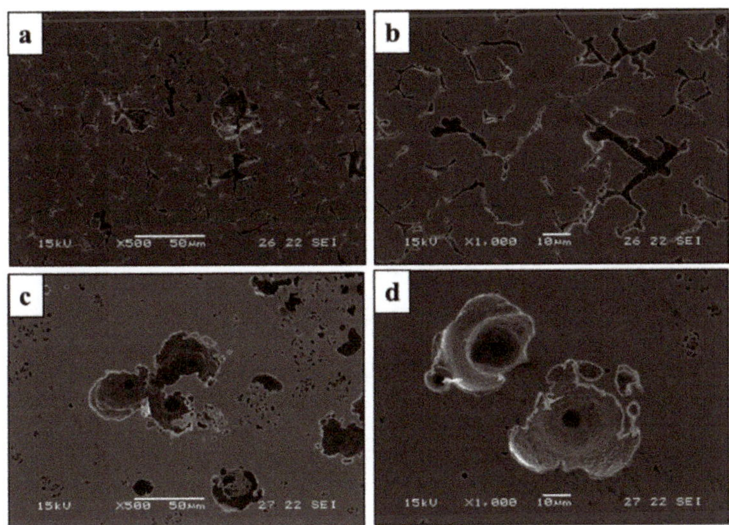

Figure 8. (**a**,**b**) Close-up views of the corrosion attack at surface of printed alloy, (**c**,**d**) close-up views of the corrosion attack at surface of counterpart AISI alloy.

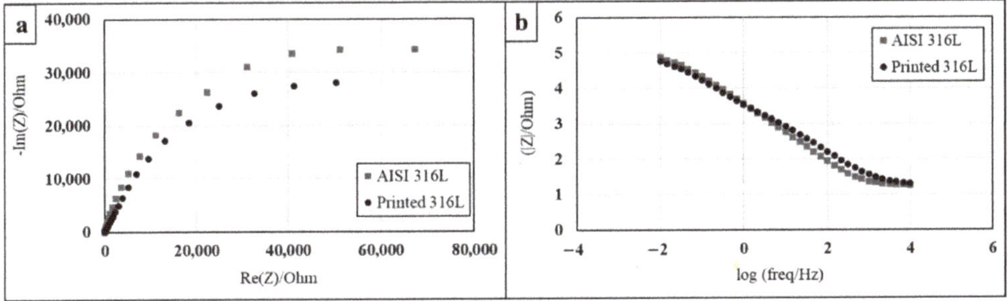

Figure 9. Electrochemical impedance spectroscopy analysis of printed and AISI 316L in 3.5% NaCl solution: (**a**) Nyquist diagram, (**b**) Bode diagram.

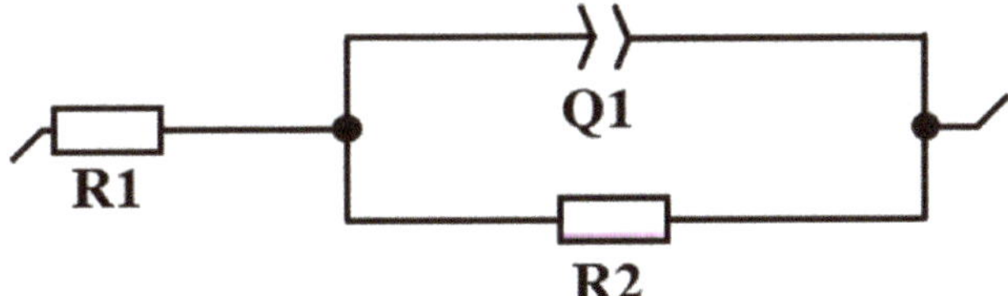

Figure 10. Electrical equivalent circuits for the EIS analysis shown in Figure 9.

Table 5. Corresponding fitting parameters for the EIS analysis shown in Figure 9.

Material	R1 (Ohm)	Q1 (F·s^{a-1})	a	R2 (Ohm)
Printed 316L	15.6	7.31×10^{-5}	0.705	92,178
AISI 316L	16.98	6.07×10^{-5}	0.825	83,863

The stress corrosion behavior of the printed and counterpart AISI 316L in terms of SSRT in 3.5% NaCl solution, are shown in Figures 8–10. Although the stress corrosion performance of the two alloys was similar, as reflected by nearly equal time to failure vs. strain rate (Figure 11), the two alloys maintain their inherent UTS and elongation properties (Figures 12 and 13, respectively). This similarity could also be seen by the fitting equations ($\sigma_{UTS} = C \times \dot{\varepsilon}^m$) of UTS vs. strain rate According to these fittings, the stain rate sensitively factors (m) of the two alloys were very close: 0.007 and 0.009 for the printed and AISI 316L, respectively. Fractography analysis of the two alloys (Figure 14a–d) clearly demonstrates that both alloys showed ductile failure behavior in the form of "cap and cone" fractures, as expected from 316L stainless steel alloy.

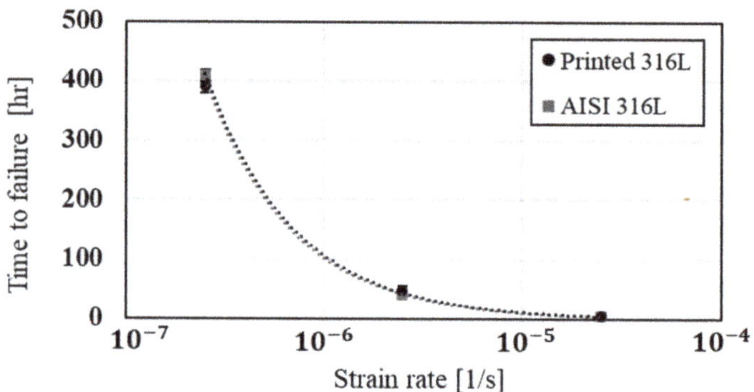

Figure 11. The effect of strain rate on time to failure of 316L produced by WAAM process in comparison with conventional wrought alloy AISI 316L.

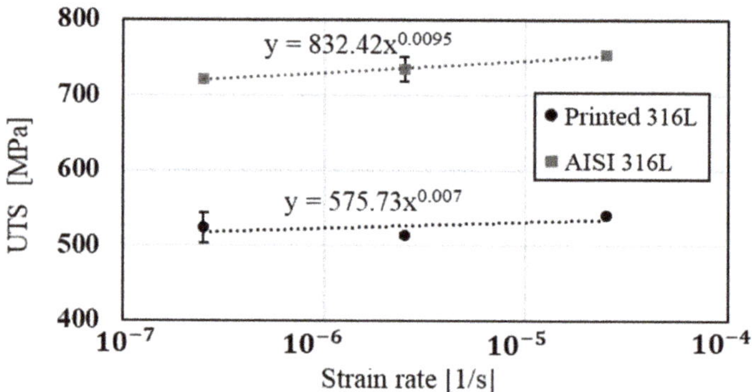

Figure 12. Ultimate tensile strength (UTS) versus strain rate of 316L produced by WAAM process compared to its counterpart wrought alloy AISI 316L.

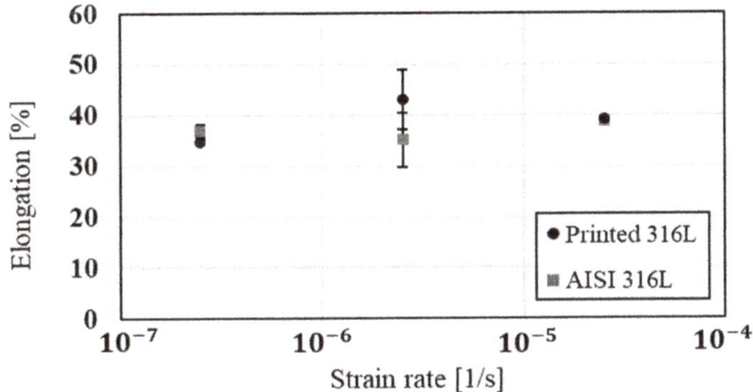

Figure 13. Ductility in terms of elongation versus strain rate of 316L produced by WAAM process compared to wrought alloy AISI 316L.

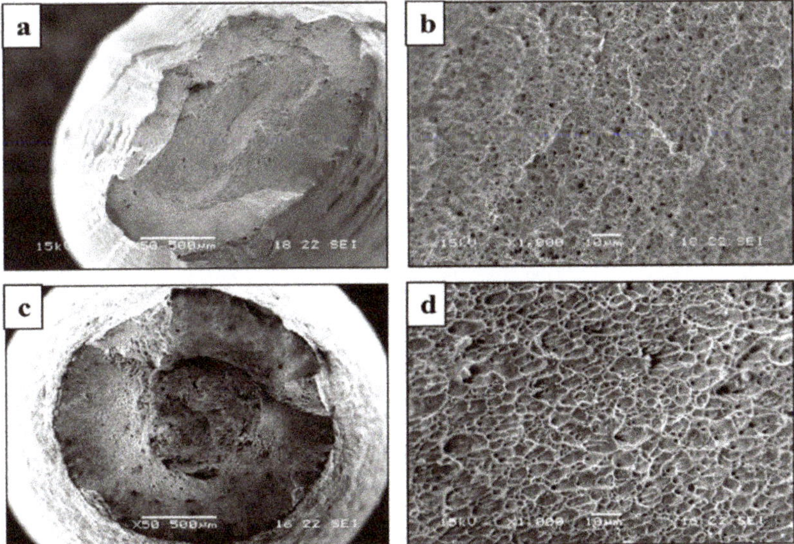

Figure 14. Fracture surface after SSRT at a strain rate of 2.5 × 10^{-7} (1/S) in 3.5% NaCl solution. (**a,b**) printed alloy, (**c,d**) counterpart AISI 316L.

4. Discussion

In spite of the differences between the microstructure and mechanical properties of WAAM 316L alloy and its counterpart AISI 316L, their corrosion performance in 3.5% NaCl solution was quite similar. This was strongly supported by the results of potentiodynamic polarization, EIS and stress corrosion analysis in terms of SSRT examination. Nevertheless, the corrosion mechanism of the two alloys was slightly different. This was clearly demonstrated by the surface corrosion attack shown in Figure 8. According to this figure, the localized corrosion attack in the printed alloy was mainly located at the interface between the austenitic matrix and the secondary ferritic phase (Figure 8a,b). This was mainly attributed to the relatively reduced corrosion resistance of the ferritic phase compared to the austenitic phase, which can induce micro-galvanic corrosion. In the case of the counterpart AISI 316L, the localized corrosion attack was in the form of pitting corrosion that was uniformly scattered on the surface (Figure 8c,d).

Regarding the effect of strain rate on UTS and elongation under a corrosive environment, both the printed and the counterpart AISI 316L displayed a similar response, according to their inherent mechanical properties. The similarity in their stress corrosion resistance was demonstrated by nearly equal time to failure at a slow strain rate of 2.5×10^{-7} s^{-1} (Figure 11), where the environmental effect was most dominant. This similarity was also manifested by the fractography analysis of the two alloys (Figure 14a–d) that clearly showed ductile failure characteristics in the form of "cap and cone" fractures, as can be expected from 316L stainless steel alloy [41].

As a final remark it should be pointed out that the similar corrosion performance of printed WAAM 316L alloy and its counterpart AISI 316L in 3.5% NaCl solution cannot be simply extrapolated to any different environment. This is mainly due to the inherent differences between the microstructure of the two alloys that can affect their corrosion behavior primarily in a more aggressive environment.

5. Conclusions

The stress corrosion behavior of additively manufactured austenitic stainless steel (316L alloy) produced by WAAM process in terms of stress corrosion susceptibility and electrochemical performance was similar to that of its counterpart wrought alloy. This similarity was obtained in spite of the inherent differences in microstructure and mechanical properties of the two alloys. The corrosion attack in the printed alloy was mainly located at the interface between the austenitic matrix and the ferritic phase, while that of the counterpart alloy composed of a single austenite phase was in the form of pitting corrosion uniformly scattered on the surface. The fractography analysis of the two alloys in post-SSRT experiments revealed that both alloys showed ductile failure in the form of "cap and cone" fractures, as expected from austenitic stainless steel.

Author Contributions: E.A., A.S. and T.R. conceived, designed and performed the experiments; A.L. and O.D. assisted in analyzing the data; E.A. and T.R. writing—original draft preparation and writing—review and editing. All authors have read and agreed to the published version of the manuscript.

Funding: This research received no external funding.

Institutional Review Board Statement: Not applicable.

Informed Consent Statement: Not applicable.

Data Availability Statement: Data sharing is not applicable to this article.

Acknowledgments: The authors thank the Kreitman School for Advanced Studies at Ben-Gurion University of the Negev for their financial contribution in support of this research.

Conflicts of Interest: The authors declare no conflict of interest.

References

1. Leon, A.; Levy, G.K.; Ron, T.; Shirizly, A.; Aghion, E. The effect of strain rate on stress corrosion performance of Ti6Al4V alloy produced by additive manufacturing process. *J. Mater. Res. Technol.* **2020**. [CrossRef]
2. Rodrigues, T.A.; Duarte, V.; Avila, J.A.; Santos, T.G.; Miranda, R.; Oliveira, J. Wire and arc additive manufacturing of HSLA steel: Effect of thermal cycles on microstructure and mechanical properties. *Addit. Manuf.* **2019**, *27*, 440–450. [CrossRef]
3. Ding, D.; Pan, Z.; Cuiuri, D.; Li, H. Wire-feed additive manufacturing of metal components: Technologies, developments and future interests. *Int. J. Adv. Manuf. Technol.* **2015**, *81*, 465–481. [CrossRef]
4. Leon, A.; Shirizly, A.; Aghion, E. Corrosion Behavior of AlSi10Mg Alloy Produced by Additive Manufacturing (AM) vs. Its Counterpart Gravity Cast Alloy. *Metals* **2016**, *6*, 148. [CrossRef]
5. Herzog, D.; Seyda, V.; Wycisk, E.; Emmelmann, C. Additive manufacturing of metals. *Acta Mater.* **2016**, *117*, 371–392. [CrossRef]
6. Haghdadi, N.; Laleh, M.; Moyle, M.; Primig, S. Additive manufacturing of steels: A review of achievements and challenges. *J. Mater. Sci.* **2020**, *56*, 64–107. [CrossRef]
7. Leon, A.; Levy, G.K.; Ron, T.; Shirizly, A.; Aghion, E. The effect of hot isostatic pressure on the corrosion performance of Ti6Al4V produced by an electron-beam melting additive manufacturing process. *Addit. Manuf.* **2020**, *33*, 101039. [CrossRef]
8. Zakay, A.; Aghion, E. Effect of Post-heat Treatment on the Corrosion Behavior of AlSi10Mg Alloy Produced by Additive Manufacturing. *JOM* **2019**, *71*, 1150–1157. [CrossRef]

9. Rodrigues, T.A.; Duarte, V.; Miranda, R.M.; Santos, T.G.; Oliveira, J.P. Current Status and Perspectives on Wire and Arc Additive Manufacturing (WAAM). *Materials* **2019**, *12*, 1121. [CrossRef]

10. Ding, D.; Pan, Z.; Cuiuri, D.; Li, H. A multi-bead overlapping model for robotic wire and arc additive manufacturing (WAAM). *Robot. Comput. Manuf.* **2015**, *31*, 101–110. [CrossRef]

11. Gisario, A.; Kazarian, M.; Martina, F.; Mehrpouya, M. Metal additive manufacturing in the commercial aviation industry: A review. *J. Manuf. Syst.* **2019**, *53*, 124–149. [CrossRef]

12. Donoghue, J.M.; Antonysamy, A.A.; Martina, F.; Colegrove, P.; Williams, S.W.; Prangnell, P. The effectiveness of combining rolling deformation with Wire–Arc Additive Manufacture on β-grain refinement and texture modification in Ti-6Al-4V. *Mater. Charact.* **2016**, *114*, 103–114. [CrossRef]

13. Busachi, A.; Erkoyuncu, J.; Colegrove, P.; Martina, F.; Watts, C.; Drake, R. A review of Additive Manufacturing technology and Cost Estimation techniques for the defence sector. *CIRP J. Manuf. Sci. Technol.* **2017**, *19*, 117–128. [CrossRef]

14. Bekker, A.C.M.; Verlinden, J.C.; Galimberti, G. Challenges in assessing the sustainability of wire+ arc additive manufacturing for large structures. In Proceedings of the Solid Freeform Fabrication Symposium, Austin, TX, USA, 8–10 August 2016; pp. 406–418.

15. Cunningham, C.R.; Wikshåland, S.; Xu, F.; Kemakolam, N.; Shokrani, A.; Dhokia, V.; Newman, S. Cost Modelling and Sensitivity Analysis of Wire and Arc Additive Manufacturing. *Procedia Manuf.* **2017**, *11*, 650–657. [CrossRef]

16. Ron, T.; Levy, G.K.; Dolev, O.; Leon, A.; Shirizly, A.; Aghion, E. Environmental Behavior of Low Carbon Steel Produced by a Wire Arc Additive Manufacturing Process. *Metals* **2019**, *9*, 888. [CrossRef]

17. Bandari, Y.K.; Williams, S.W.; Ding, J.; Martina, F. Additive manufacture of large structures: Robotic or CNC systems. In Proceedings of the 26th international solid freeform fabrication symposium, Austin, TX, USA, 10–12 August 2015; pp. 12–14.

18. Williams, S.W.; Martina, F.; Addison, A.C.; Ding, J.; Pardal, G.; Colegrove, P.A. Wire + Arc Additive Manufacturing. *Mater. Sci. Technol.* **2016**, *32*, 641–647. [CrossRef]

19. Ron, T.; Levy, G.K.; Dolev, O.; Leon, A.; Shirizly, A.; Aghion, E. The Effect of Microstructural Imperfections on Corrosion Fatigue of Additively Manufactured ER70S-6 Alloy Produced by Wire Arc Deposition. *Metals* **2020**, *10*, 98. [CrossRef]

20. Shirizly, A.; Dolev, O. From Wire to Seamless Flow-Formed Tube: Leveraging the Combination of Wire Arc Additive Manufacturing and Metal Forming. *JOM* **2018**, *71*, 709–717. [CrossRef]

21. Derekar, K.S. A review of wire arc additive manufacturing and advances in wire arc additive manufacturing of aluminium. *Mater. Sci. Technol.* **2018**, *34*, 895–916. [CrossRef]

22. Wu, B.; Pan, Z.; Ding, D.; Cuiuri, D.; Li, H.; Xu, J.; Norrish, J. A review of the wire arc additive manufacturing of metals: Properties, defects and quality improvement. *J. Manuf. Process.* **2018**, *35*, 127–139. [CrossRef]

23. Pan, Z.; Ding, D.; Wu, B.; Cuiuri, D.; Li, H.; Norrish, J. *Arc Welding Processes for Additive Manufacturing: A Review*; Springer Science and Business Media LLC: Berlin, Germany, 2017; pp. 3–24.

24. Wang, L.; Xue, J.; Wang, Q. Correlation between arc mode, microstructure, and mechanical properties during wire arc additive manufacturing of 316L stainless steel. *Mater. Sci. Eng. A* **2019**, *751*, 183–190. [CrossRef]

25. Chen, X.; Li, J.; Cheng, X.; He, B.; Wang, H.; Huang, Z. Microstructure and mechanical properties of the austenitic stainless steel 316L fabricated by gas metal arc additive manufacturing. *Mater. Sci. Eng. A* **2017**, *703*, 567–577. [CrossRef]

26. Kaya, A.; Uzan, P.; Eliezer, D.; Aghion, E. Electron microscopical investigation of as cast AZ91D alloy. *Mater. Sci. Technol.* **2000**, *16*, 1001–1006. [CrossRef]

27. Aghion, E.; Gueta, Y.; Moscovitch, N.; Bronfin, B. Effect of yttrium additions on the properties of grain-refined Mg–3%Nd alloy. *J. Mater. Sci.* **2008**, *43*, 4870–4875. [CrossRef]

28. Available online: https://www.astm.org/Standards/G129 (accessed on 21 December 2020).

29. Arnon, A.; Aghion, E. Stress Corrosion Cracking of Nano/Sub-micron E906 Magnesium Alloy. *Adv. Eng. Mater.* **2008**, *10*, 742–745. [CrossRef]

30. Hakimi, O.; Aghion, E.; Goldman, J. Improved stress corrosion cracking resistance of a novel biodegradable EW62 magnesium alloy by rapid solidification, in simulated electrolytes. *Mater. Sci. Eng. C* **2015**, *51*, 226–232. [CrossRef]

31. Guo, P.; Zou, B.; Huang, C.; Gao, H. Study on microstructure, mechanical properties and machinability of efficiently additive manufactured AISI 316L stainless steel by high-power direct laser deposition. *J. Mater. Process. Technol.* **2017**, *240*, 12–22. [CrossRef]

32. Chen, X.; Li, J.; Cheng, X.; Wang, H.; Huang, Z. Effect of heat treatment on microstructure, mechanical and corrosion properties of austenitic stainless steel 316L using arc additive manufacturing. *Mater. Sci. Eng. A* **2018**, *715*, 307–314. [CrossRef]

33. Xiong, J.; Tan, M.Y.; Forsyth, M. The corrosion behaviors of stainless steel weldments in sodium chloride solution observed using a novel electrochemical measurement approach. *Desalination* **2013**, *327*, 39–45. [CrossRef]

34. Koseki, T.; Flemings, M.C. Solidification of undercooled Fe-Cr-Ni alloys: Part I. Thermal behavior. *Met. Mater. Trans. A* **1995**, *26*, 2991–2999. [CrossRef]

35. Itzhak, D.; Aghion, E. Corrosion behaviour of hot-pressed austenitic stainless steel in H2SO4 solutions at room temperature. *Corros. Sci.* **1983**, *23*, 1085–1094. [CrossRef]

36. Itzhak, D.; Aghion, E. An anodic behaviour study of an analogical sintered system of austenitic stainless steel in H2SO4 solution. *Corros. Sci.* **1984**, *24*, 145–149. [CrossRef]

37. Garcia-Cabezon, C.; Martín, F.; Blanco, Y.; De Tiedra, P.; Aparicio, M. Corrosion behaviour of duplex stainless steels sintered in nitrogen. *Corros. Sci.* **2009**, *51*, 76–86. [CrossRef]

38. Al Saadi, S.; Yi, Y.; Cho, P.; Jang, C.; Beeley, P. Passivity breakdown of 316L stainless steel during potentiodynamic polarization in NaCl solution. *Corros. Sci.* **2016**, *111*, 720–727. [CrossRef]
39. Leon, A.; Aghion, E. Effect of surface roughness on corrosion fatigue performance of AlSi10Mg alloy produced by Selective Laser Melting (SLM). *Mater. Charact.* **2017**, *131*, 188–194. [CrossRef]
40. Kafri, A.; Ovadia, S.; Yosafovich-Doitch, G.; Aghion, E. The Effects of 4%Fe on the Performance of Pure Zinc as Biodegradable Implant Material. *Ann. Biomed. Eng.* **2019**, *47*, 1400–1408. [CrossRef]
41. Kubík, P.; Šebek, F.; Petruška, J.; Hůlka, J.; Park, N.; Huh, H. Comparative investigation of ductile fracture with 316L austenitic stainless steel in small punch tests: Experiments and simulations. *Theor. Appl. Fract. Mech.* **2018**, *98*, 186–198. [CrossRef]

 *materials*

Communication

Molar Ratio Effect of Sodium to Chloride Ions on the Electrochemical Corrosion of Alloy 600 and SA508 in HCl + NaOH Mixtures

Do Haeng Hur [1,*], **Jeoh Han** [1] **and Jun Choi** [1,2]

[1] Korea Atomic Energy Research Institute, 989-111, Deadeok-daero, Yuseong-gu, Daejeon 34057, Korea; jeohhan@kaeri.re.kr (J.H.); jun@dsv.co.kr (J.C.)

[2] Daesung Machinery Co., Ltd., 3Da-401, Sihwa Industrial Complex, 334, Gongdan 1-daero, Gyeonggi-do, Siheung 15106, Korea

* Correspondence: dhhur@kaeri.re.kr; Tel.: +82-42-868-8388; Fax: +82-42-868-8696

Received: 24 March 2020; Accepted: 20 April 2020; Published: 23 April 2020

Abstract: This study aims to investigate the molar ratio effect of sodium to chloride ions on the corrosion of an Alloy 600 steam generator tube and an SA508 tubesheet material. The corrosion behavior was evaluated in solutions with three different molar ratios of sodium to chloride ions using a potentiodynamic polarization method. The corrosion potentials and corrosion rates of both the two materials were significantly decreased as the molar ratio increased from 0.1 to 10. Therefore, it is recommended that the molar ratio control to a value of 1 is beneficial only when the crevice chemistry has a low molar ratio with an acidic pH. The corrosion potentials and corrosion rates were little affected by the total sodium and chloride ion concentrations. SA508 acted as an anode and its corrosion rate was accelerated by galvanic coupling with Alloy 600.

Keywords: molar ratio index; steam generator; tubesheet; crevice chemistry; galvanic couple

1. Introduction

Stress corrosion cracking (SCC), intergranular attack (IGA) and pitting have been major degradation modes of Alloy 600 steam generator (SG) tubes in the secondary side water environments of pressurized water reactors (PWRs) [1–4]. Corrosion damage of the Alloy 600 tube materials is accelerated in both acidic and alkaline environments resulting from the impurity concentration processes. The formation of these corrosive environments is closely associated with local boiling within tube to tube support crevices and tube to tubesheet crevices in which sludge is accumulated [5,6]. The laboratory and field experience data have indicated that both IGA and SCC of Alloy 600 steam generator tubes are minimized at a near neutral pH [7,8]. Therefore, it is expected that maintaining a crevice pH near neutral reduces the corrosion damage of Alloy 600.

Based on the above backgrounds, the molar ratio control program was developed by the Electric Power Research Institute (EPRI), of which the goal is to maintain the crevice chemistry at near neutral pH values [7]. A molar ratio index (MRI) for managing the secondary water chemistry in PWRs was also proposed as the following equation [7]:

$$\text{MRI} = \frac{Na + K}{Cl + excess\ SO_4} \tag{1}$$

To maintain a desired MRI, several methods has been implemented, such as sodium source reduction, chloride injection, and ion exchange resin manipulation [9]. This index was derived from a viewpoint of the corrosion behavior of SG tube material itself. SG tubes are equipped in a tubesheet by expanding the both end parts of the tubes within drilled-holes of the tubesheet. The SG tubing is made of nickel-based alloys, while the tubesheet is SA508 low alloy steels. Therefore, an SG tube and a tubesheet are galvanically contacted in the tube to tubesheet crevice region of a SG. In addition, corrosion of the tubesheet material induces denting damage of the tubes at the top of the tubesheet, accelerating SCC of the tubes [10,11]. Therefore, the corrosion behavior of the tubesheet material should also be considered in the application of the molar ratio control. In addition, since the MRI is a simple molar ratio of cations to anions, the total concentration of the cations and the anions is not considered.

Potentiodynamic polarization tests provide useful information such as corrosion rate and susceptibility of materials to corrosion in aqueous solutions, as well as pitting corrosion. Furthermore, it has been reported that polarization behaviors of Alloy 600 and stainless steels are associated with IGA and SCC [12–14]. In this case, however, additional IGA and SCC tests should be performed to find a correlation at applied potentials selected from a polarization curve. In this work, an electrochemical polarization method was used to survey the effect of the MRI on the polarization responses of an Alloy 600 SG tube and an SA508 tubesheet material at approximately 25 °C and to provide bases for comparison with the results at higher temperatures, which will be obtained in the next phase. The effect of impurity concentration at a constant MRI was also evaluated.

2. Experimental Methods

2.1. Specimen and Solution Preparation

Alloy 600 and SA508 were used as a tube material and a tubesheet material of a SG, respectively. Chemical compositions of the materials are given in Tables 1 and 2. Alloy 600 had an average grain size of about 48.9 mm with chromium carbides at grain boundaries, which satisfied the EPRI specification [15]. SA508 was heat-treated according to the American Society of Mechanical Engineers (ASME) standard specification [16] and its microstructure was typical bainite. Because the material microstructure has a significant impact on corrosion behavior [17–19], all specimens were prepared from a single heat of each material.

Table 1. Chemical composition of Alloy 600 material (wt.%).

C	Cr	Fe	Si	Mn	Ti	Al	Ni
0.02	15.7	10.0	0.1	0.3	0.1	0.06	Bal.

Table 2. Chemical composition of SA508 material (wt.%).

C	Si	Mn	P	S	Ni	Cr	Mo	Fe
0.199	0.051	1.52	0.005	0.006	0.987	0.232	0.582	Bal.

Specimens were cut into a size of $10 \times 5 \times 1$ mm^3 for the electrochemical corrosion tests. They were ground using silicon carbide paper down to 1000-grit and then ultrasonically cleaned in acetone for 5 min.

To prepare a working electrode for the electrochemical test, an Alloy 600 specimen was spot-welded to an Alloy 600 wire, while an SA508 specimen was spot-welded to a pure iron wire. The lead wire was then shielded with a polytetrafluoroethylene (PTFE) tube for electrical insulation. A resin was coated around the spot-weld to prevent the test solution from penetrating into any remaining crevice there. After curing the resin, the specimen was ultrasonically cleaned in acetone for 1 min. When subtracting the weld-junction area, the surface area exposed to the solutions during the electrochemical tests was 1.28 cm^2.

The MRIs of sodium ions to chloride ions in the test solutions were controlled to be 0.1, 1 and 10 by the addition of HCl and NaOH into demineralized water with the resistivity near 18 MΩ·cm, as shown in Table 3. Any other chemical species were not included to simplify the Equation (1). The total ion concentrations of sodium and chloride ions were fixed to be 0.011 and 0.11 M at each MRI. Regardless of the total ion concentrations, the measured solution pH was dependent on the MRI and was about 2 at the MRI 0.1, 7 at the MRI 1, and 12 at the MRI 10.

Table 3. Experimental conditions for the electrochemical corrosion tests.

Na^+ (M)	Cl^- (M)	Total Concentration of Na^+ and Cl^- (M)	MRI (Na^+/Cl^-)	pH
0.001	0.01	0.011	0.1	2
0.0055	0.0055	0.011	1	7
0.01	0.001	0.011	10	12
0.01	0.1	0.11	0.1	2
0.055	0.055	0.11	1	7
0.1	0.01	0.11	10	12

2.2. Electrochemical Corrosion Test

Potentiodynamic polarization tests were performed in each solution at 25 °C by using a PAR273 potentiostat (EG&G Princeton Applied Research, Berwyn, PA, USA) with Power-Suite software (version 2.58, Ametek, Berwyn, PA, USA) and a conventional corrosion cell with three electrodes. A saturated calomel electrode was used as the reference electrode, and a platinum wire was used as the counter electrode. The test solutions were deaerated by bubbling ultra-high purity (99.999%) nitrogen gas at a rate of 300 mL/min. The open circuit potential (OCP) of a working electrode reached a stable value within 1 h. After that, the potential was scanned either to the positive direction for the anodic curve, or to the negative direction for the cathodic curve at a rate of 20 mV/min under continuous blowing of nitrogen. Each anodic and cathodic polarization curve was finally combined in one graph. The polarization curves were obtained at least three times to ensure their reproducibility using fresh specimens and solutions.

The corrosion current density (i_{corr}) of the materials at the OCPs was calculated by using the Tafel extrapolation method of cathodic polarization curves. The galvanic corrosion potential and the galvanic current density between Alloy 600 and SA508 were determined by the application of the mixed potential theory.

3. Results and Discussion

Figure 1 shows the potentiodynamic polarization behaviors of Alloy 600 in the solutions of MRI 0.1, 1, and 10 at a total sodium and chloride concentration of 0.011 M. The corrosion potentials (E_{corr}) and corrosion rates (i_{corr}) of Alloy 600 were significantly decreased as the MRI increased from 0.1 to 10. The active–passive transition appeared at the MRI 0.1, but the alloy was passivated without active dissolution at the MRI 1 and 10.

Figure 2 shows the SEM micrographs of corroded surfaces after the anodic polarization scans. The surfaces exposed to the MRI 0.1 and 1 solutions showed extensive pitting, whereas the surface at the MRI 10 was uniformly corroded without pitting corrosion. Therefore, the transpassivity showing an abrupt increase of current at about 0.260 and 0.390 V in the solutions of the MRI 0.1 and 1, respectively, was due to pitting, whereas the current increase at 0.600 V in the solution of the MRI 10 was owing to oxygen evolution, the reaction of which can be expressed by the following equation [20]:

$$2H_2O = O_2 + 4H^+ + 4e \tag{2}$$

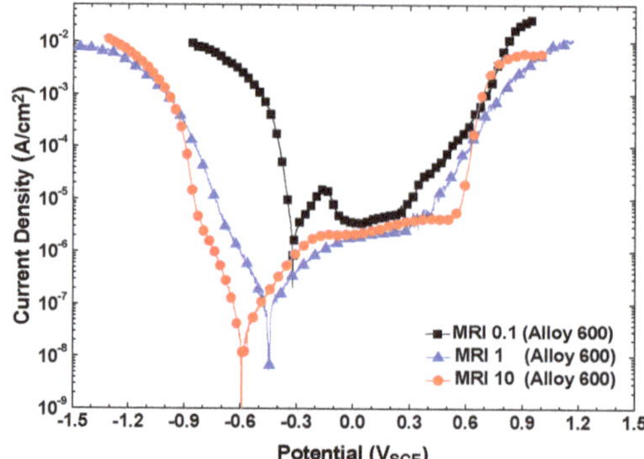

Figure 1. Polarization curves of Alloy 600 in the solutions of the MRI 0.1, 1 and 10 at a total sodium and chloride ion concentration of 0.011 M.

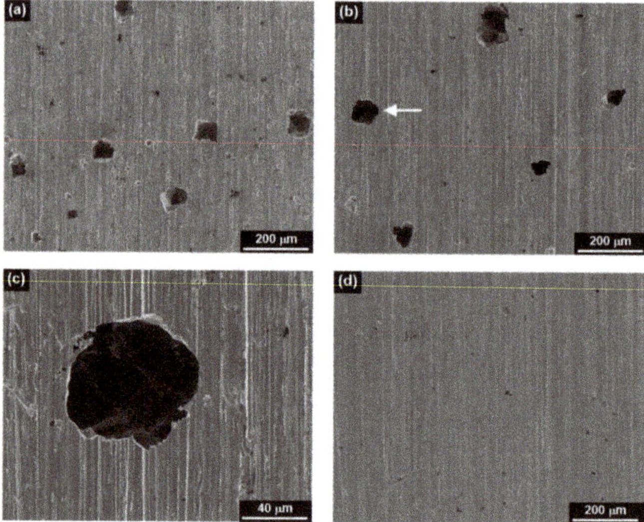

Figure 2. Scanning electron microscopy (SEM) micrographs showing the corroded surfaces of Alloy 600 after polarization scans at (**a**) the MRI 0.1, (**b**) the MRI 1, (**c**) magnification of the pit denoted by the white arrow in (**b**), and (**d**) the MRI 10.

As shown in Figure 3, the corrosion potentials of SA508 were also significantly decreased as the MRI increased from 0.1 to 10. SA508 actively dissolved at high corrosion rates without any passivation at the MRI 0.1 and 1, whereas the alloy showed the lowest corrosion rate with a passive behavior in a potential range of −0.190 to 0.600 V at the MRI 10. SA508 also showed an abrupt increase of current density due to oxygen evolution near a potential of 0.600 V at the MRI 10, as did Alloy 600.

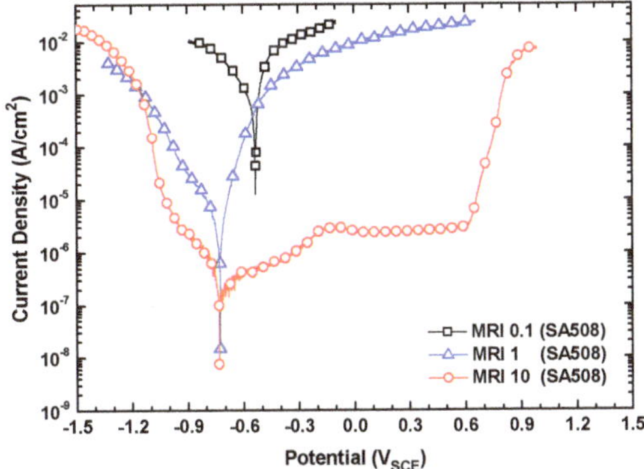

Figure 3. Polarization curves of SA508 in the solutions of the MRI 0.1, 1 and 10 at a total sodium and chloride ion concentration of 0.011 M.

Figure 4 shows the SEM micrographs of SA508 surfaces after the anodic polarization scans. The surface exposed to the MRI 0.1 was severely and uniformly corroded enough to dissolve out the grinding marks, which was made by emery paper during the surface finishing process. The surface at the MRI 1 was also corroded uniformly, but less severely than at the MRI 0.1. On the contrary, it can still clearly be seen the grinding marks on the surface exposed to the MRI 10, indicating that the anodic dissolution rate was very low in the MRI 10 solution. Therefore, the morphologies of these corroded surfaces were in good agreement with the polarization behaviors shown in Figure 3.

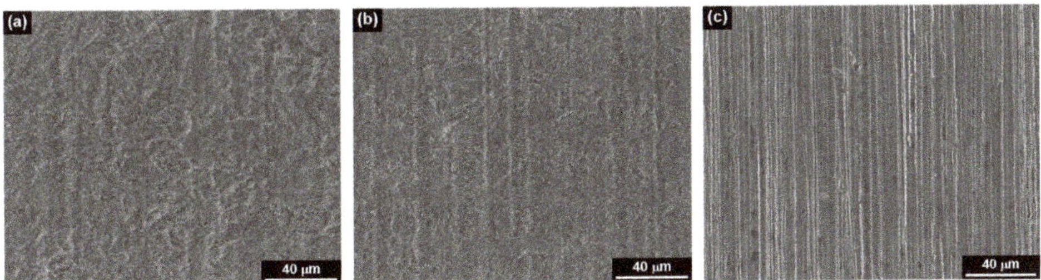

Figure 4. SEM micrographs showing the corroded surfaces of SA508 after polarization scans at (**a**) the MRI 0.1, (**b**) the MRI 1, and (**c**) the MRI 10.

The important corrosion parameters from Figures 1 and 3 are summarized in Table 4. Alloy 600 showed the highest corrosion rate at the MRI 0.1, while the corrosion rates at the MRIs 1 and 10 were nearly similar. In case of SA508, this alloy also showed the highest corrosion rate at the MRI 0.1, but the corrosion rate at the MRIs 10 was rather smaller than that at the MRI 1. Consequently, this result indicates that the molar ratio control method is beneficial only when the crevice chemistry has a low MRI and pH.

Table 4. Corrosion potentials and corrosion rates of Alloy 600 and SA508 obtained from the polarization tests.

MRI	Alloy 600		SA508	
	E_{corr} (V_{SCE})	i_{corr} ($\mu A/cm^2$)	E_{corr} (V_{SCE})	i_{corr} ($\mu A/cm^2$)
0.1	−0.314	3.2	−0.532	520
1	−0.446	0.088	−0.732	1.8
10	−0.593	0.064	−0.728	0.19

As shown in Tables 1 and 2, Alloy 600 is a high-alloyed steel containing 15.7 wt.% Cr and 73.7 wt.% Ni. Thus, this alloy has an excellent resistance to corrosion in overall pH ranges from acidic to alkaline. Therefore, the difference between the corrosion rates (i_{corr}) at the acidic MRI 0.1 and at the alkaline MRI 10 is not so large. However, SA508 is an iron-based steel containing only 0.23 wt.% Cr and 0.58 wt.% Mo and thus has a basically poor corrosion resistance, especially in acidic solutions. From Table 4, the corrosion rates (i_{corr}) of Alloy 600 were always significantly lower than those of SA508 in all the test conditions. In addition, there was a significant decrease of i_{corr} for SA508 at the MRI 10 in comparison with the MRI 1 as well as the MRI 0.1. The reason for this can be attributed to the fact that the solubility of magnetite is significantly dependent on the pH of a solution [21,22]. The solubility of magnetite at pH 3 is about 5×10^4 times higher than that at pH 12 in water at 100 °C [21]. Therefore, the corrosion rate of SA508 increases significantly in low pH solutions (i.e., low MRI solutions) with a high solubility of magnetite because the corroding surface of the alloy cannot be protected by the magnetite film. Conversely, the alloy showed a passive behavior with a low corrosion current in a potential range of −0.190 to 0.600 V at the MRI 10 solution of pH 12, owing to a significantly low solubility of magnetite.

Figure 5 shows the potentiodynamic polarization behaviors of Alloy 600 and SA508 in the solutions with total sodium and chloride ion concentration of 0.011 M and 0.11 M at a constant MRI 1. The corrosion potentials of Alloy 600 and SA508 were approximately −0.450 V and −0.730 V at both concentrations, respectively, indicating that the corrosion potentials of the two materials were not affected by a change in the total ion concentration. The cathodic and anodic current density of Alloy 600 was also little affected by an increase of the ion concentration from 0.011 M to 0.11 M. However, the pitting potentials of Alloy 600 decreased from 0.390 V in 0.011 M to 0.170 V in 0.11 M. In case of SA508, the polarization current density was nearly same in the solutions of 0.011 M and 0.11 M, when this alloy was polarized around the corrosion potential. The above results mean that the electrochemical corrosion behavior of these materials in a region near the corrosion potentials does not depend on the total sodium and chloride ion concentrations if the sodium to chloride molar ratio in a solution is the same. Similar behaviors were also observed at the MRI 0.1 and 10.

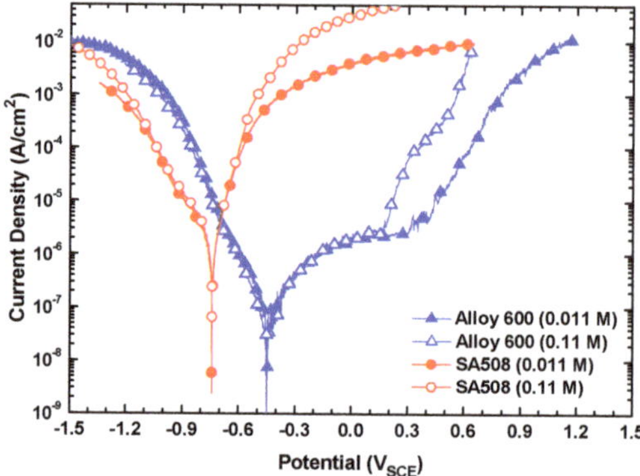

Figure 5. Polarization curves of Alloy 600 and SA508 in the 0.011 and 0.11 M solutions at a constant MRI 1.

From Figure 1, Figure 3, and Figure 5, it is clear that the corrosion potential of SA508 is always lower than that of Alloy 600 in each test condition. The anodic curve of SA508 also intersects with the cathodic curve of Alloy 600. This result demonstrates that SA508 is an anodic member of the galvanic couple and its corrosion rate is accelerated, when SA508 and Alloy 600 are electrically contacted. When the two materials are coupled in equal area, the galvanic current density (i_{couple}) of SA508, acting as an anode, is determined at the intersection of the anodic curve of SA508 and the cathodic curve of Alloy 600. Figure 6 shows the effect of the MRIs on the galvanic corrosion of SA508, based on the polarization curves. Upon coupling to an equal area of Alloy 600, the current density (i_{couple}) of the coupled SA508 was increased by about 2~6 times compared to that (i_{SA508}) before coupling. However, the area of the tubesheet around a tube is much smaller than that of the tube in actual SGs, because the tubes are densely inserted into the tubesheet to increase the heat transfer area. Consequently, the corrosion rate of SA508 would be more accelerated by the effect of small anode (SA508) and large cathode (Alloy 600). In addition, the galvanic corrosion rate of SA508 was little changed by the total ion concentration at a fixed MRI as shown in Figure 6. This is because the polarization current density of the two materials was not affected by an increase in the total ion concentration from 0.011 M to 0.11 M, as shown in Figure 5.

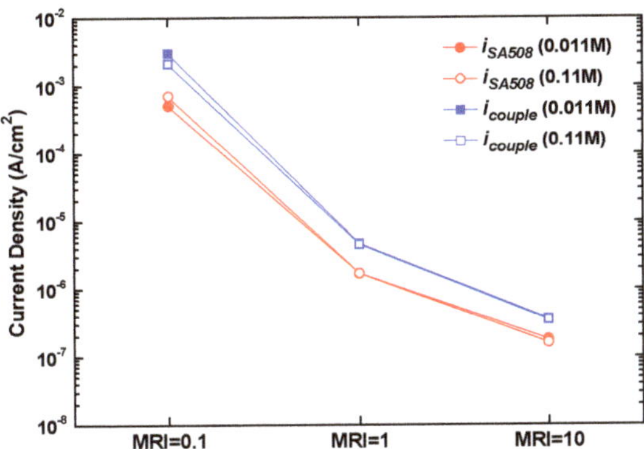

Figure 6. Effect of the MRIs on the galvanic corrosion rate of SA508 coupled to an equal area of Alloy 600.

An SG tube can be slowly deformed by volume expansion of corrosion products due to corrosion of the tube support materials adjacent to and around the tube, which is called denting. The denting was attributed to concentration of chlorides and oxidants such as copper, in the crevices, leading to rapid corrosion of the tube support materials [2,23]. SG tubes are expanded into the tubesheet of SA508, a dissimilar metal. Therefore, based on the results obtained in this work, it is worth mentioning that corrosion of the tubesheet is accelerated by the galvanic coupling itself without concentration of chemical impurities in the crevices.

4. Conclusions

The electrochemical corrosion behavior of Alloy 600 and SA508 was investigated in solutions with three different molar ratios of sodium to chloride ions. The corrosion potentials and corrosion rates of both materials were significantly decreased as the molar ratio increased from 0.1 to 10. Therefore, it is expected that the molar ratio control method is beneficial only when the crevice chemistry has a low molar ratio with an acidic pH. The corrosion potentials and corrosion rates were little affected by the total sodium and chloride ion concentrations if the alloys were polarized not far from their corrosion potentials. Alloy 600 and SA508 acted as a cathode and an anode, respectively, when they were electrically coupled. Therefore, this result indicates that the corrosion rate of the SA508 tubesheet material is accelerated owing to the galvanic coupling effect itself without concentration of chemical impurities in the crevices.

Author Contributions: D.H.H. conceived the experiments and wrote the paper; J.H. contributed analysis tools; J.C. performed the experiments. All authors have read and agreed to the published version of the manuscript.

Funding: This work was supported by the National Research Foundation (NRF) grant funded by the government of the Republic of Korea (NRF-2017M2A8A4015159).

Conflicts of Interest: The authors declare no conflict of interest.

References

1. Gorman, J. Corrosion problems affecting steam generator tubes in commercial water-cooled nuclear power plants. In *Steam Generators for Nuclear Power Plants*; Elsevier BV: Amsterdam, The Netherlands, 2017; pp. 155–181.
2. Staehle, R.W.; Gorman, J.A. Quantitative assessment of submodes of stress corrosion cracking on the secondary side of steam generator tubing in pressurized water reactors: Part 1. *Corrosion* **2003**, *59*, 931–944. [CrossRef]
3. Hur, D.H.; Choi, M.S.; Lee, D.H.; Song, M.H.; Han, J.H. Pitting corrosion and its countermeasures for pressurized water reactor steam generator tubes. *Corrosion* **2006**, *62*, 905–910. [CrossRef]
4. Hur, D.H.; Choi, M.S.; Lee, D.H.; Song, M.H.; Han, J.H. Root causes of intergranular attack in an operating nuclear steam generator tube. *J. Nucl. Mater.* **2008**, *375*, 382–387. [CrossRef]
5. Bahn, C.B.; Oh, S.H.; Park, B.K.; Hwang, I.S.; Rhee, I.H.; Kim, U.C.; Na, J.W. Impurity concentration behaviors in a boiling tubesheet crevice Part I. Open crevice. *Nucl. Eng. Des.* **2003**, *225*, 129–144. [CrossRef]
6. Engelhardt, G.R.; Macdonald, D.D.; Millett, P.J. Transport processes in steam generator crevices—I. General corrosion model. *Corros. Sci.* **1999**, *41*, 2165–2190. [CrossRef]
7. Millett, P.J. *PWR Molar Ratio Control Application Guidelines. Volume 1: Summary*; TR-104811-V1; Electric Power Research Institute: Palo Alto, Houston, CA, USA, 1995; Available online: https://www.epri.com/#/pages/product/TR-104811-V1/?lang=en- (accessed on 20 March 2020).
8. Kawamura, H.; Hirano, H.; Koike, M.; Suda, M. Inhibitory effect of boric acid on intergranular attack and stress corrosion cracking of Alloy 600 in high-temperature water. *Corrosion* **2002**, *58*, 941–952. [CrossRef]
9. Fruzzetti, K. *Pressurized Water Reactor Secondary Water Chemistry Guidelines-Revision 8*; 3002010645; Electric Power Research Institute: Palo Alto, Houston, CA, USA, 2017; Available online: https://www.epri.com/#/pages/product/3002010645/ (accessed on 20 March 2020).
10. Choi, S.; Marks, C.; Wolfe, R. PWR steam generator tube denting at top of tubesheet. In Proceedings of the International Conference on Nuclear Power Chemistry, Sapporo, Japan, 26–31 October 2014; p. 10137.
11. Hur, D.H.; Choi, M.S.; Lee, D.H.; Han, J.H.; Shim, H.S. Corrosion inhibition of steam generator tubesheet by Alloy 690 cladding in secondary side environments. *J. Nucl. Mater.* **2013**, *442*, 326–329. [CrossRef]
12. Hsu, S.S.; Tai, S.C.; Kai, J.J.; Tai, C.H. SCC behavior and anodic dissolution of Inconel 600 in low concentration thiosulfate. *J. Nucl. Mater.* **1991**, *184*, 97–106. [CrossRef]
13. Bandy, R.; Roberge, R.; van Rooyen, D. Intergranular failures of Alloy 600 in high temperature caustic environments. *Corrosion* **1985**, *41*, 142–150. [CrossRef]
14. Lin, L.F.; Cragnolino, G.; Szklarska-Smialowska, Z.; Macdonald, D.D. Stress corrosion cracking of sensitized Type 304 stainless steel in high temperature chloride solutions. *Corrosion* **1981**, *37*, 616–627. [CrossRef]
15. EPRI. *Guidelines for PWR Steam Generator Tubing Specifications and Repair*; TR-016743-V2R1; Electric Power Research Institute: Palo Alto, Houston, CA, USA, 1999; Available online: https://www.epri.com/#/pages/product/TR-016743-V2R1 (accessed on 27 February 2020).
16. ASME. *Specification for Quenched and Tempered Vacuum-Treated Carbon and Alloy Steel Forgings for Pressure Vssels*; SA-508/SA-508M: New York, NY, USA, 1995.
17. Gnedenkov, A.S.; Sinebryukhov, S.L.; Mashtalyar, D.V.; Imshinetskiy, I.M.; Vyaliy, I.E.; Gnedenkov, S.V. Effect of microstructure on the corrosion resistance of TIG welded 1579 alloy. *Materials* **2019**, *12*, 2615. [CrossRef] [PubMed]
18. Dutta, R.S.; Tewari, R.; De, P.K. Effects of heat-treatment on the extent of chromium depletion and caustic corrosion resistance of Alloy 690. *Corros. Sci.* **2007**, *49*, 303–318. [CrossRef]
19. Briant, C.L.; O'Toole, C.S.; Hall, E.L. The effect of microstructure on the corrosion and stress corrosion cracking of Alloy 600 in acidic and neutral environments. *Corrosion* **1986**, *42*, 15–27. [CrossRef]
20. Pouraix, M. *Atlas of Electrochemical Equilibria in Aqueous Solutions*; NACE: Houston, TX, USA, 1974.
21. Tremaine, P.R.; LeBlanc, J.C. The solubility of magnetite and the hydrolysis and oxidation of Fe^{2+} in water to 300 °C. *J. Solut. Chem.* **1980**, *9*, 415–442. [CrossRef]

22. Macdonald, D.D.; Shierman, G.R.; Butler, P. *The Thermodynamics of Metal-Water Systems at Elevated Temperatures, Part 2: The Iron-Water System*; AECL-4137; Atomic Energy of Canada Ltd.: Pinawa, MB, Canada, 1972.
23. Fernández-Saavedra, R.; Fernández-Díaz, M.; Gómez-Mancebo, M.B.; de Diego, G.; Quejido-Cabezas, A.J.; Gómez-Briceño, D. Hard sludge and denting on the secondary side of PWR steam generators. In Proceedings of the International Conference on Nuclear Power Chemistry, Sapporo, Japan, 26–31 October 2014.

© 2020 by the authors. Licensee MDPI, Basel, Switzerland. This article is an open access article distributed under the terms and conditions of the Creative Commons Attribution (CC BY) license (http://creativecommons.org/licenses/by/4.0/).

 *materials*

Article

Corrosion Prediction of Weathered Galvanised Structures Using Machine Learning Techniques

Marta Terrados-Cristos *, Francisco Ortega-Fernández, Guillermo Alonso-Iglesias, Marina Díaz-Piloneta and Ana Fernández-Iglesias

Project Engineering Department, University of Oviedo, 33004 Oviedo, Spain; fdeasis@uniovi.es (F.O.-F.); guillermo.alonso@api.uniovi.es (G.A.-I.); marina.diaz@api.uniovi.es (M.D.-P.); fernandeziana@uniovi.es (A.F.-I.)
* Correspondence: marta.terrados@api.uniovi.es

Abstract: Galvanised steel atmospheric corrosion is a complex multifactorial phenomenon that globally affects many structures, equipment, and sectors. Moreover, the International Organization of Standardization (ISO) standards require specific pollutant depositions values for any atmosphere classification or corrosion loss prediction result. The aim of this research is to develop predictive models to estimate corrosion loss based on easily worldwide available parameters. Experimental data from internationally validated studies were used for the data mining process, basing their characterisation on seven globally accessible qualitative and quantitative variables. Self-Organising Maps including both supervised and unsupervised layers were used to predict first-year corrosion loss, its corrosivity categories, and an uncertainty range. Additionally, a formula optimised with Newton's method has been proposed for extrapolating these results to long-term results. The predictions obtained were compared with real values using Euclidean distances to know its similarity degree, offering high prediction performance. Specifically, evaluation results showed an average saving of up to 16% in coatings using these predictions. Therefore, using the proposed models reduces the uncertainty of the final structures state by predicting their material loss, avoiding initial over-dimensioning of structures, and meeting the principles of efficiency and sustainability, thus reducing costs.

Keywords: weathered galvanised steel; corrosion; predictive models; optimisation

Citation: Terrados-Cristos, M.; Ortega-Fernández, F.; Alonso-Iglesias, G.; Díaz-Piloneta, M.; Fernández-Iglesias, A. Corrosion Prediction of Weathered Galvanised Structures Using Machine Learning Techniques. *Materials* **2021**, *14*, 3906. https://doi.org/10.3390/ma14143906

Academic Editor: Marián Palcut

Received: 10 June 2021
Accepted: 12 July 2021
Published: 13 July 2021

Publisher's Note: MDPI stays neutral with regard to jurisdictional claims in published maps and institutional affiliations.

Copyright: © 2021 by the authors. Licensee MDPI, Basel, Switzerland. This article is an open access article distributed under the terms and conditions of the Creative Commons Attribution (CC BY) license (https://creativecommons.org/licenses/by/4.0/).

1. Introduction

Multiple metallic structures and equipment operate in outdoor conditions [1]. In such cases, one of the main problems related to their stability and durability is corrosion [2,3]. World Corrosion Organization (WCO) estimates the world direct cost of corrosion to be between 1.3 and 1.4 trillion EUR, 3.1% to 3.5% of a nation's GDP annually [4].

Corrosion is a very complex phenomenon based on the degradation of a material or its properties due to its reaction with the environment [5]. Multiple factors [6], particles [7], and variables [8,9] are involved. The character of the attack and the corrosion rate are consequences of the system formed by metallic materials, atmospheric environment, technical parameters, and operating conditions [10]. Corrective factors are introduced in the design phases to guarantee the structure's integrity during its useful life [11]. However, the difficulty of quantifying the material loss causes unnecessary over-dimensioning, leading to superfluous costs and resources consumption [12]. Proper management of this complex multifactorial phenomenon is key to sustainable development [13].

To ensure the integrity of the outer layer, structures are designed with physical protection. Historically, metallic zinc has provided excellent corrosion protection of steel structures [14]. Unfortunately, corrosion damage also occurs in such systems [15]. Since corrosion leads to a mass loss, an excess thickness is often considered to ensure service life. This not only increases manufacturing cost but also does not satisfy the principles of sustainable engineering efficiency [16]. Therefore, lacking an automated monitoring system or predictive model, routine thickness monitoring would be required [17]. These phenomena have drawn increasing attention in recent decades due to the resulting catastrophic accidents [18] and the growing demand for sustainable designs [19]. For an optimal selection of materials, atmospheric aggressiveness must be considered. Depending on this, coating needs can be set.

The current regulation regarding galvanised metallic structures (ISO 9223:2012 [20]) groups the corrosivity level of an atmosphere into six categories. After studying the effect of corrosion on standard samples during 1 year of weathering exposure, the level of corrosion rates achieved can be established by measuring weight losses for different materials. This material's loss due to corrosion is commonly used as an initial measure for determining coating requirements. However, material loss margins are allowed within these categories, and coating thickness designs based on them are not fixed. These margins imply variability in the amount of material that can be translated into increased costs.

According to [20], two methods are proposed to classify the corrosivity of atmospheric environments, depending on the availability of experimental data. When experimental data are available, dose–response functions can be used. However, when no experimental data are available, corrosivity category estimation using the informative procedure is recommended, and as stated in the norm, it is based on the comparison of local environmental conditions with the description of typical atmospheric environments, which may cause misinterpretations [21]. Finding the optimum point between efficiency and competitive price, while remaining within limits, is therefore challenging given the lack of characterisation of the specific construction site.

The objective of this work is to develop machine learning models that, by analysing real cases, predict corrosion mass loss of zinc coatings over time. The aim is to characterise an environment without requiring long testing periods and sampling and generalising it to any location worldwide, with the data available from international studies. This considerably increases the existing knowledge about coated steel structure corrosion and extends it to the full diversity of atmospheres, thereby reducing the uncertainty of its final state.

This paper starts with a state-of-the-art analysis. Then, it explains the creation of the database through the characterisation of each sample. Next, the applied methodology is explained, and modelling and evaluation techniques are defined. Finally, results are discussed, and the conclusions obtained in this research are proposed.

2. Literature Review

There is a wide range of corrosion problems in the industry, resulting from the different combinations of materials, environments, and service conditions [22]. Therefore, the concern about corrosion is not new. The science of atmospheric corrosion started with Faraday in the nineteenth century [23]. Another important contribution was made by Vernon who began systematic experiments in atmospheric corrosion in the 1920s [24]. In 1986, Benarie and Lipfert published their work on atmospheric corrosion [25], relating this phenomenon to the concentration of certain pollutants and pH of the rain. Subsequently, Feliu et al. developed regression equations for mild steel, zinc, copper, and aluminium [26].

There are several kinetic corrosion models that attempt to predict atmospheric corrosion over time: the general linear model [27], the power function models [28], and the power-linear models [29]. However, the corrosion process is influenced by multiple environmental factors [30]. Therefore, these corrosion kinetic models are valid at specific locations. When the environmental condition changes, the model may no longer be applicable [31]. It would be interesting to classify the aggressiveness of different atmospheres, which would allow preventive measures to be taken. Therefore, it is important to introduce the interaction parameters between environmental factors and corrosion rates for their efficient prediction.

In accordance with this approach, the ISOCORRAG program was launched in 1986 [32]. The ISO 156 technical committee developed this project with the intention of obtaining sufficient information to standardise atmospheric corrosion on metals and alloys. Four international standards were created as a result of this project: ISO 9223 [21], ISO 9224 [33], ISO 9225 [20], and ISO 9226 [34]. Since then, these standards have served as practical guidelines and aids for the design of both structures and their corrosion protection. In September 1987, the Executive Body for the Convention on Long-Range Transboundary Air Pollution (CLRTAP) decided to launch an International Cooperation Program with the United Nations European Economic Commission (ICP/UNECE) [35] whose objective was to carry out a quantitative assessment of the effect of pollutants on atmospheric corrosion [6]. In addition, a third cooperative program was launched, named MICAT [36] (Ibero-American Atmospheric Corrosivity Map). Its objective was to understand the mechanisms that take place when this phenomenon occurs, to generate, with the data obtained, mathematical models to calculate corrosion as a function of climate condition or pollutant levels [13]. The three projects evaluated corrosion by measuring mass loss and were based on what was indicated in the standard for measuring SO_2 or Cl^- levels and other pollutant concentrations.

In 1992, the ASTM (American Society for Testing and Materials) published a study discussing an alternative method for measuring corrosion penetration, with models that are tighter and more rational than the traditional potential model [37]. In 2003, several workers compiled atmospheric exposure data from many research reports and journal articles [38]. R.E. Melchers, an engineer at Newcastle University, focused on studying the corrosion of metals in marine atmospheres in his studies in 2008 [39] and 2013 [40]. Later, Morcillo et al. [27] made a comprehensive compilation in the scientific literature on weathering steel atmospheric corrosion [6]. In addition, they developed Damage Functions to know the damage that a metallic structure can suffer depending on weathering conditions. In the subsequent years, there have been local experimental studies to characterise this phenomenon, such as those in Greece [41] and the Czech Republic [42].

The dose–response function is the most widely used. It directly correlates the influencing environmental factors with the corrosion parameters [43]. The basic form of this function follows the simple linear [36,44] or logarithmic–linear relationships [45]. However, many researchers also started to depart from judging the effect of each environmental factor separately and established a new multi-factor combination model [46,47]. A response surface model (RSM) takes into account the interactive effect and the non-linearity of the atmospheric corrosion process and allows a better approximation compared to conventional dose–response function models [48]. The models offer a closer approximation of corrosion rate by introducing different input variables. Temperature, humidity, sulphur dioxide concentration, and chloride concentration are typically used.

In conclusion, there are different options to predict corrosion rates of metals based on experimental input data. However, for the cases when pollutants' concentration is unknown, the options are limited. Time and cost constraints make the development of these measurements difficult as they would be unrepresentative when only completed at a specific point in time. As the environmental conditions continuously change, it is necessary to know their distribution over larger distances and longer periods of time. All corrosion related research carried out so far showed that there are certain factors that clearly influence the corrosion process. Regarding atmospheric corrosion, the factors include temperature, relative humidity, precipitation level, and pollutant concentrations (SO_x, Cl^-, etc.) [49,50]. A combination of parameters, such as Time of Wetness (TOW), is also used. TOW represents the fraction of time when relative humidity exceeds 80% and ambient temperature is above 0 °C (h/year) [51].

Climate has a significant influence on corrosion since some of the factors mentioned above depend on the climatic zone. A Köppen–Geiger classification [52] is the most popular technique for climate characterisation. According to this method, six precipitation levels can be distinguished [52]: desert (0), steppe (1), totally humid (2), summer dry (3), winter dry (4), and monsoon (5). Temperature and relative humidity are easily analysable climatic variables, and their values are generally accessible. There are also additional factors besides climate, mainly derived from human activities, whose importance is also significant. It is evident that the most populated and most-developed areas with accumulations of vehicles and high industrial activity have greater corrosive potential. It is also known that materials situated in areas closer to the sea tend to have a worse corrosion performance. Therefore, it is necessary to include these additional factors as well as they are critical for the successful operation of the model.

3. Materials and Methods

3.1. Data

This work seeks a more practical approach to characterise the environment. After a complete analysis of the data from existing experimental studies, it has been concluded that ISOCORRAG program data [32] should be used as it also analysed the corrosion in helical samples. Corrosion rates on helical samples have higher average corrosion rate values and do not limit corrosion loss to a single direction. This approach is useful in our case, as it more closely relates to galvanised structures used in civil engineering. Besides, it includes enough helical specimens distributed globally to represent a wide variety of cases. The project was carried out at more than 50 different locations in Asia, Europe, and America (Figure 1). During the ISOCORRAG program, the exposed specimens were used to determine the first-year corrosion rate. Nevertheless, some of the specimens were also used to study long term corrosion exposure. Grouped in different sets, triplicate samples were exposed every 6 months, and left for up to 1 year. The monitoring process lasted from 1986 to mid-1989.

Figure 1. ISOCORRAG program sample's location.

ISO 9223 and ISO 9224 standards are highlighted for this project. First, ISO 9223:2012 [20] divides the corrosivity of atmospheres into 6 categories. Each of these categories corresponds to a different corrosion level. For zinc, data are shown in Table 1.

Table 1. Corrosion rates of zinc for first-year exposure for different corrosivity categories according to ISO 9223:2012.

Corrosivity Category	Corrosivity	Unit	Zinc
C1	Very low	µm/year	$r_{corr} \leq 0.1$
C2	Low	µm/year	$0.1 < r_{corr} \leq 0.7$
C3	Medium	µm/year	$0.7 < r_{corr} \leq 2.1$
C4	High	µm/year	$2.1 < r_{corr} \leq 4.2$
C5	Very high	µm/year	$4.2 < r_{corr} \leq 8.4$
CX	Extreme	µm/year	$8.4 < r_{corr} \leq 25$

Second, ISO 9224:2012 proposes a relationship for long-term corrosion exposures. This relationship is based on the power function according to the following equation:

$$D = r_{corr}\, t^b \tag{1}$$

In Equation (1), r_{corr} is the first-year corrosion rate, t is the number of years to be analysed, and b is the environment and metal-specific time exponent.

3.1.1. Variables

Willing to characterise any location worldwide, its atmospheric corrosivity and climate need to be considered. For this work, three specific types of atmospheric environments have been introduced as binary synthetic variables, trying to represent the behaviour of sulphates-related pollution and chlorides deposition:

- Industrial/Non-industrial: industrial are areas with fossil fuel combustion industries (refineries, thermal power plants, etc.).
- Marine/Non-marine: this characterisation has been made according to the distance from the coast, considering as Marine any location within 15 km from the seashore [53,54].
- Urban/Rural: locations with more than 5000 inhabitants or 300 inhabitants per square kilometre have been considered urban locations [55].

Regarding the climate characterisation, temperature, relative humidity, TOW, and Köppen–Geiger level of precipitation were the main characteristics, unified in a simple, accessible, and complete way. Therefore, a total of seven numeric predictor variables were set for the model: mean annual temperature, mean annual relative humidity, TOW, precipitation, industrial, marine, and urban. The variable to be predicted was the zinc corrosion loss during first-year exposure, directly taken from experimental studies, and its atmospheric corrosivity category, based on the standard. Each sample was characterised, following the rules mentioned above, as explained in Figure 2.

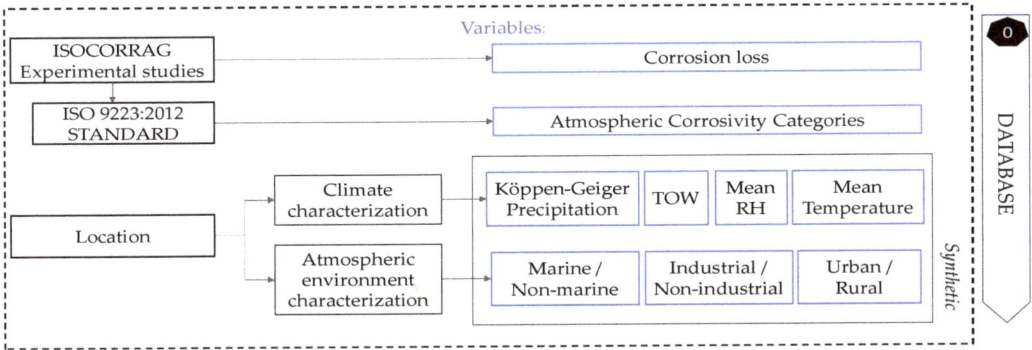

Figure 2. Flow chart for database creation and future locations characterisation.

A summary of variables is shown in Table 2. The mean annual temperature is represented as T_annual and mean annual relative humidity as RH_annual in the table.

Table 2. Information on new continuous and discrete variables added.

Continuous Variables					Discrete Variables	
Variable	Unit	Min	Avg	Max	Variable	Range
T_annual	°C	−15	14.5	29.1	Marine	0 (Non-Marine)–1 (Marine)
RH_annual	%	33	74.7	98	Industrial	0 (Non-industrial)–1 (Industrial)
TOW	h/year	37	2723	6350	Rural	0 (Urban)–1 (Rural)
					Precipitation	1–5

3.1.2. Data Analysis

Data quality and representativeness are crucial for modelling; otherwise, the results obtained would be inconsistent. Frequency distributions of the 4 discrete variables are shown in Figure 3. All possible combinations between different environment types (Rural/Urban, Industrial, Marine) have been observed. In addition, colours show the number of samples in each of the 5 possible precipitation levels. All precipitation levels were represented; however, there some combinations were represented more often than others (urban, industrial, and marine zone).

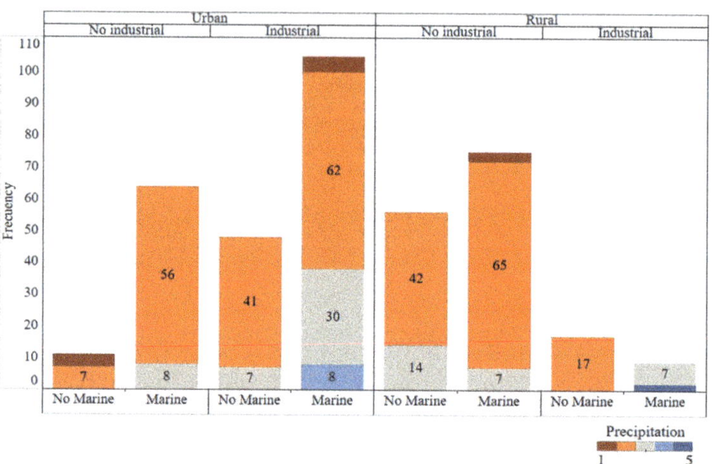

Figure 3. Frequency graphical analysis of the categorical variables. All possible atmospheric environment combinations are represented and coloured by precipitation type.

Regarding continuous variables, Figure 4 shows the geographical distribution of temperature and mean annual relative humidity in each location, according to the numerical values obtained. The data are obtained from web services that use weather stations spread all over the world. Worldwide distribution of cases has been achieved.

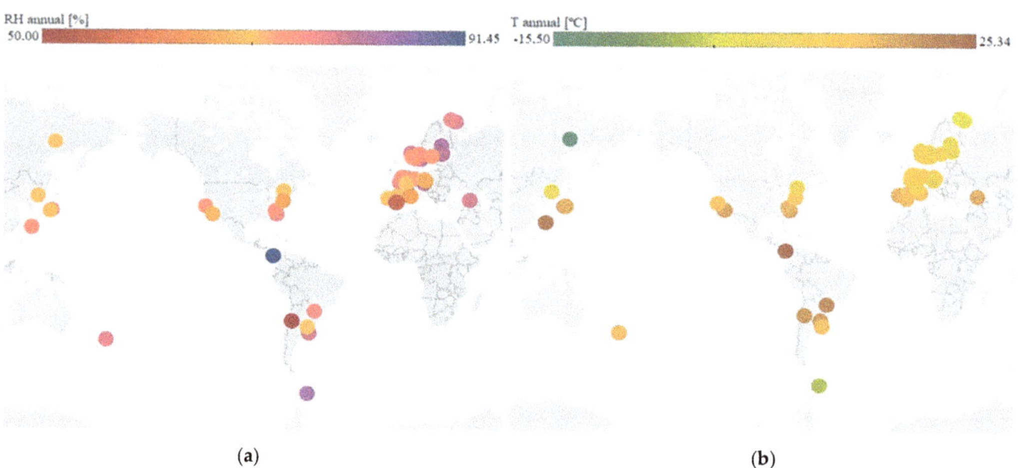

(a)   (b)

Figure 4. Analysis of continuous variables at each location. (a) Distribution of mean annual relative humidity. (b) Distribution of mean annual temperature.

3.2. Methodology

The methodology followed in this paper consisted of 6 phases (Figure 5). The preparatory stage (stage zero) in the previous subsection was concluded with the creation of the database. Then, the remaining five phases included modelling and testing. The first step for data pre-processing was to identify input variable's importance for better understanding their behaviour and obtaining additional information regarding their usefulness in the final model. This was completed using Multivariate Adaptive Regression Splines (MARS, Step 1). Then, the next phase was to define the first-year corrosion loss of galvanised steel. Self-Organising Maps (SOM) were used, including various layers (supersom) of both supervised and unsupervised learning. The next two steps used the result of the various layers of this algorithm. The first layer has been the result of using unsupervised SOM, according to the relationships between the 7 main variables. Zinc corrosion loss during first year of exposure (Corr_Zn, in µm) was the output variable to be predicted (Step 2).

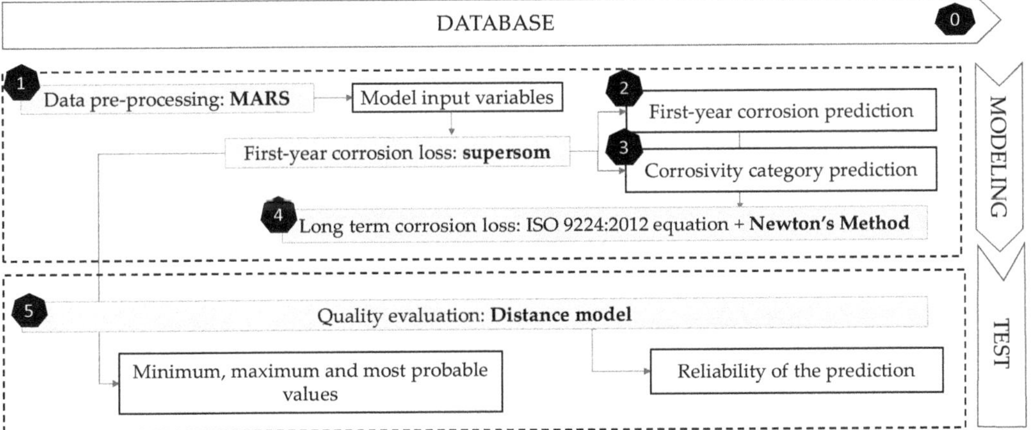

Figure 5. Flow chart showing the methodology followed in this paper. The six phases proposed are exposed as shown.

The advantage of SOM maps is that in addition to assigning an individual value, an uncertainty range is also given, obtained by adding the minimum and maximum value within each neuron. Besides, it is intended that in addition to self-organising according to the input variables, supersom networks group the data according to the various corrosivity categories. Then, the second one of the two output layers would be the result of organising corrosion in a supervised output layer that will assign the corresponding 'corrosivity category' value set to each node by the standard (Step 3). Furthermore, the corrosivity is not constant with respect to exposure time. In most cases, it decreases with increasing exposure due to accumulation of corrosion products on the surface. Step 4 includes optimising the formula that allows the extrapolation of these results to long term results. With Newton's method, a nonlinear regression of the formula used by ISO 9224 (Equation (1)) was performed to optimise the value of variable *b*.

Finally, to test the quality of the predictions, a model based on Euclidean distances was used (Step 5). This model analyses the model input variables, trying to find the most similar cases in the database to show their corrosion value and its similarity degree (quality). Then, in this fifth phase, the results obtained were compared with existing real cases to measure the quality of predictions using a Euclidean distance model. Although both supersom and distance models start from the same database and have the same inputs, their purposes are different. While supersom model gives a corrosion prediction, and a corrosivity category, the distance model sets the quality of that prediction.

Techniques

- Multivariate Adaptive Regression Splines (MARS)

One of the most widely used algorithms for solving adaptive computing problems is MARS [56]. This method consists of approximating an unknown function by the linear combination of a set of basic functions (products of the model variables) [57]. Among the key points of the algorithm, it stands out that it autonomously selects the relevant variables and interactions between them for each subregion. Thus, the dimensionality reduction of the problem is performed directly by the model, with the advantage of being locally carried out. Precisely, this benefit can be used to analyse the relevance of the variables likely to subsequently participate in the model.

- Self-Organising Maps (SOM)

The clustering model, known as SOM, is an unsupervised Artificial Neural Network (ANN) presented in 1982 by T. Kohonen [58]. This model is based on certain evidence discovered at brain level and performs a reduction of the dimensionality of the input space to produce topologically ordered maps. This type of network has competitive, unsupervised learning. The network itself is in charge of self-organising and discovering common features, regularities, correlations, or categories in the input data [59,60].

Figure 6 shows the architecture of the model and how each input neuron is connected to one of the output neurons by weights (w, according to Kohonen's notation). The output neurons will therefore have an associated vector of weights which is called the reference vector (or codebook), also constituting the average vector of the category represented by the output neuron [61,62].

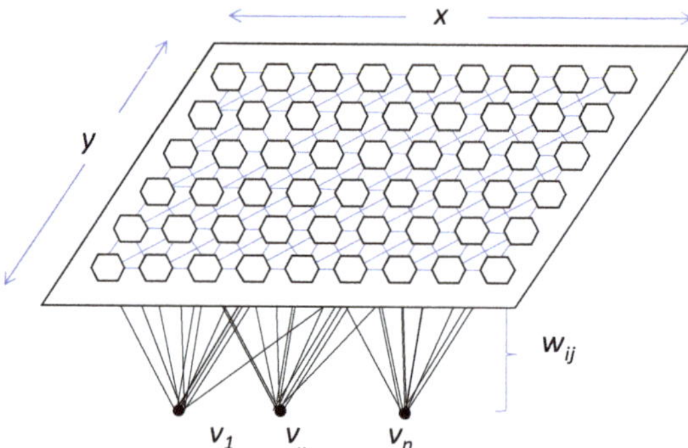

Figure 6. General example of SOM model's topography. Dimensions are expressed by x and y; v_{1-n} represent each one of the input neurons, and w_{ij} is the weight of each vector according to Kohonen's notation.

SOM's utility lies in the holistic visual interpretation of the output rather than in understanding the underlying processes [63]. Roughly speaking, the output layer (i.e., the self-organising map itself) contains neurons organised in a rectangular or hexagonal lattice to represent the entire dataset [58].

The goal of this learning is to categorise the data fed into the network. Similar values are classified into the same category and, therefore, should activate the same output neuron. Since this is an unsupervised method, classes or categories must be created by the network itself through correlations between the input data [64]. However, SOM can also be used for pattern recognition (supervised learning). The information is given at the end of the training: if classification is involved, as in this case, the winner-takes-all strategy is used. This principle can be extended to more layers, generating super-organised maps (supersom). For each layer, a similarity level is calculated, and the individual similarities are combined into a single value which is used to determine the winner node.

- Newton's method

This nonlinear regression uses Newton's Surface gradients, which is an unconstrained linear regression method based on that gradient. The gradient information is provided by analytically computed gradients. Design variables are modified, while their impact on the objective function is analysed [65].

- Euclidean distance model

The operation of this model is based on Euclidean distances (d_E). This is a non-negative function used to calculate the distance between two points $P = (p_1; p_2; \ldots; p_n)$ and $Q = (q_1; q_2; \ldots; q_n)$ on an n-dimensional space [66]. It works on the basis of the Pythagoras Theorem (Equation (2)) [67]. Results evaluation using this method involves checking that the model gives a 100 % quality in all the cases studied, i.e., that it perfectly finds its counterpart.

$$d_E(P, Q) = \sqrt{(p_1 - q_1)^2 + \cdots + (p_n - q_n)^2} = \sqrt{\sum_{i=1}^{n}(p_i - q_i)^2} \tag{2}$$

To summarise, Table 3 shows the different algorithms used in each phase of the data mining process.

Table 3. Summary of all models used.

Phase	Algorithm
Pre-processing data	MARS
Modelling	
• Corrosivity category prediction	superSOM
• First-year corrosion prediction	superSOM
• Long-term corrosion prediction	Newton method
Quality evaluation	Distance model

4. Results and Discussion

Results obtained in each of the phases are presented below.

4.1. Data Pre-Processing Using MARS

The importance of each of the variables has been analysed, assessing their influence on the variable to be predicted. Two statistics were used: generalised cross-validation criterion (GCV) and residual sum of squares (RSS). Both criteria results (blue and red lines) together with the mean of both results (light blue bars) are shown in Figure 7.

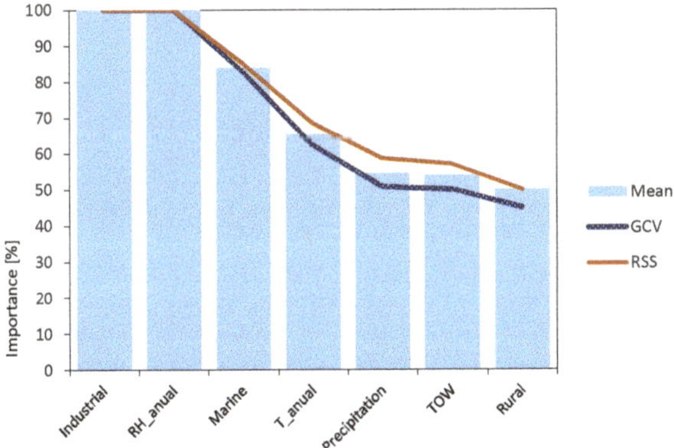

Figure 7. Variable importance analysis results, using MARS algorithm.

It is clearly evidenced that variables related to atmospheric pollutants SO_2 (Industrial) and Cl^- (Marine) are the most important factors, together with relative humidity, in agreement with what was previously described in the literature review. They can all be considered as independent variables, susceptible to providing the model with enough information to obtain valuable predictions.

4.2. First-Year Corrosion Prediction

The result of the supersom model is a mesh of 7×7 hexagonal neurons trained with the Kohonen algorithm, which provides a good representation of the sample space. The resulting trained map contains all the data in a vector structure so that the training data falls on each of the neurons (Figure 8).

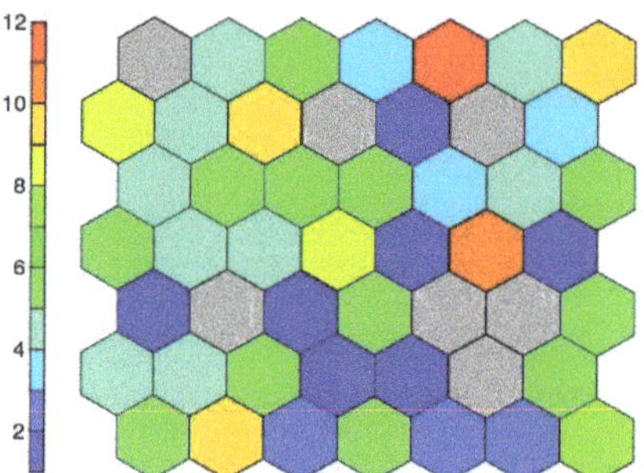

Figure 8. Number of cases on each neuron.

Each neuron, filled or not, is represented by a codebook. These neurons are arranged in such a way that nearby neurons represent points closer to each other. Analysing the result of the average corrosion values per neuron along the mesh, it can be clearly seen how the mesh is growing towards the lower right corner. Figure 9 shows this result; the larger the circle size, the higher the average corrosion. Keeping the neighbourhood properties, a uniform behaviour is shown, which indicates good training results.

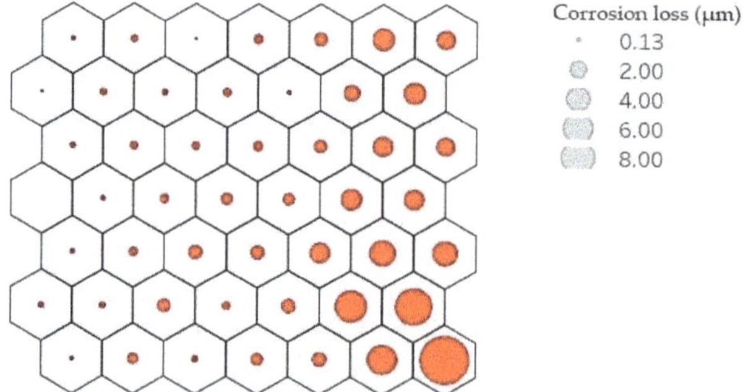

Figure 9. Mean corrosion values per neuron. Corrosion loss in μm per year is represented by circle size.

4.3. Corrosivity Category Classification

When analysing the results of both output layers, represented in each neuron by its corrosion rate value, the neurons were grouped, forming zones mostly corresponding to one type of atmosphere (Table 1). The zones division with different corrosion rates is given in Figure 10. Both C1 and CX categories were filtered out of the dataset due to a lack of consistent data. Thus, the far-left zone corresponds to C2 atmospheres, the left zones to C3, the right zones to C4, and finally, the lower-right end to C5. There is also a transition between the values so that the C5 are in contact with C4, C4 with C3, etc., demonstrating an optimal training.

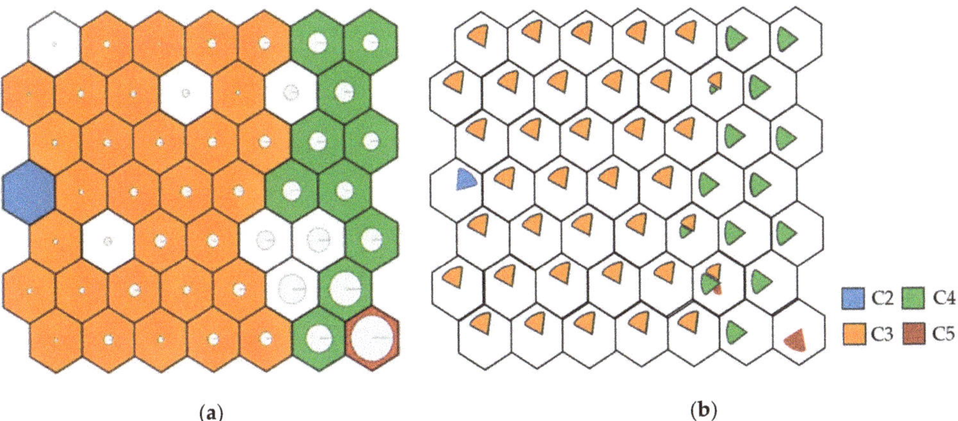

Figure 10. Corrosion zones according to the environment. (**a**) Corrosion representation (larger circle, more corrosion). (**b**) Corrosivity category representation, according to ISO 9223:2012 standard.

The predicted first-year corrosion rates using SOM trained network were compared with real values. A satisfactory correlation has been obtained (Figure 11), although not all points perfectly matched their counterpoints. The ideal situation would be if the predicted values all lied on the diagonal line. The points tend to be located on the upper side of the graph, meaning that predictions are conservative, and the decisions made based on them can provide greater safety.

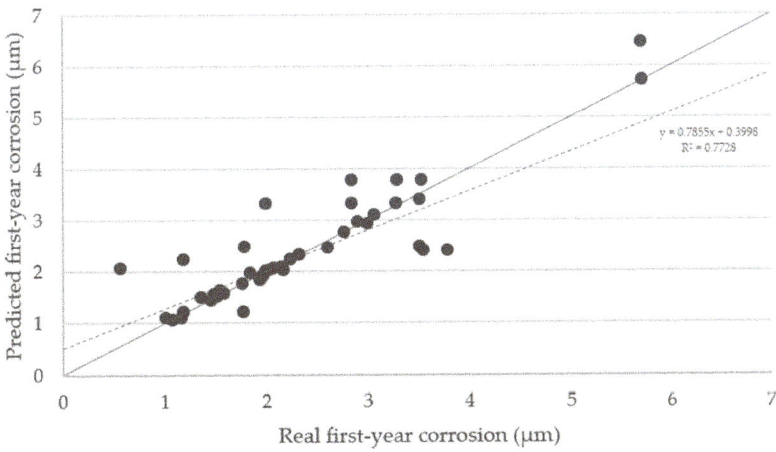

Figure 11. Predicted first-year corrosion values in micron vs. real first-year corrosion values. The dashed line is the regression line ($R^2 = 0.7728$). The points situated on the diagonal grey line represent an optimal training.

From the trained network, it is possible to determine the corrosion rate of any situation to be studied. When introducing a new case to the model, it finds the node that most closely resembles its input variables. Thus, the output of the model is the corrosion rate of that node. The uncertainty range is also given, including the minimum and maximum values within each neuron. This can be seen with the following example for a case with the characteristics defined in Table 4.

Table 4. Example of model input data.

Rural	Industrial	Marine	Precipitation	T_annual	RH_annual	TOW
0	1	0	2	11.98	72.1	3218

The case falls into the neuron indicated in Figure 12, which consists of 10 examples.

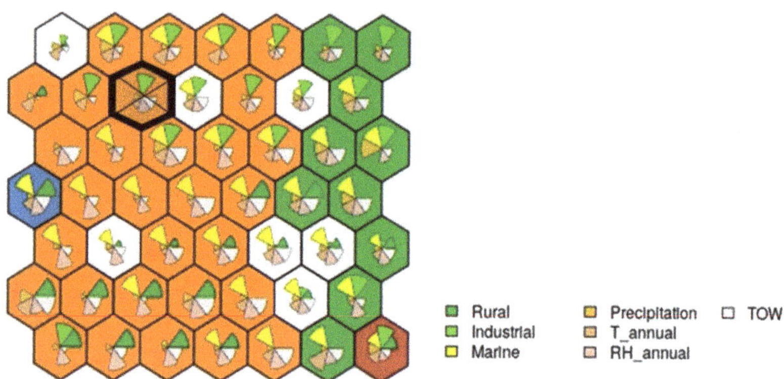

Figure 12. Case study example: the cake portions shown at each node show the contribution of each variable; the larger the size, the greater its final weight.

Table 5 shows all results obtained. Different conclusions can be made by selecting the maximum (Corr_max), minimum (Corr_min), and average (Corr_avg) values of the examples in one single neuron. As a result, when the values with the most or least corrosion occurring within the projects in the neuron are chosen, the optimistic and pessimistic predictions can be obtained. Alternatively, β-distribution is used to determine the 'most probable' rate of Corr_Zn, using the maximum, minimum, and average values. On the other hand, the category is awarded by the weighted average of the categories in each case. In this case, since all cases are C3, C3 is its category.

Table 5. Example of results for the case study.

Corr_min	Corr_avg	Corr_max	Range Given by the Model	Category	Range Given by ISO Standard
1.22	1.578	1.91	1.22–1.91	C3 100%	0.7–2.1

Comparing the range given by the model with the range given by the existing standard, it is observed that the latter represents a much higher uncertainty for each corrosivity category. Extending this comparison to the entire study scope, possible model predictions for each category, clustered on similar values and represented by boxplots, can be presented (Figure 13). Although not all categories are equally distributed, they show, in general, narrower intervals.

This study is presented as a possible alternative to the informative procedure of the ISO standard when there is no experimental data available. The results of the informative procedure regarding atmospheric categorisation provide a range of mass losses for each material. The current trend among companies and engineers, when no specific experimental information is available, is to use the highest value of each category to make their decisions. Since corrosion loss values are directly related to the required coating thickness, the higher the corrosion loss value, the more coating is required. A coating thickness can thus be directly determined by the predicted material's loss.

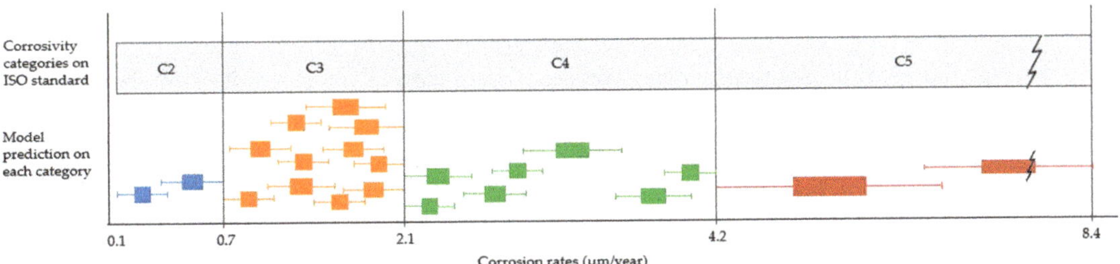

Figure 13. Comparison between each category range offered by the standard using the informative procedure and the possible mean values and uncertainties offered by the model, represented by clustered boxplots on each category.

The material requirement for coatings can be compared with the largest measurement proposed by the standard in each category and with the value predicted by the model. Following the example above, when using a Zn-coating of 1.6 µm (Corr_avg) instead of 2.1 µm (maximum in the range given by ISO), a 24% reduction in material's costs is obtained. It is then proposed to carry out this comparison for the rest of the points studied. From a more conservative perspective, comparing the maximum predicted value (Corr_max) with the maximum proposed by the standard using the informative method can also be used. In this way, uncertainties are also considered. By performing this for all data studied during the evaluation phase, an average saving of 16% in coating material is obtained.

4.4. Long-Term Corrosion Prediction

Once the first-year corrosion rate provided by the supersom model is known, the long-term loss can be identified thanks to the optimised Equation (1). Table 6 shows the different values obtained by this optimisation method for each of the corrosivity categories.

Table 6. Results obtained by Newton's method for optimised *b* coefficient.

Corrosivity Category	Value
C2–C3	0.816
C4–C5	0.704

Figure 14 compares the distribution of relative errors of both models. The nonlinear regression relative error is represented by a solid black line and the standard formula's relative error (ISO 9224) by a blue dashed line. A more uniform distribution is achieved in the nonlinear regression model.

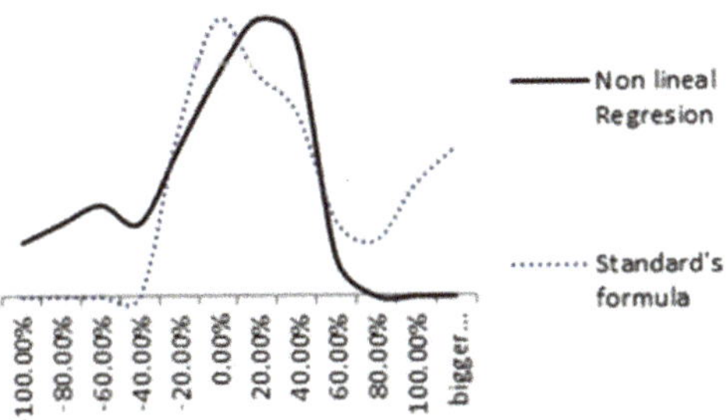

Figure 14. Comparison between Nonlinear regression and standard's formula relative errors.

4.5. Quality Evaluation

For the correct functioning of the model, data were normalised. According to the previous criteria, the most similar options are shown. The best way to show the results of this last model is using an application example, which is presented in Table 7. The quality row shows the percentage assessing the prediction's quality. The first column represents all input values of the example. The next three columns show the most similar real results in the database.

Table 7. Results of the example case, using the distance model.

Variable	Example	Result 1	Result 2	Result 3
Location	Dortmund	Bergisch Gladbach	Saint Denis	London
Quality	-	98.10%	98.00%	86.60%
Rural/Urban	Urban	Urban	Urban	Urban
Industrial	Yes	Yes	Yes	Yes
Marine	No	No	No	No
Precipitation	2	2	2	2
T_annual	11.98	11.8	12.3	12.5
RH_annual	72.1	73	73	74
TOW	3218	3149	3146	4021.3
Corr_Zn (μm/year)	-	1.60	1.48	1.67
Corrosion Category	-	C3	C3	C3

Results obtained above show high prediction reliability. Cases similar to the one under study have been found in the database. The model could also give a satisfactory result for a case that is not included in the database. Ideally, the results obtained with the proposed methodology should be compared with the results obtained with existing methods in the literature. However, since the innovative premise of this study is based on adapting the input variables to avoid the need for pollutant-specific data, such a comparison cannot be made. One of the differentiating factors of this classifier model is that to obtain a corrosion loss rate, values for pollutant concentrations are not needed. Consequently, it may be concluded that the different algorithms developed are a good alternative for technicians and engineers to make informed decisions based on their level of risk acceptance. To sum up, given a specific location and based on the available data, these models can determine the Zn-coating thickness needed for a successful short- and long-term corrosion resistance, providing the most probable, optimistic, and pessimistic predictions.

5. Conclusions

In the present work, various models for predicting galvanised coated steel corrosion damage of metal structures exposed to weathering have been developed. The following conclusions can be drawn from this research.

The application of a supersom algorithm is considered for first-year corrosion prediction, which allows categorising any environment while obtaining a predicted value, with satisfactory results. In the cases when no experimental data are available, the model can be an alternative to the conventional informative method based on pollutant input variables. The model presented in this work could help civil engineering companies to optimise the ratio between the minimum coating required and maximum service life, thus contributing to a significant lifetime extension of steel structures.

The main limitation of the model is that it lacks statistical metrics to evaluate the performance. To solve this and explore the performance and quality of the predictions, a quality model based on Euclidean distances was proposed. A long-term corrosion prediction was also optimised based on standards ISO 9224:2012 formula and the exponential coefficient with Newton's method.

To cover all different atmospheric environments, more specific characterisations are required. The future research will focus on including the development of physical variables, such as wind speed and wind direction. It is also important to feed the model with more examples from the lesser-represented categories, as there are notable differences between C3/C4 categories and the remainder of the cases. Adding new metallic materials will also be explored, following the same methodology, possibly leading to the development of new prediction models.

Author Contributions: Conceptualisation, F.O.-F. and A.F.-I.; methodology, G.A.-I. and M.D.-P; validation, G.A.-I. and M.T.-C; writing—original draft preparation, M.T.-C.; writing—review and editing, M.T.-C. and M.D.-P.; supervision, F.O.-F. All authors have read and agreed to the published version of the manuscript.

Funding: This research received no external funding.

Institutional Review Board Statement: Not applicable.

Informed Consent Statement: Not applicable.

Data Availability Statement: Data sharing is not applicable to this article.

Conflicts of Interest: The authors declare no conflict of interest.

References

1. Wang, Z.; Wang, M.; Jiang, J.; Lan, X.; Wang, F.; Geng, Z.; Tian, Q. Atmospheric Corrosion Analysis and Rust Evolution Research of Q235 Carbon Steel at Different Exposure Stages in Chengdu Atmospheric Environment of China. *Scanning* **2020**, *2020*, e9591516. [CrossRef] [PubMed]
2. Emetere, M.E.; Afolalu, S.A.; Amusan, L.M.; Mamudu, A. Role of Atmospheric Aerosol Content on Atmospheric Corrosion of Metallic Materials. *Int. J. Corros.* **2021**, *2021*, e6637499. [CrossRef]
3. Michael Schutze, R.B. *Corrosion Resistance of Steels, Nickel Alloys, and Zinc in Aqueous Media: Waste Water, Seawater, Drinking Water, High-Purity Water*; John Wiley and Sons Ltd.: Hoboken, NJ, USA, 2016; ISBN 3-527-34069-6.
4. Hays, G.F. *Now Is the Time*; World Corrosion Organization: New York, NY, USA, 2010.
5. Ahmad, Z. Chapter 2—Basic Concepts in Corrosion. In *Principles of Corrosion Engineering and Corrosion Control*; Ahmad, Z., Ed.; Butterworth-Heinemann: Oxford, UK, 2006; pp. 9–56, ISBN 978-0-7506-5924-6.
6. Chico, B.; De la Fuente, D.; Díaz, I.; Simancas, J.; Morcillo, M. Annual Atmospheric Corrosion of Carbon Steel Worldwide. An Integration of ISOCORRAG, ICP/UNECE and MICAT Databases. *Materials* **2017**, *10*, 601. [CrossRef]
7. Yin, C.; Cheng, X.; Liu, X.; Zhao, M. Identification and Classification of Atmospheric Particles Based on SEM Images Using Convolutional Neural Network with Attention Mechanism. *Complexity* **2020**, *2020*, e9673724. [CrossRef]
8. Hembrara, O.V.; Andreikiv, O.E. Effect of Hydrogenation of the Walls of Oil-and-Gas Pipelines on Their Soil Corrosion and Service Life. *Mater. Sci.* **2012**, *47*, 598–607. [CrossRef]
9. Doyle, G.; Seica, M.V.; Grabinsky, M.W. The Role of Soil in the External Corrosion of Cast Iron Water Mains in Toronto, Canada. *Can. Geotech. J.* **2003**, *40*, 225–236. [CrossRef]
10. Kusmierek, E.; Chrzescijanska, E. Atmospheric Corrosion of Metals in Industrial City Environment. *Data Brief* **2015**, *3*, 149–154. [CrossRef] [PubMed]
11. Xu, Y.; Liu, L.; Zhou, Q.; Wang, X.; Tan, M.Y.; Huang, Y. An Overview of Major Experimental Methods and Apparatus for Measuring and Investigating Erosion-Corrosion of Ferrous-Based Steels. *Metals* **2020**, *10*, 180. [CrossRef]
12. Lazorenko, G.; Kasprzhitskii, A.; Nazdracheva, T. Anti-Corrosion Coatings for Protection of Steel Railway Structures Exposed to Atmospheric Environments: A Review. *Constr. Build. Mater.* **2021**, *288*, 123115. [CrossRef]
13. Morcillo, M.; Chico, B.; Fuente, D.; Simancas, J. Looking Back on Contributions in the Field of Atmospheric Corrosion Offered by the MICAT Ibero-American Testing Network. *Int. J. Corros.* **2012**, *2012*, 824365. [CrossRef]
14. National Institute of Standars and Technology; American Bureau of Shiping; Colorado School of Mines; Mineral Management Service; Office of Pipeline Safety. *Coatings for Corrosion Protection: Offshore Oil and Gas Operation Facilities, Marine Pipeline and Ship Structures*; U.S. Department of Transportation: Washington, DC, USA, 2004.
15. Peabody, A.W. *Control of Pipeline Corrosion*; Bianchetti, R., Ed.; National Association of Corrosion Engineers (NACE): Houston, TX, USA, 2001; ISBN 1-57590-092-0.
16. Arriba-Rodriguez, L.; Villanueva-Balsera, J.; Ortega-Fernandez, F.; Rodriguez-Perez, F. Methods to Evaluate Corrosion in Buried Steel Structures: A Review. *Metals* **2018**, *8*, 334. [CrossRef]
17. Naz, M.Y.; Ismail, N.I.; Sulaiman, S.A.; Shukrullah, S. Electrochemical and Dry Sand Impact Erosion Studies on Carbon Steel. *Sci. Rep.* **2015**, *5*, 16583. [CrossRef]
18. Kubzova, M.; Krivy, V.; Kreislova, K. Influence of Chloride Deposition on Corrosion Products. *Procedia Eng.* **2017**, *192*, 504–509. [CrossRef]
19. Moins, B.; France, C.; Van den Bergh, W.; Audenaert, A. Implementing Life Cycle Cost Analysis in Road Engineering: A Critical Review on Methodological Framework Choices. *Renew. Sustain. Energy Rev.* **2020**, *133*, 110284. [CrossRef]

20. ISO. *ISO 9225:2012: Corrosion of Metals and Alloys—Corrosivity of Atmospheres—Measurement of Environmental Parameters Affecting Corrosivity of Atmospheres*; ISO/TC 156; ISO: Geneva, Switzerland, 2012.

21. ISO. *ISO 9223:2012: Corrosion of Metals and Alloys—Corrosivity of Atmospheres—Classification, Determination and Estimation*; ISO/TC 156; ISO: Geneva, Switzerland, 2012.

22. National Association of Corrosion Engineers; Baboian, R. *NACE Corrosion Engineer's Reference Book*; NACE International: Houston, TX, USA, 2016; ISBN 978-1-5231-0657-8.

23. Ahmad, Z. Chapter 1—Introduction to Corrosion. In *Principles of Corrosion Engineering and Corrosion Control*; Ahmad, Z., Ed.; Butterworth-Heinemann: Oxford, UK, 2006; pp. 1–8, ISBN 978-0-7506-5924-6.

24. Vernon, W.H.J. First (Experimental) Report to the Atmospheric Corrosion Research Committee (of the British Non-Ferrous Metals Research Association). *Nature* **1925**, *115*, 417. [CrossRef]

25. Benarie, M.; Lipfert, F.L. A General Corrosion Function in Terms of Atmospheric Pollutant Concentrations and Rain PH. *Atmos. Environ. 1967* **1986**, *20*, 1947–1958. [CrossRef]

26. Feliu, S.; Morcillo, M. The Prediction of Atmospheric Corrosion from Meteorological and Pollution Parameters. *Corros. Sci.* **1993**, *34*, 403–414. [CrossRef]

27. Morcillo, M.; Chico, B.; Díaz, I.; Cano, H.; de la Fuente, D. Atmospheric Corrosion Data of Weathering Steels. A Review. *Corros. Sci.* **2013**, *77*, 6–24. [CrossRef]

28. De la Fuente, D.; Castaño, J.G.; Morcillo, M. Long-Term Atmospheric Corrosion of Zinc. *Corros. Sci.* **2007**, *49*, 1420–1436. [CrossRef]

29. Panchenko, Y.M.; Marshakov, A.I. Long-Term Prediction of Metal Corrosion Losses in Atmosphere Using a Power-Linear Function. *Corros. Sci.* **2016**, *109*, 217–229. [CrossRef]

30. Leygraf, C.; Wallinder, I.; Tidblad, J.; Graedel, T. *Atmospheric Corrosion*, 2nd ed.; John Wiley & Sons: Hoboken, NJ, USA, 2016; p. 374.

31. Cai, Y.; Xu, Y.; Zhao, Y.; Ma, X. Atmospheric Corrosion Prediction: A Review. *Corros. Rev.* **2020**. [CrossRef]

32. Dean, S.W.; Knotkova, D.; Kreislová, K. *ISOCORRAG International Atmospheric Exposure Program: Summary of Results*; ASTM International: West Conshohocken, PA, USA, 2011.

33. ISO. *ISO 9224:2012 Corrosion of Metals and Alloys—Corrosivity of Atmospheres—Guiding Values for the Corrosivity Categories*; ISO/TC 156; ISO: Geneva, Switzerland, 2012.

34. ISO. *ISO 9226:2012 Corrosion of Metals and Alloys—Corrosivity of Atmospheres—Determination of Corrosion Rate of Standard Specimens for the Evaluation of Corrosivity*; ISO/TC 156; ISO: Geneva, Switzerland, 2012.

35. Swedish Corrosion Institute. *UN/ECE International Cooperative Programme on Effects on Materials Including Historic and Cultural Monuments*; Report no. 1: Technical Manual; Swedish Corrosion Institute: Stockholm, Sweden, 1988.

36. Morcillo, M. Atmospheric Corrosion in Ibero-America: The MICAT Project. *Atmos. Corros.* **1995**. [CrossRef]

37. McCuen, R.H.; Albrecht, P.; Cheng, J. A New Approach to Power-Model Regression of Corrosion Penetration Data. In *Corrosion Forms and Control for Infrastructure*; ASTM International: West Conshohocken, PA, USA, 1992. [CrossRef]

38. Albrecht, P.; Hall, T.T., Jr. Atmospheric Corrosion Resistance of Structural Steels. *J. Mater. Civ. Eng.* **2003**, *15*, 2–24. [CrossRef]

39. Melchers, R.E. A New Interpretation of the Corrosion Loss Processes for Weathering Steels in Marine Atmospheres. *Corros. Sci.* **2008**, *50*, 3446–3454. [CrossRef]

40. Melchers, R.E. Long-Term Corrosion of Cast Irons and Steel in Marine and Atmospheric Environments. *Corros. Sci.* **2013**, *68*, 186–194. [CrossRef]

41. Titakis, C.; Vassiliou, P. Evaluation of 4-Year Atmospheric Corrosion of Carbon Steel, Aluminum, Copper and Zinc in a Coastal Military Airport in Greece. *Corros. Mater. Degrad.* **2020**, *1*, 8. [CrossRef]

42. Kreislova, K.; Knotkova, D. The Results of 45 Years of Atmospheric Corrosion Study in the Czech Republic. *Materials* **2017**, *10*, 394. [CrossRef]

43. Tidblad, J. Atmospheric Corrosion of Metals in 2010–2039 and 2070–2099. *Atmos. Environ.* **2012**, *55*, 1–6. [CrossRef]

44. Knotkova, D.; Boschek, P.; Kreislova, K. Results of ISO CORRAG Program: Processing of One-Year Data in Respect to Corrosivity Classification. *Atmos. Corros.* **1995**. [CrossRef]

45. Panchenko, Y.M.; Marshakov, A.I.; Nikolaeva, L.A.; Kovtanyuk, V.V.; Igonin, T.N.; Andryushchenko, T.A. Comparative Estimation of Long-Term Predictions of Corrosion Losses for Carbon Steel and Zinc Using Various Models for the Russian Territory. *Corros. Eng. Sci. Technol.* **2017**, *52*, 149–157. [CrossRef]

46. Cole, I.S.; Muster, T.H.; Azmat, N.S.; Venkatraman, M.S.; Cook, A. Multiscale Modelling of the Corrosion of Metals under Atmospheric Corrosion. *Electrochim. Acta* **2011**, *56*, 1856–1865. [CrossRef]

47. Nguyen, M.N.; Wang, X.; Leicester, R.H. An Assessment of Climate Change Effects on Atmospheric Corrosion Rates of Steel Structures. *Corros. Eng. Sci. Technol.* **2013**, *48*, 359–369. [CrossRef]

48. Gomes, H.M.; Awruch, A.M. Comparison of Response Surface and Neural Network with Other Methods for Structural Reliability Analysis. *Struct. Saf.* **2004**, *26*, 49–67. [CrossRef]

49. Ahmad, Z. Chapter 10—Atmospheric Corrosion. In *Principles of Corrosion Engineering and Corrosion Control*; Ahmad, Z., Ed.; Butterworth-Heinemann: Oxford, UK, 2006; pp. 550–575. ISBN 978-0-7506-5924-6.

50. Vargel, C. Chapter C.2—The Parameters of Atmospheric Corrosion. In *Corrosion of Aluminium*; Vargel, C., Ed.; Elsevier: Amsterdam, The Netherlands, 2004; pp. 241–257, ISBN 978-0-08-044495-6.

51. Schindelholz, E.; Kelly, R.G. Wetting Phenomena and Time of Wetness in Atmospheric Corrosion: A Review. *Corros. Rev.* **2012**, *30*. [CrossRef]

52. Kottek, M.; Grieser, J.; Beck, C.; Rudolf, B.; Rubel, F. World Map of the Köppen-Geiger Climate Classification Updated. *Meteorol. Z.* **2006**, *15*, 259–263. [CrossRef]

53. Committee MT-014 (Corrosion Of Metals). *AS 4312-2008 Atmospheric Corrosivity Zones in Australia*; Standards Australia: Sydney, NSW, Australia, 2008.

54. Chico, B.; Otero, E.; Mariaca, L.; Morcillo, M. La Corrosión En Atmósferas Marinas. Efecto de La Distancia a La Costa. *Rev. Metal.* **1998**, *34*. [CrossRef]

55. Goerlich, G.F.J.; Cantarino, M.I. Estimaciones de la población rural y urbana a nivel municipal. *Estad. Esp.* **2015**, *57*, 5–28.

56. Friedman, J.H. Multivariate Adaptive Regression Splines. *Ann. Stat.* **1991**, *19*, 1–67. [CrossRef]

57. Vanegas, J.; Vásquez, F. Multivariate Adaptive Regression Splines (MARS), Una Alternativa Para El Análisis de Series de Tiempo. *Gac. Sanit.* **2017**, *31*, 235–237. [CrossRef]

58. Oja, E.; Kaski, S. *Kohonen Maps*, 1st ed.; Elsevier Science: Amsterdam, The Netherlands, 1999.

59. Wehrens, R.; Buydens, L. Self- and Super-Organizing Maps in R: The Kohonen Package. *J. Stat. Softw.* **2007**, *21*, 1–19. [CrossRef]

60. Villmann, T.; Bauer, H.-U. Applications of the Growing Self-Organizing Map11This Work Has Been Supported by Deutsche Forschungsgemeinschaft, SFB 185 "Nichtlineare Dynamik", TP E6. *Neurocomputing* **1998**. [CrossRef]

61. Diazaraque, J.M.M. Los Mapas Auto-Organizados de Kohonen (SOM). Available online: https://docplayer.es/9172924-Los-mapas-auto-organizados-de-kohonen-som.html (accessed on 15 April 2021).

62. Pachghare, V.; Kulkarni, P.; Nikam, D. Intrusion Detection System Using Self Organizing Maps. In Proceedings of the 2009 International Conference on Intelligent Agent & Multi-Agent Systems (IAMA 2009), Chennai, India, 22–24 July 2009; pp. 1–5.

63. Heasley, E.L.; Millington, J.D.A.; Clifford, N.J.; Chadwick, M.A. A Waterbody Typology Derived from Catchment Controls Using Self-Organising Maps. *Water* **2020**, *12*, 78. [CrossRef]

64. Kohonen, T. *Self-Organizing Maps*, 3rd ed.; Springer Series in Information Sciences; Springer: Berlin/Heidelberg, Germany, 2001; ISBN 978-3-540-67921-9.

65. Shanno, D.F. Conditioning of Quasi-Newton Methods for Function Minimization. *Math. Comput.* **1970**, *24*, 647–656. [CrossRef]

66. Bronshtein, I.; Semendiaev, K. *Manual de Matemáticas para Ingenieros y Estudiantes*; Mir: Moscow, Russia; Rubiños-1860: Madrid, Spain, 1993; ISBN 978-84-8041-022-9.

67. Bourbaki, N. *Topological Vector Spaces: Chapters 1–5*; Springer: Berlin/Heidelberg, Germany, 2002; ISBN 978-3-540-42338-6.

 *materials* 

Article

Oxidation of Al-Co Alloys at High Temperatures

Patrik Šulhánek [1], Marián Drienovský [1], Ivona Černičková [1], Libor Ďuriška [1], Ramūnas Skaudžius [2], Žaneta Gerhátová [1] and Marián Palcut [1,*]

[1]   Faculty of Materials Science and Technology in Trnava, Slovak University of Technology in Bratislava, J. Bottu 24, 91724 Trnava, Slovakia; patrik.sulhanek@stuba.sk (P.Š.); marian.drienovsky@stuba.sk (M.D.); ivona.cernickova@stuba.sk (I.Č.); libor.duriska@stuba.sk (L.Ď.); zaneta.gerhatova@stuba.sk (Ž.G.)

[2]   Faculty of Chemistry and Geosciences, Vilnius University, Naugarduko g. 24, 01513 Vilnius, Lithuania; ramunas.skaudzius@chgf.vu.lt

*   Correspondence: marian.palcut@stuba.sk

Received: 20 June 2020; Accepted: 13 July 2020; Published: 15 July 2020

Abstract: In this work, the high temperature oxidation behavior of $Al_{71}Co_{29}$ and $Al_{76}Co_{24}$ alloys (concentration in at.%) is presented. The alloys were prepared by controlled arc-melting of Co and Al granules in high purity argon. The as-solidified alloys were found to consist of several different phases, including structurally complex $m\text{-}Al_{13}Co_4$ and $Z\text{-}Al_3Co$ phases. The high temperature oxidation behavior of the alloys was studied by simultaneous thermal analysis in flowing synthetic air at 773–1173 K. A protective Al_2O_3 scale was formed on the sample surface. A parabolic rate law was observed. The rate constants of the alloys have been found between 1.63×10^{-14} and 8.83×10^{-12} g cm^{-4} s^{-1}. The experimental activation energies of oxidation are 90 and 123 kJ mol^{-1} for the $Al_{71}Co_{29}$ and $Al_{76}Co_{24}$ alloys, respectively. The oxidation mechanism of the Al-Co alloys is discussed and implications towards practical applications of these alloys at high temperatures are provided.

Keywords: Al alloy; Co alloy; complex intermetallic; oxidation kinetics; oxide scale

---

1. Introduction

Co-based superalloys are promising materials for high temperature structural applications because of their high melting points and favorable mechanical properties [1–3]. Applications of these alloys include gas turbines, aircraft engines, and chemical reactors [4–6]. The Co-based superalloys are often alloyed with chromium to provide oxidation resistance [7,8]. The superalloys alloyed with Cr form a compact chromia scale (Cr_2O_3) on their surface. Nevertheless, at high temperatures and high oxygen partial pressures, the Cr_2O_3 scale is prone to degradation. During long-term oxidation, volatile high-valent oxides of Cr, such as CrO_2 and CrO_3, start to form at the expense of Cr_2O_3. This effect is called "chromia evaporation" and is often pronounced in humid atmospheres [9]. The loss of protective chromia scale leads to a reduced life span of the Co-based superalloys. Several authors have, therefore, investigated the possibility of improving the high temperature oxidation stability of Co superalloys by alloying with Al [10–12]. Al-based alloys form a protective oxide scale composed of alumina (Al_2O_3). Al_2O_3 has a lower growth rate compared to Cr_2O_3 and is non-volatile. Furthermore, Al is a non-transition element. It has a smaller tendency to form complex oxides with transition metals compared to Cr, thereby reducing the risk of scale spallation over time. The formation of alumina scales may be achieved by pack aluminizing the alloy's surface [13–15]. The application of Al-Co coatings could significantly extend the alloy's lifetime [16–22].

Aluminides are aluminum-based intermetallic compounds with transition metals. Cobalt aluminides are interesting for high temperature applications since they possess a combination of high melting points and good corrosion resistance. At ~18–30 at.% Co, different structurally complex aluminides in the Al-Co binary system have been observed (Figure 1, [23,24]). These include Al_9Co_2 (P2$_1$/C), Al_5Co_2 (P6$_3$/mmc), Z-Al$_3$Co (P2/m) and family of $Al_{13}Co_4$ phases containing m-$Al_{13}Co_4$ (C2/m), O-$Al_{13}Co_4$ (Pmn2$_1$), O'-$Al_{13}Co_4$ (Pnma), Y$_1$-$Al_{13}Co_4$ (C2/m) and Y$_2$-$Al_{13}Co_4$ (Immm) [25–34]. Although the individual cobalt aluminides are brittle [35], the Al-Co precipitates may significantly strengthen the Al alloys [36,37]. The presence of Al_5Co_2 and $Al_{13}Co_4$ intermetallic compounds (IMCs) may also be beneficial in Co-based alloys as they may greatly improve the alloy's wear resistance [38]. Most aluminides form protective alumina scales with large resistance against corrosion [39].

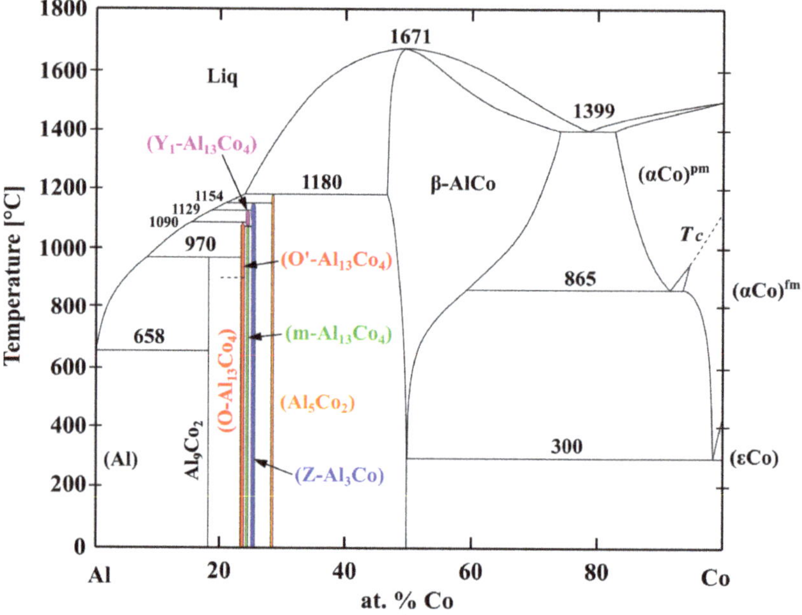

Figure 1. Phase diagram of the Al-Co binary system, redrawn from [23,24].

High temperature corrosion studies of cobalt aluminides are limited. Metal oxidation is a heterogeneous reaction taking place in several elementary steps [40]. In the first step, a gaseous oxygen molecule is transported to the metal surface. Upon approaching the solid phase, the adsorption of oxygen molecules to the metal substrate occurs. Subsequently, the oxygen is dissociated to atoms and later reduced to the O^{2-} anions. In parallel, oxidation of metal atoms takes place at the metal-oxide interface. Recently, M. Wardé et al. studied the adsorption of oxygen on the Al_9Co_2 (001) and $Al_{13}Co_4$ (100) surfaces at high temperatures and reduced oxygen pressures [41]. At the surfaces, only Al–O bonding was observed. Al–O distances were also calculated from first principles [41,42]. The Al–O lengths were shorter in comparison with Co–O distances. The obtained Al–O distances were in agreement with the typical distances of oxygen adsorption on the Al (111) surface, as well as with the Al–O distances in Al_2O_3.

Al-rich Al-Co alloys belong to a relatively new group of complex metallic alloys (CMA, [43,44]). These materials contain structurally complex phases including quasicrystals. The structurally complex phases have non-periodically ordered atomic arrangements [45,46]. Consequently, the properties of CMA are different from those observed in traditional materials [43]. The quasicrystalline surfaces have a good adhesion and low coefficient of friction [47]. Owing to their high hardness and good oxidation resistance, the Al-TM alloys (TM = transition metal) are suitable for high temperature coatings [48–51]. The Al-Co CMAs are also interesting for catalytic and hydrogen generation applications [52–55]. The corrosion studies of Al-Co alloys are limited. Lekatou et al. investigated the corrosion behavior of an $Al_{82}Co_{18}$ (metal concentrations are given in at.%) alloy in saline solution [56]. Three methods of alloy preparation were investigated: casting, arc-melting and free sintering. The alloy prepared by arc-melting was found to be the most corrosion-resistant. The $Al_{82}Co_{18}$ alloy was composed of (Al), Al_9Co_2 and m-$Al_{13}Co_4$. The complex intermetallic m-$Al_{13}Co_4$ in the alloy was found to have the highest corrosion resistance. Recently, we have investigated the corrosion behavior of as-solidified and near equilibrium Al-Co alloys in various environments [24,57,58]. The alloys were composed of various intermetallic phases. In HCl and NaCl solutions, a pitting corrosion occurred. A higher corrosion resistance of structurally complex Z-Al_3Co phase in Cl-containing electrolytes was observed. The difference in the corrosion behavior could be ascribed to the strong covalent character of metallic bonds in the structurally complex Al-Co phases which prevents aluminum diffusion. The studies also suggest that the existence of an electrical contact between different alloy phases play an important role in the overall alloy's corrosion behavior.

High temperature oxidation studies of Al-Co alloys have been limited to Co-rich alloys only. Zhang et al. investigated the oxidation behavior of Co-5 at.% Al and Co-10 at.% Al alloys at 973 and 1073 K [59,60]. The oxide scales were primarily composed of cobalt oxide and cobalt-aluminum oxide. The oxides grown on these alloys were relatively thick (>10 μm after 24 h). Only a limited amount of Al_2O_3 was found at the inner side of the scales. Irving et al. studied the oxidation behavior of Co-xAl alloys (0 < x < 32.4 at.%) at 1073–1273 K [61]. The authors found that 20–25 at.% Al is necessary to form a continuous alumina scale. As such, larger Al concentrations are needed to improve the corrosion resistance of Al-Co alloys.

To our best knowledge, oxidation studies of Al-rich Al-Co alloys have not been reported yet. In the present work, we aim to study the oxidation behavior of Al-rich Al-Co alloys in air at 773–1173 K. Two alloys, $Al_{71}Co_{29}$ and $Al_{76}Co_{24}$ (composition in at.%) were prepared by arc-melting. The composition of the $Al_{71}Co_{29}$ alloy was chosen close to Al_5Co_2 (71.4 at.%). The composition of the $Al_{76}Co_{24}$ alloy was close to $Al_{13}Co_4$ (76.5 at.%). The high temperature oxidation of the alloys was studied with the aim to identify the role of the alloy's chemical composition and microstructure on the overall corrosion behavior.

2. Materials and Methods

The alloys with nominal compositions $Al_{71}Co_{29}$ and $Al_{76}Co_{24}$ (metal concentrations in at.%) were prepared from Co and Al lumps (purity of 99.95%, smart-elements.com) by arc-melting. The melting was conducted in MAM-1 arc melter (Edmund Buehler Ltd., Bodelshausen, Germany) in high purity argon. A piece of Ti (oxygen getter) was melted first to remove residual traces of oxygen in argon. The Co and Al lumps were placed in the center of Cu mold and rapidly melted by striking an arc from tungsten cathode. The homogeneity of the molten samples was improved by repeated arc-melting. Subsequently, the melts were solidified on a water-cooled Cu mold to form button ingots. The as-solidified alloys were removed from the arc-melter, cut into smaller specimens by diamond blade, and prepared for oxidation experiments. The samples were ground with grade 1200 abrasive paper and polished with diamond suspensions down to 1 μm surface roughness.

The oxidation behavior of the polished alloys was studied in flowing synthetic air (20 vol. % O_2 and 80% vol. % N_2). The air flow rate was 20 mL/min. Isothermal oxidation experiments were performed at 773, 973 and 1173 K. The polished samples were placed in an alumina crucible and heated from room temperature to peak temperature by heating rate 20 K/min. The oxidation time was 30 h. The mass gain of the samples was recorded continuously in a chamber of NETZCH STA 409 CD thermogravimeter (NETZSCH-Gerätebau GmbH, Selb, Germany). After oxidation, the samples were cooled down, mounted in epoxy resin, and cut perpendicularly to the reaction interface. The cross-sections of the specimens were prepared by wet grinding and polishing and subjected to microscopy observation.

The microstructure and chemical composition of the alloys was studied by scanning electron microscopy. During experiments, a JEOL JSM-7600F microscope (JEOL Ltd., Tokyo, Japan), equipped with an energy-dispersive x-ray spectrometer X-max (EDS), was used. The EDS was operated by

INCA software (Oxford Instruments Nanoanalysis, Bucks, UK). Regimes of secondary electrons (SE) and backscatter electrons (BSE) were used during imaging. The accelerating voltage of the electron beam was 20 kV and working distance was 15 mm. Furthermore, a Panalytical Empyrean PIXCel 3D diffractometer was used for the phase analysis (Malvern Panalytical Ltd., Malvern, UK). The diffractometer was working with Bragg–Brentano geometry and used $CoK_{\alpha 1}$ radiation. The X-ray beam was generated at 40 kV and 50 mA.

3. Results and Discussion

3.1. Alloy Microstructure and Constitution before Oxidation

The $Al_{71}Co_{29}$ and $Al_{76}Co_{24}$ alloys were prepared by arc-melting. The microstructure of the as-solidified $Al_{71}Co_{29}$ alloy is presented in Figure 2. In this alloy, three different microstructure constituents have been found. The image was acquired in backscatter electron mode to provide element resolution. The chemical composition of the constituents, measured by EDS point analysis, is provided in Table 1. The dendritic constituents have a significantly higher Co concentration (44.8 at.%) compared to the remainder of the alloy. The other two constituents have 28.1 and 25.4 at.% Co, respectively. The black areas, found in the microstructure, are pores.

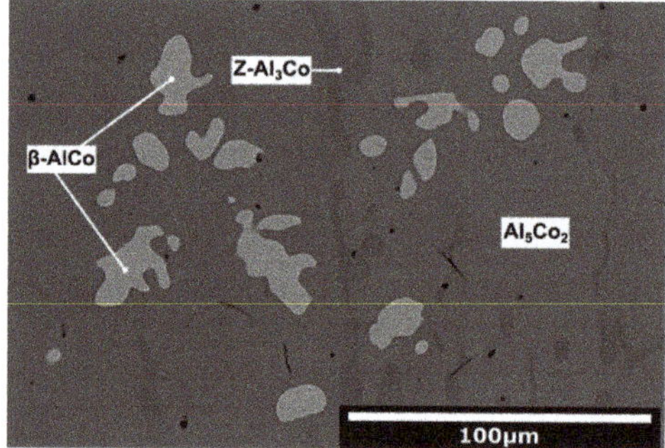

Figure 2. Microstructure of the as-solidified $Al_{71}Co_{29}$ alloy.

Table 1. Chemical compositions of microstructure constituents observed in the as-solidified $Al_{71}Co_{29}$ and $Al_{76}Co_{24}$ alloys.

Al₇₁Co₂₉ Alloy				
Microstructure Constituent	**Phase Identified [57]**	**Al [at.%]**	**Co [at.%]**	**Volume Fraction [%]**
Light-grey	β-AlCo	55.2 ± 0.9	44.8 ± 0.9	8
Medium-grey	Al_5Co_2	71.9 ± 0.4	28.1 ± 0.4	86
Dark-grey	Z-Al_3Co	74.6 ± 0.4	25.4 ± 0.4	6
Al₇₆Co₂₄ Alloy				
Microstructure Constituent	**Phase Identified [58]**	**Al [at.%]**	**Co [at.%]**	**Volume Fraction [%]**
Light-grey	Z-Al_3Co	74.4 ± 0.1	25.6 ± 0.1	5
Medium-grey	m-$Al_{13}Co_4$	75.2 ± 0.2	24.8 ± 0.2	83
Dark-grey	Al_9Co_2	81.5 ± 0.1	18.5 ± 0.1	12

The microstructure of the as-solidified $Al_{76}Co_{24}$ alloy is presented in Figure 3 in BSE imaging mode. In this alloy, three different microstructure constituents have been found (light grey, medium grey and dark grey). The light grey constituent in the $Al_{76}Co_{24}$ alloy has 74.4 at.% Al and 25.6 at.% Co (Table 1). As such, its chemical composition is comparable to chemical compositions of the dark grey constituent observed in the as-solidified $Al_{71}Co_{29}$ alloy (Figure 2). The medium grey constituent contains 75.2 at.%

Al and 24.8 at.% Co. The dark grey constituent of the as-solidified $Al_{76}Co_{24}$ alloy has 81.5 at.% Al and 18.5 at.% Co. The black areas located within the dark grey constituent are pores (Figure 3).

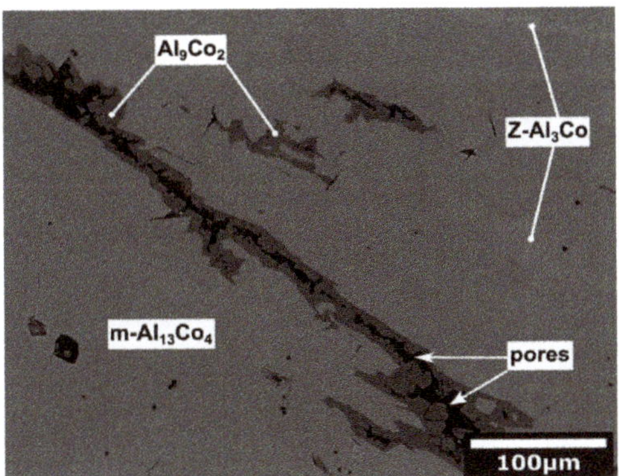

Figure 3. Microstructure of the as-solidified $Al_{76}Co_{24}$ alloy.

A phase assignment of the alloy's microstructure constituents is presented in Table 1. The assignment has been made based on the experimental chemical composition of the constituents obtained by EDS and crystal structure of the phases identified by XRD in our previous work [57,58]. In the studied alloys, altogether five different phases have been identified: β-AlCo, Al_5Co_2, Z-Al₃Co, m-$Al_{13}Co_4$ and Al_9Co_2 (Table 1).

The presence of different phases in the alloys indicates that non-equilibrium processes had been taking place during rapid solidification. The dendritic shape of β-AlCo in the as-solidified $Al_{71}Co_{29}$ alloy suggests that it solidified first, directly from the melt. The dendritic β-AlCo is located inside the Al_5Co_2 phase. Therefore, Al_5Co_2 was probably formed by partial transformation of β. The Al_5Co_2 phase is located next to Z-Al₃Co (Figure 2). As such, Z-Al₃Co was probably formed by peritectic reaction of Al_5Co_2 with the surrounding melt.

The as-solidified $Al_{76}Co_{24}$ alloy was found to consist of Z-Al₃Co, m-$Al_{13}Co_4$ and Al_9Co_2, respectively (Figure 3). The m-$Al_{13}Co_4$ phase is located next to Z-Al₃Co. As such, it was probably formed by peritectic reaction of Z-Al₃Co with the melt. The Z-Al₃Co phase has a lower Al concentration compared to m-$Al_{13}Co_4$ (Table 1). Al_9Co_2, observed in the as-solidified $Al_{76}Co_{24}$ alloy, is located next to m-$Al_{13}Co_4$ (Figure 3). As such, Al_9Co_2 was probably formed by peritectic reaction of m-$Al_{13}Co_4$ with the remaining melt. It was observed to be porous (Figure 3). The pores are usually formed by vacancy migration, with sub-grains and natural surfaces serving as sinks [62,63]. In the present case, the pores were found in the interior of Al_9Co_2. The preferential pore formation indicates that the pores could be a result of rapid transformation of the liquid Al_9Co_2 into solid in the final step of solidification.

3.2. Oxidation Behavior

The oxidation behavior of the as-solidified $Al_{76}Co_{24}$ and $Al_{71}Co_{29}$ alloys was studied in flowing synthetic air at 773, 973 and 1173 K. The microstructure and chemical composition of the oxide scale were investigated by SEM/EDS. The cross-section of the $Al_{71}Co_{29}$ alloy after oxidation at 1173 K is presented in Figure 4. The oxide scale was homogeneous. EDS element maps are included in Figure 4b–d. The scale was found to be composed of aluminum oxide.

The cross-section image of the $Al_{76}Co_{24}$ alloy after oxidation at 1173 K is presented in Figure 5. The thickness of the oxide scale was approximately 1 µm after 30 h of oxidation. The chemical composition of the oxide scale was studied by EDS analysis. Results presented in Figure 5b show that the scale was predominantly composed of Al_2O_3. A small amount of Co (~ 2 at.%) was also detected in the scale, however, this result is attributable to a possible interference of the bulk alloy signal.

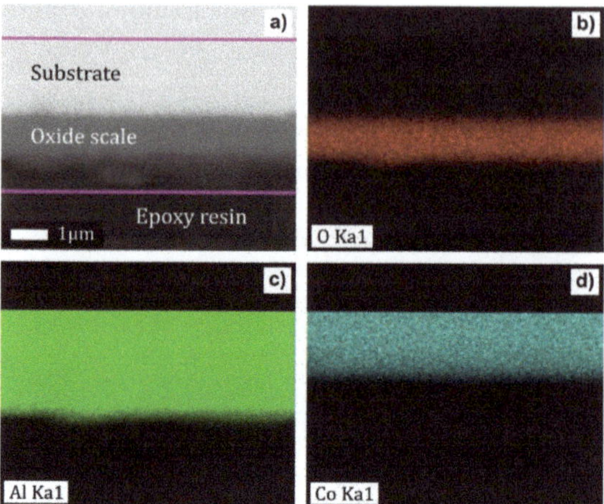

Figure 4. Cross section of the Al$_{71}$Co$_{29}$ alloy after oxidation at 1173 K for 30 h (**a**) and energy-dispersive x-ray spectrometer X-max (EDS) element maps for O (**b**), Al (**c**) and Co (**d**).

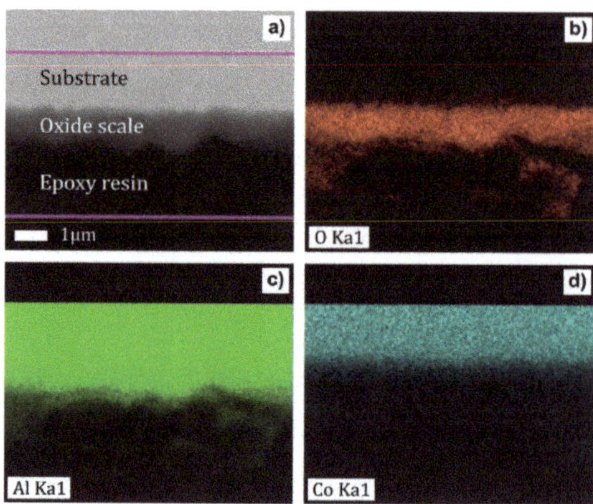

Figure 5. Cross section of the Al$_{76}$Co$_{24}$ alloy after oxidation at 1173 K for 30 h (**a**) and EDS element maps for O (**b**), Al (**c**) and Co(**d**).

The phase constitution of the oxide scale was studied by room temperature X-ray diffraction. The diffraction patterns of the Al$_{71}$Co$_{29}$ alloy are presented in Figure 6. In the alloy, peaks corresponding to θ-Al$_2$O$_3$ have been identified. θ-Al$_2$O$_3$ is a metastable alumina phase [64,65]. θ-Al$_2$O$_3$ structures are based on a cubic close packing of oxygen anions. The cubic close packing of oxygen anions of θ-Al$_2$O$_3$ is deformed monoclinic. The thermodynamically stable α-Al$_2$O$_3$ adopts a corundum structure. θ-Al$_2$O$_3$ is a transition phase. It transforms into stable forms of alumina during long term annealing [65]. Metastable θ-Al$_2$O$_3$ has been found in oxidized Al-Cu-Fe alloys studied previously [66]. It was formed initially with an orientational relationship to the substrate. At 1173 K, θ-Al$_2$O$_3$ was found to slowly transform into α-Al$_2$O$_3$ with an increasing oxidation time (70 h, [66]).

At 1173 K, an orientation of θ-Al$_2$O$_3$ in (002) crystallographic plane has been found (Figure 6). The same behavior was also observed for the Al$_{76}$Co$_{24}$ alloy (Figure 6). This observation indicates a preferential crystal growth. The morphology of alumina scale formed on the Al$_{71}$Co$_{29}$ and Al$_{76}$Co$_{24}$ alloys after oxidation at 1173 K for 30 h is given in Figure 7. The scale had a blade-like structure.

The platelet-like scale morphology is indicative of rapid outwards growth. Alumina scales grow by counter-diffusion of aluminum and oxygen [67]. The ions, however, diffuse faster in polycrystalline alumina at near-atmospheric oxygen partial pressures. The scale morphology is indicative of rapid diffusion through the scale. A grain boundary diffusion was probably the preferred transport path for the Al^{3+} and O^{2-} ions in the scale.

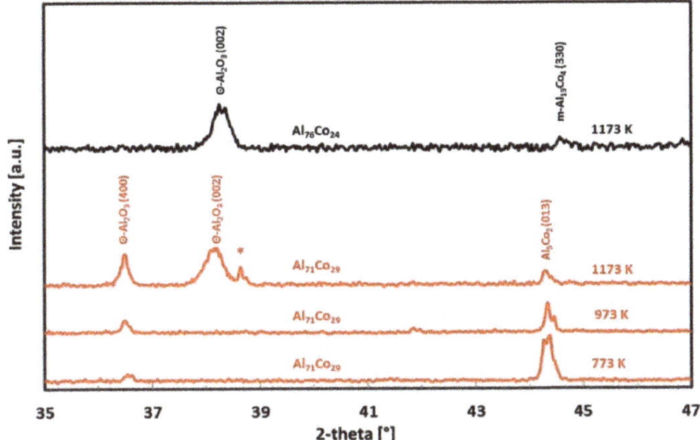

Figure 6. Room temperature XRD patterns of the scales formed on the $Al_{71}Co_{29}$ and $Al_{76}Co_{24}$ alloys. For the discussion of the peak marked with an asterisk (*), please refer to the article text.

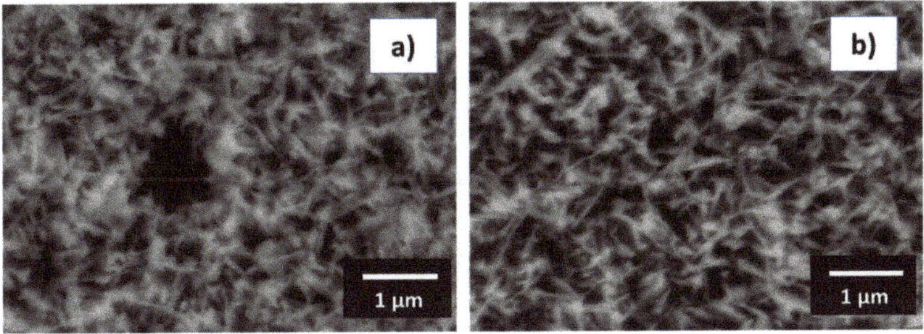

Figure 7. Blade-like morphology of alumina scale formed on the oxidized $Al_{71}Co_{29}$ alloy (**a**) and $Al_{76}Co_{24}$ alloy (**b**) during oxidation at 1173 K.

The un-indexed peak next to θ-Al_2O_3 (002), marked with an asterisk in Figure 6, is an alumina peak. The closest match was found for hexagonal form of Al_2O_3 (reference code 98-017-3713, [68]). Nevertheless, it should also be mentioned that θ-Al_2O_3 has a disordered structure [69–72]. As such, the peak could also be a result of stacking faults (twinning) or other structural defects in θ-Al_2O_3 [73–75]. The precise peak assignment was not possible, owing to the difficulty to unambiguously distinguish the various Al_2O_3 polymorphs by XRD technique alone.

The alumina scale was well adherent to the substrate (Figures 4a and 5a). Nevertheless, locally, a detachment of the scale on the $Al_{71}Co_{29}$ alloy was observed (Figure 8). A scale delamination was found preferentially around β-AlCo dendrites. The situation is shown in Figure 8c. An explanation of the layer spallation could reside in a mechanical stress developed during oxide growth. The stress is formed due to different molar volumes of the oxide and the underlying original metal substrate [76]. The stress generated in the oxide during excessive growth may lead to crack formation in the scale. The pores, cracks and other defects facilitate the access of molecular oxygen to the metal substrate.

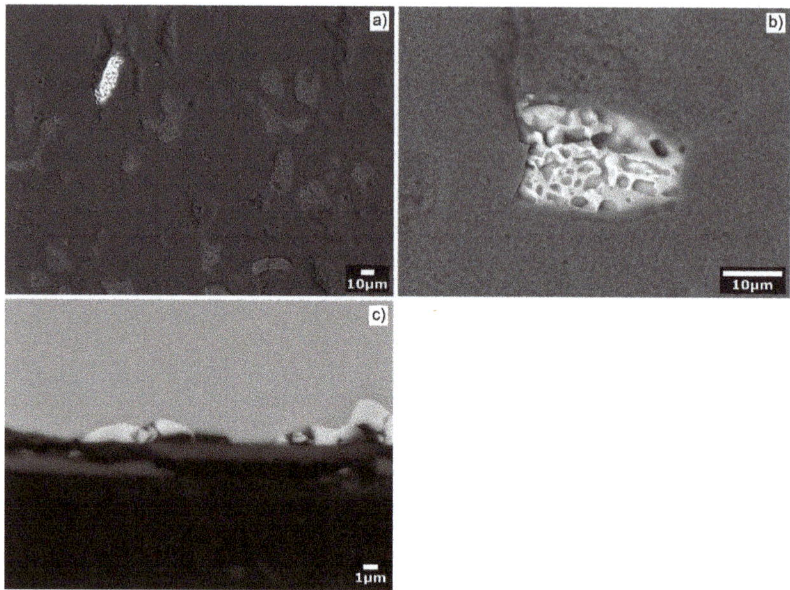

Figure 8. Scale microstructure around β-AlCo dendrites of the $Al_{71}Co_{29}$ alloy: (**a**) an overview; (**b**) a detailed view; (**c**) cross section image of the scale.

Another interesting feature was the porosity of β-AlCo dendrites located beneath the spalled scale of the $Al_{71}Co_{29}$ alloy (Figure 8b). The voids in β-AlCo were not observed before oxidation (Figure 2). The voids were thus formed during reaction, probably because of rapid aluminum outward diffusion from the metal surface. During oxidation, metallic species diffuse out from the alloy bulk to the alloy/oxide interface. The Al atoms, however, leave behind their vacant sites. The vacancies may have coalesced into larger defects, giving rise to the observed macroscopic porosity. The oxide spallation has not been observed on the $Al_{76}Co_{24}$ alloy. The $Al_{76}Co_{24}$ alloy is composed of structurally complex intermetallic phases (Z-Al_3Co, m-$Al_{13}Co_4$ and Al_9Co_2, Figure 3). The surfaces of these phases are typically Al-rich [77]. Aluminum necessary for the scale growth was readily available at the surface of complex intermetallics. As such, the diffusion of Al atoms from these phases was less rapid compared to β-AlCo.

The scale of the $Al_{76}Co_{24}$ alloy did not show any spallation. It was well adherent to the substrate and had a wave-like morphology (Figure 5). The wave-like morphology of the scale could be indicative of epitaxial growth. The XRD pattern showed a preferential orientation of θ-Al_2O_3 grains in (002) crystallographic plane (Figure 6). The θ-Al_2O_3 (400) peak was not observed. The $Al_{76}Co_{24}$ alloy was primarily composed of m-$Al_{13}Co_4$, with small amounts of Z-Al_3Co and Al_9Co_2 (Table 1). A preferred orientation is typical for m-$Al_{13}Co_4$ phase [26,78]. It is therefore likely that the θ-Al_2O_3 phase was formed with an orientation relationship to the m-$Al_{13}Co_4$ phase. This may be reflected in the preferential orientation of the θ-Al_2O_3 grains, as a result of which the intensities of the peaks of this phase change and some may disappear [79,80].

The oxide layer on the $Al_{71}Co_{29}$ alloy, on the other hand, was more uniform (Figure 4). The preferential layer growth is a self-limited process driven by atomic diffusion, and surface energy minimization [81,82]. It requires a certain concentration and certain mobility of Al atoms on the surface. The $Al_{71}Co_{29}$ alloy was primarily composed of Al_5Co_2. The $Al_{76}Co_{24}$ alloy, on the other hand, was mainly composed of m-$Al_{13}Co_4$ (Table 1). The surface of Al_5Co_2 is terminated at specific bulk layers (Al-rich puckered layers) where various fractions of specific sets of Al atoms are missing [83]. The surface of m-$Al_{13}Co_4$ has a higher density of Al atoms when compared to Al_5Co_2. Previous investigations on $Al_{13}Co_4$ (100) showed that it was terminated by a dense aluminum topmost layer [84]. Therefore, the conditions for the atomic diffusion and subsequent oxide growth on the $Al_{76}Co_{24}$ and $Al_{71}Co_{29}$ alloys were different. The preferential oxide growth was less favored on the $Al_{71}Co_{29}$ alloy.

The mass gain of the samples was recorded by simultaneous thermogravimetry (TGA). Thermogravimetric curves of the $Al_{71}Co_{29}$ and $Al_{76}Co_{24}$ alloys are presented in Figures 9 and 10. The mass gain of the samples was increasing with increasing time. The kinetic curves obeyed a parabolic behavior.

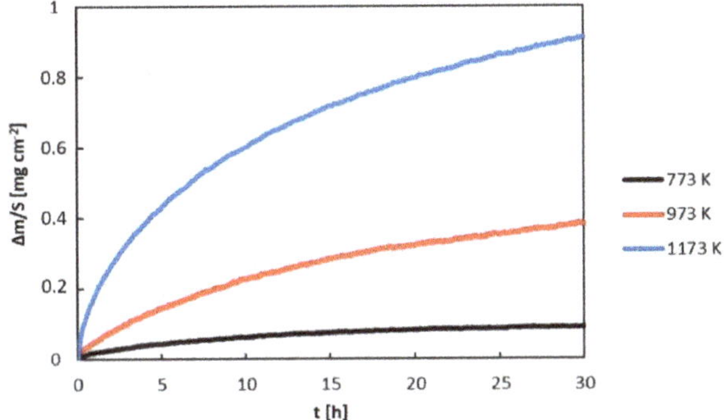

Figure 9. Mass gain of the $Al_{71}Co_{29}$ alloy during isothermal oxidation in flowing synthetic air.

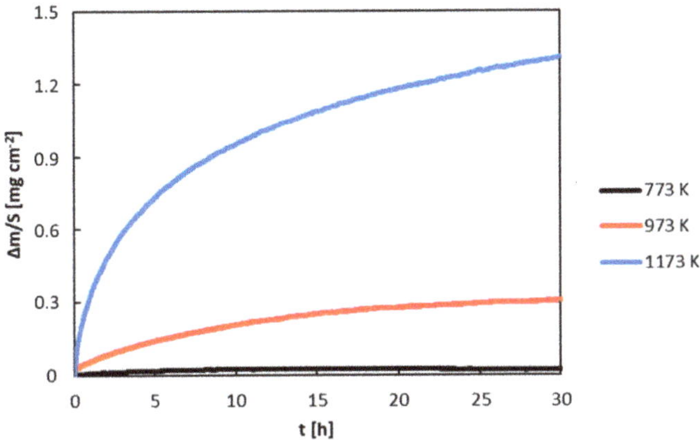

Figure 10. Mass gain of the $Al_{76}Co_{24}$ alloy during isothermal oxidation in flowing synthetic air.

The parabolic oxidation could be described by the following equation

$$\left(\frac{\Delta m}{S}\right)^2 = k_p t + C \tag{1}$$

In this equation, $\frac{\Delta m}{S}$ represents the specific mass gain (mass increase per unit area), k_p is the parabolic constant, t is the annealing time and C is the integration constant. The plot of $\left(\frac{\Delta m}{S}\right)^2$ versus t was linear and the slope of the line represented the parabolic rate constant. The k_p values are collected in Table 2. The rate constants of the alloys are found between 1.6×10^{-14} and 2.5×10^{-11} g^2 cm^{-4} s^{-1}. The obedience of the parabolic behavior shows that the oxidation process of the $Al_{71}Co_{29}$ and $Al_{74}Co_{26}$ alloys was controlled by ionic diffusion in the scale.

Table 2. Parabolic rate constants of the $Al_{71}Co_{29}$ and $Al_{76}Co_{24}$ alloys in air.

k_p [g^2 cm^{-4} s^{-1}]		
$Al_{71}Co_{29}$ Alloy		
773 K	**973 K**	**1173 K**
1.04×10^{-13} (0–15 h)	1.64×10^{-12} (0–15 h)	9.71×10^{-12} (0–15 h)
4.20×10^{-14} (20–30 h)	1.23×10^{-12} (15–30 h)	5.81×10^{-12} (15–30 h)
$Al_{76}Co_{24}$ Alloy		
773 K	**973 K**	**1173 K**
1.63×10^{-14}	1.21×10^{-12} (0–15 h)	2.54×10^{-11} (0–10 h)
	5.43×10^{-13} (15–30 h)	8.83×10^{-12} (20–30 h)

Oxidation is a thermally activated process. The parabolic rate constants increase with increasing temperature. The activation energy of oxidation thus could be obtained from the following equation

$$\log k_p = \log A - 0.434\frac{E_A}{RT} \tag{2}$$

In this equation, A is a constant, E_A is the activation energy, R is the molar gas constant (8.3144 J K^{-1} mol^{-1}) and T is the absolute temperature (in K). The plot of rate constants at different temperatures is presented in Figure 11. The rate constants follow the Arrhenius-type behavior. Activation energy has been found from the slope of lines given in Figure 11. The activation energy of the $Al_{71}Co_{29}$ alloy is 90 kJ mol^{-1} and the activation energy of the $Al_{76}Co_{24}$ alloy is 129 kJ mol^{-1}. These activation energies are comparable to activation energies for Al oxidation reported by previous studies [85,86].

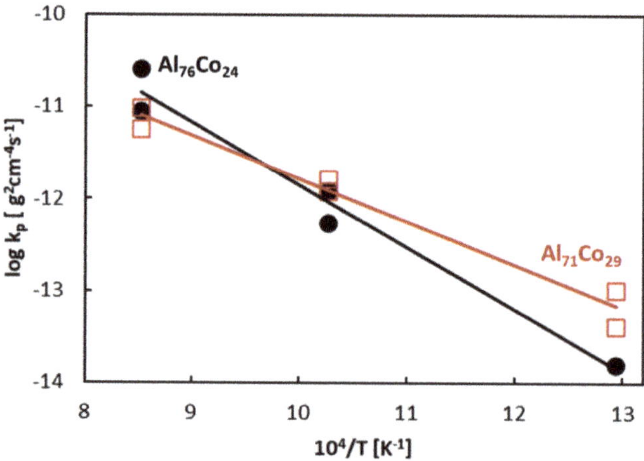

Figure 11. Temperature dependence of parabolic rate constants of the Al-Co alloys.

Results presented above show that a protective scale has been formed on the surface of complex metallic alloys. The rate constants were relatively low, and a thin alumina scale was found on the surface (Figures 4 and 5). Alumina is formed by the following reaction

$$\frac{4}{3}Al + O_2 \rightarrow \frac{2}{3}Al_2O_3 \tag{3}$$

The Gibbs energy (ΔG) of reaction (3) is given in Figure 12. $\Delta G(Al_2O_3)$ is very low which indicates a strong affinity of aluminum towards oxygen. In principle, cobalt oxidation in the Al-Co alloys is also possible. This reaction can be given by the following equation

$$2Co + O_2 \rightarrow 2CoO \tag{4}$$

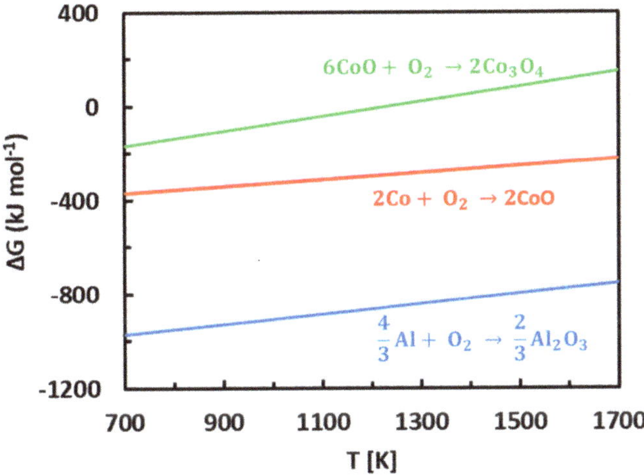

Figure 12. Gibbs free energies of metal oxidation reactions at elevated temperatures, redrawn from [54].

Nevertheless, ΔG reaction (4) is considerably larger compared to Gibbs free energy for aluminum oxidation (Figure 12). A further oxidation of CoO is even more energetically demanding [87,88]. Therefore, CoO tends to decompose in reaction with Al. The reaction can be expressed by the following equation

$$\frac{4}{3}Al + 2CoO \rightarrow \frac{2}{3}Al_2O_3 + Co \tag{5}$$

In this disproportionation reaction, CoO is reduced, and Al oxidized. The Gibbs energy of reaction (5) is negative. Therefore, the selective oxidation of aluminum in the Al-Co alloys is thermodynamically possible.

The oxidation of Co-rich Al-Co alloys was previously studied by Irving et al. [61]. The alloys were studied in the as-cast state. The authors studied several Co-xAl alloys with x = 0–32 at.%. Alloys with small Al concentration (< 10 at.%) formed a single CoO layer. Cobalt oxide layer grew with a considerably higher corrosion rate compared to Al_2O_3. At intermediate Al concentrations (10–20 at.%), the authors found that an inner layer of Al_2O_3 started to form below the outer CoO scale. With increasing aluminum concentration, a continuous Al_2O_3 scale has been developed. The comparison of the present results with those from literature is given in Figure 13. Our data complement the previous results of Irving et al. The continuous Al_2O_3 scale forms a barrier to cobalt diffusion. It hinders the nucleation and growth of cobalt oxides. Irving et al. found that a protective alumina scale can be formed when Al concentration 24 at.% at 1173 K is reached. Comparable minimum Al concentrations required to form the external alumina scale were also found for the Ni-Al and Fe-Al alloys [89].

The comparison of the present results with previously studied complex metallic alloys is provided in Figure 14. Kinetics of oxidation of Al-Cu-Fe and Al-Pd-Mn quasicrystal surfaces was studied in synthetic air [66,90]. High temperature oxidation kinetics of Al-Cr-Fe complex metallic alloys was studied in pure oxygen [91]. Our data are comparable to Al-Cu-Fe and Al-Pd-Mn alloys. The parabolic rate constants of the Al-Cr-Fe complex metallic alloys are lower compared to the remainder of the alloys. The oxidation resistance of the Al-Fe-TM (TM = Cr, Cu) alloys is related to the chemical composition of the oxide scale. The scale found in Al-Cu-Fe alloys was alumina. The scale formed in Al-Cr-Fe complex metallic alloys, however, was composed of Al_2O_3 and $(Al_{0.9}Cr_{0.1})_2O_3$. The second scale component provided an additional barrier against corrosion. Chromium as a third alloying element may improve the overall oxidation resistance of the alloy. The corrosion resistance of alumina forming alloys alloyed with chromium is higher compared to alloys without Cr. When a sufficient Cr concentration is available, a complete chromia scale can be formed on top of the alumina scale [92,93]. The duplex Al_2O_3/Cr_2O_3 scale has an outstanding corrosion resistance.

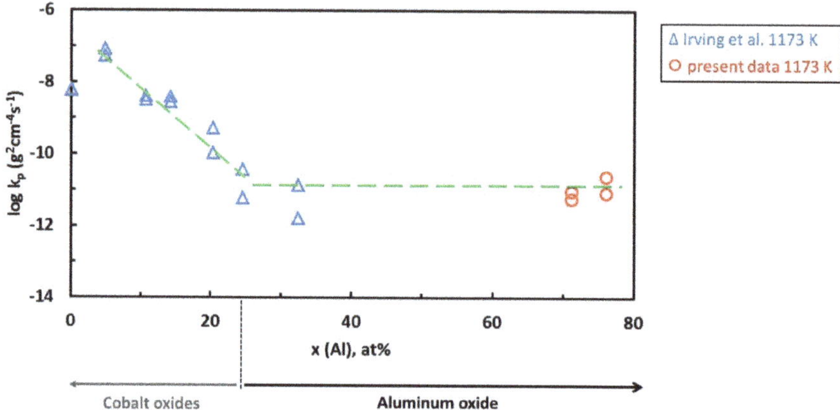

Figure 13. Variation of parabolic rate constants of Al-Co alloys with increasing aluminum atomic fraction.

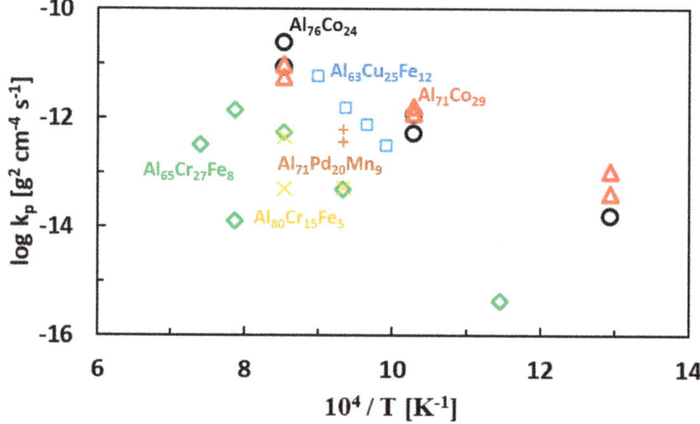

Figure 14. Parabolic rate constants for metal oxidation of Al-TM complex metallic alloys.

Previous authors also studied the microstructure evolution of the oxide scale. In early oxidation stages, γ-Al_2O_3 on the $Al_{63}Cu_{25}Fe_{12}$ alloy was formed with an orientational relationship to the underlying Al-Cu-Fe quasicrystal [66]. γ-Al_2O_3 continued to grow as θ-Al_2O_3 until the oxide layer of several hundred nanometers has been formed. θ-Al_2O_3 was later transformed into the thermodynamically stable α-Al_2O_3. α-Al_2O_3 continued to grow with a nodular morphology. The oxide-nodule formation changed the growth mechanism. A protective layer formation was no longer observed. A massive spallation occurred after few days of oxidation [66]. A spallation of Al_2O_3

scale from the oxidized Al-Cr-Fe surfaces at high temperatures was also observed [91]. The massive oxide spallation has not been observed in the present study. However, the oxide spallation was observed locally on β-AlCo dendrites (Figure 8). It is possible that further stresses in the scale could develop during long term annealing. Therefore, further experiments on the Al-Co complex metallic alloys are required to study the effects of long-term annealing and/or thermal cycling on the oxidation behavior. These studies could shed further light into the long-term oxidation resistance for practical applications of the Al-Co alloys at high temperatures.

4. Conclusions

In the present work, the oxidation behavior of the $Al_{71}Co_{29}$ and $Al_{76}Co_{24}$ alloys (concentration in at.%) was studied at 773–1173 K in air. The alloys were prepared by rapid solidification of Al and Co lumps in argon. The alloys were studied in as-solidified state. The following conclusions can be drawn:

1. The alloys were composed of different microstructure constituents. The $Al_{76}Co_{24}$ alloy was composed of Al_9Co_2, m-$Al_{13}Co_4$ and Z-Al_3Co. The $Al_{71}Co_{29}$ alloy consisted of Z-Al_3Co, Al_5Co_2 and β-AlCo. The precipitation sequences of the constituents were explained based on the equilibrium Al-Co phase diagram and rapid solidification processes taking place during casting.
2. During oxidation in air, aluminum in the alloys was selectively oxidized and a protective alumina scale was formed on the alloy surfaces. The oxidation kinetics followed a parabolic rate law. The rate constants of the alloys were between 1.63×10^{-14} and 8.83×10^{-12} g cm^{-4} s^{-1}, depending on the annealing temperature. The activation energies of oxidation were 90 kJ mol^{-1} for the $Al_{71}Co_{29}$ alloy and 123 kJ mol^{-1} for the $Al_{76}Co_{24}$ alloy, respectively.
3. The scale of the $Al_{76}Co_{24}$ alloy was adherent to the substrate and had a wave-like morphology. At 1173 K, a preferential orientation of θ-Al_2O_3 in (002) crystallographic plane was found. The scale on the $Al_{71}Co_{29}$ alloy was more uniform and spallation was observed locally on the dendritic β-AlCo.
4. The oxidation kinetics of the $Al_{71}Co_{29}$ and $Al_{76}Co_{24}$ alloys was comparable to previously studied $Al_{24}Co_{76}$ and $Al_{32}Co_{68}$ alloys where a continuous alumina scale had been formed. The increased Al concentration contributes to the alloy's corrosion resistance. The continuous Al_2O_3 scale forms a barrier to cobalt diffusion, and it hinders the nucleation and growth of cobalt oxides.

Author Contributions: Conceptualization, M.P., and M.D.; methodology, M.P., M.D., and I.Č.; formal analysis, P.Š.; investigation, P.Š., M.D., I.Č., L.Ď., R.S., Ž.G., and M.P.; resources, M.D., and M.P.; data curation, P.Š., M.P., M.D., I.Č., and L.Ď.; writing—original draft preparation, M.P.; writing—review and editing, M.D., L.Ď., Ž.G., and I.Č.; supervision, M.D., R.S., L.Ď., I.Č., and M.P.; project administration, M.P.; funding acquisition, M.P. All authors have read and agreed to the published version of the manuscript.

Funding: This research was supported by the Grant Agency VEGA of the Slovakian Ministry of Education, Research, Science and Sport project no. 1/0490/18 and by the Slovak Research and Development Agency project no. APVV-15-0049.

Conflicts of Interest: The authors declare no conflict of interest.

References

1. Sato, J.; Omori, T.; Oikawa, K.; Ohnuma, I.; Kainuma, R.; Ishida, K. Cobalt-Base High-Temperature Alloys. *Science* **2006**, *312*, 90–91. [CrossRef]
2. Ma, Q.; Wang, W.; Dong, C. Co-Al-W-based superalloys: A summary of current knowledge. *Cailiao Daobao/ Mater. Rep.* **2020**, *34*, 3157–3164. [CrossRef]
3. Pollock, T.M.; Dibbern, J.; Tsunekane, M.; Zhu, J.; Suzuki, A. New Co-based γ-γ' high-temperature alloys. *JoM* **2010**, *62*, 58–63. [CrossRef]
4. Betteridge, W.; Shaw, S.W.K. Development of superalloys. *Mater. Sci. Technol.* **1987**, *3*, 682–694. [CrossRef]
5. Coutsouradis, D.; Davin, A.; Lamberigts, M. Cobalt-based superalloys for applications in gas turbines. *Mater. Sci. Eng.* **1987**, *88*, 11–19. [CrossRef]
6. Klarstrom, D.L. Wrought cobalt- base superalloys. *J. Mater. Eng. Perform.* **1993**, *2*, 523–530. [CrossRef]

7. Moffat, J.P.; Whitfield, T.; Christofidou, K.A.; Pickering, E.; Jones, N.; Stone, H.J. The Effect of Heat Treatment on the Oxidation Resistance of Cobalt-Based Superalloys. *Metals* **2020**, *10*, 248. [CrossRef]

8. Liu, L.; Wu, S.-S.; Dong, Y.; Lü, S. Effects of alloyed Mn on oxidation behaviour of a Co-Ni-Cr-Fe alloy between 1050 and 1250 °C. *Corros. Sci.* **2016**, *104*, 236–247. [CrossRef]

9. Holcomb, G.R. Steam Oxidation and Chromia Evaporation in Ultrasupercritical Steam Boilers and Turbines. *J. Electrochem. Soc.* **2009**, *156*, C292–C297. [CrossRef]

10. Forsik, S.A.J.; Rosas, A.O.P.; Wang, T.; Colombo, G.A.; Zhou, N.; Kernion, S.J.; Epler, M.E. High-Temperature Oxidation Behavior of a Novel Co-Base Superalloy. *Met. Mater. Trans. A* **2018**, *49*, 4058–4069. [CrossRef]

11. Klein, L.; Bauer, A.; Neumeier, S.; Göken, M.; Virtanen, S. High temperature oxidation of γ/γ′-strengthened Co-base superalloys. *Corros. Sci.* **2011**, *53*, 2027–2034. [CrossRef]

12. Yan, H.-Y.; Vorontsov, V.; Dye, D. Effect of alloying on the oxidation behaviour of Co-Al-W superalloys. *Corros. Sci.* **2014**, *83*, 382–395. [CrossRef]

13. Goward, G.W.; Cannon, L.W. Pack Cementation Coatings for Superalloys: A Review of History, Theory, and Practice. *J. Eng. Gas Turbines Power* **1988**, *110*, 150–154. [CrossRef]

14. Goward, G. Protective coatings—Purpose, role, and design. *Mater. Sci. Technol.* **1986**, *2*, 194–200. [CrossRef]

15. Goward, G. Progress in coatings for gas turbine airfoils. *Surf. Coat. Technol.* **1998**, *108*, 73–79. [CrossRef]

16. Streiff, R.; Boone, D.H. Corrosion resistant modified aluminide coatings. *J. Mater. Eng.* **1988**, *10*, 15–26. [CrossRef]

17. Meier, G.; Pettit, F. High-temperature corrosion of alumina-forming coatings for superalloys. *Surf. Coat. Technol.* **1989**, *39*, 1–17. [CrossRef]

18. Liu, P.; Liang, K.; Gu, S. High-temperature oxidation behavior of aluminide coatings on a new cobalt-base superalloy in air. *Corros. Sci.* **2001**, *43*, 1217–1226. [CrossRef]

19. Rahman, A.; Jayaganthan, R.; Chandra, R.; Ambardar, R. Microstructural Characterization and Cyclic Hot Corrosion Behaviour of Sputtered Co-Al Nanostructured Coatings on Superalloy. *Oxid. Met.* **2011**, *76*, 307–330. [CrossRef]

20. Rahman, A.; Jayaganthan, R.; Chandra, R.; Ambardar, R. High temperature degradation behavior of sputtered nanostructured Co-Al coatings on superalloy. *Appl. Surf. Sci.* **2013**, *265*, 10–23. [CrossRef]

21. Rahman, A.; Chawla, V.; Jayaganthan, R.; Chandra, R.; Ambardar, R. Degradation behaviour of sputtered Co-Al coatings on superalloy. *Mater. Chem. Phys.* **2013**, *138*, 49–62. [CrossRef]

22. Jiang, J.; Zhou, T.; Shao, W.; Zhou, C. Interdiffusion behavior and lifetime prediction of Co-Al coating on Ni-based superalloy. *J. Alloys Compd.* **2019**, *786*, 920–929. [CrossRef]

23. Stein, F.; He, C.; Dupin, N. Melting behaviour and homogeneity range of B2 CoAl and updated thermodynamic description of the Al-Co system. *Intermetallics* **2013**, *39*, 58–68. [CrossRef]

24. Priputen, P.; Palcut, M.; Babinec, M.; Mišík, J.; Černičková, I.; Janovec, J. Correlation between Microstructure and Corrosion Behavior of Near-Equilibrium Al-Co Alloys in Various Environments. *J. Mater. Eng. Perform.* **2017**, *26*, 3970–3976. [CrossRef]

25. Grushko, B.; Wittenberg, R.; Bickmann, K.; Freiburg, C. The constitution of aluminum-cobalt alloys between Al5Co2 and Al9Co2. *J. Alloys Compd.* **1996**, *233*, 279–287. [CrossRef]

26. Priputen, P.; Kusy, M.; Drienovský, M.; Janičkovič, D.; Čička, R.; Černičková, I.; Janovec, J. Experimental reinvestigation of Al-Co phase diagram in vicinity of Al13Co4 family of phases. *J. Alloys Compd.* **2015**, *647*, 486–497. [CrossRef]

27. Cooper, M.J. The electron distribution in the phases CoAl and NiAl. *Philos. Mag.* **1963**, *8*, 811–821. [CrossRef]

28. Burkhardt, U.; Ellner, M.; Grin, Y.; Baumgartner, B. Powder diffraction refinement of the Co2Al5 structure. *Powder Diffr.* **1998**, *13*, 159–162. [CrossRef]

29. Ma, X.L.; Li, X.Z.; Kuo, K.H. A family of τ-inflated monoclinic Al13Co4 phases. *Acta Crystallogr. Sect. B Struct. Sci.* **1995**, *51*, 36–43. [CrossRef]

30. Freiburg, C.; Grushko, B.; Wittenberg, R.; Reichert, W. Once More about Monoclinic Al13Co4. *Mater. Sci. Forum* **1996**, *228*, 583–586. [CrossRef]

31. Grin, J.; Burkhardt, U.; Ellner, M.; Peters, K. Crystal structure of orthorhombic Co4Al13. *J. Alloys Compd.* **1994**, *206*, 243–247. [CrossRef]

32. Fleischer, F.; Weber, T.; Jung, D.; Steurer, W. o′-Al13Co4, a new quasicrystal approximant. *J. Alloys Compd.* **2010**, *500*, 153–160. [CrossRef]

33. Sugiyama, K.; Genba, M.; Hiraga, K.; Waseda, Y. The structure of Y-Al13-xCo4 (x = 0.8) analyzed by single crystal X-ray diffraction coupled with anomalous X-ray scattering. *J. Alloys Compd.* **2010**, *494*, 98–101. [CrossRef]

34. Boström, M.; Rosner, H.; Prots, Y.; Burkhardt, U.; Grin, Y. The Co2Al9 Structure Type Revisited. *Z. Anorg. Allg. Chem.* **2005**, *631*, 534–541. [CrossRef]

35. Heggen, M.; Deng, D.; Feuerbacher, M. Plastic deformation properties of the orthorhombic complex metallic alloy phase Al13Co4. *Intermetallics* **2007**, *15*, 1425–1431. [CrossRef]

36. Samuel, F.H. A study of the microstructure, mechanical properties and failure behaviour of Al-Li-Co melt-spun ribbons: Effect of Al9Co2 phase particle precipitation. *J. Mater. Sci.* **1986**, *21*, 3097–3107. [CrossRef]

37. Lekatou, A.G.; Sfikas, A.; Karantzalis, A.E. The influence of the fabrication route on the microstructure and surface degradation properties of Al reinforced by Al 9 Co 2. *Mater. Chem. Phys.* **2017**, *200*, 33–49. [CrossRef]

38. Wolf, W.; Schulz, R.; Savoie, S.; Bolfarini, C.; Kiminami, C.S.; Botta, W. Structural, mechanical and thermal characterization of an Al-Co-Fe-Cr alloy for wear and thermal barrier coating applications. *Surf. Coat. Technol.* **2017**, *319*, 241–248. [CrossRef]

39. Hagel, W.C. The oxidation of iron, nickel and cobalt-base alloys containing aluminum. *Corrosion* **1965**, *21*, 316–326. [CrossRef]

40. Young, D.J. Oxidation of Pure Metals. In *High Temperature Oxidation and Corrosion of Metals*, 2nd ed.; Elsevier: Amsterdam, The Netherlands, 2016; Chapter 3; pp. 85–144. [CrossRef]

41. Warde, M.; Herinx, M.; Ledieu, J.; Loli, L.S.; Fournée, V.; Gille, P.; Le Moal, S.; Barthés-Labrousse, M.-G. Adsorption of O2 and C2H (n = 2, 4, 6) on the Al9Co2(0 0 1) and o-Al13Co4(1 0 0) complex metallic alloy surfaces. *Appl. Surf. Sci.* **2015**, *357*, 1666–1675. [CrossRef]

42. Villaseca, S.A.; Loli, L.N.S.; Ledieu, J.; Fournée, V.; Gille, P.; Dubois, J.-M.; Gaudry, É. Oxygen adsorption on the Al9Co2(001) surface: First-principles and STM study. *J. Phys. Condens. Matter* **2013**, *25*, 355003. [CrossRef] [PubMed]

43. Dubois, J.-M. Properties- and applications of quasicrystals and complex metallic alloys. *Chem. Soc. Rev.* **2012**, *41*, 6760. [CrossRef] [PubMed]

44. Wolf, W.; Bolfarini, C.; Kiminami, C.S.; Botta, W. Designing new quasicrystalline compositions in Al-based alloys. *J. Alloys Compd.* **2020**, *823*, 153765. [CrossRef]

45. Steurer, W. Twenty years of structure research on quasicrystals. Part I. Pentagonal, octagonal, decagonal and dodecagonal quasicrystals. *Z. Krist. Cryst. Mater.* **2004**, *219*, 391–446. [CrossRef]

46. Barber, E.M. Chemical Bonding and Physical Properties in Quasicrystals and Their Related Approximant Phases: Known Facts and Current Perspectives. *Appl. Sci.* **2019**, *9*, 2132. [CrossRef]

47. Rabson, D. Toward theories of friction and adhesion on quasicrystals. *Prog. Surf. Sci.* **2012**, *87*, 253–271. [CrossRef]

48. Balbyshev, V.; King, D.; Khramov, A.; Kasten, L.; Donley, M. Investigation of quaternary Al-based quasicrystal thin films for corrosion protection. *Thin Solid Films* **2004**, *447*, 558–563. [CrossRef]

49. Moskalewicz, T.; Dubiel, B.; Wendler, B. AlCuFe(Cr) and AlCoFeCr coatings for improvement of elevated temperature oxidation resistance of a near-α titanium alloy. *Mater. Charact.* **2013**, *83*, 161–169. [CrossRef]

50. Parsamehr, H.; Chang, S.-Y.; Lai, C.-H. Mechanical and surface properties of aluminum-copper-iron quasicrystal thin films. *J. Alloy Compd.* **2018**, *732*, 952–957. [CrossRef]

51. Parsamehr, H.; Chen, T.-S.; Wang, D.-S.; Leu, M.-S.; Han, I.; Xi, Z.; Tsai, A.-P.; Shahani, A.J.; Lai, C.-H. Thermal spray coating of Al-Cu-Fe quasicrystals: Dynamic observations and surface properties. *Materialia* **2019**, *8*, 100432. [CrossRef]

52. Chatelier, C.; Garreau, Y.; Piccolo, L.; Vlad, A.; Resta, A.; Ledieu, J.; Fournée, V.; De Weerd, M.-C.; Picca, F.-E.; De Boissieu, M.; et al. From the Surface Structure to Catalytic Properties of Al5Co2(2$\overline{1}$0): A Study Combining Experimental and Theoretical Approaches. *J. Phys. Chem. C* **2020**, *124*, 4552–4562. [CrossRef]

53. Piccolo, L.; Chatelier, C.; De Weerd, M.-C.; Morfin, F.; Ledieu, J.; Fournée, V.; Gille, P.; Gaudry, E. Catalytic properties of Al13TM4 complex intermetallics: Influence of the transition metal and the surface orientation on butadiene hydrogenation. *Sci. Technol. Adv. Mater.* **2019**, *20*, 557–567. [CrossRef] [PubMed]

54. Meier, M.; Ledieu, J.; Fournée, V.; Gaudry, E. Semihydrogenation of Acetylene on Al5Co2 Surfaces. *J. Phys. Chem. C* **2017**, *121*, 4958–4969. [CrossRef]

55. Soler, L.; Macanás, J.; Muñoz, M.; Casado, J. Aluminum and aluminum alloys as sources of hydrogen for fuel cell applications. *J. Power Sources* **2007**, *169*, 144–149. [CrossRef]

56. Lekatou, A.G.; Sfikas, A.; Karantzalis, A.E.; Sioulas, D. Microstructure and corrosion performance of Al-32%Co alloys. *Corros. Sci.* **2012**, *63*, 193–209. [CrossRef]

57. Palcut, M.; Priputen, P.; Kusý, M.; Janovec, J. Corrosion behaviour of Al-29at%Co alloy in aqueous NaCl. *Corros. Sci.* **2013**, *75*, 461–466. [CrossRef]

58. Palcut, M.; Priputen, P.; Šalgó, K.; Janovec, J. Phase constitution and corrosion resistance of Al-Co alloys. *Mater. Chem. Phys.* **2015**, *166*, 95–104. [CrossRef]

59. Zhang, H.H.; Xiang, J.; Wang, W.; Shi, Y.-J. The Oxidation of Co-5Al Alloys in 1 Atm of Pure O2 at 700 and 800 °C. *Adv. Mater. Res.* **2011**, *295*, 319–322. [CrossRef]

60. Zhang, H.H.; Xiang, J.H.; Xu, X.C.; Wang, C. Comparison of Oxidation Behavior of Binary Co-10X (x = Al, Si, Cr) Alloys at 973 and 1073K. *Appl. Mech. Mater.* **2013**, *395*, 238–242. [CrossRef]

61. Irving, G.N.; Stringer, J.; Whittle, D.P. The high-temperature oxidation resistance of Co-Al alloys. *Oxid. Met.* **1975**, *9*, 427–440. [CrossRef]

62. Agliozzo, S.; Brunello, E.; Klein, H.; Mancini, L.; Hartwig, J.; Baruchel, J.; Gastaldi, J. Extended investigation of porosity in quasicrystals by synchrotron X-ray phase contrast radiography—I: In icosahedral AlPdMn grains. *J. Cryst. Growth* **2005**, *281*, 623–638. [CrossRef]

63. Brunello, E.; Agliozzo, S.; Klein, H.; Mancini, L.; Härtwig, J.; Baruchel, J.; Gastaldi, J. Extended investigation of porosity in quasicrystals by synchrotron X-ray phase contrast radiography: Part II, in grains of other alloys and structures than AlPdMn. *J. Cryst. Growth* **2005**, *282*, 228–235. [CrossRef]

64. Prescott, R.; Graham, M.J. The oxidation of iron-aluminum alloys. *Oxid. Met.* **1992**, *38*, 73–87. [CrossRef]

65. Chevalier, S. Formation and growth of protective alumina scales. In *Developments in High Temperature Corrosion and Protection of Materials*, 1st ed.; Woodhead Publishing: Abington, Cambridge, UK, 2008; Chapter 10; pp. 290–328. [CrossRef]

66. Wehner, B.I.; Köster, U. Microstructural Evolution of Alumina Layers on an Al-Cu-Fe Quasicrystal during High-Temperature Oxidation. *Oxid. Met.* **2000**, *54*, 445–456. [CrossRef]

67. Tolpygo, V.; Clarke, D.R. Microstructural evidence for counter-diffusion of aluminum and oxygen during the growth of alumina scales. *Mater. High Temp.* **2003**, *20*, 261–271. [CrossRef]

68. Dan'Ko, A.J.; Rom, M.A.; Sidelnikova, N.S.; Nizhankovskiy, S.V.; Budnikov, A.T.; Grin', L.A.; Kaltaev, K.S.-O. Transformation of the corundum structure upon high-temperature reduction. *Crystallogr. Rep.* **2008**, *53*, 1112–1118. [CrossRef]

69. Trunov, M.A.; Schoenitz, M.; Zhu, X.; Dreizin, E.L. Effect of polymorphic phase transformations in Al2O3 film on oxidation kinetics of aluminum powders. *Combust. Flame* **2005**, *140*, 310–318. [CrossRef]

70. Zhou, R.S.; Snyder, R.L. Structures and transformation mechanisms of the η, γ and θ transition aluminas. *Acta Crystallogr. Sect. B Struct. Sci.* **1991**, *47*, 617–630. [CrossRef]

71. Kovarik, L.; Bowden, M.; Shi, D.; Washton, N.M.; Andersen, A.; Hu, J.Z.; Lee, J.; Szanyi, J.; Kwak, J.-H.; Peden, C.H.F. Unraveling the Origin of Structural Disorder in High Temperature Transition Al2O3: Structure of θ-Al2O3. *Chem. Mater.* **2015**, *27*, 7042–7049. [CrossRef]

72. Jbara, A.S.; Othaman, Z.; Saeed, M. Structural, morphological and optical investigations of θ-Al2O3 ultrafine powder. *J. Alloys Compd.* **2017**, *718*, 1–6. [CrossRef]

73. Wang, Y.; Bronsveld, P.; De Hosson, J.T.M.; Djuričić, B.; McGarry, D.; Pickering, S. Twinning in θ Alumina Investigated with High Resolution Transmission Electron Microscopy. *J. Eur. Ceram. Soc.* **1998**, *18*, 299–304. [CrossRef]

74. Kovarik, L.; Bowden, M.; Andersen, A.; Jaegers, N.R.; Washton, N.; Szanyi, J. Quantification of High Temperature Transitions and Their Phase Transformations. Available online: https://chemrxiv.org/articles/preprint/Quantification_of_High_Temperature_Transition_Al2O3_and_Their_Phase_Transformations/12584783 (accessed on 4 July 2020).

75. Boumaza, A.; Favaro, L.; Ledion, J.; Sattonnay, G.; Brubach, J.; Berthet, P.; Huntz, A.; Roy, P.; Tétot, R. Transition alumina phases induced by heat treatment of boehmite: An X-ray diffraction and infrared spectroscopy study. *J. Solid State Chem.* **2009**, *182*, 1171–1176. [CrossRef]

76. McCafferty, E. High-Temperature Gaseous Oxidation. In *Introduction to Corrosion Science*; Springer: New York, NY, USA, 2009; pp. 453–476. [CrossRef]

77. Ledieu, J.; Gaudry, E.; Fournée, V. Surfaces of Al-based complex metallic alloys: Atomic structure, thin film growth and reactivity. *Sci. Technol. Adv. Mater.* **2014**, *15*, 34802. [CrossRef] [PubMed]

78. Korte-Kerzel, S.; Schnabel, V.; Clegg, W.; Heggen, M. Room temperature plasticity in m-Al13Co4 studied by microcompression and high resolution scanning transmission electron microscopy. *Scr. Mater.* **2018**, *146*, 327–330. [CrossRef]

79. Wang, W.; Yang, W.; Liu, Z.; Lin, Y.; Zhou, S.; Qian, H.; Wang, H.; Lin, Z.; Li, G. Epitaxial growth and characterization of high-quality aluminum films on sapphire substrates by molecular beam epitaxy. *CrystEngComm* **2014**, *16*, 7626–7632. [CrossRef]

80. Alessandri, M.; Piagge, R.; Caniatti, M.; Del Vitto, A.; Wiemer, C.; Pavia, G.; Alberici, S.; Bellandi, E.; Nale, A. Structural and Chemical Investigation of Annealed Al2O3 Films for Interpoly Dielectric Application in Flash Memories. *ECS Trans.* **2006**, *3*, 183–192. [CrossRef]

81. Queraltó, A.; De La Mata, M.; Arbiol, J.; Obradors, X.; Puig, T. Disentangling Epitaxial Growth Mechanisms of Solution Derived Functional Oxide Thin Films. *Adv. Mater. Interfaces* **2016**, *3*, 1600392. [CrossRef]

82. Mattox, D.M. Atomistic Film Growth and Some Growth-Related Film Properties. In *Handbook of Physical Vapor Deposition (PVD) Processing*, 2nd ed.; Elsevier: Oxford, UK, 2010; pp. 333–398. [CrossRef]

83. Meier, M.; Ledieu, J.; De Weerd, M.-C.; Huang, Y.-T.; Abreu, G.J.P.; Pussi, K.; Diehl, R.; Mazet, T.; Fournée, V.; Gaudry, E. Interplay between bulk atomic clusters and surface structure in complex intermetallic compounds: The case study of the Al5Co2(001) surface. *Phys. Rev. B* **2015**, *91*, 085414. [CrossRef]

84. Anand, K.; Fournée, V.; Prevot, G.; Ledieu, J.; Gaudry, E. Nonwetting Behavior of Al-Co Quasicrystalline Approximants Owing to Their Unique Electronic Structures. *ACS Appl. Mater. Interfaces* **2020**, *12*, 15793–15801. [CrossRef]

85. Gulbransen, E.A.; Wysong, W.S. Thin Oxide Films on Aluminum. *J. Phys. Chem.* **1947**, *51*, 1087–1103. [CrossRef]

86. Smeltzer W, W. Oxidation of An Aluminum-3 Per Cent Magnesium Alloy in the Temperature Range 200–550 °C. *J. Electrochem. Soc.* **1958**, *105*, 67–71. [CrossRef]

87. Sabat, K.C.; Paramguru, R.K.; Pradhan, S.; Mishra, B.K. Reduction of Cobalt Oxide (Co3O4) by Low Temperature Hydrogen Plasma. *Plasma Chem. Plasma Process.* **2014**, *35*, 387–399. [CrossRef]

88. Chattopadhyay, B.; Wood, G.C. The transient oxidation of alloys. *Oxid. Met.* **1970**, *2*, 373–399. [CrossRef]

89. Stott, F.H. Influence of alloy additions on oxidation. *Mater. Sci. Technol.* **1989**, *5*, 734–740. [CrossRef]

90. Wehner, B.; Köster, U.; Rüdiger, A.; Pieper, C.; Sordelet, D. Oxidation of Al–Cu–Fe and Al–Pd–Mn quasicrystals. *Mater. Sci. Eng. A* **2000**, *294*, 830–833. [CrossRef]

91. Prud'Homme, N.; Ribot, P.; Herinx, M.; Gille, P.; Bauer, B.; De Weerd, M.-C.; Barthés-Labrousse, M.-G. High temperature oxidation of AlCrFe complex metallic alloys. *Corros. Sci.* **2014**, *89*, 118–126. [CrossRef]

92. Irving, G.N.; Stringer, J.; Whittle, D.P. The Oxidation Behavior of Co-Cr-Al Alloys at 1000 °C. *Corrosion* **1977**, *33*, 56–60. [CrossRef]

93. Wood, G.C.; Stott, F.H. The influence of aluminum additions on the oxidation of Co-Cr alloys at 1000 and 1200 °C. *Oxid. Met.* **1971**, *3*, 365–398. [CrossRef]

© 2020 by the authors. Licensee MDPI, Basel, Switzerland. This article is an open access article distributed under the terms and conditions of the Creative Commons Attribution (CC BY) license (http://creativecommons.org/licenses/by/4.0/).

 *materials* 

Article

Effect of Mo Addition on the Chemical Corrosion Process of SiMo Cast Iron

Marcin Stawarz [1,*] and Paweł M. Nuckowski [2]

1   Department of Foundry Engineering, Silesian University of Technology, 7 Towarowa Street, 44-100 Gliwice, Poland

2   Department of Engineering Materials and Biomaterials, Silesian University of Technology, 18A Konarskiego Street, 44-100 Gliwice, Poland; pawel.nuckowski@polsl.pl

*   Correspondence: marcin.stawarz@polsl.pl; Tel.: +48-32-338-5532

Received: 24 March 2020; Accepted: 7 April 2020; Published: 9 April 2020

Abstract: The study was carried out to evaluate five SiMo cast iron grades and their resistance to chemical corrosion at elevated temperature. Corrosion tests were carried out under conditions of an actual cyclic operation of a retort coal-fired boiler. The duration of the study was 3840 h. The range of temperature changes during one cycle was in the range of 300–650 °C. Samples of SiMo cast iron with Si content at the level of 5% and variable Mo content in the range 0%–2.5% were used as the material for the study. The examined material was subjected to preliminary metallographic analysis using scanning microscopy and an Energy dispersive spectroscopy (EDS) system. The chemical composition was determined on the basis of a Leco spectrometer and a Leco carbon and sulfur analyzer. The examination of the oxide layer was carried out with the use of Scanning electron microscope (SEM), EDS, and X-ray diffraction (XRD) methods. It was discovered that, in the analyzed alloys, oxide layers consisting of Fe_2O_3, Fe_3O_4, SO_2, and Fe_2SiO_4 were formed. The analyzed oxide layers were characterized by high adhesion to the substrate material, and their total thickness was about 20 μm.

Keywords: chemical corrosion; SiMo cast iron; fayalite; hematite; magnetite; maghemite; sulfur dioxide

---

1. Introduction

Corrosive behavior of SiMo cast iron in the air and flue gases was presented in the works [1–5]. When we subject pure iron to the corrosion process at elevated temperatures and in the ambient atmospheric air, a multilayer oxide structure composed of FeO, Fe_3O_4, and Fe_2O_3 is formed [5]. For ductile cast iron, an oxide layer is formed, located both in the material and in the surface layer as a result of migration of Fe atoms [5]. After introducing an alloying element in the form of Si into cast iron, a SiO_2 compound is formed at the metal–oxide layer point of contact, which constitutes a barrier to further oxidation processes [5]. Simultaneously, SiO_2 can react with O, Fe, and FeO. The result of these reactions may be the formation of fayalite, Fe_2SiO_4 [6–8]. In the paper [9], the author writes that the oxide layer on the surface of SiMo cast iron is composed of the following sub-layers situated from the outside to the inside of the material: Fe_2O_3, Fe_3O_4, FeO, FeO + Fe_2SiO_4.

The oxide layer adheres well to the base material and the inner layer consisting of FeO + Fe_2SiO_4 [9]. The higher the silicon content in the base material, the faster the oxide layer forms. A number of studies concerning the corrosion resistance of SiMo cast iron focus on a relatively short time of exposure to oxidation (500 h on average) [10]. These studies are conducted mainly in terms of the use of SiMo cast iron in automotive castings, as described by Rouczka [11] and many other authors [12–17]. SiMo cast iron is an increasingly popular material, and research on this material is also conducted with a focus on optimizing the manufacturing process. In their work, Guzik et al. [18] write about the method of introducing two flexible hoses with the diameter of Ø 9 mm; one filled with a FeSi + Mg mixture, and the other with a graphitizing modifier for the treatment drum ladle. Guzik et al. [18] describe it as a new method of secondary treatment of ferritic cast iron production of SiMo type. This method can be used for the production of ductile iron melted in an induction furnace [18,19].

SiMo cast iron can also be successfully used in other areas of industry: exhaust parts for combustion engines, turbocharger housings and rotors, gas turbine components, molds for the glass industry, molds for aluminum alloys, zinc, forging dies, heat treatment furnace components, aluminum melting furnace components, and waste incineration furnaces. This happens wherever elevated operating temperatures and gases resulting from the combustion, e.g., of solid fuels are involved. A good example of such a system is a coal-fired retort furnace. Nyashina and her team write about the problems related to the emission of pollutants during the combustion process [20]. Released into the atmosphere with exhaust gases, nitrogen oxides (mainly NO and NO_2) are the main reason why photochemical smog appears, which reaches the stratosphere to act as a catalyst for ozone layer depletion. Rapid oxidation of NOx and SOx and their interaction with water vapor in the atmosphere generates tiny droplets of sulfuric (H_2SO_4) and nitric (HNO_3) acids [20]. Sulfuric acid causes significant losses in the ecosystem, which has been mentioned by many authors [21]. It is important to optimize the combustion process of solid fuels in boilers by improving the materials from which these boilers are built. For the above reasons, in this work, studies of resistance to chemical corrosion of SiMo cast iron were carried out during actual operation of a retort boiler. The duration of the study was 3840 hours. To date, the corrosion resistance of SiMo cast iron during the operation of a retort boiler has not been described in the literature. Due to its properties, it can be successfully used for manufacturing furnace elements fired with solid, liquid, or gaseous fuels.

2. Methods and Materials

Experimental melts were conducted in an induction furnace (PI25, ELKON Sp. z o.o., Rybnik, Poland) with medium frequency and a capacity of 25 kg. The charge consisted of steel scrap with low sulphur content. Other ingredients added during the melting were ferrosilicon FeSi75, synthetic graphite of carbon content above 99.35%, and FeMo65-rich alloy. The spheroidization process of cast iron was conducted at the bottom of the ladle after covering the nodulizing agent with pieces of steel scrap. The magnesium-rich alloy used in the studies was FeSiMg5RE. The studies were carried out under conditions of cyclic temperature changes in the range of 300–650 °C. The duration of the study was 3840 h. The full cycle time of heating and cooling was 12 min 30 s. A total of 18,432 full cycles of heat load were carried out. The length of the test cycle was selected so that the samples would reach the assumed minimum and maximum temperatures. Temperature measurement of the samples was performed with a NiCr-Ni thermocouple, with no recording of temperature changes in time, and the measurement of surface temperature of the samples was performed to determine the minimum and maximum temperature for the test cycle. The tests were carried out under conditions of a reverberatory furnace (the scheme of a single retort stoker is shown in Figure 1).

The fuel used was bituminous coal with a calorific value of 26–28 MJ/kg, a combustion heat of 29 MJ/kg, a granulation of 5–25 mm, a humidity of <10%, a maximum ash content of 7%, and a maximum sulfur content of 0.6%. The fuel used was certified by Główny Instytut Górnictwa (Central Mining Institute).

In the studies, samples of SiMo cast iron with Si content of 5% and Mo content of 0%–2.5% were used. The chemical composition was determined on the basis of a Leco spectrometer (Model No 607-500, Leco Corporation, 3000 Lakeview Ave, St. Joseph, MI, USA) and a CS-125 Leco carbon and sulfur analyzer (Leco Corporation, 3000 Lakeview Ave, St. Joseph, MI, USA). The chemical composition of the tested samples is presented in Table 1.

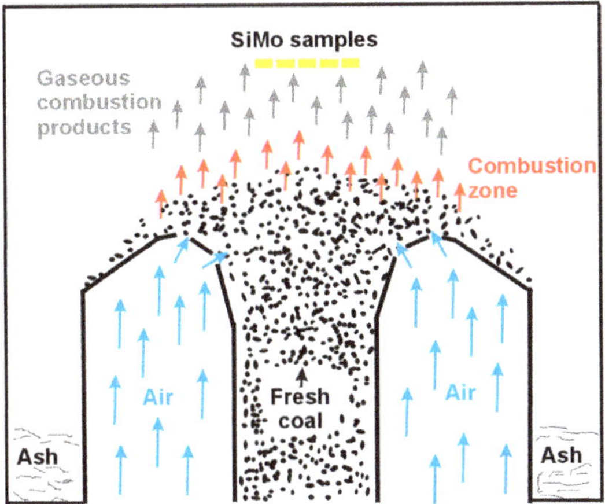

Figure 1. Scheme of a single retort with (yellow) SiMo samples placed above the stoker.

Table 1. Chemical composition of the tested SiMo cast iron.

Melt Number	Chemical Composition, % of Weight						
	C	Si	Mo	P	S	Mg	Fe (Balance)
SiMo 1	3.02	5.03	0.01	0.021	0.009	0.032	91.878
SiMo 2	2.94	5.14	0.47	0.020	0.005	0.029	91.396
SiMo 3	3.04	4.94	1.09	0.022	0.005	0.031	90.872
SiMo 4	2.71	5.17	1.92	0.018	0.009	0.062	90.111
SiMo 5	2.74	4.42	2.51	0.022	0.007	0.033	90.268

In order to determine the phase composition of the studied material, X-ray diffraction analyses were carried out with the use of an X'Pert Pro multipurpose x-ray diffractometer by Panalytical (Almelo, The Netherlands). The measurements were conducted utilizing filtered radiation of a cobalt anode lamp ($\lambda K\alpha = 0.179$ nm) as well as a PIXcel 3D detector on the diffracted beam axis. The diffraction lines were recorded in the Bragg–Brentano geometry in the angular scope of 10°–120° (2θ), with the step of 0.026° and the step time of 100 s. Furthermore, to obtain more precise information from the surface oxide layer, grazing incidence diffraction (GID) geometry with a proportional detector on the diffracted beam axis was used. In this geometry, a primary X-ray beam was set at a constant, low angle (1.5°) related to the sample plane, which affected the corresponding slope of diffraction vectors related to the normal to surface. This allowed us to obtain during the measurement a constant penetration depth of the X-ray beam, limited mainly to the surface oxide layer. The analysis of the obtained diffraction patterns was made in the Panalytical High Score Plus software (ver.: 3.0e), with the dedicated Panalytical Inorganic Crystal Structure Database (PAN-ICSD)

The analysis of the structure and the chemical composition was performed on a Phenom ProX scanning microscope (Phenom-World Eindhoven, Noord-Brabant, Netherlands) equipped with an energy-dispersive X-ray spectrometer (EDS).

3. Results

3.1. SEM Analysis

Figure 2 shows photos of SiMo cast iron samples after the chemical corrosion resistance test cycle. The microstructure of the cast iron consisted of a ferritic matrix, graphite nodules, molybdenum carbide (Mo$_2$C), a carburized zone (Figure 2e), and the passive layer and the loose oxide layer on the top surface of the samples. Microstructure components are highlighted in Figure 2.

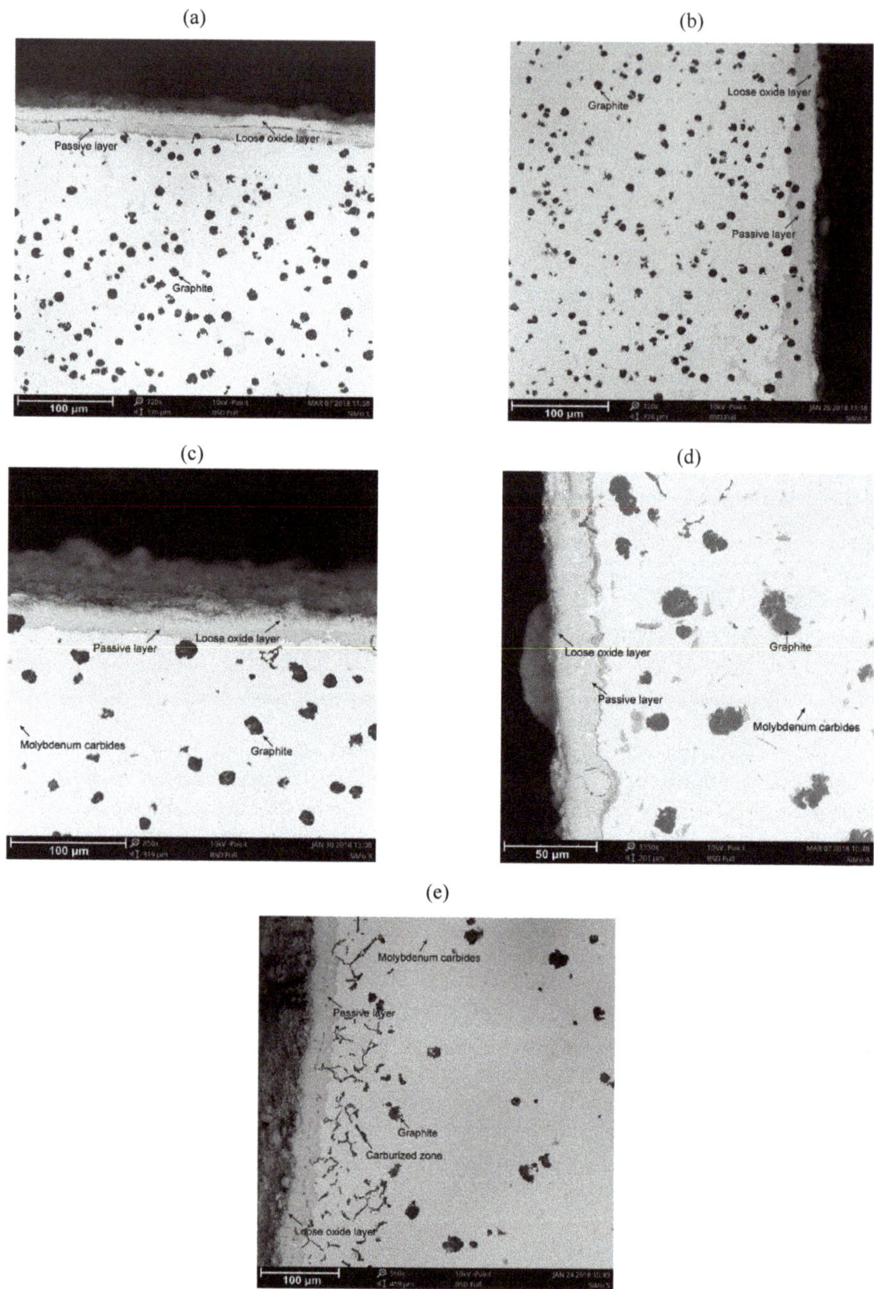

Figure 2. Microstructure of SiMo cast iron after the corrosion test cycle. SiMo cast iron with addition of: (**a**) 0.01% Mo, (**b**) 0.47% Mo, (**c**) 1.09% Mo, (**d**) 1.92% Mo, (**e**) 2.51% Mo. SEM.

In all cases (SiMo samples 1–5), the thickness of passive layers is around 10 µm. For elevated molybdenum content (Figure 2e, SiMo melt 5, 2.51% Mo) in the near-surface layer, the carburized zone in the form of black inclusions distributed in the vicinity of Mo_2C carbide precipitates is clearly visible. All the cases considered are characterized by a cohesive passive layer tightly adhering to the sample. No cracks nor defects in the passive layer were observed, even after the process of preparing metallographic sections (cutting, grinding, and polishing).

3.2. EDS Analysis

Figures 3 and 4 show the collective results of metallographic studies for selected alloys using the EDS system. The presented results indicate the presence of a loose oxide layer (blue), which is adjacent to the passive layer. The passive layer is an area also marked in blue, where the area has an increased silicon content (intense yellow bands on the maps—see Figure 3e, Figure 4e, Figure 5e). The increased Si content in the passive layer results from the diffusion of iron atoms from this layer to the loose oxide layer, which makes the area richer in Si. Penetrating oxygen reaches the area enriched with silicon and forms compounds with it (e.g., SiO_2) creating a tight barrier to the propagation of corrosive phenomena. Of course, a SiO_2 compound can react with Fe, O, and FeO, forming a fayalite Fe_2SiO_4 [8,17].

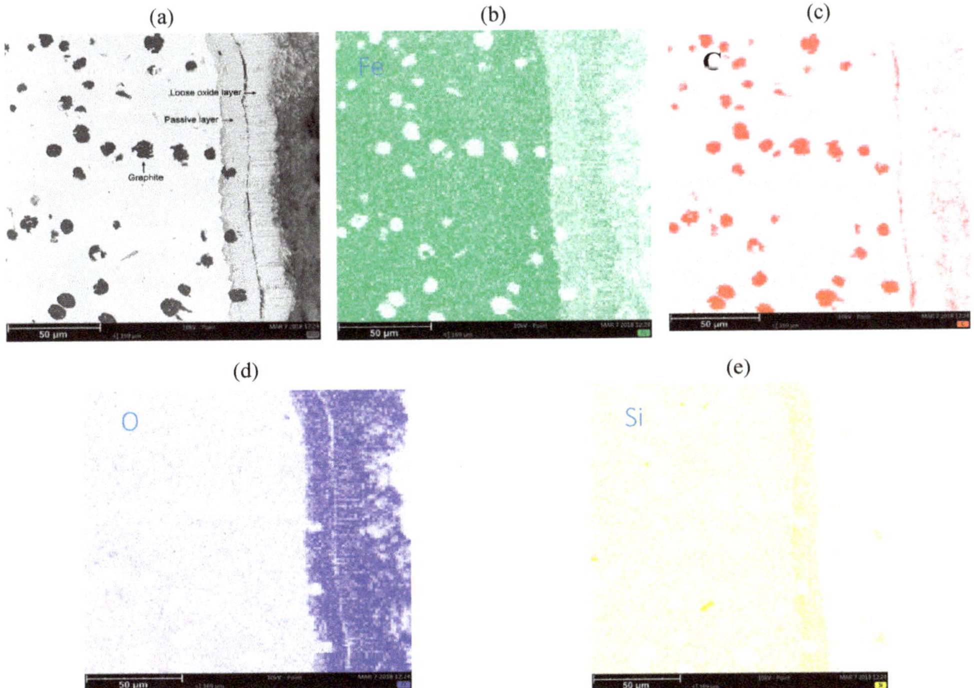

Figure 3. Microstructure of SiMo cast iron, 0.01% Mo (**a**). Elements decomposition maps. Map for (**b**) Fe, (**c**) C, (**d**) O, and (**e**) Si.

The characteristic phenomenon of these alloys is their ability to perform so-called "self-healing". Removing the passive layer creates a new one in its place. The average thickness of the loose oxide layer was 10 µm for the tested samples, while the thickness of the passive layer was also around 10 µm. Of course, the thickness of this layer depends on the local conditions on the surface of the sample exposed to the corrosive agents.

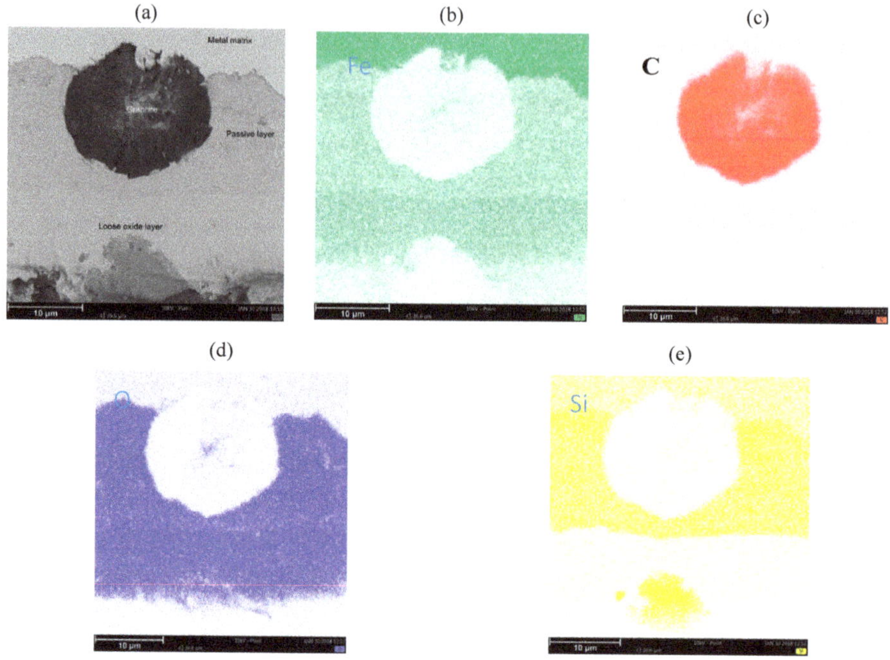

Figure 4. Microstructure of SiMo 3 cast iron, 1.09% Mo (**a**). Map for (**b**) Fe, (**c**) C, (**d**) O, and (**e**) Si.

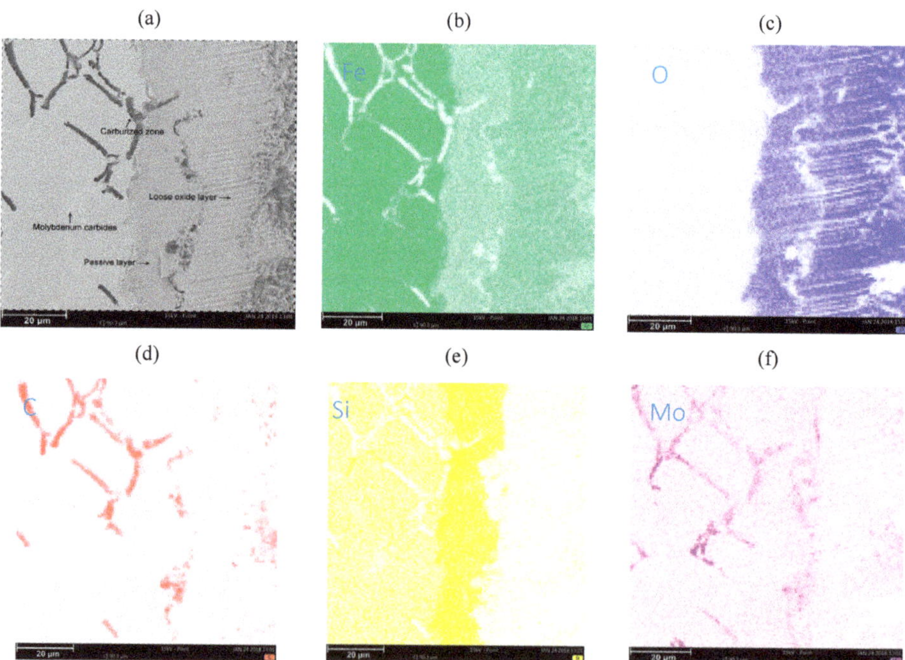

Figure 5. View of the carburized layer. SiMo sample 5. 2.51% Mo (**a**). Map for (**b**) Fe, (**c**) O, (**d**) C, (**e**) Si, and (**f**) Mo.

Figure 5 shows the analysis of the sample area (for the SiMo melt 5), where a significant degree of carburation of the casting material layer was observed. The layer is placed under the passive layer. This can be explained by the high concentration of carbon in the corrosive atmosphere of the furnace, and the penetration of carbon atoms through the passive layer towards the center of the casting.

3.3. XRD Analysis

The above SEM and EDS results were complemented with X-ray diffraction. Two measurements of the surface area of a selected sample were made (Figure 6), one with Bragg–Brentano geometry and the other one using the grazing incidence diffraction (GID) geometry. The GID geometry allows us to limit the penetration of the X-ray beam, so that diffractive information can be obtained mainly from the surface layer (in this particular case, the evidence is the almost complete disappearance of the diffraction lines from the substrate). In the figure below Figure 7, two diffractograms are superimposed on each other. The blue diffractogram was obtained using Bragg–Brentano geometry and the red diffractogram using GID geometry at a 1.5-degree tilt of the primary beam in relation to the measurement plane.

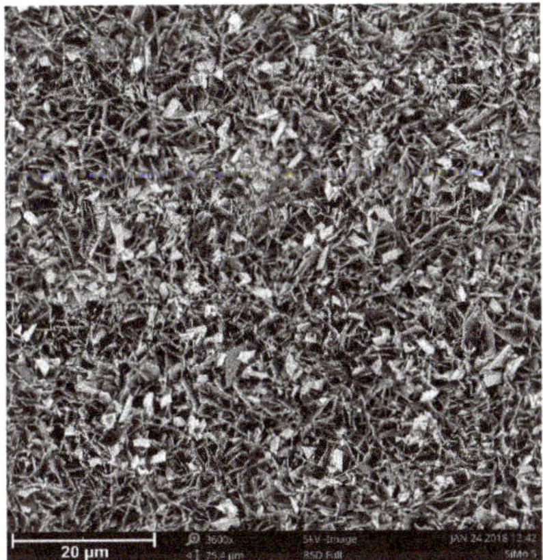

Figure 6. View of the loose oxide layer on a SiMo 5 sample. SEM.

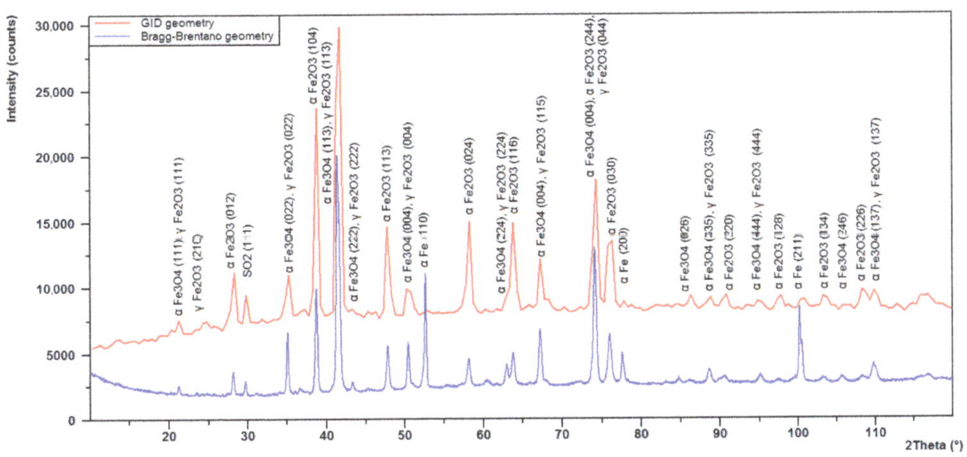

Figure 7. Diffractogram from a sample. Blue chart, Bragg–Brentano geometry; red chart, grazing incidence diffraction (GID) geometry. SiMo 3. 1.09% Mo.

From the results of the analysis presented in Figure 7, the following components of the surface oxide layer were identified. The compounds identified were, among others: α - Fe_2O_3 (hematite) with a hexagonal lattice R-3c (98-002-2505), α - Fe_3O_4 (magnetite) with a cubic lattice Fd-3m (98-026-3007), γ - Fe_2O_3 (maghemite) with a cubic lattice Fd-3m (98-017-2905), and a small share of SO_2 (an orthorhombic lattice).

4. Discussion

The presented photos of SiMo cast iron samples (Figure 2), after a series of chemical corrosion resistance tests, clearly show the microstructure of the cast iron with the passive layer and the loose oxide layer on the surface of the samples. The obtained results are analogous to the results presented by M. Ekström in his paper [5]. The results obtained by M. Ekström were obtained under different test conditions in a flue gas atmosphere. For all cases analyzed in the study, the thickness of the oxide and passive layers oscillates around the level of 20 µm. For elevated molybdenum content (SiMo melt 5, 2.51% Mo) in the near-surface layer, the carburized zone, in the form of black inclusions distributed in the vicinity of Mo_2C carbide precipitates, is clearly visible. The layer is placed under the passive layer, which is clearly visible in Figure 5. We found no description of a similar case in the available literature. The obtained effect can be explained by the high concentration of carbon in the corrosive atmosphere of the furnace; penetration of carbon atoms through the passive layer in the direction of the casting center, combined with the high concentration of molybdenum in the casting, resulted in the accumulation of carbon atoms around the precipitates rich in molybdenum (Mo_2C carbides).

The results obtained from the XRD analysis allowed for the following components of the surface oxide layer to be described, among others: α - Fe_2O_3 (hematite), α - Fe_3O_4 (magnetite), γ - Fe_2O_3 (maghemite), and SO_2 [22–24].

The X-ray diffraction patterns of magnetite and maghemite are very similar. This is related to their similar structures. Both of these two oxide phases crystallize in the cubic system and their lattice parameters are very close. For this reason, it is difficult to differentiate these structures. However, some works [25,26] report that the maghemite phase gives two additional small diffraction lines at 23.77° (210) and 26.10° (211). One of these, line (210), was identified even on a diffractogram obtained in Bragg–Brentano geometry. The formed magnetite and maghemite oxide layers do not show the much higher intensity of line (113) in relation to the highest hematite line (104). This result can be explained by the similar crystal growth of these oxide phases. Also, some authors report that the formation of maghemite (γ - Fe_2O_3) is a result of oxidation of the magnetite (α - Fe_3O_4) [27,28]. One of the variables deciding the quantitative share of hematite, magnetite, and maghemite is the range of oxidation temperature described by M. Marciuš et al. [29].

5. Conclusions

Based on the conducted study, it can be stated that SiMo cast iron is fully resistant to chemical corrosion during retort furnace operations. All SiMo cast iron samples were characterized by a cohesive oxidized layer, consisting of a passive layer on the casting material side and an oxide surface layer.

The two layers adhered quite well to one another. No cracks nor defects in any of the elements of the oxidized layer were observed. The surface oxide layer was found to consist of the following compounds: Fe_2O_3, Fe_3O_4, and SO_2.

For the sample with increased Mo content, a significant carburization of the near-surface layer of the sample was observed, especially in the areas adjacent to molybdenum carbide. The carburized edge zone of the sample did not affect the corrosion resistance of SiMo cast iron.

The areas strongly enriched with silicon are the fayalite Fe_2SiO_4 resulting from the reaction of SiO_2 with O, Fe, and FeO.

Due to the relatively low operating temperature range, we suggest that the Mo content in the alloy be reduced to the range of 0%–0.5% Mo.

Author Contributions: research concept, M.S.; conducting the experiment, M.S.; microstructure studies, M.S. and P.M.N.; diffractive analysis, P.M.N.; writing—original manuscript preparation, M.S.; writing—review and edition, M.S. and P.M.N. All authors have read and agreed to the published version of the manuscript.

Funding: This research received no external funding.

Acknowledgments: This publication was financed by the statutory subsidy of the Faculty of Mechanical Engineering of the Silesian University of Technology in 2019.

Conflicts of Interest: The authors report no conflicts of interest.

References

1. Tholence, F.; Norell, M. High-Temperature Corrosion of Cast Irons and Cast Steels in Dry Air. *Mater. Sci. Forum* **2001**, *369*, 197–204. [CrossRef]
2. Tholence, F.; Norell, M. AES characterization of oxide grains formed on ductile cast irons in exhaust environments. *Surf. Interface Anal.* **2002**, *34*, 535–539. [CrossRef]
3. Tholence, F.; Norell, M. Nitride precipitation during high temperature corrosion of ductile cast irons in synthetic exhaust gases. *J. Phys. Chem. Solids* **2005**, *66*, 530–534. [CrossRef]
4. Tholence, F.; Norell, M. High Temperature Corrosion of Cast Alloys in Exhaust Environments I-Ductile Cast Irons. *Oxid. Met.* **2007**, *69*, 13–36. [CrossRef]
5. Ekström, M.; Szakálos, P.; Jonsson, S. Influence of Cr and Ni on High-Temperature Corrosion Behavior of Ferritic Ductile Cast Iron in Air and Exhaust Gases. *Oxid. Met.* **2013**, *80*, 455–466. [CrossRef]
6. Yang, Y.L.; Cao, Z.Y.; Qi, Y.; Liu, Y.B. The Study on Oxidation Resistance Properties of Ductile Cast Irons for Exhaust Manifold at High Temperatures. *Adv. Mater. Res.* **2010**, *97*, 530–533. [CrossRef]
7. Choe, K.H.; Lee, S.M.; Lee, K.W. High Temperature Oxidation Behavior of Si-Mo Ferritic Ductile Cast Iron. *Mater. Sci. Forum* **2010**, *654*, 542–545. [CrossRef]
8. Stawarz, M.; Janerka, K.; Dojka, M. Selected Phenomena of the In-Mold Nodularization Process of Cast Iron That Influence the Quality of Cast Machine Parts. *Processes* **2017**, *5*, 68. [CrossRef]
9. Henderieckx, G.D. *Silicon Cast Iron*; Gietech BV: Terneuzen, The Netherlands, 2009.
10. Brady, M.P.; Muralidharan, G.; Leonard, D.; Haynes, J.A.; Weldon, R.G.; England, R.D. Long-Term Oxidation of Candidate Cast Iron and Stainless Steel Exhaust System Alloys from 650 to 800 °C in Air with Water Vapor. *Oxid. Met.* **2014**, *82*, 359–381. [CrossRef]
11. Roucka, J.; Abramova, E.; Kana, V. Properties of type SiMo ductile irons at high temperatures. *Arch. Metall. Mater.* **2018**, *63*, 601–607.
12. Zeytin, H.K.; Kubilay, C.; Aydin, H.; Ebrinc, A.A.; Aydemir, B. Effect of microstructure on exhaust manifold cracks produced from SiMo ductile iron. *J. Iron Steel Res. Int.* **2009**, *16*, 32–36. [CrossRef]
13. Mohammadi, F.; Alfantazi, A. Corrosion of ductile iron exhaust brake housing in heavy diesel engines. *Eng. Fail. Anal.* **2013**, *31*, 248–254. [CrossRef]
14. Cygan, B.; Stawarz, M.; Jezierski, J. Heat treatment of the SiMo iron castings—Case study in the automotive foundry. *Arch. Foundry Eng.* **2018**, *18*, 103–109.
15. Li, D. Mixed graphite cast iron for automotive exhaust component applications. *China Foundry* **2017**, *14*, 519–524. [CrossRef]
16. Unkić, F.; Glavaš, Z.; Terzić, K. The influence of elevated temperatures on microstructure of cast irons for automotive engine thermo-mechanical loaded parts. *Mater. Geoenvironment* **2009**, *56*, 9–23.
17. Ibrahim, M.; Nofal, A.; Mourad, M.M. Microstructure and Hot Oxidation Resistance of SiMo Ductile Cast Irons Containing Si-Mo-Al. *Met. Mater. Trans. A* **2016**, *48*, 1149–1157. [CrossRef]
18. Guzik, E.; Kopyciński, D.; Wierzchowski, D. Manufacturing of Ferritic Low-Silicon and Molybdenum Ductile Cast Iron with the Innovative 2PE-9 Technique. *Arch. Met. Mater.* **2014**, *59*, 687–691. [CrossRef]
19. Vaško, A.; Belan, J.; Tillová, E. Static and Dynamic Mechanical Properties of Nodular Cast Irons. *Arch. Metall. Mater.* **2019**, *64*, 185–190.
20. Nyashina, G.S.; Kuznetsov, G.V.; Strizhak, P. Energy efficiency and environmental aspects of the combustion of coal-water slurries with and without petrochemicals. *J. Clean. Prod.* **2018**, *172*, 1730–1738. [CrossRef]
21. Bhadra, B.N.; Jhung, S.H. Oxidative desulfurization and denitrogenation of fuels using metal-organic framework-based/-derived catalysts. *Appl. Catal. B Environ.* **2019**, *259*, 118021. [CrossRef]

22. Avvakumov, E.G.; Golovin, A.V.; Paukshtis, E.A.; Ivanov, V.P.; Kolomiichuk, V.N.; Kustova, G.N.; Burgina, E.B.; Litvak, G.S.; Cherepanova, S.V.; Tsybulya, S.V.; et al. *Golden Book of Phase Transitions*, 1st ed.; Tomaszewski P.E.: Wroclaw, Poland, 2002; pp. 1–123.

23. Ju, S.; Cai, T.; Lu, H.-S.; Gong, C.-D. Pressure-Induced Crystal Structure and Spin-State Transitions in Magnetite (Fe3O4). *J. Am. Chem. Soc.* **2012**, *134*, 13780–13786. [CrossRef] [PubMed]

24. Post, B.; Schwartz, R.S.; Fankuchen, I. The crystal structure of sulfur dioxide. *Acta Crystallogr.* **1952**, *5*, 372–374. [CrossRef]

25. Kim, W.; Suh, C.-Y.; Cho, S.-W.; Roh, K.-M.; Kwon, H.; Song, K.; Shon, I.-J. A new method for the identificationand quantification of magnetite-maghemite mixtureusing conventional X-ray diffraction technique. *Talanta* **2012**, *94*, 348–352. [CrossRef] [PubMed]

26. Darezereshki, E.; Ranjbar, M.; Bakhtiari, F. One-step synthesis of maghemite (γ-Fe2O3) nano-particles by wet chemical method. *J. Alloy. Compd.* **2010**, *502*, 257–260. [CrossRef]

27. Long, R.; Zhou, S.; Wiley, B.J.; Xiong, Y. Oxidative etching for controlled synthesis of metal nanocrystals: Atomic addition and subtraction. *Chem. Soc. Rev.* **2014**, *43*, 6288–6310. [CrossRef]

28. Li, C.; Wei, Y.; Liivat, A.; Zhu, Y.; Zhu, J. Microwave-solvothermal synthesis of Fe3O4 magnetic nanoparticles. *Mater. Lett.* **2013**, *107*, 23–26. [CrossRef]

29. Marciuš, M.; Ristić, M.; Ivanda, M.; Music, S. Formation of Iron Oxides by Surface Oxidation of Iron Plate. *Croat. Chem. Acta* **2012**, *85*, 117–124. [CrossRef]

 © 2020 by the authors. Licensee MDPI, Basel, Switzerland. This article is an open access article distributed under the terms and conditions of the Creative Commons Attribution (CC BY) license (http://creativecommons.org/licenses/by/4.0/).

MDPI

St. Alban-Anlage 66

4052 Basel

Switzerland

Tel. +41 61 683 77 34

Fax +41 61 302 89 18

www.mdpi.com

Materials Editorial Office

E-mail: materials@mdpi.com

www.mdpi.com/journal/materials

www.ingramcontent.com/pod-product-compliance
Lightning Source LLC
LaVergne TN
LVHW071609170726
843515LV00008B/2358